LINEAR ALGEBRA

CORE TOPICS FOR THE FIRST COURSE

LINEAR ALGEBRA

CORE TOPICS FOR THE FIRST COURSE

Dragu Atanasiu
University of Borås, Sweden

Piotr Mikusiński
University of Central Florida, USA

NEW JERSEY • LONDON • SINGAPORE • BEIJING • SHANGHAI • HONG KONG • TAIPEI • CHENNAI • TOKYO

Published by

World Scientific Publishing Co. Pte. Ltd.
5 Toh Tuck Link, Singapore 596224
USA office: 27 Warren Street, Suite 401-402, Hackensack, NJ 07601
UK office: 57 Shelton Street, Covent Garden, London WC2H 9HE

Library of Congress Control Number: 2020011809

British Library Cataloguing-in-Publication Data
A catalogue record for this book is available from the British Library.

LINEAR ALGEBRA
Core Topics for the First Course

ISBN 978-981-121-502-5 (hardcover)
ISBN 978-981-121-596-4 (paperback)
ISBN 978-981-121-503-2 (ebook for institutions)
ISBN 978-981-121-504-9 (ebook for individuals)

For any available supplementary material, please visit
https://www.worldscientific.com/worldscibooks/10.1142/11685#t=suppl

Typeset by Stallion Press
Email: enquiries@stallionpress.com

Printed in Singapore

To Delia and Grażyna

Preface

This is an introductory textbook on linear algebra.

The market is flooded with textbooks for the first course in linear algebra. Do we need another textbook? Is this textbook essentially different from what is already available? To answer these questions we present some highlights of the book that distinguish it from what is currently available on the market.

- The book has a chapter on examples of Jordan forms, which is rare in textbooks for a first course in linear algebra. In this chapter we keep the theory at the level of the earlier chapters because our main goal is to give students an opportunity to assimilate the standard material of the first course in linear algebra and not to give a complete treatment of Jordan forms.

  In our opinion, the presentation usually given in the first course in linear algebra does not give students enough time to grasp the difficult concepts like linear independence, basis, or linear transformation. This motivated us to write a chapter on Jordan forms. The chapter gives students many opportunities to choose a right basis, to verify that vectors are linearly independent, and to construct the matrix of a linear transformation. This allows students to see the same concepts from different angles. Students can gain better understanding of concepts such as linear independence, basis, dimension, eigenvalues, eigenvectors, linear transformation, and the matrix of a linear transformation.

  By being exposed to Jordan forms and other concepts not found in most textbooks for a first course in linear algebra, like Moore-Penrose inverse and Cholesky decomposition, students are better prepared for a second course in linear algebra.

- The general vector spaces are introduced at the end of the book. This gives students a possibility to concentrate on the difficulties of linear algebra in the more concrete context of the vector space $\mathbb{R}^n$.

- Whenever possible and natural, our proofs are consequences of the Gauss elimination process or the properties of the inverse matrices. It has been our experience that students find it easier to follow such arguments.

- The chapter on solving linear algebra problems with a computer algebra system not only shows how to use CAS to facilitate calculations, but also to improve understanding of linear algebra.

- The book has a companion book, A Bridge to Linear Algebra, which is a study book that presents matrices, Gauss elimination and solving systems of linear equations in an elementary way and all important ideas from the present book in the context of $\mathbb{R}^2$ and $\mathbb{R}^3$. Since it also has a substantial amount of geometry, rotations, a careful presentation of cross product, and some applications, it offers effective help for students studying the present book.

Now we turn to the description of the content of the book. The first six chapters cover what a traditional first course in linear algebra should include. The next three chapters offer an opportunity to practice the methods introduced in the first six chapters and, at same time, introduce some new concepts. These chapters should be thought of as applications of the first six chapters.

Chapter 1 is about matrices, elementary matrices, Gauss elimination and systems of linear equations, with special emphasis on the understanding of Gauss elimination.

Chapter 2 is about subspaces of $\mathbb{R}^n$. Here we introduce the essential concepts of linear algebra like linear independence and basis. Most of our proofs are given for particular cases and are consequences of Gauss elimination. The essential results on dimensions are obtained using inverse matrices.

Chapter 3 is about orthogonality. Here we first calculate projections on subspaces without using an orthogonal basis. With this approach we can introduce the normal equation and the least square method in a more natural and intuitive way.

A relatively short Chapter 4 introduces the determinants so that they can be used for the calculations of eigenvalues in Chapter 5. We do not discuss cofactor expansions.

Chapter 5 is about eigenvalues and eigenvectors. Here we present diagonalization of general square matrices and the spectral decomposition of a symmetric matrix.

In Chapter 6 we discuss linear transformations between subspaces of $\mathbb{R}^n$. We give a detailed presentation of the matrix associated with a linear transformation.

In Chapter 7 we present examples of Jordan forms for 2×2, 3×3, and 4×4 matrices. Some examples involving 5×5 matrices are given in the exercises. Because we want students to practice the concepts and methods specific to a first course in linear algebra we only use the definition of the characteristic polynomial already introduced in Chapter 5.

Chapter 8 is about singular value decomposition which is obtained from the outer product expansion. Here we give more examples than most books at this level because the calculation of outer product expansion involves many essential ideas from the first six chapters. At the end of the chapter we present the Moore-Penrose inverse.

Chapter 9 is about quadratic forms. Here we classify the quadratic forms and introduce the Cholesky decomposition of a positive definite matrix.

Chapter 10 is about general vector spaces and inner product spaces. It revisits some of the main ideas introduced in Chapters 1-6 in the context of vector spaces of matrices and functions. It is treated as a preparation for a second course in linear algebra.

In Chapter 11 we present 33 problems that are solved using Maple. Here our goal is not only to show how Maple works and how it can be used to facilitate calculations with matrices, but also to improve understanding of basic concepts of linear algebra.

We hope that students using this book will find it easier to learn linear algebra, will gain better understanding of the fundamental ideas, and will be better prepared for a second course in linear algebra.

Acknowledgments

We would like to take this opportunity to express our gratitude to those who helped us in different ways while we were writing this book. We were influenced and benefited from discussions with Joseph Brennan, Alvaro Islas, Heath Martin, and Sona Swanson from the University of Central Florida and Magnus Lundin from the University of Borås. We thank students from these two universities for their feedback when parts of the book were used in their classes.

The first author would like to thank Xin Li, the department chair, for giving him an opportunity to teach linear algebra at the University of Central Florida.

We are grateful to the World Scientific Publishing team, including Rochelle Kronzek-Miller, Tan Rok Ting, Ng Chin Choon and Peng Ah Huay, for their support and assistance with this project.

Contents

Chapter 1

Matrices

In this chapter we study algebraic properties of matrices, including operations on matrices, Gaussian elimination, and inverse matrices.

1.1 Matrix algebra

Basic definitions

Definition 1.1. By an $m \times n$ *matrix* we mean an array of numbers

$$A = \begin{bmatrix} a_{11} & a_{12} & \dots & a_{1n} \\ a_{21} & a_{22} & \dots & a_{2n} \\ \vdots & \vdots & & \vdots \\ a_{m1} & a_{m2} & \dots & a_{mn} \end{bmatrix}.$$

The matrices

$$\begin{bmatrix} a_{i1} & a_{i2} & \dots & a_{in} \end{bmatrix},$$

where $1 \le i \le m$, are called the *rows* of the matrix A. The matrices

$$\begin{bmatrix} a_{1j} \\ a_{2j} \\ \vdots \\ a_{mj} \end{bmatrix},$$

where $1 \le j \le n$, are called the *columns* of the matrix A. The numbers a_{ij} are referred to as *entries* of the matrix. The entry in the i-th row and j-th column will be referred to as the (i, j) *entry*.

An $m \times n$ matrix has m rows, n columns, and mn entries.

Example 1.2. The matrix

$$A = \begin{bmatrix} a_{11} & a_{12} & a_{13} & a_{14} \\ a_{21} & a_{22} & a_{23} & a_{24} \\ a_{31} & a_{32} & a_{33} & a_{34} \end{bmatrix}$$

has three rows

$$\begin{bmatrix} a_{11} & a_{12} & a_{13} & a_{14} \end{bmatrix}, \begin{bmatrix} a_{21} & a_{22} & a_{23} & a_{24} \end{bmatrix}, \begin{bmatrix} a_{31} & a_{32} & a_{33} & a_{34} \end{bmatrix},$$

four columns

$$\begin{bmatrix} a_{11} \\ a_{21} \\ a_{31} \end{bmatrix}, \begin{bmatrix} a_{12} \\ a_{22} \\ a_{32} \end{bmatrix}, \begin{bmatrix} a_{13} \\ a_{23} \\ a_{33} \end{bmatrix}, \begin{bmatrix} a_{14} \\ a_{24} \\ a_{34} \end{bmatrix},$$

and 12 entries.

Matrices of the form $\begin{bmatrix} a_1 \\ a_2 \\ \vdots \\ a_n \end{bmatrix}$, that is, $n \times 1$ matrices, are called *vectors*. To indicate the size of a vector we will say a *vector in* $\mathbb{R}^n$. For example, when we say that $\mathbf{x}$ is a vector in $\mathbb{R}^3$, we mean that $\mathbf{x} = \begin{bmatrix} x_1 \\ x_2 \\ x_3 \end{bmatrix}$ for some real numbers x_1, x_2, and x_3. It is natural to identify 1×1 matrices with real numbers.

While vectors are matrices, it is customary to use bold Roman lower case letters to denote vectors, that is,

$$\mathbf{a}, \mathbf{b}, \mathbf{c}, \dots, \mathbf{x}, \mathbf{y}, \mathbf{z}.$$

As we will see, vectors play a very important role in linear algebra and in applications of linear algebra.

A matrix

$$A = \begin{bmatrix} a_{11} & a_{12} & \dots & a_{1n} \\ a_{21} & a_{22} & \dots & a_{2n} \\ \vdots & \vdots & & \vdots \\ a_{m1} & a_{m2} & \dots & a_{mn} \end{bmatrix}$$

can be thought of as built from n vectors, namely,

$$\begin{bmatrix} a_{11} \\ a_{21} \\ \vdots \\ a_{m1} \end{bmatrix}, \begin{bmatrix} a_{12} \\ a_{22} \\ \vdots \\ a_{m2} \end{bmatrix}, \begin{bmatrix} a_{13} \\ a_{23} \\ \vdots \\ a_{m3} \end{bmatrix}, \dots, \begin{bmatrix} a_{1n} \\ a_{21n} \\ \vdots \\ a_{mn} \end{bmatrix}.$$

It turns out that interpreting a matrix in this way is often beneficial. If it is not necessary to specify the entries of an $m \times n$ matrix A, we can write

$$A = \begin{bmatrix} \mathbf{a}_1 & \mathbf{a}_2 & \dots & \mathbf{a}_n \end{bmatrix},$$

where $\mathbf{a}_1, \mathbf{a}_2, \dots, \mathbf{a}_n$ are vectors in $\mathbb{R}^m$.

Two matrices are equal if they have the same size and the corresponding entries are the same. More precisely, two $m \times n$ matrices

$$A = \begin{bmatrix} a_{11} & a_{12} & \dots & a_{1n} \\ a_{21} & a_{22} & \dots & a_{2n} \\ \vdots & \vdots & & \vdots \\ a_{m1} & a_{m2} & \dots & a_{mn} \end{bmatrix} \quad \text{and} \quad B = \begin{bmatrix} b_{11} & b_{12} & \dots & b_{1n} \\ b_{21} & b_{22} & \dots & b_{2n} \\ \vdots & \vdots & & \vdots \\ b_{m1} & b_{m2} & \dots & b_{mn} \end{bmatrix}$$

are equal if and only if $a_{ij} = b_{ij}$ for every $1 \le i \le m$ and every $1 \le j \le n$.

Example 1.3. The matrices

$$\begin{bmatrix} 1 & 1 & 1 \\ 1 & 1 & 1 \end{bmatrix} \quad \text{and} \quad \begin{bmatrix} 1 & 1 \\ 1 & 1 \\ 1 & 1 \end{bmatrix}$$

are not equal, because the one on the left is a 2×3 matrix and the one on the right is a 3×2 matrix. The matrices

$$\begin{bmatrix} 1 & -1 & 2 & 5 \\ 0 & \pi & -\frac{1}{2} & 7 \\ \sqrt{2} & 0 & -0.77 & \frac{1}{3} \end{bmatrix} \quad \text{and} \quad \begin{bmatrix} 1 & -1 & 2 & 5 \\ 0 & \pi & -\frac{1}{2} & 7 \\ \sqrt{2} & 0 & 0.77 & \frac{1}{3} \end{bmatrix}$$

are not equal for a different reason; one of the entries is different.

Sum of matrices

Definition 1.4. If A and B are two $m \times n$ matrices, the *sum* $A+B$ is the matrix with (i, j) entry $a_{ij} + b_{ij}$, where a_{ij} is the (i, j) entry of the matrix A and b_{ij} is the (i, j) entry of the matrix B. If

$$A = \begin{bmatrix} a_{11} & a_{12} & \dots & a_{1n} \\ a_{21} & a_{22} & \dots & a_{2n} \\ \vdots & \vdots & & \vdots \\ a_{m1} & a_{m2} & \dots & a_{mn} \end{bmatrix} \quad \text{and} \quad B = \begin{bmatrix} b_{11} & b_{12} & \dots & b_{1n} \\ b_{21} & b_{22} & \dots & b_{2n} \\ \vdots & \vdots & & \vdots \\ b_{m1} & b_{m2} & \dots & b_{mn} \end{bmatrix},$$

then

$$A+B = \begin{bmatrix} a_{11}+b_{11} & a_{12}+b_{12} & \dots & a_{1n}+b_{1n} \\ a_{21}+b_{21} & a_{22}+b_{22} & \dots & a_{2n}+b_{2n} \\ \vdots & \vdots & & \vdots \\ a_{m1}+b_{m1} & a_{m2}+b_{m2} & \dots & a_{mn}+b_{mn} \end{bmatrix}.$$

In other words, we add matrices by adding the corresponding entries.

Note that $A+B$ does not make sense if the matrices A and B are not of the same size.

Example 1.5.

$$\begin{bmatrix} 3 & 4 & -1 \\ 2 & 5 & 1 \end{bmatrix} + \begin{bmatrix} 1 & 0 & 3 \\ 4 & -3 & 7 \end{bmatrix} = \begin{bmatrix} 4 & 4 & 2 \\ 6 & 2 & 8 \end{bmatrix}$$

$$\begin{bmatrix} a_{11} & a_{12} \\ a_{21} & a_{22} \\ a_{31} & a_{32} \end{bmatrix} + \begin{bmatrix} b_{11} & b_{12} \\ b_{21} & b_{22} \\ b_{31} & b_{32} \end{bmatrix} = \begin{bmatrix} a_{11}+b_{11} & a_{12}+b_{12} \\ a_{21}+b_{21} & a_{22}+b_{22} \\ a_{31}+b_{31} & a_{32}+b_{32} \end{bmatrix}.$$

Definition 1.6. The $m \times n$ matrix whose entries are all 0,

$$\begin{bmatrix} 0 & 0 & \dots & 0 \\ 0 & 0 & \dots & 0 \\ \vdots & \vdots & & \vdots \\ 0 & 0 & \dots & 0 \end{bmatrix},$$

is called the $m \times n$ *zero matrix*. The $m \times n$ zero matrix will be denoted by $\mathbf{0}_{m,n}$ or simply by $\mathbf{0}$ when the dimension of the matrix is clear from the context.

If A, B, and C are three $m \times n$ matrices and $\mathbf{0}$ is the $m \times n$ zero matrix, then

$$A + \mathbf{0} = \mathbf{0} + A = A,$$

$$A + B = B + A,$$

$$(A + B) + C = A + (B + C).$$

The above properties are immediate consequences of the definition of addition of matrices and the corresponding properties of addition of real numbers.

Scalar multiplication

Definition 1.7. If A is a $m \times n$ matrix and t is a real number, then tA is the matrix whose (i, j) entry is ta_{ij}, where a_{ij} is the (i, j) entry of the matrix A:

$$t \begin{bmatrix} a_{11} & a_{12} & \dots & a_{1n} \\ a_{21} & a_{22} & \dots & a_{2n} \\ \vdots & \vdots & & \vdots \\ a_{m1} & a_{m2} & \dots & a_{mn} \end{bmatrix} = \begin{bmatrix} ta_{11} & ta_{12} & \dots & ta_{1n} \\ ta_{21} & ta_{22} & \dots & ta_{2n} \\ \vdots & \vdots & & \vdots \\ ta_{m1} & ta_{m2} & \dots & ta_{mn} \end{bmatrix}.$$

In other words, to multiply a matrix A by a number t we multiply every entry of A by t. The operation of multiplication of a matrix by a number is referred to as *scalar multiplication.*

Example 1.8.

$$5 \begin{bmatrix} 3 & 4 \\ 2 & 5 \end{bmatrix} = \begin{bmatrix} 15 & 20 \\ 10 & 25 \end{bmatrix}$$

$$t \begin{bmatrix} a_{11} & a_{12} \\ a_{21} & a_{22} \\ a_{31} & a_{32} \end{bmatrix} = \begin{bmatrix} ta_{11} & ta_{12} \\ ta_{21} & ta_{22} \\ ta_{31} & ta_{32} \end{bmatrix}.$$

If A and B are $m \times n$ matrices and s and t are real numbers, then

$$(s + t)A = sA + tA,$$

$$(st)A = s(tA),$$

$$t(A + B) = tA + tB.$$

The above properties are immediate consequences of the definitions of scalar multiplication and addition of matrices and the corresponding properties of addition and multiplication of real numbers.

Product of matrices

Products of matrices play a fundamental role in linear algebra. First we define

$$\begin{bmatrix} a_1 & a_2 \end{bmatrix} \begin{bmatrix} b_1 \\ b_2 \end{bmatrix} = a_1 b_1 + a_2 b_2.$$

This product can be easily extended to higher dimensions:

$$\begin{bmatrix} a_1 & a_2 & a_3 \end{bmatrix} \begin{bmatrix} b_1 \\ b_2 \\ b_3 \end{bmatrix} = a_1 b_1 + a_2 b_2 + a_3 b_3,$$

and, in general,

$$\begin{bmatrix} a_1 & a_2 & \dots & a_m \end{bmatrix} \begin{bmatrix} b_1 \\ b_2 \\ \vdots \\ b_m \end{bmatrix} = a_1 b_1 + a_2 b_2 + \dots + a_m b_m.$$

Two matrices can be multiplied, if the number of columns of the one on the left is the same as the number of rows of the one on the right. So, if A is a $k \times l$ matrix and B is a $m \times n$ matrix, then the product AB is well-defined only if $l = m$.

Definition 1.9. The *product* of a $k \times m$ matrix A and an $m \times n$ matrix B is the $k \times n$ matrix AB such that the (i, j) entry is

$$\begin{bmatrix} a_{i1} & a_{i2} & \dots & a_{im} \end{bmatrix} \begin{bmatrix} b_{1j} \\ b_{2j} \\ \vdots \\ b_{mj} \end{bmatrix},$$

where $\begin{bmatrix} a_{i1} & a_{i2} & \dots & a_{im} \end{bmatrix}$ is the i-th row of the matrix A and $\begin{bmatrix} b_{1j} \\ b_{2j} \\ \vdots \\ b_{mj} \end{bmatrix}$ is the j-th column of the matrix B.

For example,

$$\begin{bmatrix} a_1 & a_2 & a_3 & a_4 \\ c_1 & c_2 & c_3 & c_4 \\ f_1 & f_2 & f_3 & f_4 \end{bmatrix} \begin{bmatrix} b_1 & d_1 \\ b_2 & d_2 \\ b_3 & d_3 \\ b_4 & d_4 \end{bmatrix} = \begin{bmatrix} \begin{bmatrix} a_1 & a_2 & a_3 & a_4 \end{bmatrix} \begin{bmatrix} b_1 \\ b_2 \\ b_3 \\ b_4 \end{bmatrix} & \begin{bmatrix} a_1 & a_2 & a_3 & a_4 \end{bmatrix} \begin{bmatrix} d_1 \\ d_2 \\ d_3 \\ d_4 \end{bmatrix} \\ \begin{bmatrix} c_1 & c_2 & c_3 & c_4 \end{bmatrix} \begin{bmatrix} b_1 \\ b_2 \\ b_3 \\ b_4 \end{bmatrix} & \begin{bmatrix} c_1 & c_2 & c_3 & c_4 \end{bmatrix} \begin{bmatrix} d_1 \\ d_2 \\ d_3 \\ d_4 \end{bmatrix} \\ \begin{bmatrix} f_1 & f_2 & f_3 & f_4 \end{bmatrix} \begin{bmatrix} b_1 \\ b_2 \\ b_3 \\ b_4 \end{bmatrix} & \begin{bmatrix} f_1 & f_2 & f_3 & f_4 \end{bmatrix} \begin{bmatrix} d_1 \\ d_2 \\ d_3 \\ d_4 \end{bmatrix} \end{bmatrix}$$

$$= \begin{bmatrix} a_1b_1 + a_2b_2 + a_3b_3 + a_4b_4 & a_1d_1 + a_2d_2 + a_3d_3 + a_4d_4 \\ c_1b_1 + c_2b_2 + c_3b_3 + c_4b_4 & c_1d_1 + c_2d_2 + c_3d_3 + c_4d_4 \\ f_1b_1 + f_2b_2 + f_3b_3 + f_4b_4 & f_1d_1 + f_2d_2 + f_3d_3 + f_4d_4 \end{bmatrix}.$$

Here are some concrete examples of products of matrices.

Example 1.10.

$$\begin{bmatrix} 1 & 3 & 0 & 2 \\ 2 & 0 & 1 & 0 \end{bmatrix} \begin{bmatrix} 4 & 1 & 1 \\ 0 & 2 & 1 \\ 1 & 1 & 6 \\ 0 & 1 & 1 \end{bmatrix} = \begin{bmatrix} 4 & 9 & 6 \\ 9 & 3 & 8 \end{bmatrix}$$

$$\begin{bmatrix} 1 & 2 & 0 \\ 0 & 3 & 1 \\ 1 & -1 & 1 \end{bmatrix} \begin{bmatrix} 2 & 2 & 3 & 6 \\ 3 & 1 & 1 & 6 \\ 2 & 0 & 1 & 0 \end{bmatrix} = \begin{bmatrix} 8 & 4 & 5 & 18 \\ 11 & 3 & 4 & 18 \\ 1 & 1 & 3 & 0 \end{bmatrix}$$

$$\begin{bmatrix} a \\ b \end{bmatrix} \begin{bmatrix} c & d & e \end{bmatrix} = \begin{bmatrix} ac & ad & ae \\ bc & bd & be \end{bmatrix}.$$

The following observation is often useful in proofs and practical calculations.

If A is a $k \times m$ matrix and B is a $m \times n$ matrix, then the j-th column of the matrix AB is $A\mathbf{b}_j$, where $\mathbf{b}_j$ is the j-th column of the matrix B. In other words, we have

$$AB = A\begin{bmatrix} \mathbf{b}_1 & \mathbf{b}_2 & \dots & \mathbf{b}_n \end{bmatrix} = \begin{bmatrix} A\mathbf{b}_1 & A\mathbf{b}_2 & \dots & A\mathbf{b}_n \end{bmatrix}.$$

Theorem 1.11. *If A is an $m \times n$ matrix and B and C are $n \times p$ matrices, then*

$$A(B+C) = AB + AC.$$

Proof. We illustrate the method of the proof by considering the case when $n = 3$.
The (i, j) entry in $A(B+C)$ is

$$\begin{aligned}
\begin{bmatrix} a_{i1} & a_{i2} & a_{i3} \end{bmatrix} \left(\begin{bmatrix} b_{1j} \\ b_{2j} \\ b_{3j} \end{bmatrix} + \begin{bmatrix} c_{1j} \\ c_{2j} \\ c_{3j} \end{bmatrix} \right) &= \begin{bmatrix} a_{i1} & a_{i2} & a_{i3} \end{bmatrix} \begin{bmatrix} b_{1j} + c_{1j} \\ b_{2j} + c_{2j} \\ b_{3j} + c_{3j} \end{bmatrix} \\
&= a_{i1}(b_{1j} + c_{1j}) + a_{i2}(b_{2j} + c_{2j}) + a_{i3}(b_{3j} + c_{3j}) \\
&= a_{i1}b_{1j} + a_{i2}b_{2j} + a_{i3}b_{3j} + a_{i1}c_{1j} + a_{i2}c_{2j} + a_{i3}c_{3j} \\
&= \begin{bmatrix} a_{i1} & a_{i2} & a_{i3} \end{bmatrix} \begin{bmatrix} b_{1j} \\ b_{2j} \\ b_{3j} \end{bmatrix} + \begin{bmatrix} a_{i1} & a_{i2} & a_{i3} \end{bmatrix} \begin{bmatrix} c_{1j} \\ c_{2j} \\ c_{3j} \end{bmatrix},
\end{aligned}$$

which is the (i, j) entry of the matrix $AB + AC$. □

Theorem 1.12. *If A and B are $m \times n$ matrices and C is an $n \times p$ matrix, then*

$$(A+B)C = AC + BC.$$

Proof. To illustrate the method of the proof we consider the case when $n = 3$.
The (i, j) entry in $(A+B)C$ is

$$\begin{aligned}
\left(\begin{bmatrix} a_{i1} & a_{i2} & a_{i3} \end{bmatrix} + \begin{bmatrix} b_{i1} & b_{i2} & b_{i3} \end{bmatrix} \right) \begin{bmatrix} c_{1j} \\ c_{2j} \\ c_{3j} \end{bmatrix} &= \begin{bmatrix} a_{i1} + b_{i1} & a_{i2} + b_{i2} & a_{i3} + b_{i3} \end{bmatrix} \begin{bmatrix} c_{1j} \\ c_{2j} \\ c_{3j} \end{bmatrix} \\
&= (a_{i1} + b_{i1})c_{1j} + (a_{i2} + b_{i2})c_{2j} + (a_{i3} + b_{i3})c_{3j} \\
&= \begin{bmatrix} a_{i1} & a_{i2} & a_{i3} \end{bmatrix} \begin{bmatrix} c_{1j} \\ c_{2j} \\ c_{3j} \end{bmatrix} + \begin{bmatrix} b_{i1} & b_{i2} & b_{i3} \end{bmatrix} \begin{bmatrix} c_{1j} \\ c_{2j} \\ c_{3j} \end{bmatrix},
\end{aligned}$$

which is the (i, j) entry of the matrix $AC + AC$. □

Theorem 1.13. *If A is an $m \times n$ matrix, B is an $n \times p$ matrix, and t is a real number, then*

$$tAB = (tA)B = A(tB).$$

Proof. We illustrate the method of the proof by considering the case $n = 3$.

We calculate the (i, j) entry in $t(AB)$, $(tA)B$, and $A(tB)$:

$$t\left(\begin{bmatrix} a_{i1} & a_{i2} & a_{i3} \end{bmatrix} \begin{bmatrix} b_{1j} \\ b_{2j} \\ b_{3j} \end{bmatrix}\right) = t(a_{i1}b_{1j} + a_{i2}b_{2j} + a_{i3}b_{3j}),$$

$$\begin{bmatrix} ta_{i1} & ta_{i2} & ta_{i3} \end{bmatrix} \begin{bmatrix} b_{1j} \\ b_{2j} \\ b_{3j} \end{bmatrix} = ta_{i1}b_{1j} + ta_{i2}b_{2j} + ta_{i3}b_{3j},$$

$$\begin{bmatrix} a_{i1} & a_{i2} & a_{i3} \end{bmatrix} \begin{bmatrix} tb_{1j} \\ tb_{2j} \\ tb_{3j} \end{bmatrix} = a_{i1}tb_{1j} + a_{i2}tb_{2j} + a_{i3}tb_{3j}.$$

Clearly, all three expressions on the right are equal. □

From Theorems 1.11 and 1.13 we obtain the following useful properties of products of matrices and vectors. In Chapter 6 we will see that, in some sense, these two properties characterize $m \times n$ matrices.

Corollary 1.14. *Let A be a $m \times n$ matrix. Then*

$$A(\mathbf{x}+\mathbf{y}) = A\mathbf{x} + A\mathbf{y} \quad \text{and} \quad A(t\mathbf{x}) = tA\mathbf{x}$$

for any vectors $\mathbf{x}$ and $\mathbf{y}$ in $\mathbb{R}^n$ and any real number t.

In the next theorem we show that the product of matrices is associative, that is, $A(BC) = (AB)C$ for any matrices A, B, and C such that the products AB and BC are defined.

Theorem 1.15. *If A is an $m \times n$ matrix, B is an $n \times p$ matrix, and C is a $p \times q$ matrix, then we have*

$$A(BC) = (AB)C.$$

Proof of a particular case. The identity can be obtained by direct calculations. We illustrate the idea of the proof by considering the case when $p = 3$ and $q = 2$. Note that Corollary 1.14 is used in the proof.

Let

$$B = \begin{bmatrix} \mathbf{b}_1 & \mathbf{b}_2 & \mathbf{b}_3 \end{bmatrix} \quad \text{and} \quad C = \begin{bmatrix} c_{11} & c_{12} \\ c_{21} & c_{22} \\ c_{31} & c_{32} \end{bmatrix}.$$

Then

$$\begin{aligned}
A(BC) &= A\left(\begin{bmatrix}\mathbf{b}_1 & \mathbf{b}_2 & \mathbf{b}_3\end{bmatrix}\begin{bmatrix}c_{11} & c_{12}\\ c_{21} & c_{22}\\ c_{31} & c_{32}\end{bmatrix}\right)\\
&= A\begin{bmatrix}c_{11}\mathbf{b}_1 + c_{21}\mathbf{b}_2 + c_{31}\mathbf{b}_3 & c_{12}\mathbf{b}_1 + c_{22}\mathbf{b}_2 + c_{32}\mathbf{b}_3\end{bmatrix}\\
&= \begin{bmatrix}A(c_{11}\mathbf{b}_1 + c_{21}\mathbf{b}_2 + c_{31}\mathbf{b}_3) & A(c_{12}\mathbf{b}_1 + c_{22}\mathbf{b}_2 + c_{32}\mathbf{b}_3)\end{bmatrix}\\
&= \begin{bmatrix}c_{11}A\mathbf{b}_1 + c_{21}A\mathbf{b}_2 + c_{31}A\mathbf{b}_3 & c_{12}A\mathbf{b}_1 + c_{22}A\mathbf{b}_2 + c_{32}A\mathbf{b}_3\end{bmatrix}\\
&= \begin{bmatrix}A\mathbf{b}_1 & A\mathbf{b}_2 & A\mathbf{b}_3\end{bmatrix}\begin{bmatrix}c_{11} & c_{12}\\ c_{21} & c_{22}\\ c_{31} & c_{32}\end{bmatrix}\\
&= \left(A\begin{bmatrix}\mathbf{b}_1 & \mathbf{b}_2 & \mathbf{b}_3\end{bmatrix}\right)\begin{bmatrix}c_{11} & c_{12}\\ c_{21} & c_{22}\\ c_{31} & c_{32}\end{bmatrix}\\
&= (AB)C.
\end{aligned}$$

□

Definition 1.16. The $n \times n$ matrix with 1's on the main diagonal, that is the diagonal in the matrix below, and 0's everywhere else is called a *unit matrix* or an *identity matrix* and denoted by I_n:

$$I_n = \begin{bmatrix}1 & 0 & \dots & 0\\ 0 & 1 & \dots & 0\\ \vdots & \vdots & \ddots & \vdots\\ 0 & 0 & \dots & 1\end{bmatrix}.$$

I_n is called an identity matrix because when we multiply a matrix A by the identity matrix of the appropriate size, then the result is the original matrix A.

Theorem 1.17. *Let A be an $m \times n$ matrix. Then*

$$I_m A = A \quad \textit{and} \quad AI_n = A.$$

Proof of a particular case. We verify the result for 4×3 matrices.

First observe that

$$\begin{bmatrix}1 & 0 & 0 & 0\\ 0 & 1 & 0 & 0\\ 0 & 0 & 1 & 0\\ 0 & 0 & 0 & 1\end{bmatrix}\begin{bmatrix}a_1\\ a_2\\ a_3\\ a_4\end{bmatrix} = \begin{bmatrix}a_1\\ a_2\\ a_3\\ a_4\end{bmatrix}.$$

If $A = \begin{bmatrix} \mathbf{a}_1 & \mathbf{a}_2 & \mathbf{a}_3 \end{bmatrix}$, where $\mathbf{a}_1$, $\mathbf{a}_2$, and $\mathbf{a}_3$ are vectors in $\mathbb{R}^4$, then

$$\begin{bmatrix} 1 & 0 & 0 & 0 \\ 0 & 1 & 0 & 0 \\ 0 & 0 & 1 & 0 \\ 0 & 0 & 0 & 1 \end{bmatrix} \begin{bmatrix} \mathbf{a}_1 & \mathbf{a}_2 & \mathbf{a}_3 \end{bmatrix} = \begin{bmatrix} \begin{bmatrix} 1 & 0 & 0 & 0 \\ 0 & 1 & 0 & 0 \\ 0 & 0 & 1 & 0 \\ 0 & 0 & 0 & 1 \end{bmatrix} \mathbf{a}_1 & \begin{bmatrix} 1 & 0 & 0 & 0 \\ 0 & 1 & 0 & 0 \\ 0 & 0 & 1 & 0 \\ 0 & 0 & 0 & 1 \end{bmatrix} \mathbf{a}_2 & \begin{bmatrix} 1 & 0 & 0 & 0 \\ 0 & 1 & 0 & 0 \\ 0 & 0 & 1 & 0 \\ 0 & 0 & 0 & 1 \end{bmatrix} \mathbf{a}_3 \end{bmatrix}$$

$$= \begin{bmatrix} \mathbf{a}_1 & \mathbf{a}_2 & \mathbf{a}_3 \end{bmatrix}.$$

Moreover, since

$$\begin{bmatrix} a_{11} & a_{12} & a_{13} \\ a_{21} & a_{22} & a_{23} \\ a_{31} & a_{32} & a_{33} \\ a_{41} & a_{42} & a_{43} \end{bmatrix} \begin{bmatrix} 1 \\ 0 \\ 0 \end{bmatrix} = \begin{bmatrix} a_{11} \\ a_{21} \\ a_{31} \\ a_{41} \end{bmatrix}, \quad \begin{bmatrix} a_{11} & a_{12} & a_{13} \\ a_{21} & a_{22} & a_{23} \\ a_{31} & a_{32} & a_{33} \\ a_{41} & a_{42} & a_{43} \end{bmatrix} \begin{bmatrix} 0 \\ 1 \\ 0 \end{bmatrix} = \begin{bmatrix} a_{12} \\ a_{22} \\ a_{32} \\ a_{42} \end{bmatrix},$$

and

$$\begin{bmatrix} a_{11} & a_{12} & a_{13} \\ a_{21} & a_{22} & a_{23} \\ a_{31} & a_{32} & a_{33} \\ a_{41} & a_{42} & a_{43} \end{bmatrix} \begin{bmatrix} 0 \\ 0 \\ 1 \end{bmatrix} = \begin{bmatrix} a_{13} \\ a_{23} \\ a_{33} \\ a_{43} \end{bmatrix},$$

we have

$$\begin{bmatrix} \mathbf{a}_1 & \mathbf{a}_2 & \mathbf{a}_3 \end{bmatrix} \begin{bmatrix} 1 & 0 & 0 \\ 0 & 1 & 0 \\ 0 & 0 & 1 \end{bmatrix} = \begin{bmatrix} \mathbf{a}_1 & \mathbf{a}_2 & \mathbf{a}_3 \end{bmatrix}.$$

□

The following simple theorem is often quite useful in arguments when it is necessary to show that two matrices are equal.

Theorem 1.18. *Let A and B be $m \times n$ matrices. If*

$$A\mathbf{x} = B\mathbf{x}$$

for every vector $\mathbf{x}$ in $\mathbb{R}^n$, then $A = B$.

Proof of a particular case. If the equality

$$\begin{bmatrix} \mathbf{a}_1 & \mathbf{a}_2 & \mathbf{a}_3 \end{bmatrix} \begin{bmatrix} x_1 \\ x_2 \\ x_3 \end{bmatrix} = \begin{bmatrix} \mathbf{b}_1 & \mathbf{b}_2 & \mathbf{b}_3 \end{bmatrix} \begin{bmatrix} x_1 \\ x_2 \\ x_3 \end{bmatrix},$$

or equivalently

$$x_1\mathbf{a}_1 + x_2\mathbf{a}_2 + x_3\mathbf{a}_3 = x_1\mathbf{b}_1 + x_2\mathbf{b}_2 + x_3\mathbf{b}_3,$$

holds for every vector $\mathbf{x} = \begin{bmatrix} x_1 \\ x_2 \\ x_3 \end{bmatrix}$ in $\mathbb{R}^3$, then it must hold for the vectors $\begin{bmatrix} 1 \\ 0 \\ 0 \end{bmatrix}$, $\begin{bmatrix} 0 \\ 1 \\ 0 \end{bmatrix}$, and $\begin{bmatrix} 0 \\ 0 \\ 1 \end{bmatrix}$. Consequently, we have

$$\mathbf{a}_1 = \mathbf{b}_1, \quad \mathbf{a}_2 = \mathbf{b}_2, \quad \text{and} \quad \mathbf{a}_3 = \mathbf{b}_3.$$

But this means that $A = B$. □

Transpose of a matrix and symmetric matrices

Definition 1.19. The *transpose* of a matrix

$$A = \begin{bmatrix} a_{11} & a_{12} & \dots & a_{1n} \\ a_{21} & a_{22} & \dots & a_{2n} \\ \vdots & \vdots & & \vdots \\ a_{m1} & a_{m2} & \dots & a_{mn} \end{bmatrix}$$

is the matrix denoted by A^T whose rows are the columns of A in the same order, that is

$$A^T = \begin{bmatrix} a_{11} & a_{21} & \dots & a_{m1} \\ a_{12} & a_{22} & \dots & a_{m2} \\ \vdots & \vdots & & \vdots \\ a_{1n} & a_{2n} & \dots & a_{mn} \end{bmatrix}.$$

The first row of A^T is the same as the first column of A, the second row of A^T is the same as the second column of A, and so on. Note that the columns of A^T are the same as the rows of A. If A is an $m \times n$ matrix, then A^T is an $n \times m$ matrix.

Example 1.20. Here are some examples of transposes of matrices of different sizes:

$$\begin{bmatrix} 1 & 3 \\ 2 & 0 \\ 3 & 4 \end{bmatrix}^T = \begin{bmatrix} 1 & 2 & 3 \\ 3 & 0 & 4 \end{bmatrix}$$

$$\begin{bmatrix} 1 & -1 & 0 \\ 2 & 3 & 5 \\ -3 & 4 & -7 \end{bmatrix}^T = \begin{bmatrix} 1 & 2 & -3 \\ -1 & 3 & 4 \\ 0 & 5 & -7 \end{bmatrix}$$

$$\begin{bmatrix} 1 \\ 2 \\ 3 \\ 4 \end{bmatrix}^T = \begin{bmatrix} 1 & 2 & 3 & 4 \end{bmatrix}.$$

The products AA^T and A^TA are always defined. If A is an $m \times n$ matrix, then AA^T is a square $m \times m$ matrix and A^TA is a square $n \times n$ matrix.

Example 1.21. We have

$$\begin{bmatrix} 1 & 3 \\ 2 & 0 \\ 3 & 4 \end{bmatrix} \begin{bmatrix} 1 & 3 \\ 2 & 0 \\ 3 & 4 \end{bmatrix}^T = \begin{bmatrix} 1 & 3 \\ 2 & 0 \\ 3 & 4 \end{bmatrix} \begin{bmatrix} 1 & 2 & 3 \\ 3 & 0 & 4 \end{bmatrix} = \begin{bmatrix} 10 & 2 & 15 \\ 2 & 4 & 6 \\ 15 & 6 & 25 \end{bmatrix}$$

and

$$\begin{bmatrix} 1 & 3 \\ 2 & 0 \\ 3 & 4 \end{bmatrix}^T \begin{bmatrix} 1 & 3 \\ 2 & 0 \\ 3 & 4 \end{bmatrix} = \begin{bmatrix} 1 & 2 & 3 \\ 3 & 0 & 4 \end{bmatrix} \begin{bmatrix} 1 & 3 \\ 2 & 0 \\ 3 & 4 \end{bmatrix} = \begin{bmatrix} 14 & 15 \\ 15 & 25 \end{bmatrix}.$$

Example 1.22. We have

$$\begin{bmatrix} 3 \\ 1 \\ 5 \\ -1 \end{bmatrix} \begin{bmatrix} 3 \\ 1 \\ 5 \\ -1 \end{bmatrix}^T = \begin{bmatrix} 3 \\ 1 \\ 5 \\ -1 \end{bmatrix} \begin{bmatrix} 3 & 1 & 5 & -1 \end{bmatrix} = \begin{bmatrix} 9 & 3 & 15 & -3 \\ 3 & 1 & 5 & -1 \\ 15 & 5 & 25 & -5 \\ -3 & -1 & -5 & 1 \end{bmatrix}$$

and

$$\begin{bmatrix} 3 \\ 1 \\ 5 \\ -1 \end{bmatrix}^T \begin{bmatrix} 3 \\ 1 \\ 5 \\ -1 \end{bmatrix} = \begin{bmatrix} 3 & 1 & 5 & -1 \end{bmatrix} \begin{bmatrix} 3 \\ 1 \\ 5 \\ -1 \end{bmatrix} = 3 \cdot 3 + 1 \cdot 1 + 5 \cdot 5 + (-1) \cdot (-1) = 36.$$

The next result is an immediate consequence of the definition of the transpose.

Theorem 1.23. *If A and B are $m \times n$ matrices, then*

$$(A+B)^T = A^T + B^T.$$

Theorem 1.24. *If A is an $m \times n$ matrix then*

$$(A^T)^T = A.$$

Proof. The (i, j) entry of the matrix $(A^T)^T$ is the (j, i) entry of the matrix A^T which is the (i, j) entry of the matrix A. □

Example 1.25.

$$\begin{bmatrix} a_{11} & a_{12} & a_{13} \\ a_{21} & a_{22} & a_{23} \end{bmatrix}^T = \begin{bmatrix} a_{11} & a_{21} \\ a_{12} & a_{22} \\ a_{13} & a_{23} \end{bmatrix}$$

and

$$\left(\begin{bmatrix} a_{11} & a_{12} & a_{13} \\ a_{21} & a_{22} & a_{23} \end{bmatrix}^T\right)^T = \begin{bmatrix} a_{11} & a_{21} \\ a_{12} & a_{22} \\ a_{13} & a_{23} \end{bmatrix}^T = \begin{bmatrix} a_{11} & a_{12} & a_{13} \\ a_{21} & a_{22} & a_{23} \end{bmatrix}.$$

Theorem 1.26. *If A is an $m \times n$ matrix and B is an $n \times p$ matrix, then*

$$(AB)^T = B^T A^T.$$

Proof for $n = 5$. The (i, j) entry of the matrix $B^T A^T$ is

$$\begin{bmatrix} b_{1i} & b_{2i} & b_{3i} & b_{4i} & b_{5i} \end{bmatrix} \begin{bmatrix} a_{j1} \\ a_{j2} \\ a_{j3} \\ a_{j4} \\ a_{j5} \end{bmatrix} = a_{j1}b_{1i} + a_{j2}b_{2i} + a_{j3}b_{3i} + a_{j4}b_{4i} + a_{j5}b_{5i},$$

which is the (j, i) entry of the matrix AB which is the (i, j) entry of the matrix $(AB)^T$. □

Note that the order of matrices in the above equality changes. Since B^T is a $p \times n$ matrix and A^T is an $n \times m$, the product $B^T A^T$ makes sense. For example, we have

$$\left(\begin{bmatrix} a_{11} & a_{12} & a_{13} \\ a_{21} & a_{22} & a_{23} \end{bmatrix}\begin{bmatrix} b_{11} & b_{12} \\ b_{21} & b_{22} \\ b_{31} & b_{32} \end{bmatrix}\right)^T = \begin{bmatrix} b_{11} & b_{12} \\ b_{21} & b_{22} \\ b_{31} & b_{32} \end{bmatrix}^T \begin{bmatrix} a_{11} & a_{12} & a_{13} \\ a_{21} & a_{22} & a_{23} \end{bmatrix}^T$$

$$= \begin{bmatrix} b_{11} & b_{21} & b_{31} \\ b_{12} & b_{22} & b_{32} \end{bmatrix}\begin{bmatrix} a_{11} & a_{21} \\ a_{12} & a_{22} \\ a_{13} & a_{23} \end{bmatrix}.$$

Definition 1.27. An $n \times n$ matrix is called a *square matrix.* A square matrix A is called *symmetric* if $A^T = A$.

Example 1.28. Here are some examples of symmetric matrices:

$$\begin{bmatrix} a & b \\ b & c \end{bmatrix}, \quad \begin{bmatrix} a & b & c \\ b & d & e \\ c & e & f \end{bmatrix}, \quad \begin{bmatrix} 5 & 3 & 7 & 9 \\ 3 & 4 & 8 & 5 \\ 7 & 8 & -1 & 4 \\ 9 & 5 & 4 & 3 \end{bmatrix}.$$

Both matrices AA^T and $A^T A$ found in Example 1.21 are symmetric. It is not difficult to show that, for an arbitrary matrix A, the matrices AA^T and $A^T A$ are symmetric.

Exercises 1.1

Find the sum of the given matrices.

1. $\begin{bmatrix} a & b & c \end{bmatrix}, \begin{bmatrix} d & e & f \end{bmatrix}$

2. $\begin{bmatrix} a \\ b \\ c \\ d \end{bmatrix}, \begin{bmatrix} e \\ f \\ g \\ h \end{bmatrix}$

3. $\begin{bmatrix} a_{11} & a_{12} \\ a_{21} & a_{22} \end{bmatrix}, \begin{bmatrix} b_{11} & b_{12} \\ b_{21} & b_{22} \end{bmatrix}$

4. $\begin{bmatrix} 7 & 4 & 3 \\ 1 & -5 & 0 \end{bmatrix}, \begin{bmatrix} -4 & 8 & 5 \\ 9 & -12 & 19 \end{bmatrix}$

Write the following expressions as single matrices.

5. $2\begin{bmatrix} a & b & c \end{bmatrix} + 5\begin{bmatrix} d & e & f \end{bmatrix}$

6. $7\begin{bmatrix} a \\ b \\ c \\ e \end{bmatrix} - 3\begin{bmatrix} e \\ f \\ g \\ h \end{bmatrix}$

7. $s\begin{bmatrix} 2 & 0 & 2 \\ 0 & 1 & 3 \\ 5 & 0 & 0 \end{bmatrix} + t\begin{bmatrix} 1 & 0 & 4 \\ 1 & 1 & 0 \\ 0 & 2 & -1 \end{bmatrix}$

8. $s\begin{bmatrix} 3 & 1 & 1 \\ 2 & 4 & 1 \end{bmatrix} + t\begin{bmatrix} 0 & 4 & -3 \\ 2 & -5 & 0 \end{bmatrix} + u\begin{bmatrix} 1 & -3 & 7 \\ 0 & 1 & 0 \end{bmatrix}$

Find the following products.

9. $\begin{bmatrix} 1 & 1 & 0 & 1 \\ 0 & 0 & 1 & 1 \end{bmatrix}\begin{bmatrix} 2 & 1 & 0 \\ 1 & 0 & 2 \\ 0 & 1 & 0 \\ 0 & 0 & 1 \end{bmatrix}$

10. $\begin{bmatrix} 1 & 1 & 0 \\ 0 & 0 & 1 \\ 2 & 0 & 1 \\ 1 & 1 & 1 \end{bmatrix}\begin{bmatrix} 2 & 1 \\ 1 & 0 \\ 0 & 1 \end{bmatrix}$

11. $\begin{bmatrix} a \\ b \end{bmatrix}\begin{bmatrix} c & d \end{bmatrix}$

12. $\begin{bmatrix} a & b & c \end{bmatrix}\begin{bmatrix} e \\ f \\ g \end{bmatrix}$

13. $\begin{bmatrix} 7 \\ 4 \\ 5 \end{bmatrix}\begin{bmatrix} 4 & 3 \end{bmatrix}$

14. $\begin{bmatrix} 5 \\ 2 \end{bmatrix}\begin{bmatrix} a & b & c \end{bmatrix}$

15. $\begin{bmatrix} a_1 & a_2 \end{bmatrix}\begin{bmatrix} b_{11} & b_{12} & b_{13} \\ b_{21} & b_{22} & b_{23} \end{bmatrix}$

16. $\begin{bmatrix} b_{11} & b_{12} & b_{13} \\ b_{21} & b_{22} & b_{23} \end{bmatrix}\begin{bmatrix} c_1 \\ c_2 \\ c_3 \end{bmatrix}$

Explain why the following products are not defined.

17. $\begin{bmatrix} a & b & c \end{bmatrix}\begin{bmatrix} d \\ e \\ f \\ g \end{bmatrix}$

18. $\begin{bmatrix} a & b & c \end{bmatrix}\begin{bmatrix} d \\ f \end{bmatrix}$

19. $\begin{bmatrix} 1 & 1 & 3 & 1 & 1 \\ 7 & 3 & 1 & 1 & 3 \end{bmatrix}\begin{bmatrix} 2 & 1 & 3 \\ 1 & 5 & 1 \\ 4 & 1 & 2 \end{bmatrix}$

20. $\begin{bmatrix} 4 & 0 \\ 4 & 2 \\ 5 & 1 \\ 1 & 3 \end{bmatrix}\begin{bmatrix} 3 & 2 \\ 4 & 3 \\ 1 & 1 \end{bmatrix}$

21. Show that

$$\begin{bmatrix} a_1 & a_2 \end{bmatrix}\left(\begin{bmatrix} b_{11} & b_{12} & b_{13} \\ b_{21} & b_{22} & b_{23} \end{bmatrix}\begin{bmatrix} c_1 \\ c_2 \\ c_3 \end{bmatrix}\right) = \left(\begin{bmatrix} a_1 & a_2 \end{bmatrix}\begin{bmatrix} b_{11} & b_{12} & b_{13} \\ b_{21} & b_{22} & b_{23} \end{bmatrix}\right)\begin{bmatrix} c_1 \\ c_2 \\ c_3 \end{bmatrix}.$$

22. Show, without using Theorem 1.18, that if $\begin{bmatrix} a_{11} & a_{12} \\ a_{21} & a_{22} \end{bmatrix}\begin{bmatrix} x_1 \\ x_2 \end{bmatrix} = \begin{bmatrix} b_{11} & b_{12} \\ b_{21} & b_{22} \end{bmatrix}\begin{bmatrix} x_1 \\ x_2 \end{bmatrix}$ for every $\begin{bmatrix} x_1 \\ x_2 \end{bmatrix}$, then $\begin{bmatrix} a_{11} & a_{12} \\ a_{21} & a_{22} \end{bmatrix} = \begin{bmatrix} b_{11} & b_{12} \\ b_{21} & b_{22} \end{bmatrix}$.

Find A^T for the given matrix A.

23. $A = \begin{bmatrix} 5 & 7 & 9 \\ 1 & 2 & 3 \\ 2 & 4 & 8 \end{bmatrix}$

24. $A = \begin{bmatrix} 1 & 7 & 9 \\ 7 & 2 & 3 \\ 9 & 3 & 5 \end{bmatrix}$

25. $A = \begin{bmatrix} a & b & c \end{bmatrix}$

26. $A = \begin{bmatrix} a \\ b \\ c \\ d \end{bmatrix}$

27. $A = \begin{bmatrix} a & b \\ c & d \\ e & f \end{bmatrix}$

28. $A = \begin{bmatrix} a & b & c & d \\ e & f & g & h \end{bmatrix}$

29. $A = \begin{bmatrix} 2 & 3 & 1 & 4 \\ 7 & 1 & 5 & 2 \\ 9 & 8 & 3 & 3 \end{bmatrix}$

30. $A = \begin{bmatrix} 3 & 1 & 7 \\ 3 & 5 & 2 \\ 8 & 4 & 4 \\ 1 & 0 & -3 \end{bmatrix}$

31. Prove Theorem 1.18 for $n = 2$.

32. Prove Theorem 1.18 for $n = 4$.

33. If A is an $m \times n$ matrix, B is an $n \times p$ matrix, C is a $p \times q$ matrix, and D is a $q \times r$ matrix, show that
$$(ABC)D = (AB)(CD) = A(BCD).$$

34. If A is an $m \times n$ matrix, B is an $n \times p$ matrix, C is a $p \times q$ matrix, D is a $q \times r$ matrix, and E is a $r \times s$ matrix, show that
$$(ABC)(DE) = A(BC)DE = A(BCD)E.$$

35. If A is an $m \times n$ matrix, B is an $n \times p$ matrix, and C is a $p \times q$ matrix, show that
$$(ABC)^T = C^T B^T A^T.$$

36. If A is an $m \times n$ matrix, B is an $n \times p$ matrix, C is a $p \times q$ matrix, and D is a $q \times r$ matrix, show that
$$(ABCD)^T = D^T C^T B^T A^T.$$

37. If A is an $n \times p$ matrix and B is an $m \times n$ matrix, show that the matrix $BAA^T B^T$ is symmetric.

38. If A is an $m \times n$ matrix, B is an $n \times p$ matrix, and C is a $p \times q$ matrix, show that the matrix $ABCC^T B^T A^T$ is symmetric.

39. Show that
$$\begin{bmatrix} a_{11} & a_{12} \\ a_{21} & a_{22} \end{bmatrix} \begin{bmatrix} b_{11} & b_{12} \\ b_{21} & b_{22} \end{bmatrix} = \begin{bmatrix} a_{11} \\ a_{21} \end{bmatrix} \begin{bmatrix} b_{11} & b_{12} \end{bmatrix} + \begin{bmatrix} a_{12} \\ a_{22} \end{bmatrix} \begin{bmatrix} b_{21} & b_{22} \end{bmatrix}.$$

40. Show that

$$\begin{bmatrix} a_{11} & a_{12} & a_{13} \\ a_{21} & a_{22} & a_{23} \end{bmatrix} \begin{bmatrix} b_{11} & b_{12} \\ b_{21} & b_{22} \\ b_{31} & b_{32} \end{bmatrix} = \begin{bmatrix} a_{11} \\ a_{21} \end{bmatrix} \begin{bmatrix} b_{11} & b_{12} \end{bmatrix} + \begin{bmatrix} a_{12} \\ a_{22} \end{bmatrix} \begin{bmatrix} b_{21} & b_{22} \end{bmatrix} + \begin{bmatrix} a_{13} \\ a_{23} \end{bmatrix} \begin{bmatrix} b_{31} & b_{32} \end{bmatrix}.$$

1.2 Gaussian elimination

By a *linear equation* we mean any equation that can be written in the form

$$a_1 x_1 + a_2 x_2 + \cdots + a_n x_n = b \tag{1.1}$$

where $a_1, a_2, \ldots, a_n$ and b are constants (real or complex numbers) and $x_1, x_2, \ldots, x_n$ are *variables*. The numbers $a_1, a_2, \ldots, a_n$ are called the *coefficients* of the linear equation. By a *solution* of the equation (1.1) we mean any n-tuple of numbers $s_1, s_2, \ldots, s_n$ such that if we substitute $x_1 = s_1, x_2 = s_2, \ldots, x_n = s_n$ in (1.1) then we get a valid equality.

Example 1.29. The equation

$$x + 2y - 3z = 4$$

is a linear equation. It is easy to check that $x = 5, y = 4, z = 3$ is a solution of the equation, while $x = 3, y = 4, z = 5$ is not. This equation has infinitely many solutions. For arbitrary real numbers s and t,

$$x = 4 - 2s + 3t,\ y = s,\ z = t$$

is a solution of the equation. As we will learn later, every solution of the equation $x + 2y - 3z = 4$ is of this form for some numbers s and t. For example, the solution $x = 5, y = 4, z = 3$ is obtained if we take $s = 4$ and $t = 3$.

By a *system of linear equations* we mean a collection of any finite number of linear equations with the same variables:

$$\begin{cases} a_{11}x_1 + a_{12}x_2 + \cdots + a_{1n}x_n = b_1 \\ a_{21}x_1 + a_{22}x_2 + \cdots + a_{2n}x_n = b_2 \\ \quad\vdots \qquad\quad \vdots \qquad\quad \vdots \qquad\quad \vdots \\ a_{m1}x_1 + a_{m2}x_2 + \ldots + a_{mn}x_n = b_m \end{cases} \tag{1.2}$$

By a *solution of a system of equations* of the system (1.2) we mean any n-tuple of numbers $s_1, s_2, \ldots, s_n$ such that it is a solution of every equation in the system.

Example 1.30. It is easy to verify that $x = 0, y = 1, z = 1$ is a solution of the system

$$\begin{cases} 2x + 3y - 2z = \ \ 1 \\ \ \ x + 2y + \ \ z = \ \ 3 \\ 3x + 4y - 6z = -2 \end{cases}.$$

While verifying that some numbers are a solution of a given system of linear equations is easy, finding a solution for a system with many equations and many variables can be quite difficult and time-consuming. Solving systems of linear equations is one of the fundamental problems in linear algebra. One method is based on the observation that solving a system of the form

$$\begin{cases} a_{1,1}x_1 + a_{1,2}x_2 + a_{1,3}x_3 + \cdots + a_{1,n-1}x_{n-1} + a_{1,n}x_n = b_1 \\ \qquad\quad\ a_{2,2}x_2 + a_{2,3}x_3 + \cdots + a_{2,n-1}x_{n-1} + a_{2,n}x_n = b_2 \\ \qquad\qquad\qquad\qquad\qquad\qquad\qquad \vdots \qquad\qquad\quad \vdots \\ \qquad\qquad\qquad\qquad\qquad a_{m-1,n-1}x_{n-1} + a_{m-1,n}x_n = b_{m-1} \\ \qquad\qquad\qquad\qquad\qquad\qquad\qquad\qquad\quad a_{m,n}x_n = b_m \end{cases} \tag{1.3}$$

is relatively easy.

Example 1.31. Solve the system

$$\begin{cases} 2x + 3y + 2z = \ \ 6 \\ \qquad\ \ y + 4z = \ \ 3 \\ \qquad\qquad 2z = -1 \end{cases}.$$

Solution. From the last equation we get $z = -\frac{1}{2}$. Now we can substitute the known value of z in the second equation which gives us

$$y + 4 \cdot \left(-\frac{1}{2}\right) = 3$$

and consequently $y = 5$. Finally, we substitute the known values of y and z in the first equation and get

$$2x + 15 - 1 = 6$$

and thus $x = -4$. The solution is

$$x = -4, \quad y = 5, \quad \text{and} \quad z = -\frac{1}{2}.$$

□

It turns out that any system (1.2) can be transformed to a system of the form (1.3) that has the same solutions as the original system. First we note that we will not affect the solution of the system of linear equations if we multiply one of the equations by a number different from 0 or multiply one of the equations by a number and then add the result to another equation. Moreover, if necessary, we can always change the order of equations in the system. Then we observe that by manipulating any system (1.2) using such operations we can eventually change it to the form (1.3) or a similar form that makes solving the system quite easy. This process is referred to as *Gaussian elimination.* In this section we discuss the process of Gaussian elimination in detail and examine different possible outcomes of the process.

The process of solving systems of linear equations using the Gaussian elimination method can be simplified if we use matrices. We observe that the complete information about a system of linear equations

$$\begin{cases} a_{11}x_1 + a_{12}x_2 + \cdots + a_{1n}x_n = b_1 \\ a_{21}x_1 + a_{22}x_2 + \cdots + a_{2n}x_n = b_2 \\ \quad\vdots \qquad\quad \vdots \qquad \vdots \qquad\quad \vdots \\ a_{m1}x_1 + a_{m2}x_2 + \ldots + a_{mn}x_n = b_m \end{cases} \tag{1.4}$$

is contained in the matrix

$$\begin{bmatrix} a_{11} & a_{12} & \ldots & a_{1n} & b_1 \\ a_{21} & a_{22} & \ldots & a_{2n} & b_2 \\ \vdots & \vdots & & \vdots & \vdots \\ a_{m1} & a_{m2} & \ldots & a_{mn} & b_m \end{bmatrix}.$$

It suffices to remember that the first column corresponds to the variable x_1, the second column to the variable x_2, and so on until we reach the n-th column. The last column contains the numbers on the other side of the = sign. This matrix is called the *augmented matrix* of the system. In some books the last column is separated from the remaining columns by a vertical line, that is,

$$\left[\begin{array}{cccc|c} a_{11} & a_{12} & \ldots & a_{1n} & b_1 \\ a_{21} & a_{22} & \ldots & a_{2n} & b_2 \\ \vdots & \vdots & & \vdots & \vdots \\ a_{m1} & a_{m2} & \ldots & a_{mn} & b_m \end{array}\right],$$

but we are not going to do that. When we use the augmented matrix of a system of equations it is easy to remember what the last column represents. Adding a vertical line may suggest that an augmented matrix is not an ordinary matrix, while it is.

We will now describe how to solve a system of linear equations by first converting it to the augmented matrix of the system and then performing elementary operations on rows of that matrix, that is, interchange two rows of the matrix, multiply a row of the matrix by a nonzero constant, and multiply a row of the matrix by a constant and then add the result to another row of the matrix.

Example 1.32. Solve the system

$$\begin{cases} 2x + 3y - 2z = \ \ 1 \\ \ \ x + 2y + \ \ z = \ \ 3 \\ 3x + 4y - 6z = -2 \end{cases}.$$

Solution. We solve the system by first converting it to an augmented matrix and then performing elementary operations on rows of that matrix until we obtain a matrix of the form

$$\begin{bmatrix} 1 & * & * & * \\ 0 & 1 & * & * \\ 0 & 0 & 1 & * \end{bmatrix},$$

where each $*$ indicates an arbitrary number. Then we convert the obtained matrix back to a system of linear equations that is easy to solve.

The augmented matrix of the system is

$$\begin{bmatrix} 2 & 3 & -2 & 1 \\ 1 & 2 & 1 & 3 \\ 3 & 4 & -6 & -2 \end{bmatrix}.$$

Now apply the indicated elementary row operations:

interchange rows 1 and 2

$$\begin{bmatrix} 1 & 2 & 1 & 3 \\ 2 & 3 & -2 & 1 \\ 3 & 4 & -6 & -2 \end{bmatrix}$$

multiply row 1 by -2 and then add to row 2

$$\begin{bmatrix} 1 & 2 & 1 & 3 \\ 0 & -1 & -4 & -5 \\ 3 & 4 & -6 & -2 \end{bmatrix}$$

multiply row 1 by -3 and then add to row 3

$$\begin{bmatrix} 1 & 2 & 1 & 3 \\ 0 & -1 & -4 & -5 \\ 0 & -2 & -9 & -11 \end{bmatrix}$$

multiply row 2 by -1

$$\begin{bmatrix} 1 & 2 & 1 & 3 \\ 0 & 1 & 4 & 5 \\ 0 & -2 & -9 & -11 \end{bmatrix}$$

multiply row 2 by 2 and then add to equation 3

$$\begin{bmatrix} 1 & 2 & 1 & 3 \\ 0 & 1 & 4 & 5 \\ 0 & 0 & -1 & -1 \end{bmatrix}$$

multiply row 3 by -1

$$\begin{bmatrix} 1 & 2 & 1 & 3 \\ 0 & 1 & 4 & 5 \\ 0 & 0 & 1 & 1 \end{bmatrix}.$$

The last matrix corresponds to the system

$$\begin{cases} x + 2x + \quad z = 3 \\ \qquad\quad y + 4z = 5 \\ \qquad\qquad\quad z = 1 \end{cases}.$$

Proceeding as in Example 1.31 we obtain

$$x = 0, \quad y = 1, \quad \text{and} \quad z = 1.$$

□

We note that the two operations
multiply row 1 by -2 and then add to row 2
multiply row 1 by -3 and then add to row 3
in the above example do not depend on each other in any way and the order of these two operations does not matter. We can combine them together. More precisely, from the matrix

$$\begin{bmatrix} 1 & 2 & 1 & 3 \\ 2 & 3 & -2 & 1 \\ 3 & 4 & -6 & -2 \end{bmatrix}$$

by applying the following operations
multiply row 1 by -2 and then add to row 2
multiply row 1 by -3 and then add to row 3
we get the matrix

$$\begin{bmatrix} 1 & 2 & 1 & 3 \\ 0 & -1 & -4 & -5 \\ 0 & -2 & -9 & -11 \end{bmatrix}.$$

The Gaussian elimination algorithm

In the previous example we illustrated the Gaussian elimination method in the context of solving a system of three linear equations with three variables. The process can be applied to any matrix and has applications other than solving systems of linear equations. Now we will carefully describe the general process of Gaussian elimination. First we need to introduce some new terminology.

Definition 1.33. By a *leading entry* in a row of a matrix we mean the first nonzero entry. In other words, a leading entry in a row is a nonzero entry in that row such that all entries to the left of it are 0. If a leading entry is equal to 1, then we call it a *leading* 1.

Example 1.34. Here are some examples of matrices with the leading entries enclosed in boxes. Each leading 1 is indicated by a red box.

$$\begin{bmatrix} \boxed{1} & -1 & 0 & 9 \\ 0 & \boxed{2} & 0 & 2 \\ 0 & 0 & \boxed{-1} & 5 \end{bmatrix}, \begin{bmatrix} \boxed{1} & 3 & -2 & 0 & 3 \\ 0 & 0 & 0 & \boxed{1} & 7 \\ 0 & 0 & 0 & 0 & 0 \\ 0 & 0 & 0 & 0 & 0 \end{bmatrix}, \begin{bmatrix} \boxed{3} & 5 & 0 & 2 & 3 & 2 & 0 \\ 0 & 0 & \boxed{-2} & 7 & 2 & 4 & 0 \\ 0 & 0 & 0 & 0 & 0 & 0 & \boxed{3} \\ 0 & 0 & 0 & 0 & 0 & 0 & 0 \end{bmatrix}.$$

Definition 1.35. A matrix is in a *reduced row echelon form* (or *Gauss-Jordan form*) if it satisfies all of the following conditions:

(a) All leading entries of the matrix are equal to 1;

(b) In each column with a leading 1 all other entries are equal to 0;

(c) Each leading 1 is in a column to the right of the leading 1 in any row above it;

(d) Each row whose entries are all 0 is below all rows with some nonzero entries.

Example 1.36. Here are examples of matrices in a reduced row echelon form.

$$\begin{bmatrix} 1 & 0 & 3 & 0 \\ 0 & 1 & 8 & 0 \\ 0 & 0 & 0 & 1 \end{bmatrix}, \begin{bmatrix} 1 & 3 & -2 & 0 & 3 \\ 0 & 0 & 0 & 1 & 7 \\ 0 & 0 & 0 & 0 & 0 \\ 0 & 0 & 0 & 0 & 0 \end{bmatrix}, \begin{bmatrix} 1 & 0 & 4 & 0 & 3 & 1 & 0 & 2 \\ 0 & 1 & 7 & 0 & 2 & 0 & 0 & 2 \\ 0 & 0 & 0 & 1 & 8 & 3 & 0 & 7 \\ 0 & 0 & 0 & 0 & 0 & 0 & 1 & 5 \end{bmatrix}.$$

Example 1.37. Here are examples of matrices that are not in a reduced row echelon form.

1. The matrix
$$\begin{bmatrix} 1 & 0 & 0 & 9 \\ 0 & 1 & 0 & 2 \\ 0 & 0 & \boxed{2} & 5 \end{bmatrix}$$
is not in a reduced row echelon form because condition (a) is not satisfied. The leading entry in the third row is not 1.

2. The matrix
$$\begin{bmatrix} 1 & 0 & 0 \\ 0 & 1 & \boxed{-1} \\ 0 & 0 & 1 \end{bmatrix}$$
is not in a reduced row echelon form because condition (b) is not satisfied. The third column has a leading 1 and another nonzero entry.

3. The matrix
$$\begin{bmatrix} 1 & 3 & 0 & 0 & 3 \\ 0 & 0 & 0 & \boxed{1} & 7 \\ 0 & 0 & \boxed{1} & 0 & 0 \\ 0 & 0 & 0 & 0 & 0 \end{bmatrix}$$
is not in a reduced row echelon form because condition (c) is not satisfied. The leading 1 in the third row is not in a column to the right of the leading 1 in the row above it.

4. The matrix
$$\begin{bmatrix} 1 & 5 & 0 & 2 & 3 & 2 & 0 & 5 \\ 0 & 0 & 1 & 7 & 2 & 4 & 0 & 3 \\ 0 & 0 & 0 & 0 & 0 & 0 & 0 & 0 \\ 0 & 0 & 0 & 0 & 0 & 0 & \boxed{1} & 0 \end{bmatrix}$$
is not in a reduced row echelon form because condition (d) is not satisfied. All entries in the third row are 0's, but the fourth row has a nonzero entry.

Definition 1.38. Every entry in a matrix where a leading 1 is located is called a *pivot position*. Every column that contains a pivot position is called a *pivot column*.

Example 1.39. All pivot positions in the matrix
$$\begin{bmatrix} \boxed{1} & 5 & 0 & 2 & 3 & 2 & 0 & 5 \\ 0 & 0 & \boxed{1} & 7 & 2 & 4 & 0 & 3 \\ 0 & 0 & 0 & 0 & 0 & 0 & \boxed{1} & 2 \\ 0 & 0 & 0 & 0 & 0 & 0 & 0 & 0 \end{bmatrix}$$

are marked. Columns 1, 3, and 7 are the pivot columns of this matrix.

Definition 1.40. By *elementary row operations* on a matrix we mean the following three operations:

- **Row interchange**: Interchange two rows of the matrix.
- **Row scaling**: Multiply a row of the matrix by a nonzero constant.
- **Row replacement**: Multiply a row of the matrix by a constant and then add the result to another row of the matrix.

Definition 1.41. If a matrix A can be transformed into a matrix B by elementary operations, we write

$$A \sim B$$

and say that matrices A and B are *equivalent*.

Example 1.42. Note that in Example 1.32 we have shown that

$$\begin{bmatrix} 2 & 3 & -2 & 1 \\ 1 & 2 & 1 & 3 \\ 3 & 4 & -6 & -2 \end{bmatrix} \sim \begin{bmatrix} 1 & 2 & 1 & 3 \\ 0 & 1 & 4 & 5 \\ 0 & 0 & 1 & 1 \end{bmatrix}.$$

The general algorithm for obtaining the Gauss-Jordan form of any matrix is based on three basic ideas used in this process:

- If there is a nonzero entry in a column, we can always move it to a desired position in that column by applying an appropriate **row interchange**.
- Any nonzero entry a can be changed to 1 by an appropriate **scaling**.
- If a column has a entry equal to 1, then any other nonzero entry b in that column can be changed to 0 by an appropriate **row replacement**.

Lemma 1.43. *If an $m \times n$ matrix A has at least one nonzero entry, then there is an integer $1 \le p \le n$ such that*

$$A \sim \begin{bmatrix} B & C \end{bmatrix}$$

where B is an $m \times p$ matrix such that the $(1,p)$ entry is 1 and all other entries are 0 and C is a $m \times (n-p)$ matrix, that is,

$$A \sim \begin{bmatrix} 0 & 0 & \cdots & 0 & 1 & * & \cdots & * \\ 0 & 0 & \cdots & 0 & 0 & * & \cdots & * \\ \vdots & \vdots & & \vdots & \vdots & \vdots & & \vdots \\ 0 & 0 & \cdots & 0 & 0 & * & \cdots & * \end{bmatrix}.$$
$$\underbrace{\qquad\qquad}_{p\ columns} \quad \underbrace{\qquad\qquad}_{n-p\ columns}$$

If $p = n$, then

$$A \sim \begin{bmatrix} 0 & 0 & \cdots & 0 & 1 \\ 0 & 0 & \cdots & 0 & 0 \\ \vdots & \vdots & & \vdots & \vdots \\ 0 & 0 & \cdots & 0 & 0 \end{bmatrix}.$$

Proof of a particular case. We consider the matrix

$$G = \begin{bmatrix} a_1 & b_1 & c_1 & d_1 & e_1 & f_1 \\ a_2 & b_2 & c_2 & d_2 & e_2 & f_2 \\ a_3 & b_3 & c_3 & d_3 & e_3 & f_3 \\ a_4 & b_4 & c_4 & d_4 & e_4 & f_4 \end{bmatrix}$$

and we suppose that that the first two columns have no nonzero entries, that is,

$$G = \begin{bmatrix} 0 & 0 & c_1 & d_1 & e_1 & f_1 \\ 0 & 0 & c_2 & d_2 & e_2 & f_2 \\ 0 & 0 & c_3 & d_3 & e_3 & f_3 \\ 0 & 0 & c_4 & d_4 & e_4 & f_4 \end{bmatrix},$$

and that $c_3 \neq 0$.

Now we apply the following elementary operations to the matrix A:

interchange rows 1 and 3

$$G \sim \begin{bmatrix} 0 & 0 & c_3 & d_3 & e_3 & f_3 \\ 0 & 0 & c_2 & d_2 & e_2 & f_2 \\ 0 & 0 & c_1 & d_1 & e_1 & f_1 \\ 0 & 0 & c_4 & d_4 & e_4 & f_4 \end{bmatrix}$$

multiply row 1 by $\frac{1}{c_3}$

$$G \sim \begin{bmatrix} 0 & 0 & 1 & \frac{d_3}{c_3} & \frac{e_3}{c_3} & \frac{f_3}{c_3} \\ 0 & 0 & c_2 & d_2 & e_2 & f_2 \\ 0 & 0 & c_1 & d_1 & e_1 & f_1 \\ 0 & 0 & c_4 & d_4 & e_4 & f_4 \end{bmatrix} = \begin{bmatrix} 0 & 0 & 1 & d'_3 & e'_3 & f'_3 \\ 0 & 0 & c_2 & d_2 & e_2 & f_2 \\ 0 & 0 & c_1 & d_1 & e_1 & f_1 \\ 0 & 0 & c_4 & d_4 & e_4 & f_4 \end{bmatrix}$$

multiply row 1 by $-c_2$ and then add to row 2
multiply row 1 by $-c_1$ and then add to row 3
multiply row 1 by $-c_4$ and then add to row 4

$$G \sim \begin{bmatrix} 0 & 0 & 1 & d'_3 & e'_3 & f'_3 \\ 0 & 0 & 0 & d_2 - c_2 d'_3 & e_2 - c_2 e'_3 & f_2 - c_2 f'_3 \\ 0 & 0 & 0 & d_1 - c_1 d'_3 & e_1 - c_2 e'_1 & f_1 - c_1 f'_3 \\ 0 & 0 & 0 & d_4 - c_4 d'_3 & e_4 - c_4 e'_3 & f_4 - c_4 f'_3 \end{bmatrix} = \begin{bmatrix} 0 & 0 & 1 & d'_3 & e'_3 & f'_3 \\ 0 & 0 & 0 & d'_2 & e'_2 & f'_2 \\ 0 & 0 & 0 & d'_1 & e'_1 & f'_1 \\ 0 & 0 & 0 & d'_4 & e'_4 & f'_4 \end{bmatrix}.$$

□

Below we describe the process of obtaining the matrix defined in Lemma 1.43. Note that it follows the method of the proof.

Obtaining the first pivot column.

Step 1 Identify the first (from the left) nonzero column. (This is a pivot column.)

Step 2 If necessary, move a row with a nonzero entry in the pivot column to the top using an appropriate row interchange. (The position at the top of the pivot column is now in a pivot position.)

Step 3 If necessary, change the entry in the pivot position to 1 using an appropriate scaling. (The 1 in the pivot position is a leading 1.)

Step 4 Replace, if necessary, every entry below the leading 1 by 0 using appropriate **row replacements**.

Example 1.44. We consider the matrix

$$\begin{bmatrix} 0 & 0 & 1 & -1 & 0 & 1 & 2 \\ 0 & 0 & 3 & 0 & 3 & -2 & 0 \\ 0 & 2 & 0 & 1 & 3 & -1 & 3 \\ 0 & -1 & 2 & 0 & 1 & 0 & 5 \end{bmatrix}.$$

Step 1 The second column is the first one that has nonzero entries. This is a pivot column.

Step 2 We interchange rows 1 and 4 to get the nonzero entry -1 at the top of the pivot column:

$$\begin{bmatrix} 0 & -1 & 2 & 0 & 1 & 0 & 5 \\ 0 & 0 & 3 & 0 & 3 & -2 & 0 \\ 0 & 2 & 0 & 1 & 3 & -1 & 3 \\ 0 & 0 & 1 & -1 & 0 & 1 & 2 \end{bmatrix}.$$

Step 3 We multiply the first row by -1 to get 1 in a pivot position:

$$\begin{bmatrix} 0 & 1 & -2 & 0 & -1 & 0 & -5 \\ 0 & 0 & 3 & 0 & 3 & -2 & 0 \\ 0 & 2 & 0 & 1 & 3 & -1 & 3 \\ 0 & 0 & 1 & -1 & 0 & 1 & 2 \end{bmatrix}.$$

Step 4 We add the first row multiplied by -2 to the third row to replace the 2 in the pivot column by 0:

$$\begin{bmatrix} 0 & 1 & -2 & 0 & -1 & 0 & -5 \\ 0 & 0 & 3 & 0 & 3 & -2 & 0 \\ 0 & 0 & 4 & 1 & 5 & -1 & 13 \\ 0 & 0 & 1 & -1 & 0 & 1 & 2 \end{bmatrix}.$$

Theorem 1.45. *Let A be an $m \times n$ matrix that is not in the reduced row echelon form. If*

(a) *the first k columns of A form an $m \times k$ matrix that is in the reduced row echelon form,*

(b) *the k-th column is a pivot column, and*

(c) *the (j,k) entry is a leading 1,*

then there is an $m \times n$ matrix A' and an integer k' such that

(a') $k < k' \leq n$,

(b') *the first k' columns of A' form an $m \times k'$ matrix that is in the reduced row echelon form,*

(c') *k'-th column of A' is a pivot column,*

(d') *the $(j+1,k')$ entry of A' is a leading 1, and*

(e') $A \sim A'$.

Now we illustrate the above by a couple of particular cases.

Example 1.46. We consider a matrix of the form

$$A = \begin{bmatrix} 0 & 1 & a_1 & b_1 & c_1 \\ 0 & 0 & 0 & b_2 & c_2 \\ 0 & 0 & 0 & b_3 & c_3 \end{bmatrix}.$$

We suppose that $\begin{bmatrix} b_2 \\ b_3 \end{bmatrix} \neq \begin{bmatrix} 0 \\ 0 \end{bmatrix}$. By applying Lemma 1.43 to the matrix $\begin{bmatrix} 0 & b_2 & c_2 \\ 0 & b_3 & c_3 \end{bmatrix}$ we get

$$A \sim \begin{bmatrix} 0 & 1 & a_1 & b_1 & c_1 \\ 0 & 0 & 0 & 1 & c_2' \\ 0 & 0 & 0 & 0 & c_3' \end{bmatrix}.$$

Next we

multiply row 2 by $-b_1$ and then add to row 1

$$A \sim \begin{bmatrix} 0 & 1 & a_1 & 0 & c_1 - b_1 c_2' \\ 0 & 0 & 0 & 1 & c_2' \\ 0 & 0 & 0 & 0 & c_3' \end{bmatrix} = \begin{bmatrix} 0 & 1 & a_1 & 0 & c_1' \\ 0 & 0 & 0 & 1 & c_2' \\ 0 & 0 & 0 & 0 & c_3' \end{bmatrix} = A'.$$

Example 1.47. We consider the matrix

$$A = \begin{bmatrix} 0 & 1 & 0 & a_1 & 0 & b_1 & c_1 & d_1 & e_1 & f_1 \\ 0 & 0 & 1 & a_2 & 0 & b_2 & c_2 & d_2 & e_2 & f_2 \\ 0 & 0 & 0 & 0 & 1 & b_3 & c_3 & d_3 & e_3 & f_3 \\ 0 & 0 & 0 & 0 & 0 & 0 & c_4 & d_4 & e_4 & f_4 \\ 0 & 0 & 0 & 0 & 0 & 0 & c_5 & d_5 & e_5 & f_5 \\ 0 & 0 & 0 & 0 & 0 & 0 & c_6 & d_6 & e_6 & f_6 \\ 0 & 0 & 0 & 0 & 0 & 0 & c_7 & d_7 & e_7 & f_7 \end{bmatrix}.$$

We suppose that $\begin{bmatrix} c_4 \\ c_5 \\ c_6 \\ c_7 \end{bmatrix} \neq \begin{bmatrix} 0 \\ 0 \\ 0 \\ 0 \end{bmatrix}$.

By applying Lemma 1.43 to the matrix $\begin{bmatrix} 0 & c_4 & d_4 & e_4 & f_4 \\ 0 & c_5 & d_5 & e_5 & f_5 \\ 0 & c_6 & d_6 & e_6 & f_6 \\ 0 & c_7 & d_7 & e_7 & f_7 \end{bmatrix}$ we get

$$A \sim \begin{bmatrix} 0 & 1 & 0 & a_1 & 0 & b_1 & c_1 & d_1 & e_1 & f_1 \\ 0 & 0 & 1 & a_2 & 0 & b_2 & c_2 & d_2 & e_2 & f_2 \\ 0 & 0 & 0 & 0 & 1 & b_3 & c_3 & d_3 & e_3 & f_3 \\ 0 & 0 & 0 & 0 & 0 & 0 & 1 & d_4' & e_4' & f_4' \\ 0 & 0 & 0 & 0 & 0 & 0 & 0 & d_5' & e_5' & f_5' \\ 0 & 0 & 0 & 0 & 0 & 0 & 0 & d_6' & e_6' & f_6' \\ 0 & 0 & 0 & 0 & 0 & 0 & 0 & d_7' & e_7' & f_7' \end{bmatrix}.$$

Next we

multiply row 4 by $-c_1$ and then add to row 1
multiply row 4 by $-c_2$ and then add to row 2
multiply row 4 by $-c_3$ and then add to row 3

$$A \sim \begin{bmatrix} 0 & 1 & 0 & a_1 & 0 & b_1 & 0 & d_1 - c_1 d_4' & e_1 - c_1 e_4' & f_1 - c_1 f_4' \\ 0 & 0 & 1 & a_2 & 0 & b_2 & 0 & d_2 - c_2 d_4' & e_2 - c_2 e_4' & f_2 - c_2 f_4' \\ 0 & 0 & 0 & 0 & 1 & b_3 & 0 & d_3 - c_3 d_4' & e_3 - c_3 e_4' & f_3 - c_3 f_4' \\ 0 & 0 & 0 & 0 & 0 & 0 & 1 & d_4' & e_4' & f_4' \\ 0 & 0 & 0 & 0 & 0 & 0 & 0 & d_5' & e_5' & f_5' \\ 0 & 0 & 0 & 0 & 0 & 0 & 0 & d_6' & e_6' & f_6' \\ 0 & 0 & 0 & 0 & 0 & 0 & 0 & d_7' & e_7' & f_7' \end{bmatrix},$$

so now the transformed matrix has the form

$$\begin{bmatrix} 0 & 1 & 0 & a_1 & 0 & b_1 & 0 & d_1' & e_1' & f_1' \\ 0 & 0 & 1 & a_2 & 0 & b_2 & 0 & d_2' & e_2' & f_2' \\ 0 & 0 & 0 & 0 & 1 & b_3 & 0 & d_3' & e_3' & f_3' \\ 0 & 0 & 0 & 0 & 0 & 0 & 1 & d_4' & e_4' & f_4' \\ 0 & 0 & 0 & 0 & 0 & 0 & 0 & d_5' & e_5' & f_5' \\ 0 & 0 & 0 & 0 & 0 & 0 & 0 & d_6' & e_6' & f_6' \\ 0 & 0 & 0 & 0 & 0 & 0 & 0 & d_7' & e_7' & f_7' \end{bmatrix} = A'.$$

We summarize the process described above as follows.

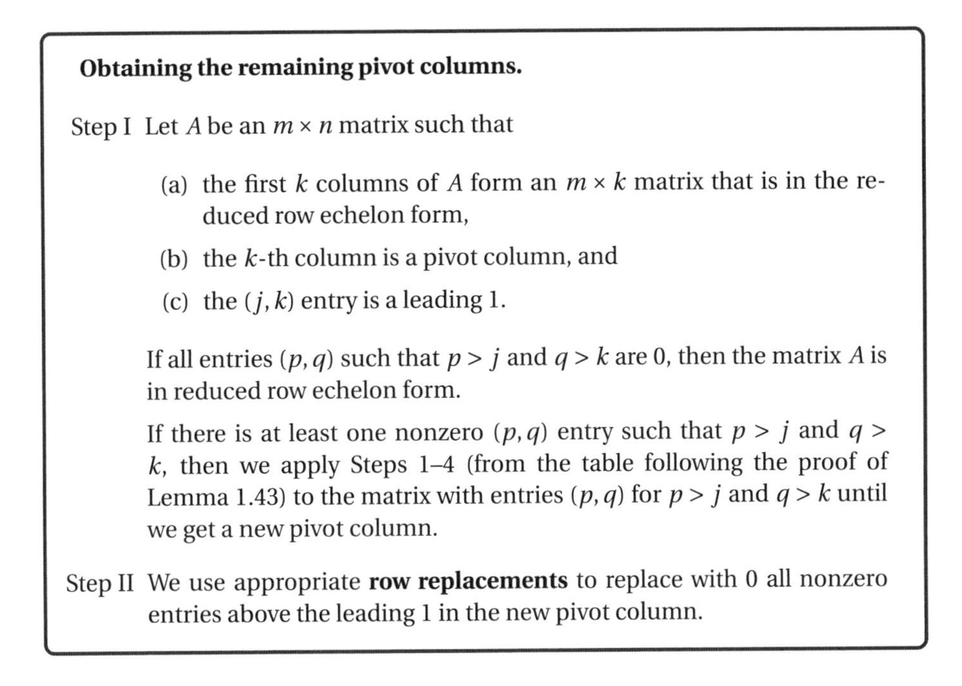

Obtaining the remaining pivot columns.

Step I Let A be an $m \times n$ matrix such that

(a) the first k columns of A form an $m \times k$ matrix that is in the reduced row echelon form,

(b) the k-th column is a pivot column, and

(c) the (j,k) entry is a leading 1.

If all entries (p,q) such that $p > j$ and $q > k$ are 0, then the matrix A is in reduced row echelon form.

If there is at least one nonzero (p,q) entry such that $p > j$ and $q > k$, then we apply Steps 1–4 (from the table following the proof of Lemma 1.43) to the matrix with entries (p,q) for $p > j$ and $q > k$ until we get a new pivot column.

Step II We use appropriate **row replacements** to replace with 0 all nonzero entries above the leading 1 in the new pivot column.

To obtain the reduced row echelon form of a matrix we first apply Steps 1–4 (in *Obtaining the first pivot column*) and then successively repeat Steps I and II above until we obtain the reduced row echelon form. We provide some specific examples to illustrate this process.

Example 1.48. Find the reduced row echelon form of the matrix

$$A = \begin{bmatrix} 2 & 4 & -1 & 2 & 1 \\ 1 & 2 & 2 & 11 & 1 \\ -1 & -2 & 1 & 1 & 0 \end{bmatrix}.$$

Solution. First we apply Steps 1–4.

interchange rows 1 and 2

$$\begin{bmatrix} 1 & 2 & 2 & 11 & 1 \\ 2 & 4 & -1 & 2 & 1 \\ -1 & -2 & 1 & 1 & 0 \end{bmatrix}$$

multiply first row by −2 and then add to row 2
multiply first row by 1 and then add to row 3

$$\begin{bmatrix} 1 & 2 & 2 & 11 & 1 \\ 0 & 0 & -5 & -20 & -1 \\ 0 & 0 & 3 & 12 & 1 \end{bmatrix}.$$

Now we are ready to apply Steps I and II.

multiply row 2 by $-\frac{1}{5}$

$$\begin{bmatrix} 1 & 2 & 2 & 11 & 1 \\ 0 & 0 & 1 & 4 & \frac{1}{5} \\ 0 & 0 & 3 & 12 & 1 \end{bmatrix}$$

multiply row 2 by -3 and then add to row 3

$$\begin{bmatrix} 1 & 2 & 2 & 11 & 1 \\ 0 & 0 & 1 & 4 & \frac{1}{5} \\ 0 & 0 & 0 & 0 & \frac{2}{5} \end{bmatrix}$$

multiply row 2 by -2 and then add to row 2

$$\begin{bmatrix} 1 & 2 & 0 & 3 & \frac{3}{5} \\ 0 & 0 & 1 & 4 & \frac{1}{5} \\ 0 & 0 & 0 & 0 & \frac{2}{5} \end{bmatrix}.$$

Now we are ready to apply Steps I and II a second time.

multiply row 3 by $\frac{5}{2}$

$$\begin{bmatrix} 1 & 2 & 0 & 3 & \frac{3}{5} \\ 0 & 0 & 1 & 4 & \frac{1}{5} \\ 0 & 0 & 0 & 0 & 1 \end{bmatrix}$$

multiply row 3 by $-\frac{3}{5}$ and then add to first row

multiply row 3 by $-\frac{1}{5}$ and then add to row 2

$$\begin{bmatrix} 1 & 2 & 0 & 3 & 0 \\ 0 & 0 & 1 & 4 & 0 \\ 0 & 0 & 0 & 0 & 1 \end{bmatrix}.$$

This is the reduced row echelon form of our matrix. The matrix has three pivot columns, first, third, and fifth, and three pivot positions, $(1,1)$, $(2,3)$, and $(3,5)$. □

Example 1.49. Determine the reduced row echelon form of the matrix

$$\begin{bmatrix} 3 & 2 & 12 & 1 & 7 \\ 2 & 1 & 7 & 2 & 7 \\ -1 & 1 & 1 & 1 & 2 \\ 1 & 1 & 5 & 1 & 4 \end{bmatrix}.$$

Solution. First we apply Steps 1–4.

interchange rows 1 and 4

$$\begin{bmatrix} 1 & 1 & 5 & 1 & 4 \\ 2 & 1 & 7 & 2 & 7 \\ -1 & 1 & 1 & 1 & 2 \\ 3 & 2 & 12 & 1 & 7 \end{bmatrix}$$

multiply first row by -2 and then add to row 2
multiply first row by 1 and then add to row 3
multiply first row by -3 and then add to row 4

$$\begin{bmatrix} 1 & 1 & 5 & 1 & 4 \\ 0 & -1 & -3 & 0 & -1 \\ 0 & 2 & 6 & 2 & 6 \\ 0 & -1 & -3 & -2 & -5 \end{bmatrix}.$$

Now we are ready to apply Steps I and II.
multiply row 2 by -1

$$\begin{bmatrix} 1 & 1 & 5 & 1 & 4 \\ 0 & 1 & 3 & 0 & 1 \\ 0 & 2 & 6 & 2 & 6 \\ 0 & -1 & -3 & -2 & -5 \end{bmatrix}$$

multiply row 2 by -2 and then add to row 3
multiply row 2 by 1 and then add to row 4

$$\begin{bmatrix} 1 & 1 & 5 & 1 & 4 \\ 0 & 1 & 3 & 0 & 1 \\ 0 & 0 & 0 & 2 & 4 \\ 0 & 0 & 0 & -2 & -4 \end{bmatrix}$$

multiply row 2 by -1 and then add to first row

$$\begin{bmatrix} 1 & 0 & 2 & 1 & 3 \\ 0 & 1 & 3 & 0 & 1 \\ 0 & 0 & 0 & 2 & 4 \\ 0 & 0 & 0 & -2 & -4 \end{bmatrix}.$$

Now we apply Steps I and II a second time.
multiply row 3 by $\frac{1}{2}$

$$\begin{bmatrix} 1 & 0 & 2 & 1 & 3 \\ 0 & 1 & 3 & 0 & 1 \\ 0 & 0 & 0 & 1 & 2 \\ 0 & 0 & 0 & -2 & -4 \end{bmatrix}$$

multiply row 3 by 2 and then add to row 4

$$\begin{bmatrix} 1 & 0 & 2 & 1 & 3 \\ 0 & 1 & 3 & 0 & 1 \\ 0 & 0 & 0 & 1 & 2 \\ 0 & 0 & 0 & 0 & 0 \end{bmatrix}.$$

Finally we

multiply row 3 by -1 and then add to first row

$$\begin{bmatrix} 1 & 0 & 2 & 0 & 1 \\ 0 & 1 & 3 & 0 & 1 \\ 0 & 0 & 0 & 1 & 2 \\ 0 & 0 & 0 & 0 & 0 \end{bmatrix}.$$

This is the reduced row echelon form of our matrix. The matrix has three pivot columns, first, second, and fourth, and three pivot positions, $(1,1)$, $(2,2)$, and $(3,4)$.
□

While it is important to understand the presented algorithm, it would not make much sense to try to memorize it. It is important to understand what the reduced row echelon form is and what "moves" are allowed. We can think of it as a game. Its goal is to reach a certain form of the matrix. We should try to reach that goal quickly, but always following the rules. Following exactly the process described above will always produce the desired result, but it may not be the quickest way. As an example of such a situation we show a different way of obtaining the reduced row echelon form of the matrix in Example 1.48, that is, the matrix

$$\begin{bmatrix} 2 & 4 & -1 & 2 & 1 \\ 1 & 2 & 2 & 11 & 1 \\ -1 & -2 & 1 & 1 & 0 \end{bmatrix}.$$

First we obtain the matrix

$$\begin{bmatrix} 1 & 2 & 2 & 11 & 1 \\ 0 & 0 & 1 & 4 & \frac{1}{5} \\ 0 & 0 & 0 & 0 & \frac{2}{5} \end{bmatrix}$$

proceeding as in Example 1.48. Then we continue in the following way.

multiply row 3 by $\frac{5}{2}$

$$\begin{bmatrix} 1 & 2 & 2 & 11 & 1 \\ 0 & 0 & 1 & 4 & \frac{1}{5} \\ 0 & 0 & 0 & 0 & 1 \end{bmatrix}.$$

Now we replace by 0 entries in the last column that are above the leading 1, that is, entries $(1,5)$ and $(2,5)$. Note that this does not change the remaining columns.

multiply row 3 by -1 and then add to row 1

multiply row 3 by $-\frac{1}{5}$ and then add to row 2

$$\begin{bmatrix} 1 & 2 & 2 & 11 & 0 \\ 0 & 0 & 1 & 4 & 0 \\ 0 & 0 & 0 & 0 & 1 \end{bmatrix}.$$

Next we replace by 0 the entry that is above the leading 1 in the second row, that is, entry $(1,3)$. Note that this does not change columns 1, 2, and 5.

multiply row 2 by -2 and then add to row 1

$$\begin{bmatrix} 1 & 2 & 0 & 3 & 0 \\ 0 & 0 & 1 & 4 & 0 \\ 0 & 0 & 0 & 0 & 1 \end{bmatrix}.$$

This is, as expected, the same reduced row echelon form obtained in Example 1.48.

The method presented above is based on the following observation.

> If entry (p,q) is a leading 1, then the columns $1,\dots,q-1$ do not change if we multiply row p by any number and then add to any other row.

When we construct the reduced row echelon form of a matrix by first applying Steps 1–4 (in *Obtaining the first pivot column*) and then successively repeating Steps I and II, we "move" through the columns of the matrix from left to right. When a column is already in the desired form, it does not change any more when we work on columns to the right of that column. This has some advantages. For example, when we are done with the first k columns of a matrix A, then these first k columns form a matrix that is the reduced row echelon form of the first k columns of the original matrix A.

Example 1.50. In Example 1.49 we obtain the reduced row echelon form of the matrix

$$\begin{bmatrix} 3 & 2 & 12 & 1 & 7 \\ 2 & 1 & 7 & 2 & 7 \\ -1 & 1 & 1 & 1 & 2 \\ 1 & 1 & 5 & 1 & 4 \end{bmatrix}.$$

When we are finished with the first three columns we obtain the matrix

$$\begin{bmatrix} 1 & 0 & 2 & 1 & 3 \\ 0 & 1 & 3 & 0 & 1 \\ 0 & 0 & 0 & 2 & 4 \\ 0 & 0 & 0 & -2 & -4 \end{bmatrix}.$$

Note that the first three columns of this matrix form a matrix that is the reduced row echelon form of the first three columns of the original matrix, that is, the reduced row echelon form of the matrix

$$\begin{bmatrix} 3 & 2 & 12 \\ 2 & 1 & 7 \\ -1 & 1 & 1 \\ 1 & 1 & 5 \end{bmatrix}$$

is

$$\begin{bmatrix} 1 & 0 & 2 \\ 0 & 1 & 3 \\ 0 & 0 & 0 \\ 0 & 0 & 0 \end{bmatrix}.$$

Elementary matrices

Definition 1.51. By an *elementary matrix* we mean a matrix obtained from an identity matrix by one elementary row operation.

Example 1.52. The matrix

$$\begin{bmatrix} 1 & 0 & 0 \\ 0 & 0 & 1 \\ 0 & 1 & 0 \end{bmatrix}$$

is an elementary matrix since it can be obtained from the unit matrix $\begin{bmatrix} 1 & 0 & 0 \\ 0 & 1 & 0 \\ 0 & 0 & 1 \end{bmatrix}$ by interchanging rows 1 and 2.

The matrix

$$\begin{bmatrix} 1 & 0 & 0 & 0 \\ 0 & 1 & 0 & 0 \\ 0 & 0 & k & 0 \\ 0 & 0 & 0 & 1 \end{bmatrix}$$

is an elementary matrix since it can be obtained from the unit matrix

$$\begin{bmatrix} 1 & 0 & 0 & 0 \\ 0 & 1 & 0 & 0 \\ 0 & 0 & 1 & 0 \\ 0 & 0 & 0 & 1 \end{bmatrix}$$

by multiplication of row 3 by k.

The matrix

$$\begin{bmatrix} 1 & 0 & 0 & 0 & 0 \\ 0 & 1 & 0 & k & 0 \\ 0 & 0 & 1 & 0 & 0 \\ 0 & 0 & 0 & 1 & 0 \\ 0 & 0 & 0 & 0 & 1 \end{bmatrix}$$

is an elementary matrix since it can be obtained from the unit matrix

$$\begin{bmatrix} 1 & 0 & 0 & 0 & 0 \\ 0 & 1 & 0 & 0 & 0 \\ 0 & 0 & 1 & 0 & 0 \\ 0 & 0 & 0 & 1 & 0 \\ 0 & 0 & 0 & 0 & 1 \end{bmatrix}$$

by multiplying row 4 by k and adding it to row 2.

The following observation is very useful in calculations involving elementary matrices.

> The result of applying an elementary row operation to a matrix A is equivalent to multiplication of A by an elementary matrix obtained from the unit matrix by the same elementary row operation.

Example 1.53. Multiplying a 3×4 matrix by the elementary matrix

$$\begin{bmatrix} 1 & 0 & 0 \\ 0 & 0 & 1 \\ 0 & 1 & 0 \end{bmatrix}$$

is equivalent to the elementary row operation of interchanging rows 2 and 3:

$$\begin{bmatrix} 1 & 0 & 0 \\ 0 & 0 & 1 \\ 0 & 1 & 0 \end{bmatrix} \begin{bmatrix} a_1 & b_1 & c_1 & d_1 \\ a_2 & b_2 & c_2 & d_2 \\ a_3 & b_3 & c_3 & d_3 \end{bmatrix} = \begin{bmatrix} a_1 & b_1 & c_1 & d_1 \\ a_3 & b_3 & c_3 & d_3 \\ a_2 & b_2 & c_2 & d_2 \end{bmatrix}.$$

Multiplying a 4×2 matrix by the matrix

$$\begin{bmatrix} 1 & 0 & 0 & 0 \\ 0 & 1 & 0 & 0 \\ 0 & 0 & k & 0 \\ 0 & 0 & 0 & 1 \end{bmatrix}$$

is equivalent to the elementary row operation of multiplication of row 3 by k:

$$\begin{bmatrix} 1 & 0 & 0 & 0 \\ 0 & 1 & 0 & 0 \\ 0 & 0 & k & 0 \\ 0 & 0 & 0 & 1 \end{bmatrix} \begin{bmatrix} a_1 & b_1 \\ a_2 & b_2 \\ a_3 & b_3 \\ a_4 & b_4 \end{bmatrix} = \begin{bmatrix} a_1 & b_1 \\ a_2 & b_2 \\ ka_3 & kb_3 \\ a_4 & b_4 \end{bmatrix}.$$

Multiplying a 5×3 matrix by the matrix

$$\begin{bmatrix} 1 & 0 & 0 & 0 & 0 \\ 0 & 1 & 0 & k & 0 \\ 0 & 0 & 1 & 0 & 0 \\ 0 & 0 & 0 & 1 & 0 \\ 0 & 0 & 0 & 0 & 1 \end{bmatrix}$$

is equivalent to the elementary row operation of multiplying row 4 by k and adding it to row 2:

$$\begin{bmatrix} 1 & 0 & 0 & 0 & 0 \\ 0 & 1 & 0 & k & 0 \\ 0 & 0 & 1 & 0 & 0 \\ 0 & 0 & 0 & 1 & 0 \\ 0 & 0 & 0 & 0 & 1 \end{bmatrix} \begin{bmatrix} a_1 & b_1 & c_1 \\ a_2 & b_2 & c_2 \\ a_3 & b_3 & c_3 \\ a_4 & b_4 & c_4 \\ a_5 & b_5 & c_5 \end{bmatrix} = \begin{bmatrix} a_1 & b_1 & c_1 \\ a_2 + ka_4 & b_2 + kb_4 & c_2 + kc_4 \\ a_3 & b_3 & c_3 \\ a_4 & b_4 & c_4 \\ a_5 & b_5 & c_5 \end{bmatrix}.$$

Theorem 1.54. *Let A and B be an $m \times n$ matrices. We have*

$$A \sim B,$$

if and only if there are elementary $m \times m$ matrices $E_1, E_2, \ldots, E_{k-1}, E_k$ such that

$$E_k E_{k-1} \ldots E_2 E_1 A = B.$$

Proof. This result is a consequence of the fact that each step in the Gaussian elimination process is equivalent to multiplication by an elementary matrix. □

Corollary 1.55. *Let A be an $m \times n$ matrix. If*

$$\begin{bmatrix} A & I_m \end{bmatrix} \sim \begin{bmatrix} R & P \end{bmatrix},$$

where R is the reduced row echelon form of the matrix A, then $R = PA$.

Proof. There are elementary matrices $E_1, E_2, \dots, E_{k-1}, E_k$ such that

$$E_k E_{k-1} \dots E_2 E_1 \begin{bmatrix} A & I_m \end{bmatrix} = \begin{bmatrix} R & P \end{bmatrix}.$$

Since

$$E_k E_{k-1} \dots E_2 E_1 \begin{bmatrix} A & I_m \end{bmatrix} = \begin{bmatrix} E_k E_{k-1} \dots E_2 E_1 A & E_k E_{k-1} \dots E_2 E_1 \end{bmatrix} = \begin{bmatrix} R & P \end{bmatrix},$$

we have

$$R = E_k E_{k-1} \dots E_2 E_1 A = PA.$$

□

Example 1.56. Find a matrix P such that the product PA is the reduced row echelon form of the matrix A, where

$$A = \begin{bmatrix} 1 & 2 & 3 & -1 \\ 1 & 3 & 1 & 1 \\ 3 & 5 & 11 & -5 \end{bmatrix}.$$

Solution. The reduced row echelon form of the matrix

$$\begin{bmatrix} 1 & 2 & 3 & -1 & 1 & 0 & 0 \\ 1 & 3 & 1 & 1 & 0 & 1 & 0 \\ 3 & 5 & 11 & -5 & 0 & 0 & 1 \end{bmatrix}$$

is

$$\begin{bmatrix} 1 & 0 & 7 & -5 & 0 & -5/4 & 3/4 \\ 0 & 1 & -2 & 2 & 0 & 3/4 & -1/4 \\ 0 & 0 & 0 & 0 & 1 & -1/4 & -1/4 \end{bmatrix}.$$

Hence

$$P = \begin{bmatrix} 0 & -5/4 & 3/4 \\ 0 & 3/4 & -1/4 \\ 1 & -1/4 & -1/4 \end{bmatrix}.$$

□

Exercises 1.2

Calculate the products

1. $\begin{bmatrix} 1 & 0 & 0 \\ 0 & 9 & 0 \\ 0 & 0 & 1 \end{bmatrix} \begin{bmatrix} a_1 & b_1 & c_1 & d_1 \\ a_2 & b_2 & c_2 & d_2 \\ a_3 & b_3 & c_3 & d_3 \end{bmatrix}$

2. $\begin{bmatrix} 1 & 0 & 0 \\ 0 & 1 & 0 \\ 0 & 0 & 7 \end{bmatrix} \begin{bmatrix} a_1 & b_1 & c_1 & d_1 \\ a_2 & b_2 & c_2 & d_2 \\ a_3 & b_3 & c_3 & d_3 \end{bmatrix}$

3. $\begin{bmatrix} 0 & 0 & 1 \\ 0 & 1 & 0 \\ 1 & 0 & 0 \end{bmatrix} \begin{bmatrix} a_1 & b_1 & c_1 & d_1 \\ a_2 & b_2 & c_2 & d_2 \\ a_3 & b_3 & c_3 & d_3 \end{bmatrix}$

4. $\begin{bmatrix} 0 & 1 & 0 \\ 1 & 0 & 0 \\ 0 & 0 & 1 \end{bmatrix} \begin{bmatrix} a_1 & b_1 & c_1 & d_1 \\ a_2 & b_2 & c_2 & d_2 \\ a_3 & b_3 & c_3 & d_3 \end{bmatrix}$

5. $\begin{bmatrix} 1 & 0 & 0 & 0 & 0 \\ 0 & 1 & 0 & 0 & 0 \\ 0 & 0 & 1 & 0 & 0 \\ 0 & -\frac{5}{3} & 0 & 1 & 0 \\ 0 & 0 & 0 & 0 & 1 \end{bmatrix} \begin{bmatrix} 0 \\ 3 \\ 0 \\ 5 \\ 0 \end{bmatrix}$

6. $\begin{bmatrix} 1 & 0 & 0 & 0 & 0 \\ 0 & 1 & 0 & 0 & 0 \\ 0 & 0 & 1 & 0 & 0 \\ 0 & 0 & 0 & 1 & 0 \\ 0 & 0 & -4 & 0 & 1 \end{bmatrix} \begin{bmatrix} a_1 \\ a_2 \\ a_3 \\ a_4 \\ a_5 \end{bmatrix}$

7. $\begin{bmatrix} 1 & 0 & 0 & 0 \\ 0 & 1 & 0 & 0 \\ 0 & 7 & 1 & 0 \\ 0 & 0 & 0 & 1 \end{bmatrix} \begin{bmatrix} a_1 & b_1 \\ a_2 & b_2 \\ a_3 & b_3 \\ a_4 & b_4 \end{bmatrix}$

8. $\begin{bmatrix} 1 & 0 & 0 & 0 \\ 0 & 1 & 8 & 0 \\ 0 & 0 & 1 & 0 \\ 0 & 0 & 0 & 1 \end{bmatrix} \begin{bmatrix} a_1 & b_1 & c_1 \\ a_2 & b_2 & c_2 \\ a_3 & b_3 & c_3 \\ a_4 & b_4 & c_4 \end{bmatrix}$

9. $\begin{bmatrix} 1 & 0 & -1 \\ 0 & 1 & 0 \\ 0 & 0 & 1 \end{bmatrix} \begin{bmatrix} 1 & 0 & 0 \\ 3 & 1 & 0 \\ 0 & 0 & 1 \end{bmatrix} \begin{bmatrix} 1 & 0 & 0 \\ 0 & 1 & 0 \\ -2 & 0 & 1 \end{bmatrix} \begin{bmatrix} 1 & 4 & 0 \\ 0 & 1 & 0 \\ 0 & 0 & 1 \end{bmatrix}$

10. $\begin{bmatrix} 1 & 0 & 0 \\ 0 & 1 & 2 \\ 0 & 0 & 1 \end{bmatrix} \begin{bmatrix} 1 & 0 & 0 \\ 0 & 1 & 0 \\ 4 & 0 & 1 \end{bmatrix} \begin{bmatrix} 1 & 0 & 0 \\ 5 & 1 & 0 \\ 0 & 0 & 1 \end{bmatrix} \begin{bmatrix} 1 & 0 & 0 \\ 0 & 1 & 0 \\ 0 & 7 & 1 \end{bmatrix}$

11. $\begin{bmatrix} 1 & 0 & 3 \\ 0 & 1 & 0 \\ 0 & 0 & 1 \end{bmatrix} \begin{bmatrix} 1 & 0 & 0 \\ 0 & 1 & 0 \\ 0 & 0 & 5 \end{bmatrix} \begin{bmatrix} 1 & 0 & 0 \\ 0 & 0 & 1 \\ 0 & 1 & 0 \end{bmatrix} \begin{bmatrix} 1 & 0 & 0 \\ 0 & 1 & 0 \\ 0 & -2 & 1 \end{bmatrix} \begin{bmatrix} 2 & 3 & 1 & -1 \\ 1 & 1 & 3 & 2 \\ 4 & 1 & 1 & 3 \end{bmatrix}$

12. $\begin{bmatrix} 0 & 0 & 1 \\ 0 & 1 & 0 \\ 1 & 0 & 0 \end{bmatrix} \begin{bmatrix} 1 & 0 & 0 \\ 0 & -2 & 0 \\ 0 & 0 & 1 \end{bmatrix} \begin{bmatrix} 0 & 0 & 1 \\ 1 & 0 & 0 \\ 0 & 1 & 0 \end{bmatrix} \begin{bmatrix} 1 & 0 & 3 \\ 0 & 1 & 0 \\ 0 & 0 & 1 \end{bmatrix} \begin{bmatrix} 3 & 0 & 1 & 1 \\ 2 & 1 & 2 & 4 \\ 3 & 2 & 1 & 0 \end{bmatrix}$

Find the reduced row echelon form of the following matrices.

13. $\begin{bmatrix} 2 & 2 & 14 & 0 \\ 3 & 1 & 11 & -2 \\ 2 & 1 & 9 & -1 \end{bmatrix}$

14. $\begin{bmatrix} 2 & -2 & 2 & 10 \\ 3 & -5 & 1 & 11 \\ 2 & -3 & 1 & 8 \end{bmatrix}$

15. $\begin{bmatrix} 3 & 2 & 1 & 10 \\ 2 & 1 & 0 & 4 \\ -1 & 1 & 1 & 4 \\ 1 & 0 & 1 & 4 \end{bmatrix}$

16. $\begin{bmatrix} 1 & 2 & 1 & 4 \\ 2 & 1 & -1 & 2 \\ -1 & 1 & 1 & 1 \\ 1 & -1 & 1 & 1 \end{bmatrix}$

17. $\begin{bmatrix} 1 & 0 & 1 & 1 \\ 0 & -a-1 & 1 & 1 \\ 1 & 1 & a & 0 \\ 1 & 1 & 0 & 1 \end{bmatrix}$

18. $\begin{bmatrix} 1 & 1 & 1 & -1 \\ 2 & -1 & a & 1 \\ 1 & 1 & -2 & 2 \\ 1 & 1 & -1 & 1 \end{bmatrix}$

19. $\begin{bmatrix} 2 & 2 & 2p+2q & 4q+2p \\ 3 & 1 & 3p+q & 4q+p \\ 2 & 1 & 2p+q & 3q+p \end{bmatrix}$

20. $\begin{bmatrix} 2 & 2 & 2p+2q & 4q+2p \\ 3 & 1 & 3p+q & 4q+p \\ 2 & 1 & 2p+q & 3q+p \end{bmatrix}$

21. Write all reduced row echelon forms of 4×4 matrices with 2 pivots.

22. Write all reduced row echelon forms of 4×4 matrices with 3 pivots.

Find a matrix P such that the product PA is the reduced row echelon form of the given matrix A.

23. $A = \begin{bmatrix} 3 & 2 & 10 \\ -1 & 5 & -9 \end{bmatrix}$

24. $\begin{bmatrix} 1 & 2 & 12 \\ 1 & 5 & 27 \end{bmatrix}$

25. $A = \begin{bmatrix} 15 \\ 18 \\ 50 \end{bmatrix}$

26. $A = \begin{bmatrix} 5 \\ 7 \\ 15 \end{bmatrix}$

27. $A = \begin{bmatrix} 1 & 2 \\ 1 & 5 \\ 2 & 3 \end{bmatrix}$

28. $A = \begin{bmatrix} 3 & 1 \\ 2 & -1 \\ 1 & 2 \end{bmatrix}$

29. $A = \begin{bmatrix} 1 \\ 3 \\ 2 \\ 5 \end{bmatrix}$

30. $A = \begin{bmatrix} 4 \\ 1 \\ 1 \\ 2 \end{bmatrix}$

1.3 The inverse of a matrix

The inverse of a matrix plays a special role in linear algebra. There are two basic questions in connection with invertible matrices: how can we check if a matrix is invertible and how do we find the inverse of an invertible matrix.

Invertible matrices

Definition 1.57. An $n \times n$ matrix A is *invertible* if there is a matrix B such that

$$AB = BA = I_n,$$

where I_n is the $n \times n$ identity matrix.

Theorem 1.58. *If an $n \times n$ matrix A is invertible, then there is a unique matrix B such that $AB = BA = I_n$.*

Proof. We need to show that, if

$$AB = BA = I_n \quad \text{and} \quad AC = CA = I_n,$$

then $B = C$. Indeed, if $BA = I_n$ and $AC = I_n$, then

$$B = BI_n = B(AC) = (BA)C = I_nC = C.$$

□

Definition 1.59. Let A be an invertible $n \times n$ matrix. The unique matrix B such that $AB = BA = I_n$ is called the *inverse* of the matrix A and is denoted by A^{-1}.

All elementary matrices are invertible and the inverse of an elementary matrix is an elementary matrix.

Example 1.60. The inverse of an elementary matrix corresponding to the interchange of two rows is the same matrix:

$$\begin{bmatrix} 0 & 1 & 0 \\ 1 & 0 & 0 \\ 0 & 0 & 1 \end{bmatrix}^{-1} = \begin{bmatrix} 0 & 1 & 0 \\ 1 & 0 & 0 \\ 0 & 0 & 1 \end{bmatrix},$$

$$\begin{bmatrix} 1 & 0 & 0 \\ 0 & 0 & 1 \\ 0 & 1 & 0 \end{bmatrix}^{-1} = \begin{bmatrix} 1 & 0 & 0 \\ 0 & 0 & 1 \\ 0 & 1 & 0 \end{bmatrix},$$

and

$$\begin{bmatrix} 0 & 0 & 1 \\ 0 & 1 & 0 \\ 1 & 0 & 0 \end{bmatrix}^{-1} = \begin{bmatrix} 0 & 0 & 1 \\ 0 & 1 & 0 \\ 1 & 0 & 0 \end{bmatrix}.$$

Example 1.61. The inverse of an elementary matrix corresponding to multiplication of a row by a constant $k \neq 0$ is the elementary matrix corresponding to multiplication of the same row by the constant $\frac{1}{k}$:

$$\begin{bmatrix} k & 0 & 0 \\ 0 & 1 & 0 \\ 0 & 0 & 1 \end{bmatrix}^{-1} = \begin{bmatrix} \frac{1}{k} & 0 & 0 \\ 0 & 1 & 0 \\ 0 & 0 & 1 \end{bmatrix},$$

$$\begin{bmatrix} 1 & 0 & 0 \\ 0 & k & 0 \\ 0 & 0 & 1 \end{bmatrix}^{-1} = \begin{bmatrix} 1 & 0 & 0 \\ 0 & \frac{1}{k} & 0 \\ 0 & 0 & 1 \end{bmatrix},$$

and

$$\begin{bmatrix} 1 & 0 & 0 \\ 0 & 1 & 0 \\ 0 & 0 & k \end{bmatrix}^{-1} = \begin{bmatrix} 1 & 0 & 0 \\ 0 & 1 & 0 \\ 0 & 0 & \frac{1}{k} \end{bmatrix}.$$

Example 1.62. The inverse of an elementary matrix corresponding to multiplication of a row by a constant a and then adding the result to another row of the matrix is the elementary matrix corresponding to the same operation except that we use $-a$ instead of a:

$$\begin{bmatrix} 1 & 0 & 0 \\ a & 1 & 0 \\ 0 & 0 & 1 \end{bmatrix}^{-1} = \begin{bmatrix} 1 & 0 & 0 \\ -a & 1 & 0 \\ 0 & 0 & 1 \end{bmatrix},$$

$$\begin{bmatrix} 1 & 0 & 0 \\ 0 & 1 & 0 \\ b & 0 & 1 \end{bmatrix}^{-1} = \begin{bmatrix} 1 & 0 & 0 \\ 0 & 1 & 0 \\ -b & 0 & 1 \end{bmatrix},$$

$$\begin{bmatrix} 1 & c & 0 \\ 0 & 1 & 0 \\ 0 & 0 & 1 \end{bmatrix}^{-1} = \begin{bmatrix} 1 & -c & 0 \\ 0 & 1 & 0 \\ 0 & 0 & 1 \end{bmatrix},$$

$$\begin{bmatrix} 1 & 0 & 0 \\ 0 & 1 & 0 \\ 0 & d & 1 \end{bmatrix}^{-1} = \begin{bmatrix} 1 & 0 & 0 \\ 0 & 1 & 0 \\ 0 & -d & 1 \end{bmatrix},$$

$$\begin{bmatrix} 1 & 0 & e \\ 0 & 1 & 0 \\ 0 & 0 & 1 \end{bmatrix}^{-1} = \begin{bmatrix} 1 & 0 & -e \\ 0 & 1 & 0 \\ 0 & 0 & 1 \end{bmatrix},$$

and

$$\begin{bmatrix} 1 & 0 & 0 \\ 0 & 1 & f \\ 0 & 0 & 1 \end{bmatrix}^{-1} = \begin{bmatrix} 1 & 0 & 0 \\ 0 & 1 & -f \\ 0 & 0 & 1 \end{bmatrix}.$$

The pattern described in the above three examples applies to larger matrices.

Example 1.63.

$$\begin{bmatrix} 1 & 0 & 0 & 0 & 0 \\ 0 & 0 & 0 & 0 & 1 \\ 0 & 0 & 1 & 0 & 0 \\ 0 & 0 & 0 & 1 & 0 \\ 0 & 1 & 0 & 0 & 0 \end{bmatrix}^{-1} = \begin{bmatrix} 1 & 0 & 0 & 0 & 0 \\ 0 & 0 & 0 & 0 & 1 \\ 0 & 0 & 1 & 0 & 0 \\ 0 & 0 & 0 & 1 & 0 \\ 0 & 1 & 0 & 0 & 0 \end{bmatrix}.$$

$$\begin{bmatrix} 1 & 0 & 0 & 0 & 0 \\ 0 & k & 0 & 0 & 0 \\ 0 & 0 & 1 & 0 & 0 \\ 0 & 0 & 0 & 1 & 0 \\ 0 & 0 & 0 & 0 & 1 \end{bmatrix}^{-1} = \begin{bmatrix} 1 & 0 & 0 & 0 & 0 \\ 0 & \frac{1}{k} & 0 & 0 & 0 \\ 0 & 0 & 1 & 0 & 0 \\ 0 & 0 & 0 & 1 & 0 \\ 0 & 0 & 0 & 0 & 1 \end{bmatrix},$$

for any $k \neq 0$.

$$\begin{bmatrix} 1 & 0 & 0 & c & 0 \\ 0 & 1 & 0 & 0 & 0 \\ 0 & 0 & 1 & 0 & 0 \\ 0 & 0 & 0 & 1 & 0 \\ 0 & 0 & 0 & 0 & 1 \end{bmatrix}^{-1} = \begin{bmatrix} 1 & 0 & 0 & -c & 0 \\ 0 & 1 & 0 & 0 & 0 \\ 0 & 0 & 1 & 0 & 0 \\ 0 & 0 & 0 & 1 & 0 \\ 0 & 0 & 0 & 0 & 1 \end{bmatrix},$$

for any c.

Characterization of invertible matrices

Now we address the important problem of determining whether a given matrix is invertible. We already know that elementary matrices are invertible. From the theorem below it follows that any matrix that can be written as a product of elementary matrices is also invertible.

Theorem 1.64. *If $A_1, \dots, A_m$ are invertible $n \times n$ matrices, then the product matrix $A_1 \cdots A_m$ is invertible and we have*

$$(A_1 \cdots A_m)^{-1} = A_m^{-1} \cdots A_1^{-1}.$$

Note that the order of the matrices in the equality $(A_1 \cdots A_m)^{-1} = A_m^{-1} \cdots A_1^{-1}$ is different on the left-hand side and the right-hand side.

Proof for $m = 3$. We have

$$A_1A_2A_3A_3^{-1}A_2^{-1}A_1^{-1} = A_1A_2I_nA_2^{-1}A_1^{-1} = A_1A_2A_2^{-1}A_1^{-1} = A_1I_nA_1^{-1} = A_1A_1^{-1} = I_n$$

and

$$A_3^{-1}A_2^{-1}A_1^{-1}A_1A_2A_3 = A_3^{-1}A_2^{-1}I_nA_2A_3 = A_3^{-1}A_2^{-1}A_2A_3 = A_3^{-1}I_nA_3 = A_3^{-1}A_3 = I_n.$$

It should be clear why this argument easily generalizes to any number of matrices. □

Example 1.65. Calculate the matrix

$$A = \begin{bmatrix} 1 & 0 & 0 & 0 \\ 0 & 1 & 0 & 0 \\ 7 & 0 & 1 & 0 \\ 0 & 0 & 0 & 1 \end{bmatrix} \begin{bmatrix} 1 & 0 & 0 & 0 \\ 0 & 1 & 0 & 0 \\ 0 & 0 & 2 & 0 \\ 0 & 0 & 0 & 1 \end{bmatrix} \begin{bmatrix} 0 & 0 & 1 & 0 \\ 0 & 1 & 0 & 0 \\ 1 & 0 & 0 & 0 \\ 0 & 0 & 0 & 1 \end{bmatrix} \begin{bmatrix} 1 & 0 & 0 & 0 \\ 0 & 1 & 0 & 0 \\ 0 & 0 & 1 & 0 \\ 0 & 5 & 0 & 1 \end{bmatrix}$$

and then write its inverse as a product of elementary matrices. Use the result to calculate the inverse of the matrix A.

Solution. First we find that

$$A = \begin{bmatrix} 0 & 0 & 1 & 0 \\ 0 & 1 & 0 & 0 \\ 2 & 0 & 7 & 0 \\ 0 & 5 & 0 & 1 \end{bmatrix}.$$

Now

$$\begin{aligned} A^{-1} &= \left(\begin{bmatrix} 1 & 0 & 0 & 0 \\ 0 & 1 & 0 & 0 \\ 7 & 0 & 1 & 0 \\ 0 & 0 & 0 & 1 \end{bmatrix} \begin{bmatrix} 1 & 0 & 0 & 0 \\ 0 & 1 & 0 & 0 \\ 0 & 0 & 2 & 0 \\ 0 & 0 & 0 & 1 \end{bmatrix} \begin{bmatrix} 0 & 0 & 1 & 0 \\ 0 & 1 & 0 & 0 \\ 1 & 0 & 0 & 0 \\ 0 & 0 & 0 & 1 \end{bmatrix} \begin{bmatrix} 1 & 0 & 0 & 0 \\ 0 & 1 & 0 & 0 \\ 0 & 0 & 1 & 0 \\ 0 & 5 & 0 & 1 \end{bmatrix} \right)^{-1} \\ &= \begin{bmatrix} 1 & 0 & 0 & 0 \\ 0 & 1 & 0 & 0 \\ 0 & 0 & 1 & 0 \\ 0 & -5 & 0 & 1 \end{bmatrix} \begin{bmatrix} 0 & 0 & 1 & 0 \\ 0 & 1 & 0 & 0 \\ 1 & 0 & 0 & 0 \\ 0 & 0 & 0 & 1 \end{bmatrix} \begin{bmatrix} 1 & 0 & 0 & 0 \\ 0 & 1 & 0 & 0 \\ 0 & 0 & 1/2 & 0 \\ 0 & 0 & 0 & 1 \end{bmatrix} \begin{bmatrix} 1 & 0 & 0 & 0 \\ 0 & 1 & 0 & 0 \\ -7 & 0 & 1 & 0 \\ 0 & 0 & 0 & 1 \end{bmatrix} \\ &= \begin{bmatrix} -7/2 & 0 & 1/2 & 0 \\ 0 & 1 & 0 & 0 \\ 1 & 0 & 0 & 0 \\ 0 & -5 & 0 & 1 \end{bmatrix}. \end{aligned}$$

It is easy to verify that

$$\begin{bmatrix} 0 & 0 & 1 & 0 \\ 0 & 1 & 0 & 0 \\ 2 & 0 & 7 & 0 \\ 0 & 5 & 0 & 1 \end{bmatrix} \begin{bmatrix} -7/2 & 0 & 1/2 & 0 \\ 0 & 1 & 0 & 0 \\ 1 & 0 & 0 & 0 \\ 0 & -5 & 0 & 1 \end{bmatrix} = \begin{bmatrix} 1 & 0 & 0 & 0 \\ 0 & 1 & 0 & 0 \\ 0 & 0 & 1 & 0 \\ 0 & 0 & 0 & 1 \end{bmatrix}$$

and

$$\begin{bmatrix} -7/2 & 0 & 1/2 & 0 \\ 0 & 1 & 0 & 0 \\ 1 & 0 & 0 & 0 \\ 0 & -5 & 0 & 1 \end{bmatrix} \begin{bmatrix} 0 & 0 & 1 & 0 \\ 0 & 1 & 0 & 0 \\ 2 & 0 & 7 & 0 \\ 0 & 5 & 0 & 1 \end{bmatrix} = \begin{bmatrix} 1 & 0 & 0 & 0 \\ 0 & 1 & 0 & 0 \\ 0 & 0 & 1 & 0 \\ 0 & 0 & 0 & 1 \end{bmatrix}.$$

□

We noted that any matrix that can be written as a product of elementary matrices is also invertible. It turns out that the converse is also true, that is, every invertible matrix can be written as a product of elementary matrices.

Theorem 1.66. *Let A be an $n \times n$ matrix. The following conditions are equivalent:*

(a) *A is invertible;*

(b) *The reduced row echelon form of A is the identity matrix I_n;*

(c) *$A = E_1 \cdots E_m$ for some elementary matrices $E_1, \ldots, E_m$.*

Proof. To show that (a) implies (b) we use a proof by contradiction, that is, we assume that A is invertible and that the reduced row echelon form of A is not I_n. This means that the reduced row echelon form of A has the form $\begin{bmatrix} B \\ \mathbf{0} \end{bmatrix}$, where B is an $(n-1) \times n$ matrix and $\mathbf{0}$ is the $1 \times n$ matrix with all entries 0. Since

$$A \sim \begin{bmatrix} B \\ \mathbf{0} \end{bmatrix},$$

according to Theorem 1.54, there are elementary matrices $E_1, \ldots, E_m$ such that

$$E_m \cdots E_1 A = \begin{bmatrix} B \\ \mathbf{0} \end{bmatrix}.$$

By Theorem 1.64, the matrix $E_m \cdots E_1 A$ is invertible, being a product of invertible matrices. But this is not possible because for any $n \times n$ matrix X we have

$$E_m \cdots E_1 AX = \begin{bmatrix} B \\ \mathbf{0} \end{bmatrix} X = \begin{bmatrix} Y \\ \mathbf{0} \end{bmatrix},$$

for some $(n-1) \times n$ matrix Y. In other words, since $E_m \cdots E_1 AX \neq I_n$ for any $n \times n$ matrix X, the matrix $E_m \cdots E_1 A$ cannot have an inverse. This contradiction shows that (a) implies (b).

Now we assume that the reduced row echelon form of A is the identity matrix I_n, that is,

$$A \sim I_n.$$

By Theorem 1.54, there are elementary matrices $E_1,\dots,E_m$ such that

$$E_m\cdots E_1 A = I_n.$$

Consequently,

$$A = (E_m\dots E_1)^{-1} I_n = (E_m\dots E_1)^{-1} = E_1^{-1}\dots E_m^{-1}.$$

This shows that (b) implies (c) because the inverse of a elementary matrix is an elementary matrix.

Finally, if A can be written as a product of elementary matrices, then A is invertible, because elementary matrices are invertible and the product of invertible matrices is an invertible matrix. Therefore (c) implies (a). □

Corollary 1.67. *If an $n \times n$ matrix A has a column or a row with all zero entries, then A is not invertible.*

From the proof of Theorem 1.66 we get the following useful result.

Corollary 1.68. *Let A be an $n \times n$ matrix. The following conditions are equivalent:*

(a) *A is not invertible;*

(b) *$A = E_1\cdots E_m N$ for some elementary matrices $E_1,\dots,E_m$ and a matrix N having all entries in the last row 0.*

Example 1.69. Show that the matrix $A = \begin{bmatrix} 3 & 1 & 2 & 23 \\ -1 & 0 & 1 & 3 \\ 1 & 1 & 1 & 14 \\ 4 & -1 & 1 & 6 \end{bmatrix}$ is not invertible.

Solution. The matrix A is not invertible because the reduced row echelon form of A is

$$\begin{bmatrix} 1 & 0 & 0 & 2 \\ 0 & 1 & 0 & 7 \\ 0 & 0 & 1 & 5 \\ 0 & 0 & 0 & 0 \end{bmatrix}.$$

□

Example 1.70. Write the matrix $\begin{bmatrix} 2 & 3 \\ 1 & 4 \end{bmatrix}$ as a product of elementary matrices.

Solution. Since

$$\begin{bmatrix} 0 & 1 \\ 1 & 0 \end{bmatrix}\begin{bmatrix} 2 & 3 \\ 1 & 4 \end{bmatrix} = \begin{bmatrix} 1 & 4 \\ 2 & 3 \end{bmatrix},$$

$$\begin{bmatrix} 1 & 0 \\ -2 & 1 \end{bmatrix}\begin{bmatrix} 0 & 1 \\ 1 & 0 \end{bmatrix}\begin{bmatrix} 2 & 3 \\ 1 & 4 \end{bmatrix} = \begin{bmatrix} 1 & 0 \\ -2 & 1 \end{bmatrix}\begin{bmatrix} 1 & 4 \\ 2 & 3 \end{bmatrix} = \begin{bmatrix} 1 & 4 \\ 0 & -5 \end{bmatrix},$$

$$\begin{bmatrix} 1 & 0 \\ 0 & -\frac{1}{5} \end{bmatrix}\begin{bmatrix} 1 & 0 \\ -2 & 1 \end{bmatrix}\begin{bmatrix} 0 & 1 \\ 1 & 0 \end{bmatrix}\begin{bmatrix} 2 & 3 \\ 1 & 4 \end{bmatrix} = \begin{bmatrix} 1 & 0 \\ 0 & -\frac{1}{5} \end{bmatrix}\begin{bmatrix} 1 & 4 \\ 0 & -5 \end{bmatrix} = \begin{bmatrix} 1 & 4 \\ 0 & 1 \end{bmatrix},$$

and finally

$$\begin{bmatrix} 1 & -4 \\ 0 & 1 \end{bmatrix}\begin{bmatrix} 1 & 0 \\ 0 & -\frac{1}{5} \end{bmatrix}\begin{bmatrix} 1 & 0 \\ -2 & 1 \end{bmatrix}\begin{bmatrix} 0 & 1 \\ 1 & 0 \end{bmatrix}\begin{bmatrix} 2 & 3 \\ 1 & 4 \end{bmatrix} = \begin{bmatrix} 1 & -4 \\ 0 & 1 \end{bmatrix}\begin{bmatrix} 1 & 4 \\ 0 & 1 \end{bmatrix} = \begin{bmatrix} 1 & 0 \\ 0 & 1 \end{bmatrix},$$

we have

$$\begin{bmatrix} 1 & 0 \\ 0 & -\frac{1}{5} \end{bmatrix}\begin{bmatrix} 1 & 0 \\ -2 & 1 \end{bmatrix}\begin{bmatrix} 0 & 1 \\ 1 & 0 \end{bmatrix}\begin{bmatrix} 2 & 3 \\ 1 & 4 \end{bmatrix} = \begin{bmatrix} 1 & -4 \\ 0 & 1 \end{bmatrix}^{-1}\begin{bmatrix} 1 & 0 \\ 0 & 1 \end{bmatrix} = \begin{bmatrix} 1 & -4 \\ 0 & 1 \end{bmatrix}^{-1} = \begin{bmatrix} 1 & 4 \\ 0 & 1 \end{bmatrix},$$

$$\begin{bmatrix} 1 & 0 \\ -2 & 1 \end{bmatrix}\begin{bmatrix} 0 & 1 \\ 1 & 0 \end{bmatrix}\begin{bmatrix} 2 & 3 \\ 1 & 4 \end{bmatrix} = \begin{bmatrix} 1 & 0 \\ 0 & -\frac{1}{5} \end{bmatrix}^{-1}\begin{bmatrix} 1 & 4 \\ 0 & 1 \end{bmatrix} = \begin{bmatrix} 1 & 0 \\ 0 & -5 \end{bmatrix}\begin{bmatrix} 1 & 4 \\ 0 & 1 \end{bmatrix},$$

$$\begin{bmatrix} 0 & 1 \\ 1 & 0 \end{bmatrix}\begin{bmatrix} 2 & 3 \\ 1 & 4 \end{bmatrix} = \begin{bmatrix} 1 & 0 \\ -2 & 1 \end{bmatrix}^{-1}\begin{bmatrix} 1 & 0 \\ 0 & -5 \end{bmatrix}\begin{bmatrix} 1 & 4 \\ 0 & 1 \end{bmatrix} = \begin{bmatrix} 1 & 0 \\ 2 & 1 \end{bmatrix}\begin{bmatrix} 1 & 0 \\ 0 & -5 \end{bmatrix}\begin{bmatrix} 1 & 4 \\ 0 & 1 \end{bmatrix},$$

and finally

$$\begin{bmatrix} 2 & 3 \\ 1 & 4 \end{bmatrix} = \begin{bmatrix} 0 & 1 \\ 1 & 0 \end{bmatrix}^{-1}\begin{bmatrix} 1 & 0 \\ 2 & 1 \end{bmatrix}\begin{bmatrix} 1 & 0 \\ 0 & -5 \end{bmatrix}\begin{bmatrix} 1 & 4 \\ 0 & 1 \end{bmatrix} = \begin{bmatrix} 0 & 1 \\ 1 & 0 \end{bmatrix}\begin{bmatrix} 1 & 0 \\ 2 & 1 \end{bmatrix}\begin{bmatrix} 1 & 0 \\ 0 & -5 \end{bmatrix}\begin{bmatrix} 1 & 4 \\ 0 & 1 \end{bmatrix}.$$

□

According to Definition 1.57 an $n \times n$ matrix A is invertible if there is a matrix B such that $AB = I_n$ and $BA = I_n$. In the following theorem we show that it is not necessary to check both conditions.

Theorem 1.71. *If A and B are $n \times n$ matrices such that*

$$AB = I_n,$$

then both matrices A and B are invertible and we have $A^{-1} = B$ and $B^{-1} = A$.

Proof. Assume that A and B are $n \times n$ matrices such that $AB = I_n$. If A is not invertible, then the reduced row echelon form of A is of the form $\begin{bmatrix} X \\ \mathbf{0} \end{bmatrix}$ where X is a $(n-1) \times n$ matrix and $\mathbf{0}$ is the $1 \times n$ matrix with all the entries 0. According to Theorem 1.54, there is a product P of elementary matrices such that

$$PA = \begin{bmatrix} X \\ \mathbf{0} \end{bmatrix}.$$

Hence,

$$P = PAB = \begin{bmatrix} X \\ \mathbf{0} \end{bmatrix} B = \begin{bmatrix} C \\ \mathbf{0} \end{bmatrix},$$

for some $(n-1) \times n$ matrix C. But this is not possible because the matrix P is invertible and the matrix $\begin{bmatrix} C \\ \mathbf{0} \end{bmatrix}$ is not invertible. Therefore the matrix A is invertible. Consequently, $B = A^{-1}$ and B is also invertible. □

Example 1.72. In Example 1.70 we show that

$$\begin{bmatrix} 1 & -4 \\ 0 & 1 \end{bmatrix} \begin{bmatrix} 1 & 0 \\ 0 & -\frac{1}{5} \end{bmatrix} \begin{bmatrix} 1 & 0 \\ -2 & 1 \end{bmatrix} \begin{bmatrix} 0 & 1 \\ 1 & 0 \end{bmatrix} \begin{bmatrix} 2 & 3 \\ 1 & 4 \end{bmatrix} = \begin{bmatrix} 1 & 0 \\ 0 & 1 \end{bmatrix}.$$

Consequently, by Theorem 1.71, we have

$$\begin{bmatrix} 2 & 3 \\ 1 & 4 \end{bmatrix}^{-1} = \begin{bmatrix} 1 & -4 \\ 0 & 1 \end{bmatrix} \begin{bmatrix} 1 & 0 \\ 0 & -\frac{1}{5} \end{bmatrix} \begin{bmatrix} 1 & 0 \\ -2 & 1 \end{bmatrix} \begin{bmatrix} 0 & 1 \\ 1 & 0 \end{bmatrix} = \begin{bmatrix} \frac{4}{5} & -\frac{3}{5} \\ -\frac{1}{5} & \frac{2}{5} \end{bmatrix}.$$

The above example suggests a method for finding the inverse of a matrix A: we find elementary matrices $E_1, \dots, E_m$ such that $E_m \cdots E_1 A = I_n$. Then $A^{-1} = E_m \cdots E_1$. While this method will produce the inverse of a matrix, it is not very efficient. The following theorem gives us a better way.

Theorem 1.73. *Let A be an $n \times n$ invertible matrix and let B be an $n \times p$ matrix. The reduced row echelon form of the $n \times (n+p)$ matrix*

$$\begin{bmatrix} A & B \end{bmatrix}$$

is the matrix

$$\begin{bmatrix} I_n & A^{-1}B \end{bmatrix}.$$

Proof. According to Theorem 1.66 there are elementary matrices $E_1, \dots, E_m$ such that

$$E_m \cdots E_1 A = I_n.$$

Then the reduced row echelon form of the matrix $\begin{bmatrix} A & B \end{bmatrix}$ is the matrix

$$E_m \cdots E_1 \begin{bmatrix} A & B \end{bmatrix}.$$

Since $E_m \cdots E_1 = A^{-1}$, we have

$$E_m \cdots E_1 \begin{bmatrix} A & B \end{bmatrix} = \begin{bmatrix} E_m \cdots E_1 A & E_m \cdots E_1 B \end{bmatrix} = \begin{bmatrix} I_n & A^{-1}B \end{bmatrix}.$$

□

The above result applied to the case when $B = I_n$ tells us that the reduced row echelon form of the $n \times 2n$ matrix $\begin{bmatrix} A & I_n \end{bmatrix}$ is the matrix $\begin{bmatrix} I_n & A^{-1} \end{bmatrix}$. This suggests a convenient method for finding the inverse of a matrix, as illustrated in the following example.

Example 1.74. Determine if the matrix

$$A = \begin{bmatrix} 1 & 1 & 2 & 1 \\ -1 & 1 & 1 & 1 \\ 2 & 1 & 1 & 1 \\ 3 & -1 & 1 & 1 \end{bmatrix}$$

is invertible. If A is invertible, then find its the inverse.

Solution. We find that

$$\begin{bmatrix} 1 & 1 & 2 & 1 & 1 & 0 & 0 & 0 \\ -1 & 1 & 1 & 1 & 0 & 1 & 0 & 0 \\ 2 & 1 & 1 & 1 & 0 & 0 & 1 & 0 \\ 3 & -1 & 1 & 1 & 0 & 0 & 0 & 1 \end{bmatrix} \sim \begin{bmatrix} 1 & 0 & 0 & 0 & 0 & -\frac{1}{3} & \frac{1}{3} & 0 \\ 0 & 1 & 0 & 0 & 0 & -\frac{1}{6} & \frac{2}{3} & -\frac{1}{2} \\ 0 & 0 & 1 & 0 & 1 & -\frac{1}{3} & -\frac{2}{3} & 0 \\ 0 & 0 & 0 & 1 & -1 & \frac{7}{6} & \frac{1}{3} & \frac{1}{2} \end{bmatrix}.$$

This means that the matrix A is invertible and the inverse is

$$A^{-1} = \begin{bmatrix} 0 & -\frac{1}{3} & \frac{1}{3} & 0 \\ 0 & -\frac{1}{6} & \frac{2}{3} & -\frac{1}{2} \\ 1 & -\frac{1}{3} & -\frac{2}{3} & 0 \\ -1 & \frac{7}{6} & \frac{1}{3} & \frac{1}{2} \end{bmatrix}.$$

□

A matrix equation

By a matrix equation we mean an equation of the form

$$AX = B$$

where A and B are known matrices and the matrix X is the unknown matrix. Such an equation makes sense only if the number of rows in A and B are the same. More precisely, if A is a $k \times m$ matrix and B is a $k \times n$ matrix, we want to find all $m \times n$ matrices X such that $AX = B$. As we will see, such an equation can have no solutions, exactly one solution, or infinitely many solutions. It is important to understand how we can determine for a given matrix equation which of the three possibilities occurs and how to find the solutions, if the equation has solutions. Linear algebra provides elegant and effective tools for dealing with such questions. The next theorem is an example of such a result.

Theorem 1.75. *If A is an invertible $n \times n$ matrix, then the equation*

$$AX = B \tag{1.5}$$

has a unique solution for any $n \times p$ matrix B. That solution is the $n \times p$ matrix

$$X = A^{-1}B.$$

Proof. If A is invertible, we can multiply both sides of the equation $AX = B$ by the matrix A^{-1} and get

$$A^{-1}(AX) = A^{-1}B,$$

which simplifies to

$$X = A^{-1}B. \tag{1.6}$$

This shows that, if the equation (1.5) has a solution, then it is given by the equation (1.6).

Now we need to check that $X = A^{-1}B$ is a solution. Indeed, we have

$$AX = A\left(A^{-1}B\right) = (AA^{-1})B = I_nB = B.$$

□

Example 1.76. Solve the matrix equation

$$\begin{bmatrix} 2 & 3 \\ 1 & 4 \end{bmatrix} X = \begin{bmatrix} 5 & 0 & 10 \\ -5 & 15 & 0 \end{bmatrix}. \tag{1.7}$$

Proof. In Example 1.72 we show that

$$\begin{bmatrix} 2 & 3 \\ 1 & 4 \end{bmatrix}^{-1} = \begin{bmatrix} \frac{4}{5} & -\frac{3}{5} \\ -\frac{1}{5} & \frac{2}{5} \end{bmatrix}.$$

Using this result and Theorem 1.75 we can conclude that the matrix equation (1.76) has a unique solution

$$X = \begin{bmatrix} \frac{4}{5} & -\frac{3}{5} \\ -\frac{1}{5} & \frac{2}{5} \end{bmatrix} \begin{bmatrix} 5 & 0 & 10 \\ -5 & 15 & 0 \end{bmatrix} = \begin{bmatrix} 7 & -9 & 8 \\ -3 & 6 & -2 \end{bmatrix}.$$

□

Example 1.77. Find a matrix

$$\begin{bmatrix} p & q & r \\ x & y & z \end{bmatrix}$$

such that

$$\begin{bmatrix} 3 & 4 \\ 2 & 5 \end{bmatrix} \begin{bmatrix} p & q & r \\ x & y & z \end{bmatrix} = \begin{bmatrix} a & b & c \\ d & e & f \end{bmatrix}.$$

Solution. Since

$$\begin{bmatrix} 3 & 4 & a & b & c \\ 2 & 5 & d & e & f \end{bmatrix} \sim \begin{bmatrix} 1 & 0 & -\frac{4}{7}d + \frac{5}{7}a & -\frac{4}{7}e + \frac{5}{7}b & -\frac{4}{7}f + \frac{5}{7}c \\ 0 & 1 & \frac{3}{7}d - \frac{2}{7}a & \frac{3}{7}e - \frac{2}{7}b & \frac{3}{7}f - \frac{2}{7}c \end{bmatrix},$$

we have

$$\begin{aligned} \begin{bmatrix} p & q & r \\ x & y & z \end{bmatrix} &= \begin{bmatrix} 3 & 4 \\ 2 & 5 \end{bmatrix}^{-1} \begin{bmatrix} a & b & c \\ d & e & f \end{bmatrix} \\ &= \begin{bmatrix} -\frac{4}{7}d + \frac{5}{7}a & -\frac{4}{7}e + \frac{5}{7}b & -\frac{4}{7}f + \frac{5}{7}c \\ \frac{3}{7}d - \frac{2}{7}a & \frac{3}{7}e - \frac{2}{7}b & \frac{3}{7}f - \frac{2}{7}c \end{bmatrix}, \end{aligned}$$

by Theorem 1.73. □

Systems of linear equations

Systems of linear equations can be expressed as matrix equations. This idea allows us to use tools of linear algebra to solve systems of linear equations in an elegant and efficient way.

A system of linear equations

$$\begin{cases} a_{11}x_1 + a_{12}x_2 + \cdots + a_{1n}x_n = b_1 \\ a_{21}x_1 + a_{22}x_2 + \cdots + a_{2n}x_n = b_2 \\ \quad\vdots \qquad\quad \vdots \qquad\quad \vdots \qquad\qquad \vdots \\ a_{m1}x_1 + a_{m2}x_2 + \ldots + a_{mn}x_n = b_m \end{cases}$$

can be conveniently written as a matrix equation

$$A\mathbf{x} = \mathbf{b},$$

where

$$A = \begin{bmatrix} a_{11} & a_{12} & \dots & a_{1n} \\ a_{21} & a_{22} & \dots & a_{2n} \\ \vdots & \vdots & & \vdots \\ a_{m1} & a_{m2} & \dots & a_{mn} \end{bmatrix}, \quad \mathbf{x} = \begin{bmatrix} x_1 \\ x_2 \\ \vdots \\ x_n \end{bmatrix}, \quad \text{and} \quad \mathbf{b} = \begin{bmatrix} b_1 \\ b_2 \\ \vdots \\ b_n \end{bmatrix}.$$

The following theorem describes the method of solving a system of linear equations using Gaussian elimination in terms of matrix equations.

Theorem 1.78. *Let A be an $m \times n$ matrix and let $\mathbf{b}$ be a vector in $\mathbb{R}^m$. The equation*

$$A\mathbf{x} = \mathbf{b}$$

is equivalent to the equation

$$R\mathbf{x} = \mathbf{c},$$

where $\begin{bmatrix} R & \mathbf{c} \end{bmatrix}$ is the reduced row echelon form of the augmented matrix $\begin{bmatrix} A & \mathbf{b} \end{bmatrix}$.

Proof. According to Theorem 1.54 there are elementary matrices $E_1, \dots, E_k$ such that

$$E_k \dots E_1 \begin{bmatrix} A & \mathbf{b} \end{bmatrix} = \begin{bmatrix} R & \mathbf{c} \end{bmatrix}.$$

Since the matrix $P = E_k \dots E_1$ is invertible, the equations

$$A\mathbf{x} = \mathbf{b} \quad \text{and} \quad PA\mathbf{x} = P\mathbf{b}$$

are equivalent. If we let $R = PA$ and $\mathbf{c} = P\mathbf{b}$, then the equations

$$A\mathbf{x} = \mathbf{b} \quad \text{and} \quad R\mathbf{x} = \mathbf{c}$$

are equivalent. □

Example 1.79. Solve the system

$$\begin{cases} 3x + y + 2z = 1 \\ 5x + 2y + z = 3 \\ 2x + 3y + 2z = 0 \end{cases}.$$

Solution. The augmented matrix of the system is

$$\begin{bmatrix} 3 & 1 & 2 & 1 \\ 5 & 2 & 1 & 3 \\ 2 & 3 & 2 & 0 \end{bmatrix}.$$

Since

$$\begin{bmatrix} 3 & 1 & 2 & 1 \\ 5 & 2 & 1 & 3 \\ 2 & 3 & 2 & 0 \end{bmatrix} \sim \begin{bmatrix} 1 & 0 & 0 & \frac{13}{17} \\ 0 & 1 & 0 & -\frac{2}{17} \\ 0 & 0 & 1 & -\frac{10}{17} \end{bmatrix},$$

the system has a unique solution: $x = \frac{13}{17}$, $y = -\frac{2}{17}$, and $z = -\frac{10}{17}$. □

Example 1.80. Solve the system

$$\begin{cases} p - q + 2r - s - 3t = -2 \\ 2p + q + 7r + s + 9t = -1 \\ p - 3q + s - 9t = -12 \\ 2p - q + 5r + s + t = -7 \end{cases}.$$

Solution. The augmented matrix of the system is

$$\begin{bmatrix} 1 & -1 & 2 & -1 & -3 & -2 \\ 2 & 1 & 7 & 1 & 9 & -1 \\ 1 & -3 & 0 & 1 & -9 & -12 \\ 2 & -1 & 5 & 1 & 1 & -7 \end{bmatrix}$$

and we have

$$\begin{bmatrix} 1 & -1 & 2 & -1 & -3 & -2 \\ 2 & 1 & 7 & 1 & 9 & -1 \\ 1 & -3 & 0 & 1 & -9 & -12 \\ 2 & -1 & 5 & 1 & 1 & -7 \end{bmatrix} \sim \begin{bmatrix} 1 & 0 & 3 & 0 & 2 & -1 \\ 0 & 1 & 1 & 0 & 4 & 3 \\ 0 & 0 & 0 & 1 & 1 & -2 \\ 0 & 0 & 0 & 0 & 0 & 0 \end{bmatrix}.$$

Since

$$\begin{bmatrix} 1 & 0 & 3 & 0 \\ 0 & 1 & 1 & 0 \\ 0 & 0 & 0 & 1 \\ 0 & 0 & 0 & 0 \end{bmatrix} \neq I_4,$$

the system will have infinitely many solutions. The reduced row echelon form of the augmented matrix gives us the system

$$\begin{cases} p + 3r + 2t = -1 \\ q + r + 4t = 3 \\ s + t = -2 \end{cases}.$$

The reduced row echelon form of the augmented matrix has three pivot columns, namely columns 1, 2, and 4. The variables corresponding to columns 1, 2, and 4 are p, q, and s.

The variables that correspond to the pivot columns are called *pivot variables* or *basic variables.* The other variables are called *free variables.* In the general solution of the system we let free variables be arbitrary numbers and express the basic variables in terms of the free variables.

In our example p, q, and s are basic variables of the system and r and t are free variables of the system. The solution is

$$\begin{cases} p = -1 - 3r - 2t \\ q = 3 - r - 4t \\ s = -2 - t \end{cases}$$

where r and t are arbitrary numbers. Sometimes such a solution is expressed in the following form

$$\begin{cases} p = -1 - 3r - 2t \\ q = 3 - r - 4t \\ r = r \\ s = -2 - t \\ t = t \end{cases}.$$

□

Example 1.81. Solve the system

$$\begin{cases} x - y - 2z - 3w = 2 \\ 2x + y + 2z + 9w = -1 \\ 2x - 3y + z + 10w = 1 \\ 2x - y + 2z + 11w = -3 \end{cases}.$$

Solution. The augmented matrix of the system is

$$\begin{bmatrix} 1 & -1 & -2 & -3 & 2 \\ 2 & 1 & 2 & 9 & -1 \\ 2 & -3 & 1 & 10 & 1 \\ 2 & -1 & 2 & 11 & -3 \end{bmatrix}$$

and we have

$$\begin{bmatrix} 1 & -1 & -2 & -3 & 2 \\ 2 & 1 & 2 & 9 & -1 \\ 2 & -3 & 1 & 10 & 1 \\ 2 & -1 & 2 & 11 & -3 \end{bmatrix} \sim \begin{bmatrix} 1 & 0 & 0 & 2 & 0 \\ 0 & 1 & 0 & -1 & 0 \\ 0 & 0 & 1 & 3 & 0 \\ 0 & 0 & 0 & 0 & 1 \end{bmatrix}.$$

We arrived at something that is not possible, because from the last row we get $0 = 1$. The system has no solution. □

In this section we seem to imply that the reduced row echelon form of a matrix is unique, but the question was never formally addressed. While we don't give a formal proof of the general statement, we discuss some examples that illustrate what would happen if a matrix were to have two different reduced row echelon forms.

Theorem 1.82. *The reduced row echelon form of any matrix is unique.*

Proof of a particular case. Suppose that $A \sim R$ and $A \sim Q$ where are R and Q are in the reduced row echelon form. The equations $R\mathbf{x} = \mathbf{0}$ and $Q\mathbf{x} = \mathbf{0}$ have the same solutions by the same argument as in Theorem 1.78. Now we consider two different scenarios.

Suppose that a leading 1 has different positions in R and Q, for example,

$$R = \begin{bmatrix} 1 & 0 & a & 0 & c \\ 0 & 1 & b & 0 & d \\ 0 & 0 & 0 & 1 & e \end{bmatrix} \quad \text{and} \quad Q = \begin{bmatrix} 1 & p & 0 & 0 & q \\ 0 & 0 & 1 & 0 & r \\ 0 & 0 & 0 & 1 & s \end{bmatrix}.$$

But then $\mathbf{x} = \begin{bmatrix} -p \\ 1 \\ 0 \\ 0 \\ 0 \end{bmatrix}$ is a solution of the equation $Q\mathbf{x} = \mathbf{0}$ and is not a solution of the equation $R\mathbf{x} = \mathbf{0}$.

Now suppose that R and Q have the same pivot positions, for example,

$$R = \begin{bmatrix} 1 & 0 & a & 0 & c \\ 0 & 1 & b & 0 & d \\ 0 & 0 & 0 & 1 & e \\ 0 & 0 & 0 & 0 & 0 \end{bmatrix} \quad \text{and} \quad Q = \begin{bmatrix} 1 & 0 & p & 0 & r \\ 0 & 1 & q & 0 & s \\ 0 & 0 & 0 & 1 & t \\ 0 & 0 & 0 & 0 & 0 \end{bmatrix}.$$

Since $\mathbf{x} = \begin{bmatrix} -a \\ -b \\ 1 \\ 0 \\ 0 \end{bmatrix}$ is a solution of the equation $R\mathbf{x} = \mathbf{0}$, it has to be a solution of the equation $Q\mathbf{x} = \mathbf{0}$, which gives us $p = a$ and $q = b$. In a similar way we can show that $r = c$, $s = d$, and $t = e$. □

Exercises 1.3

Calculate the matrix A and then write its inverse as a product of elementary matrices. Use the result to calculate the inverse of the matrix A.

1. $A = \begin{bmatrix} 1 & 0 \\ 0 & 2 \end{bmatrix} \begin{bmatrix} 1 & 0 \\ -5 & 1 \end{bmatrix} \begin{bmatrix} 1 & 4 \\ 0 & 1 \end{bmatrix} \begin{bmatrix} 0 & 1 \\ 1 & 0 \end{bmatrix} \begin{bmatrix} -1 & 0 \\ 0 & 1 \end{bmatrix}$

2. $A = \begin{bmatrix} 1 & 1 \\ 0 & 1 \end{bmatrix} \begin{bmatrix} 1 & 0 \\ 0 & 3 \end{bmatrix} \begin{bmatrix} 0 & 1 \\ 1 & 0 \end{bmatrix} \begin{bmatrix} 1 & 0 \\ -1 & 1 \end{bmatrix} \begin{bmatrix} 5 & 0 \\ 0 & 1 \end{bmatrix}$

3. $A = \begin{bmatrix} 1 & 0 & 3 \\ 0 & 1 & 0 \\ 0 & 0 & 1 \end{bmatrix} \begin{bmatrix} 1 & 0 & 0 \\ 0 & 1 & 4 \\ 0 & 0 & 1 \end{bmatrix} \begin{bmatrix} 1 & 0 & 0 \\ 0 & 0 & 1 \\ 0 & 1 & 0 \end{bmatrix} \begin{bmatrix} 1 & 0 & 0 \\ 0 & 1 & 2 \\ 0 & 0 & 1 \end{bmatrix}$

4. $A = \begin{bmatrix} 1 & 0 & 0 \\ 0 & 3 & 0 \\ 0 & 0 & 1 \end{bmatrix} \begin{bmatrix} 1 & 7 & 0 \\ 0 & 1 & 0 \\ 0 & 0 & 1 \end{bmatrix} \begin{bmatrix} 1 & 0 & 5 \\ 0 & 1 & 0 \\ 0 & 0 & 1 \end{bmatrix} \begin{bmatrix} 0 & 0 & 1 \\ 0 & 1 & 0 \\ 1 & 0 & 0 \end{bmatrix}$

5. $A = \begin{bmatrix} 1 & 0 & 0 & 4 \\ 0 & 1 & 0 & 0 \\ 0 & 0 & 1 & 0 \\ 0 & 0 & 0 & 1 \end{bmatrix} \begin{bmatrix} 1 & 0 & 0 & 0 \\ 0 & 0 & 0 & 1 \\ 0 & 0 & 1 & 0 \\ 0 & 1 & 0 & 0 \end{bmatrix} \begin{bmatrix} 1 & 0 & 0 & 0 \\ 0 & 1 & 0 & 0 \\ 0 & 0 & 7 & 0 \\ 0 & 0 & 0 & 1 \end{bmatrix}$

6. $A = \begin{bmatrix} 1 & 0 & 0 & 3 \\ 0 & 1 & 0 & 0 \\ 0 & 0 & 1 & 0 \\ 0 & 0 & 0 & 1 \end{bmatrix} \begin{bmatrix} 1 & 2 & 0 & 0 \\ 0 & 1 & 0 & 0 \\ 0 & 0 & 1 & 0 \\ 0 & 0 & 0 & 1 \end{bmatrix} \begin{bmatrix} 1 & 0 & 0 & 0 \\ 0 & 1 & 0 & 0 \\ 0 & 0 & 1 & 0 \\ 0 & 0 & 5 & 1 \end{bmatrix}$

Solve the given system.

7. $\begin{cases} 2x + 3y - z = 1 \\ 3x + y + 2z = 0 \\ x + 2y + 5z = 2 \end{cases}$

8. $\begin{cases} 3x + 2y + z = 1 \\ x + 3y + z = 2 \\ 2x + y + z = 1 \end{cases}$

9. $\begin{cases} x + 5y + 4z = 3 \\ 2x + 7y + 5z = 1 \\ x + 8y + 7z = 8 \end{cases}$

10. $\begin{cases} 2x + y + 3z = 4 \\ 7x + 2y + 5z = 11 \\ x + 2y + 7z = 5 \end{cases}$

11. $\begin{cases} 2x + y + 11z = 4 \\ x + 2y + 10z = 3 \\ x + y + 7z = 1 \end{cases}$

12. $\begin{cases} x + y + z = 4 \\ 5x + 2y - z = 3 \\ 2x + 3y + 4z = 1 \end{cases}$

13. $\begin{cases} 2x + y + 3z = 0 \\ x + 3y + az = 1 \\ x + 4y + 3z = 3 \end{cases}$

14. $\begin{cases} x + ay + 5z = 0 \\ x + 3y + 2z = 1 \\ 2x + 4y + z = 0 \end{cases}$

15. $\begin{cases} x + y + 2z + 13w = 7 \\ 2 + y + z + 12w = 4 \\ x + y + z + 10w = 5 \\ x - y + z = -3 \end{cases}$

16. $\begin{cases} 2x + y + 2z + 15w = 14 \\ x + 2y + 2z + 12w = 15 \\ x + y + 2z + 11w = 13 \\ x + 2y + z + 9w = 10 \end{cases}$

17. $\begin{cases} x + y + 5z + 7w = 1 \\ -x + y - z + 3w = 1 \\ 2x + y + 8z + 9w = 1 \\ 3x - y + 7z + w = -1 \end{cases}$

18. $\begin{cases} 2x + 3y + 21z + 11w = 0 \\ x + 2y + 13z + 6w = 0 \\ 3x + 2y + 19z + 14w = 0 \\ 5x + 2y + 25z + 22w = 0 \end{cases}$

Show that the given matrix is not invertible.

19. $\begin{bmatrix} 1 & 7 & -22 \\ 1 & 1 & -1 \\ 4 & 2 & 3 \end{bmatrix}$

20. $\begin{bmatrix} 2 & -1 & 5 \\ 0 & 1 & 1 \\ 1 & 1 & 4 \end{bmatrix}$

21. $\begin{bmatrix} 2 & 3 & 2 & 1 \\ 1 & 2 & 3 & 2 \\ 2 & 1 & -1 & 0 \\ 1 & 1 & 1 & 1 \end{bmatrix}$

22. $\begin{bmatrix} 1 & 1 & 0 & 1 \\ 1 & 0 & 1 & 1 \\ 0 & 1 & -1 & 0 \\ 1 & 1 & 1 & 0 \end{bmatrix}$

23. $\begin{bmatrix} 2 & 2 & 0 & 2 \\ 2 & 3 & 2 & 2 \\ 2 & 2 & 2 & 0 \\ 0 & 2 & 2 & 2 \end{bmatrix}$

24. $\begin{bmatrix} 1 & 0 & 0 & 1 \\ 1 & 0 & 1 & 0 \\ 0 & 1 & 1 & 0 \\ 0 & 1 & 0 & 1 \end{bmatrix}$

Show that the given matrix is invertible and find its inverse.

25. $\begin{bmatrix} 1 & 1 & 1 \\ 1 & 1 & -1 \\ 1 & 2 & 3 \end{bmatrix}$

26. $\begin{bmatrix} 2 & 0 & 1 \\ 3 & 1 & 1 \\ 1 & 2 & 1 \end{bmatrix}$

27. $\begin{bmatrix} 2 & 4 & 3 \\ a & 5 & 1 \\ 1 & 1 & 1 \end{bmatrix}, a \neq -3$

28. $\begin{bmatrix} 1 & 1 & 1 \\ 2 & 2 & 1 \\ a & 2 & 2 \end{bmatrix}, a \neq 2$

29. $\begin{bmatrix} 1 & 1 & 2 & 1 \\ 2 & 3 & 0 & 2 \\ 1 & 1 & a & 0 \\ 1 & 1 & 1 & 1 \end{bmatrix}$

30. $\begin{bmatrix} 4 & 1 & 1 & 1 \\ 1 & 3 & 1 & 1 \\ 1 & 1 & 2 & 1 \\ 1 & 1 & 1 & 1 \end{bmatrix}$

Write the given matrix A and its inverse as products of elementary matrices. Use the result to calculate the inverse of A.

31. $A = \begin{bmatrix} -5 & 20 \\ -2 & 5 \end{bmatrix}$

32. $A = \begin{bmatrix} 7 & 2 \\ 3 & -1 \end{bmatrix}$

33. $A = \begin{bmatrix} 2 & 1 & 0 \\ 0 & -2 & 1 \\ 1 & 1 & 0 \end{bmatrix}$

34. $A = \begin{bmatrix} 20 & 15 & 1 \\ 1 & 1 & 0 \\ 4 & 3 & 0 \end{bmatrix}$

35. Find a matrix $\begin{bmatrix} p & q & r \\ x & y & z \end{bmatrix}$ such that $\begin{bmatrix} 2 & 3 \\ 4 & 7 \end{bmatrix} \begin{bmatrix} p & q & r \\ x & y & z \end{bmatrix} = \begin{bmatrix} 1 & 1 & -1 \\ 2 & 1 & 2 \end{bmatrix}$.

36. Find a matrix $\begin{bmatrix} p & q & r \\ x & y & z \end{bmatrix}$ such that $\begin{bmatrix} 5 & 1 \\ 7 & 2 \end{bmatrix} \begin{bmatrix} p & q & r \\ x & y & z \end{bmatrix} = \begin{bmatrix} 2 & 0 & 4 \\ 1 & 3 & -2 \end{bmatrix}$.

37. Find a matrix $\begin{bmatrix} p & x \\ q & y \\ r & z \end{bmatrix}$ such that $\begin{bmatrix} 2 & 3 & 1 \\ 1 & 1 & 2 \\ 1 & 1 & 3 \end{bmatrix} \begin{bmatrix} p & x \\ q & y \\ r & z \end{bmatrix} = \begin{bmatrix} 1 & 1 \\ 1 & 2 \\ 3 & -1 \end{bmatrix}$.

38. Find a matrix $\begin{bmatrix} p & x \\ q & y \\ r & z \end{bmatrix}$ such that $\begin{bmatrix} 4 & 2 & 1 \\ 2 & 3 & 2 \\ 1 & 2 & 5 \end{bmatrix} \begin{bmatrix} p & x \\ q & y \\ r & z \end{bmatrix} = \begin{bmatrix} 2 & 0 \\ 3 & 1 \\ 1 & 1 \end{bmatrix}$.

Chapter 2

The vector space $\mathbb{R}^n$

In this chapter, we present basic topics in linear algebra, like linear independence, vector subspace, basis and dimension, that will be needed in the rest of this book.

2.1 Linear combinations of column vectors

We recall that by a vector in $\mathbb{R}^n$, where n is an integer greater than 1, we mean an $n \times 1$ matrix $\begin{bmatrix} a_1 \\ \vdots \\ a_n \end{bmatrix}$. Sometimes we refer to such vectors as *column vectors*. Since vectors are matrices, the operations of addition of vectors and multiplication of vectors by scalars are defined as for general matrices, that is,

$$\begin{bmatrix} a_1 \\ \vdots \\ a_n \end{bmatrix} + \begin{bmatrix} b_1 \\ \vdots \\ b_n \end{bmatrix} = \begin{bmatrix} a_1 + b_1 \\ \vdots \\ a_n + b_n \end{bmatrix} \quad \text{and} \quad c \begin{bmatrix} a_1 \\ \vdots \\ a_n \end{bmatrix} = \begin{bmatrix} ca_1 \\ \vdots \\ ca_n \end{bmatrix}$$

where c is an arbitrary real number.

Example 2.1. We have

$$\begin{bmatrix} 1 \\ -2 \\ 0 \\ 7 \end{bmatrix} + \begin{bmatrix} 5 \\ 4 \\ -3 \\ 1 \end{bmatrix} = \begin{bmatrix} 6 \\ 2 \\ -3 \\ 8 \end{bmatrix} \quad \text{and} \quad \frac{1}{3} \begin{bmatrix} 5 \\ -4 \\ -3 \\ 7 \end{bmatrix} = \begin{bmatrix} \frac{5}{3} \\ -\frac{4}{3} \\ -1 \\ \frac{7}{3} \end{bmatrix}.$$

Definition 2.2. Let $\mathbf{v}_1, \dots, \mathbf{v}_j$ be vectors in $\mathbb{R}^n$. A vector of the form

$$c_1\mathbf{v}_1 + \cdots + c_j\mathbf{v}_j,$$

where $c_1, \dots, c_j$ are arbitrary real numbers, is called a *linear combination* of the vectors $\mathbf{v}_1, \dots, \mathbf{v}_j$.

Example 2.3. Since

$$2\begin{bmatrix}3\\0\\-1\\2\end{bmatrix}-3\begin{bmatrix}0\\-2\\-2\\5\end{bmatrix}+\frac{1}{2}\begin{bmatrix}6\\2\\-1\\8\end{bmatrix}=\begin{bmatrix}9\\7\\\frac{7}{2}\\-7\end{bmatrix},$$

the vector $\begin{bmatrix}9\\7\\\frac{7}{2}\\-7\end{bmatrix}$ is a linear combination of vectors $\begin{bmatrix}3\\0\\-1\\2\end{bmatrix}$, $\begin{bmatrix}0\\-2\\-2\\5\end{bmatrix}$, and $\begin{bmatrix}6\\2\\-1\\8\end{bmatrix}$.

When the Gaussian elimination process is applied to a matrix A, then the resulting reduced row echelon form of A has some entries identified as leading 1's. Every position in a matrix where a leading 1 is located is called a *pivot position*. Every column that contains a pivot position is called a *pivot column*. Since the reduced row echelon form of A is unique, the pivot columns are uniquely determined. We use that name for the columns of the matrix in the reduced row echelon form as well as the corresponding columns of the original matrix.

Example 2.4. Since the reduced row echelon form of the matrix

$$A=\begin{bmatrix}2&4&-1&2&1\\1&2&2&11&1\\-1&-2&1&1&0\end{bmatrix}$$

is the matrix

$$\begin{bmatrix}1&2&0&3&0\\0&0&1&4&0\\0&0&0&0&1\end{bmatrix},$$

the matrix A has three pivot columns; the first, third, and fifth.

The columns of a matrix that are not pivot columns are referred to as *nonpivot columns*. It turns out that every nonpivot column of an arbitrary matrix A can be expressed as a linear combination of the pivot columns of A that are to the left of that nonpivot column. We formulate this result, which we will use constantly in this chapter, in the next theorem.

Theorem 2.5. *Suppose $A = \begin{bmatrix} \mathbf{a}_1 & \dots & \mathbf{a}_n \end{bmatrix}$ is a matrix with columns $\mathbf{a}_1, \dots, \mathbf{a}_n$. If $\mathbf{a}_j$ is a nonpivot column and $\mathbf{a}_{j_1}, \dots, \mathbf{a}_{j_k}$ are the pivot columns to the left of the column $\mathbf{a}_j$, then*

$$\mathbf{a}_j = b_1\mathbf{a}_{j_1} + \dots + b_k\mathbf{a}_{j_k}$$

where

$$\begin{bmatrix} b_1 \\ \vdots \\ b_k \\ 0 \\ \vdots \\ 0 \end{bmatrix}$$

is the j-th column in the reduced row echelon form of the matrix A.

Proof of a particular case. We illustrate the proof using a 4×7 matrix A. Suppose we have

$$A = \begin{bmatrix} a_{11} & a_{12} & a_{13} & a_{14} & a_{15} & a_{16} & a_{17} \\ a_{21} & a_{22} & a_{23} & a_{24} & a_{25} & a_{26} & a_{27} \\ a_{31} & a_{32} & a_{33} & a_{34} & a_{35} & a_{36} & a_{37} \\ a_{41} & a_{42} & a_{43} & b_{44} & a_{45} & a_{46} & a_{47} \end{bmatrix} \sim \begin{bmatrix} 1 & p & 0 & q & s & 0 & u \\ 0 & 0 & 1 & r & t & 0 & v \\ 0 & 0 & 0 & 0 & 0 & 1 & w \\ 0 & 0 & 0 & 0 & 0 & 0 & 0 \end{bmatrix}. \tag{2.1}$$

We will show that

$$\begin{bmatrix} a_{15} \\ a_{25} \\ a_{35} \\ a_{45} \end{bmatrix} = s \begin{bmatrix} a_{11} \\ a_{21} \\ a_{31} \\ a_{41} \end{bmatrix} + t \begin{bmatrix} a_{13} \\ a_{23} \\ a_{33} \\ a_{43} \end{bmatrix}. \tag{2.2}$$

Since

$$-s \begin{bmatrix} 1 \\ 0 \\ 0 \\ 0 \end{bmatrix} + 0 \cdot \begin{bmatrix} p \\ 0 \\ 0 \\ 0 \end{bmatrix} - t \begin{bmatrix} 0 \\ 1 \\ 0 \\ 0 \end{bmatrix} + 0 \cdot \begin{bmatrix} q \\ r \\ 0 \\ 0 \end{bmatrix} + 1 \cdot \begin{bmatrix} s \\ t \\ 0 \\ 0 \end{bmatrix} + 0 \cdot \begin{bmatrix} 0 \\ 0 \\ 1 \\ 0 \end{bmatrix} + 0 \cdot \begin{bmatrix} u \\ v \\ w \\ 0 \end{bmatrix} = \begin{bmatrix} 0 \\ 0 \\ 0 \\ 0 \end{bmatrix} \tag{2.3}$$

and

$$\begin{bmatrix} A & \mathbf{0} \end{bmatrix} = \begin{bmatrix} a_{11} & a_{12} & a_{13} & a_{14} & a_{15} & a_{16} & a_{17} & 0 \\ a_{21} & a_{22} & a_{23} & a_{24} & a_{25} & a_{26} & a_{27} & 0 \\ a_{31} & a_{32} & a_{33} & a_{34} & a_{35} & a_{36} & a_{37} & 0 \\ a_{41} & a_{42} & a_{43} & b_{44} & a_{45} & a_{46} & a_{47} & 0 \end{bmatrix} \sim \begin{bmatrix} 1 & p & 0 & q & s & 0 & u & 0 \\ 0 & 0 & 1 & r & t & 0 & v & 0 \\ 0 & 0 & 0 & 0 & 0 & 1 & w & 0 \\ 0 & 0 & 0 & 0 & 0 & 0 & 0 & 0 \end{bmatrix},$$

we have

$$-s \begin{bmatrix} a_{11} \\ a_{21} \\ a_{31} \\ a_{41} \end{bmatrix} + 0 \cdot \begin{bmatrix} a_{12} \\ a_{22} \\ a_{32} \\ a_{42} \end{bmatrix} - t \begin{bmatrix} a_{13} \\ a_{23} \\ a_{33} \\ a_{43} \end{bmatrix} + 0 \cdot \begin{bmatrix} a_{14} \\ a_{24} \\ a_{34} \\ a_{44} \end{bmatrix} + 1 \cdot \begin{bmatrix} a_{15} \\ a_{25} \\ a_{35} \\ a_{45} \end{bmatrix} + 0 \cdot \begin{bmatrix} a_{16} \\ a_{26} \\ a_{36} \\ a_{46} \end{bmatrix} + 0 \cdot \begin{bmatrix} a_{17} \\ a_{27} \\ a_{37} \\ a_{47} \end{bmatrix} = \begin{bmatrix} 0 \\ 0 \\ 0 \\ 0 \end{bmatrix}, \tag{2.4}$$

because row operations do not affect the set of solutions. From (2.4) we get (2.2). □

Example 2.6. Write all nonpivot columns of the matrix

$$A = \begin{bmatrix} 1 & 2 & 3 & 1 & 2 \\ 2 & 1 & 1 & 2 & 1 \\ 1 & -1 & -2 & 2 & 1 \end{bmatrix}$$

as linear combinations of the pivot columns of A using Theorem 2.5.

Solution. Since the reduced row echelon form of the matrix A is

$$\begin{bmatrix} 1 & 0 & -\frac{1}{3} & 0 & -2 \\ 0 & 1 & \frac{5}{3} & 0 & 1 \\ 0 & 0 & 0 & 1 & 2 \end{bmatrix},$$

we can conclude that

$$\begin{bmatrix} 3 \\ 1 \\ -2 \end{bmatrix} = -\frac{1}{3}\begin{bmatrix} 1 \\ 2 \\ 1 \end{bmatrix} + \frac{5}{3}\begin{bmatrix} 2 \\ 1 \\ -1 \end{bmatrix} \quad \text{and} \quad \begin{bmatrix} 2 \\ 1 \\ 1 \end{bmatrix} = -2\begin{bmatrix} 1 \\ 2 \\ 1 \end{bmatrix} + \begin{bmatrix} 2 \\ 1 \\ -1 \end{bmatrix} + 2\begin{bmatrix} 1 \\ 2 \\ 2 \end{bmatrix}.$$

□

As a consequence of Theorem 2.5 we get

Theorem 2.7. *Suppose* $A = \begin{bmatrix} \mathbf{a}_1 & \dots & \mathbf{a}_n \end{bmatrix}$ *is a* $m \times n$ *matrix with columns* $\mathbf{a}_1, \dots, \mathbf{a}_n$ *and* $\mathbf{b}$ *is a vector in* $\mathbb{R}^m$. *The equation*

$$A\mathbf{x} = \mathbf{b}$$

has a solution if and only if $\mathbf{b}$ *is a nonpivot column of the matrix*

$$\begin{bmatrix} A & \mathbf{b} \end{bmatrix}.$$

Consequently, the equation $A\mathbf{x} = \mathbf{b}$ *has no solution if and only if* $\mathbf{b}$ *is a pivot column of the matrix* $\begin{bmatrix} A & \mathbf{b} \end{bmatrix}$.

Writing nonpivot columns as linear combinations of pivot columns gives a new perspective on finding the inverse of a matrix as we can see in the next example.

Example 2.8. Find the inverse of the matrix

$$\begin{bmatrix} 2 & 1 & 1 \\ 1 & 3 & 1 \\ -1 & 1 & 1 \end{bmatrix}.$$

Solution. We have to find a matrix

$$\begin{bmatrix} x & p & s \\ y & q & t \\ z & r & u \end{bmatrix}$$

such that

$$\begin{bmatrix} 2 & 1 & 1 \\ 1 & 3 & 1 \\ -1 & 1 & 1 \end{bmatrix} \begin{bmatrix} x & p & s \\ y & q & t \\ z & r & u \end{bmatrix} = \begin{bmatrix} 1 & 0 & 0 \\ 0 & 1 & 0 \\ 0 & 0 & 1 \end{bmatrix}.$$

This is equivalent to solving three equations:

$$\begin{bmatrix} 2 & 1 & 1 \\ 1 & 3 & 1 \\ -1 & 1 & 1 \end{bmatrix} \begin{bmatrix} x \\ y \\ z \end{bmatrix} = \begin{bmatrix} 1 \\ 0 \\ 0 \end{bmatrix},$$

$$\begin{bmatrix} 2 & 1 & 1 \\ 1 & 3 & 1 \\ -1 & 1 & 1 \end{bmatrix} \begin{bmatrix} p \\ q \\ r \end{bmatrix} = \begin{bmatrix} 0 \\ 1 \\ 0 \end{bmatrix},$$

$$\begin{bmatrix} 2 & 1 & 1 \\ 1 & 3 & 1 \\ -1 & 1 & 1 \end{bmatrix} \begin{bmatrix} s \\ t \\ u \end{bmatrix} = \begin{bmatrix} 0 \\ 0 \\ 1 \end{bmatrix}.$$

By Theorem 2.5, since

$$\begin{bmatrix} 2 & 1 & 1 & 1 & 0 & 0 \\ 1 & 3 & 1 & 0 & 1 & 0 \\ -1 & 1 & 1 & 0 & 0 & 1 \end{bmatrix} \sim \begin{bmatrix} 1 & 0 & 0 & \frac{1}{3} & 0 & -\frac{1}{3} \\ 0 & 1 & 0 & -\frac{1}{3} & \frac{1}{2} & -\frac{1}{6} \\ 0 & 0 & 1 & \frac{2}{3} & -\frac{1}{2} & \frac{5}{6} \end{bmatrix}, \tag{2.5}$$

the solutions of the three equations are

$$\begin{bmatrix} x \\ y \\ z \end{bmatrix} = \begin{bmatrix} \frac{1}{3} \\ -\frac{1}{3} \\ \frac{2}{3} \end{bmatrix}, \quad \begin{bmatrix} p \\ q \\ r \end{bmatrix} = \begin{bmatrix} 0 \\ \frac{1}{2} \\ -\frac{1}{2} \end{bmatrix}, \quad \text{and} \quad \begin{bmatrix} s \\ t \\ u \end{bmatrix} = \begin{bmatrix} -\frac{1}{3} \\ -\frac{1}{6} \\ \frac{5}{6} \end{bmatrix}.$$

But this means that

$$\begin{bmatrix} 2 & 1 & 1 \\ 1 & 3 & 1 \\ -1 & 1 & 1 \end{bmatrix}^{-1} = \begin{bmatrix} \frac{1}{3} & 0 & -\frac{1}{3} \\ -\frac{1}{3} & \frac{1}{2} & -\frac{1}{6} \\ \frac{2}{3} & -\frac{1}{2} & \frac{5}{6} \end{bmatrix}.$$

□

Note that the last three columns of the matrix (2.5) give us the inverse of the matrix and are the nonpivot columns of this matrix.

Definition 2.9. Vectors $\mathbf{v}_1, \dots, \mathbf{v}_j$ in $\mathbb{R}^n$ are *linearly dependent* if and only if the equation

$$x_1\mathbf{v}_1 + \cdots + x_j\mathbf{v}_j = \mathbf{0}$$

has a nontrivial solution, that is, a solution such that at least one of the numbers $x_1, \dots, x_j$ is different from 0.

The following theorem gives us a practical method to determine linear dependence of vectors.

Theorem 2.10. *Vectors $\mathbf{v}_1, \dots, \mathbf{v}_j$ in $\mathbb{R}^n$ are linearly dependent if and only if the reduced row echelon form of the matrix $\begin{bmatrix} \mathbf{v}_1 & \dots & \mathbf{v}_j \end{bmatrix}$ has at least one nonpivot column.*

Proof. If the reduced row echelon form of the matrix $\begin{bmatrix} \mathbf{v}_1 & \dots & \mathbf{v}_j \end{bmatrix}$ has only pivot columns then the equation

$$x_1\mathbf{v}_1 + \cdots + x_j\mathbf{v}_j = \mathbf{0}$$

has only the trivial solution, that is, $x_1 = \cdots = x_j = 0$.

If the reduced row echelon form of the matrix $\begin{bmatrix} \mathbf{v}_1 & \dots & \mathbf{v}_j \end{bmatrix}$ has at least one nonpivot column, then it follows from the proof of Theorem 2.5 that the equation

$$x_1\mathbf{v}_1 + \cdots + x_j\mathbf{v}_j = \mathbf{0}$$

has a nontrivial solution. □

Example 2.11. Show that the vectors $\begin{bmatrix} 1 \\ 2 \\ 4 \\ -2 \end{bmatrix}, \begin{bmatrix} 2 \\ 1 \\ -1 \\ 5 \end{bmatrix}, \begin{bmatrix} 1 \\ 1 \\ 1 \\ 1 \end{bmatrix}, \begin{bmatrix} 1 \\ 1 \\ 2 \\ 1 \end{bmatrix}$ are linearly dependent.

Solution. We have

$$\begin{bmatrix} 1 & 2 & 1 & 1 \\ 2 & 1 & 1 & 1 \\ 4 & -1 & 1 & 2 \\ -2 & 5 & 1 & 1 \end{bmatrix} \sim \begin{bmatrix} 1 & 0 & \frac{1}{3} & 0 \\ 0 & 1 & \frac{1}{3} & 0 \\ 0 & 0 & 0 & 1 \\ 0 & 0 & 0 & 0 \end{bmatrix}.$$

Since the matrix $\begin{bmatrix} 1 & 0 & \frac{1}{3} & 0 \\ 0 & 1 & \frac{1}{3} & 0 \\ 0 & 0 & 0 & 1 \\ 0 & 0 & 0 & 0 \end{bmatrix}$ has a nonpivot column, namely $\begin{bmatrix} \frac{1}{3} \\ \frac{1}{3} \\ 0 \\ 0 \end{bmatrix}$, the vectors $\begin{bmatrix} 1 \\ 2 \\ 4 \\ -2 \end{bmatrix}, \begin{bmatrix} 2 \\ 1 \\ -1 \\ 5 \end{bmatrix}, \begin{bmatrix} 1 \\ 1 \\ 1 \\ 1 \end{bmatrix}, \begin{bmatrix} 1 \\ 1 \\ 2 \\ 1 \end{bmatrix}$ are linearly dependent, by Theorem 2.10. □

Theorems 2.5 and 2.10 give us the following important property of linearly dependent vectors.

Corollary 2.12. *Vectors $\mathbf{v}_1, \dots, \mathbf{v}_j$ from $\mathbb{R}^n$ are linearly dependent if and only if $\mathbf{v}_1 = \mathbf{0}$ or there is a integer $k \geq 1$ and real numbers $a_1, \dots, a_k$ such that*

$$a_1\mathbf{v}_1 + \cdots + a_k\mathbf{v}_k = \mathbf{v}_{k+1}.$$

Example 2.13. Since

$$\begin{bmatrix} 1 & 2 & 1 & 1 \\ 2 & 1 & 1 & 1 \\ 4 & -1 & 1 & 2 \\ -2 & 5 & 1 & 1 \end{bmatrix} \sim \begin{bmatrix} 1 & 0 & \frac{1}{3} & 0 \\ 0 & 1 & \frac{1}{3} & 0 \\ 0 & 0 & 0 & 1 \\ 0 & 0 & 0 & 0 \end{bmatrix},$$

we know that the vectors $\begin{bmatrix} 1 \\ 2 \\ 4 \\ -2 \end{bmatrix}, \begin{bmatrix} 2 \\ 1 \\ -1 \\ 5 \end{bmatrix}, \begin{bmatrix} 1 \\ 1 \\ 1 \\ 1 \end{bmatrix}, \begin{bmatrix} 1 \\ 1 \\ 2 \\ 1 \end{bmatrix}$ are linearly dependent and we have

$$\frac{1}{3}\begin{bmatrix} 1 \\ 2 \\ 4 \\ -2 \end{bmatrix} + \frac{1}{3}\begin{bmatrix} 2 \\ 1 \\ -1 \\ 5 \end{bmatrix} = \begin{bmatrix} 1 \\ 1 \\ 1 \\ 1 \end{bmatrix}.$$

Definition 2.14. The vectors $\mathbf{v}_1, \dots, \mathbf{v}_j$ from $\mathbb{R}^n$ are linearly independent if and only if the only solution of the equation

$$x_1\mathbf{v}_1 + \cdots + x_j\mathbf{v}_j = \mathbf{0}$$

is the trivial solution, that is $x_1 = \cdots = x_j = 0$.

From Theorem 2.10 we immediately get the following useful characterization of linearly independent vectors.

Theorem 2.15. *The vectors $\mathbf{v}_1, \dots, \mathbf{v}_k$ from $\mathbb{R}^n$ are linearly independent if the reduced row echelon form of the matrix $\begin{bmatrix} \mathbf{v}_1 & \dots & \mathbf{v}_k \end{bmatrix}$ has only pivot columns.*

Example 2.16. Show that the vectors $\begin{bmatrix} 1 \\ 1 \\ 2 \\ 2 \end{bmatrix}, \begin{bmatrix} 1 \\ 2 \\ 1 \\ 3 \end{bmatrix}, \begin{bmatrix} 1 \\ 3 \\ 3 \\ 5 \end{bmatrix}$ are linearly independent.

Proof. Since

$$\begin{bmatrix} 1 & 1 & 1 \\ 1 & 2 & 3 \\ 2 & 1 & 3 \\ 2 & 3 & 5 \end{bmatrix} \sim \begin{bmatrix} 1 & 0 & 0 \\ 0 & 1 & 0 \\ 0 & 0 & 1 \\ 0 & 0 & 0 \end{bmatrix},$$

the vectors $\begin{bmatrix} 1 \\ 1 \\ 2 \\ 2 \end{bmatrix}, \begin{bmatrix} 1 \\ 2 \\ 1 \\ 3 \end{bmatrix}, \begin{bmatrix} 1 \\ 3 \\ 3 \\ 5 \end{bmatrix}$ are linearly independent, by Theorem 2.15. □

In the next theorem we see the first indication of importance of linear independence of vectors.

Theorem 2.17. *Let $A = \begin{bmatrix} \mathbf{c}_1 & \dots & \mathbf{c}_n \end{bmatrix}$ be an $n \times n$ matrix. The following conditions are equivalent:*

(a) *The matrix A is invertible;*

(b) *The equation $A\mathbf{x} = \mathbf{0}$ has only the trivial solution, that is, $\mathbf{x} = \mathbf{0}$;*

(c) *The vectors $\mathbf{c}_1, \dots, \mathbf{c}_n$ are linearly independent.*

Proof. If the matrix A is invertible, then from the equation $A\mathbf{x} = \mathbf{0}$ we get

$$\mathbf{x} = I_n\mathbf{x} = A^{-1}A\mathbf{x} = A^{-1}\mathbf{0} = \mathbf{0}.$$

Therefore (a) implies (b).

If the equation $A\mathbf{x} = \mathbf{0}$ has only the trivial solution, then the vectors $\mathbf{c}_1, \dots, \mathbf{c}_n$ are linearly independent, by Definition 2.14. This shows that (b) implies (c).

Finally, if the vectors $\mathbf{c}_1, \dots, \mathbf{c}_n$ are linearly independent, then the reduced row echelon form of the matrix $A = \begin{bmatrix} \mathbf{c}_1 & \dots & \mathbf{c}_n \end{bmatrix}$ is I_n, because of Theorem 2.15. But then A is invertible by Theorem 1.66. Thus (c) implies (a). □

Exercises 2.1

Determine if the columns of the given matrix are linearly dependent or linearly independent.

1. $\begin{bmatrix} -3 & 4 & 1 \\ 5 & -2 & 3 \\ 2 & 5 & 7 \\ 3 & 5 & 8 \end{bmatrix}$

2. $\begin{bmatrix} 1 & 4 & 7 \\ 5 & -2 & 13 \\ 3 & -5 & 4 \\ 3 & -1 & 8 \end{bmatrix}$

3. $\begin{bmatrix} 1 & 1 & 1 \\ 1 & 1 & 1 \\ 1 & 2 & 1 \\ 2 & 1 & 1 \end{bmatrix}$

4. $\begin{bmatrix} 1 & 1 & 1 \\ 1 & 1 & 4 \\ 1 & 4 & 1 \\ 1 & 0 & 1 \end{bmatrix}$

Find a number a such that the columns of the given matrix are linearly dependent.

5. $\begin{bmatrix} -3 & 4 & 1 \\ 5 & -2 & 3 \\ 2 & 5 & a+5 \\ 1 & 1 & a \end{bmatrix}$

6. $\begin{bmatrix} 1 & 0 & a-1 \\ 2 & 1 & a \\ 1 & 2 & 0 \\ 1 & 1 & 1 \end{bmatrix}$

Find numbers a and b such that the columns of the given matrix are linearly dependent.

7. $\begin{bmatrix} 2 & 1 & 9 \\ 1 & 3 & a \\ 3 & -5 & 7 \\ 5 & -13 & b \end{bmatrix}$

8. $\begin{bmatrix} 1 & 1 & 5 \\ 2 & 1 & a \\ 1 & 3 & 11 \\ 4 & 1 & b \end{bmatrix}$

Write all nonpivot columns of the given matrix as linear combinations of the pivot columns.

9. $\begin{bmatrix} 1 & 2 & 1 & 1 \\ 1 & 1 & 3 & 2 \\ 2 & 3 & 4 & 3 \end{bmatrix}$

10. $\begin{bmatrix} 1 & 3 & 2 & 5 & 1 \\ 1 & 1 & 2 & 3 & 3 \\ 1 & 2 & 2 & 4 & 2 \end{bmatrix}$

11. $\begin{bmatrix} 1 & 2 & 4 & 0 & 5 \\ 2 & 1 & -1 & 3 & -2 \\ 1 & 1 & 1 & 1 & 1 \end{bmatrix}$

12. $\begin{bmatrix} 1 & 3 & 1 & 2 & 1 \\ 2 & 2 & 1 & 2 & 3 \\ 4 & 8 & 3 & 2 & 1 \end{bmatrix}$

13. $\begin{bmatrix} 3 & 2 & 4 & 1 & 4 \\ 2 & 1 & 4 & 1 & -2 \\ 5 & 3 & 8 & 1 & 1 \end{bmatrix}$

14. $\begin{bmatrix} 1 & 2 & 3 & 5 \\ 3 & 1 & 4 & 5 \\ 2 & -1 & 1 & 0 \\ 1 & 1 & 2 & 3 \end{bmatrix}$.

15. If $\begin{bmatrix} \mathbf{c}_1 & \mathbf{c}_2 & \mathbf{c}_3 & \mathbf{c}_4 & \mathbf{c}_5 \end{bmatrix} = \begin{bmatrix} 1 & 2 & 3 & 2 & 1 \\ 3 & 1 & 2 & 3 & 1 \\ 2 & -1 & -1 & 1 & 1 \\ 1 & 1 & 2 & 3 & 1 \end{bmatrix}$, show that the vectors $\mathbf{c}_1, \mathbf{c}_2, \mathbf{c}_3, \mathbf{c}_4$ are linearly dependent and the vectors $\mathbf{c}_1, \mathbf{c}_2, \mathbf{c}_3, \mathbf{c}_5$ are linearly independent.

16. Let $\begin{bmatrix} \mathbf{c}_1 & \mathbf{c}_2 & \mathbf{c}_3 & \mathbf{c}_4 \end{bmatrix} = \begin{bmatrix} 1 & 1 & 1 & 1 \\ 3 & 3 & 2 & 1 \\ 4 & 2 & 3 & 2 \end{bmatrix}$. Determine if

    (a) the vectors $\mathbf{c}_2, \mathbf{c}_3, \mathbf{c}_4$ are linearly independent,
    (b) the vectors $\mathbf{c}_1, \mathbf{c}_3, \mathbf{c}_4$ are linearly independent,
    (c) the vectors $\mathbf{c}_1, \mathbf{c}_2, \mathbf{c}_4$ are linearly independent,
    (d) the vectors $\mathbf{c}_1, \mathbf{c}_2, \mathbf{c}_3$ are linearly independent.

17. If $\begin{bmatrix} \mathbf{c}_1 & \mathbf{c}_2 & \mathbf{c}_3 & \mathbf{c}_4 & \mathbf{c}_5 \end{bmatrix} = \begin{bmatrix} 3 & 2 & 1 & 1 & 2 \\ 2 & 1 & 1 & 1 & -1 \\ 5 & 3 & 2 & 2 & 1 \\ 5 & 3 & 2 & 3 & 0 \end{bmatrix}$, show that

    (a) the vectors $\mathbf{c}_1$, $\mathbf{c}_2$, and $\mathbf{c}_4$ are linearly independent,
    (b) the vectors $\mathbf{c}_1$, $\mathbf{c}_2$, and $\mathbf{c}_5$ are linearly independent,
    (c) the vectors $\mathbf{c}_1$, $\mathbf{c}_3$, and $\mathbf{c}_4$ are linearly independent,
    (d) the vectors $\mathbf{c}_1$, $\mathbf{c}_3$, and $\mathbf{c}_5$ are linearly independent.

18. The matrix $\begin{bmatrix} \mathbf{c}_1 & \mathbf{c}_2 & \mathbf{c}_3 & \mathbf{c}_4 & \mathbf{c}_5 \end{bmatrix}$ has 5 rows and 5 columns. If the reduced row echelon form of the matrix is

$$\begin{bmatrix} 1 & 0 & 0 & 2 & 0 \\ 0 & 1 & 0 & 4 & 0 \\ 0 & 0 & 1 & 1 & 0 \\ 0 & 0 & 0 & 0 & 1 \end{bmatrix},$$

determine whether

(a) the vectors $\mathbf{c}_1, \mathbf{c}_2, \mathbf{c}_3, \mathbf{c}_4$ are linearly independent,

(b) the vectors $\mathbf{c}_1, \mathbf{c}_2, \mathbf{c}_3, \mathbf{c}_5$ are linearly independent,

(c) the vectors $\mathbf{c}_2, \mathbf{c}_3, \mathbf{c}_4, \mathbf{c}_5$ are linearly independent.

19. Let $A = \begin{bmatrix} \mathbf{c}_1 & \mathbf{c}_2 & \mathbf{c}_3 & \mathbf{c}_4 & \mathbf{c}_5 \end{bmatrix}$ be a 4×5 matrix such that

$$A \sim \begin{bmatrix} 1 & 0 & a & 0 & a \\ 0 & 1 & b & 0 & b \\ 0 & 0 & 0 & 1 & c \\ 0 & 0 & 0 & 0 & 0 \end{bmatrix}.$$

If all numbers a, b, and c are different from 0, determine all sets of 3 columns which are linearly independent.

20. Let $A = \begin{bmatrix} \mathbf{c}_1 & \mathbf{c}_2 & \mathbf{c}_3 & \mathbf{c}_4 & \mathbf{c}_5 \end{bmatrix}$ be a 5×5 matrix such that

$$A \sim \begin{bmatrix} 1 & 0 & 0 & 0 & 0 \\ 0 & 1 & 0 & 0 & a \\ 0 & 0 & 1 & 0 & b \\ 0 & 0 & 0 & 1 & 0 \\ 0 & 0 & 0 & 0 & 0 \end{bmatrix}.$$

If the numbers a and b are different from 0, determine all sets of 4 columns which are linearly independent.

2.2 Subspaces of $\mathbb{R}^n$

Every vector in $\mathbb{R}^3$ can be written as a linear combination of vectors $\begin{bmatrix} 1 \\ 0 \\ 0 \end{bmatrix}, \begin{bmatrix} 0 \\ 1 \\ 0 \end{bmatrix}, \begin{bmatrix} 0 \\ 0 \\ 1 \end{bmatrix}$:

$$\begin{bmatrix} a \\ b \\ c \end{bmatrix} = a \begin{bmatrix} 1 \\ 0 \\ 0 \end{bmatrix} + b \begin{bmatrix} 0 \\ 1 \\ 0 \end{bmatrix} + c \begin{bmatrix} 0 \\ 0 \\ 1 \end{bmatrix}.$$

In some sense, the three vectors $\begin{bmatrix} 1 \\ 0 \\ 0 \end{bmatrix}$, $\begin{bmatrix} 0 \\ 1 \\ 0 \end{bmatrix}$, and $\begin{bmatrix} 0 \\ 0 \\ 1 \end{bmatrix}$ are sufficient to "build" the whole space $\mathbb{R}^3$. On the other hand, if we take only two of these three vectors, say $\begin{bmatrix} 1 \\ 0 \\ 0 \end{bmatrix}$ and $\begin{bmatrix} 0 \\ 0 \\ 1 \end{bmatrix}$, then we will no longer be able to obtain the whole space $\mathbb{R}^3$. Only vectors of the form $\begin{bmatrix} a \\ 0 \\ b \end{bmatrix}$ can be obtained as linear combinations of the vectors $\begin{bmatrix} 1 \\ 0 \\ 0 \end{bmatrix}$

and $\begin{bmatrix}0\\0\\1\end{bmatrix}$. These two simple examples illustrate an idea that plays a fundamental role in linear algebra.

Definition 2.18. Let $\mathbf{v}_1,\dots,\mathbf{v}_k$ be vectors in $\mathbb{R}^n$. The set of all linear combinations of the vectors $\mathbf{v}_1,\dots,\mathbf{v}_k$, that is, the set of all possible sums of the form

$$x_1\mathbf{v}_1+\cdots+x_k\mathbf{v}_k,$$

where $x_1,\dots,x_k$ are arbitrary real numbers, is denoted by

$$\operatorname{Span}\{\mathbf{v}_1,\dots,\mathbf{v}_k\}$$

and called the linear span (or simply span) of vectors $\mathbf{v}_1,\dots,\mathbf{v}_k$.
If $\mathcal{V}=\operatorname{Span}\{\mathbf{v}_1,\dots,\mathbf{v}_k\}$, we say that $\{\mathbf{v}_1,\dots,\mathbf{v}_k\}$ is a spanning set for $\mathcal{V}$.

Example 2.19. Determine the span of the vectors $\begin{bmatrix}1\\0\\0\\1\end{bmatrix},\begin{bmatrix}1\\0\\1\\0\end{bmatrix},\begin{bmatrix}1\\1\\0\\0\end{bmatrix}$.

Solution. The span is the set of all linear combinations of the vectors, that is, all vectors of the form

$$x_1\begin{bmatrix}1\\0\\0\\1\end{bmatrix}+x_2\begin{bmatrix}1\\0\\1\\0\end{bmatrix}+x_3\begin{bmatrix}1\\1\\0\\0\end{bmatrix}=\begin{bmatrix}x_1+x_2+x_3\\x_3\\x_2\\x_1\end{bmatrix}.$$

In other words,

$$\operatorname{Span}\left\{\begin{bmatrix}1\\0\\0\\1\end{bmatrix},\begin{bmatrix}1\\0\\1\\0\end{bmatrix},\begin{bmatrix}1\\1\\0\\0\end{bmatrix}\right\}=\left\{\begin{bmatrix}x_1+x_2+x_3\\x_3\\x_2\\x_1\end{bmatrix}:x_1,x_2,x_3\text{ in }\mathbb{R}\right\}.$$

□

An important property of any span is that it is closed under multiplication by scalars and addition of vectors.

Theorem 2.20. *Let $\mathbf{v}_1,\dots,\mathbf{v}_k$ be vectors from $\mathbb{R}^n$ and let $\mathcal{V} = \operatorname{Span}\{\mathbf{v}_1,\dots,\mathbf{v}_k\}$.*

(a) *If $\mathbf{v}$ is a vector in $\mathcal{V}$ and c is a real number, then the vector $c\mathbf{v}$ is in $\mathcal{V}$.*

(b) *If $\mathbf{v}$ and $\mathbf{w}$ are vectors in $\mathcal{V}$, then the vector $\mathbf{v}+\mathbf{w}$ is in $\mathcal{V}$.*

Proof. If $\mathbf{v}$ is a vector in $\mathcal{V}$, then

$$\mathbf{v} = x_1\mathbf{v}_1 + \cdots + x_k\mathbf{v}_k,$$

for some real numbers $x_1,\dots,x_k$. For any real number c we have

$$c\mathbf{v} = c(x_1\mathbf{v}_1 + \cdots + x_k\mathbf{v}_k) = (cx_1)\mathbf{v}_1 + \cdots + (cx_k)\mathbf{v}_k,$$

which shows that the vector $c\mathbf{v}$ is in $\mathcal{V}$.

If $\mathbf{v}$ and $\mathbf{w}$ are vectors in $\mathcal{V}$, then

$$\mathbf{v} = x_1\mathbf{v}_1 + \cdots + x_k\mathbf{v}_k \quad \text{and} \quad \mathbf{w} = y_1\mathbf{v}_1 + \cdots + y_k\mathbf{v}_k,$$

for some real numbers $x_1,\dots,x_k$ and $y_1,\dots,y_k$. Since

$$\mathbf{v}+\mathbf{w} = x_1\mathbf{v}_1 + \cdots + x_k\mathbf{v}_k + y_1\mathbf{v}_1 + \cdots + y_k\mathbf{v}_k = (x_1+y_1)\mathbf{v}_1 + \cdots + (x_k+y_k)\mathbf{v}_k,$$

the vector $\mathbf{v}+\mathbf{w}$ is in $\mathcal{V}$. □

Example 2.21. The vector

$$\mathbf{v} = 3\begin{bmatrix}1\\2\\1\\5\end{bmatrix} + 7\begin{bmatrix}2\\1\\1\\3\end{bmatrix} = \begin{bmatrix}17\\13\\10\\36\end{bmatrix}$$

is in $\operatorname{Span}\left\{\begin{bmatrix}1\\2\\1\\5\end{bmatrix}, \begin{bmatrix}2\\1\\1\\3\end{bmatrix}\right\}$ and consequently the vector $2\mathbf{v} = \begin{bmatrix}34\\26\\20\\72\end{bmatrix}$ is in $\operatorname{Span}\left\{\begin{bmatrix}1\\2\\1\\5\end{bmatrix}, \begin{bmatrix}2\\1\\1\\3\end{bmatrix}\right\}$.

The vectors

$$\mathbf{v} = 5\begin{bmatrix}1\\2\\1\\5\end{bmatrix} - 3\begin{bmatrix}2\\1\\1\\3\end{bmatrix} = \begin{bmatrix}-1\\7\\2\\16\end{bmatrix} \quad \text{and} \quad \mathbf{w} = \begin{bmatrix}1\\2\\1\\5\end{bmatrix} + 2\begin{bmatrix}2\\1\\1\\3\end{bmatrix} = \begin{bmatrix}5\\4\\3\\11\end{bmatrix}$$

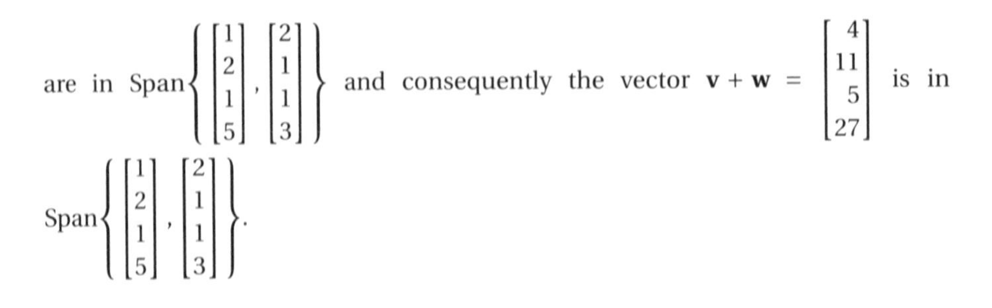

are in $\operatorname{Span}\left\{\begin{bmatrix}1\\2\\1\\5\end{bmatrix}, \begin{bmatrix}2\\1\\1\\3\end{bmatrix}\right\}$ and consequently the vector $\mathbf{v}+\mathbf{w} = \begin{bmatrix}4\\11\\5\\27\end{bmatrix}$ is in $\operatorname{Span}\left\{\begin{bmatrix}1\\2\\1\\5\end{bmatrix}, \begin{bmatrix}2\\1\\1\\3\end{bmatrix}\right\}$.

Definition 2.22. A subset $\mathcal{V}$ of $\mathbb{R}^n$ which satisfies the following two conditions

(a) If $\mathbf{v}$ is a vector in $\mathcal{V}$ and c a real number, then the vector $c\mathbf{v}$ is in $\mathcal{V}$,

(b) If $\mathbf{v}$ and $\mathbf{w}$ are vectors in $\mathcal{V}$, then the vector $\mathbf{v}+\mathbf{w}$ is in $\mathcal{V}$,

is called a *subspace* of $\mathbb{R}^n$.

For simplicity we formulate the definition of a subspace of $\mathbb{R}^n$ using conditions (a) and (b), but in practice we use any linear combination of vectors from $\mathcal{V}$.

Theorem 2.23. *Any linear combination of vectors from a subspace $\mathcal{V}$ is also in $\mathcal{V}$.*
In other words, if vectors $\mathbf{v}_1,\dots,\mathbf{v}_k$ are in a subspace $\mathcal{V}$ and $c_1,\dots,c_k$ are arbitrary real numbers, then the vector $\mathbf{v} = c_1\mathbf{v}_1 + \cdots + c_k\mathbf{v}_k$ is in $\mathcal{V}$.

Proof. The proof follows from Definition 2.22 and is left as an exercise. □

In Theorem 2.20 we prove that $\operatorname{Span}\{\mathbf{v}_1,\dots,\mathbf{v}_k\}$ is a subspace of $\mathbb{R}^n$ for any vectors $\mathbf{v}_1,\dots,\mathbf{v}_k$ from $\mathbb{R}^n$. It is because of this that $\operatorname{Span}\{\mathbf{v}_1,\dots,\mathbf{v}_k\}$ is often called the *subspace spanned by vectors* $\mathbf{v}_1,\dots,\mathbf{v}_k$.

It turns out that every subspace of $\mathbb{R}^n$ is spanned by some vectors from $\mathbb{R}^n$. Later in this chapter we will show that if $\mathcal{V}$ is a subspace of $\mathbb{R}^n$, then there are linearly independent vectors $\mathbf{v}_1,\dots,\mathbf{v}_k$ in $\mathbb{R}^n$, where $k \le n$, such that $\mathcal{V} = \operatorname{Span}\{\mathbf{v}_1,\dots,\mathbf{v}_k\}$.

Example 2.24. Show that the set $\mathcal{V}$ of all vectors of the form $\begin{bmatrix}a+3b\\b\\2a-b\\5a+2b\end{bmatrix}$, where a and b are arbitrary real numbers, is a subspace of $\mathbb{R}^4$.

Solution. Since

$$\begin{bmatrix} a+3b \\ b \\ 2a-b \\ 5a+2b \end{bmatrix} = a\begin{bmatrix} 1 \\ 0 \\ 2 \\ 5 \end{bmatrix} + b\begin{bmatrix} 3 \\ 1 \\ -1 \\ 2 \end{bmatrix},$$

we have

$$\mathcal{V} = \operatorname{Span}\left\{\begin{bmatrix} 1 \\ 0 \\ 2 \\ 5 \end{bmatrix}, \begin{bmatrix} 3 \\ 1 \\ -1 \\ 2 \end{bmatrix}\right\}.$$

□

Example 2.25. Find numbers a and b such that the vector $\begin{bmatrix} a \\ 3 \\ b \\ 4 \end{bmatrix}$ is in $\operatorname{Span}\left\{\begin{bmatrix} 1 \\ 1 \\ 3 \\ 2 \end{bmatrix}, \begin{bmatrix} 2 \\ 1 \\ 1 \\ 1 \end{bmatrix}\right\}$.

Solution. Since the reduced row echelon form of the matrix

$$\begin{bmatrix} 1 & 2 & a \\ 1 & 1 & 3 \\ 3 & 1 & b \\ 2 & 1 & 4 \end{bmatrix}$$

is

$$\begin{bmatrix} 1 & 0 & 6-a \\ 0 & 1 & a-3 \\ 0 & 0 & 2a+b-15 \\ 0 & 0 & a-5 \end{bmatrix},$$

the vector $\begin{bmatrix} a \\ 3 \\ b \\ 4 \end{bmatrix}$ is in $\operatorname{Span}\left\{\begin{bmatrix} 1 \\ 1 \\ 3 \\ 2 \end{bmatrix}, \begin{bmatrix} 2 \\ 1 \\ 1 \\ 1 \end{bmatrix}\right\}$, if

$$2a+b-15=0 \quad \text{and} \quad a-5=0.$$

Consequently, $a=b=5$. □

The basis of a subspace

To motivate the main idea of this section we first consider an example.

Example 2.26. Show that

$$\text{Span}\left\{\begin{bmatrix}1\\3\\0\\2\end{bmatrix},\begin{bmatrix}1\\1\\1\\1\end{bmatrix},\begin{bmatrix}1\\9\\-3\\5\end{bmatrix}\right\}=\text{Span}\left\{\begin{bmatrix}1\\3\\0\\2\end{bmatrix},\begin{bmatrix}1\\1\\1\\1\end{bmatrix}\right\}.$$

Solution. Clearly, every vector in $\text{Span}\left\{\begin{bmatrix}1\\3\\0\\2\end{bmatrix},\begin{bmatrix}1\\1\\1\\1\end{bmatrix}\right\}$ is in $\text{Span}\left\{\begin{bmatrix}1\\3\\0\\2\end{bmatrix},\begin{bmatrix}1\\1\\1\\1\end{bmatrix},\begin{bmatrix}1\\9\\-3\\5\end{bmatrix}\right\}$.

We need to show that, conversely, every vector from $\text{Span}\left\{\begin{bmatrix}1\\3\\0\\2\end{bmatrix},\begin{bmatrix}1\\1\\1\\1\end{bmatrix},\begin{bmatrix}1\\9\\-3\\5\end{bmatrix}\right\}$ is in

$\text{Span}\left\{\begin{bmatrix}1\\3\\0\\2\end{bmatrix},\begin{bmatrix}1\\1\\1\\1\end{bmatrix}\right\}$. Since

$$\begin{bmatrix}1&1&1\\3&1&9\\0&1&-3\\2&1&5\end{bmatrix}\sim\begin{bmatrix}1&0&4\\0&1&-3\\0&0&0\\0&0&0\end{bmatrix},$$

we have

$$\begin{bmatrix}1\\9\\-3\\5\end{bmatrix}=4\begin{bmatrix}1\\3\\0\\2\end{bmatrix}-3\begin{bmatrix}1\\1\\1\\1\end{bmatrix}.$$

Using this result we obtain

$$x_1 \begin{bmatrix} 1 \\ 9 \\ -3 \\ 5 \end{bmatrix} + x_2 \begin{bmatrix} 1 \\ 3 \\ 0 \\ 2 \end{bmatrix} + x_3 \begin{bmatrix} 1 \\ 1 \\ 1 \\ 1 \end{bmatrix} = x_1 \left(4 \begin{bmatrix} 1 \\ 3 \\ 0 \\ 2 \end{bmatrix} - 3 \begin{bmatrix} 1 \\ 1 \\ 1 \\ 1 \end{bmatrix} \right) + x_2 \begin{bmatrix} 1 \\ 3 \\ 0 \\ 2 \end{bmatrix} + x_3 \begin{bmatrix} 1 \\ 1 \\ 1 \\ 1 \end{bmatrix}$$
$$= (4x_1 + x_2) \begin{bmatrix} 1 \\ 3 \\ 0 \\ 2 \end{bmatrix} + (-3x_1 + x_3) \begin{bmatrix} 1 \\ 1 \\ 1 \\ 1 \end{bmatrix}$$
$$= y_1 \begin{bmatrix} 1 \\ 3 \\ 0 \\ 2 \end{bmatrix} + y_2 \begin{bmatrix} 1 \\ 1 \\ 1 \\ 1 \end{bmatrix}.$$

□

The observation in the example above leads to the following general theorem.

Theorem 2.27. *If* $\mathbf{v}$ *is in* $\operatorname{Span}\{\mathbf{v}_1,\dots,\mathbf{v}_k\}$, *then*

$$\operatorname{Span}\{\mathbf{v}_1,\dots,\mathbf{v}_k,\mathbf{v}\} = \operatorname{Span}\{\mathbf{v}_1,\dots,\mathbf{v}_k\}.$$

Proof. Clearly, every vector in $\operatorname{Span}\{\mathbf{v}_1,\dots,\mathbf{v}_k\}$ is also in $\operatorname{Span}\{\mathbf{v}_1,\dots,\mathbf{v}_k,\mathbf{v}\}$.

Now consider a vector $\mathbf{w}$ in $\operatorname{Span}\{\mathbf{v}_1,\dots,\mathbf{v}_k,\mathbf{v}\}$. Then there are real numbers $a_1,\dots,$ a_k, a_{k+1} such that

$$\mathbf{w} = a_1\mathbf{v}_1 + \dots + a_k\mathbf{v}_k + a_{k+1}\mathbf{v}.$$

Since $\mathbf{v}$ is in $\operatorname{Span}\{\mathbf{v}_1,\dots,\mathbf{v}_k\}$, there are real numbers $b_1,\dots, b_k$ such that

$$\mathbf{v} = b_1\mathbf{v}_1 + \dots + b_k\mathbf{v}_k.$$

Then

$$\begin{aligned} \mathbf{w} &= a_1\mathbf{v}_1 + \dots + a_k\mathbf{v}_k + a_{k+1}\mathbf{v} \\ &= a_1\mathbf{v}_1 + \dots + a_k\mathbf{v}_k + a_{k+1}(b_1\mathbf{v}_1 + \dots + b_k\mathbf{v}_k) \\ &= (a_1 + a_{k+1}b_1)\mathbf{v}_1 + \dots + (b_k + a_{k+1}b_k)\mathbf{v}_k. \end{aligned}$$

Thus $\mathbf{w}$ is in $\operatorname{Span}\{\mathbf{v}_1,\dots,\mathbf{v}_k\}$. □

The next theorem gives us a condition that can be used to check that two sets of vectors span the same subspace.

Theorem 2.28. *Let* $\mathbf{v}_1,\dots,\mathbf{v}_k$ *and* $\mathbf{w}_1,\dots,\mathbf{w}_m$ *be vectors in* $\mathbb{R}^n$. *Then*

$$\operatorname{Span}\{\mathbf{v}_1,\dots,\mathbf{v}_k\} = \operatorname{Span}\{\mathbf{w}_1,\dots,\mathbf{w}_m\}$$

if and only if

$$\mathbf{v}_1,\dots,\mathbf{v}_k \text{ are in } \operatorname{Span}\{\mathbf{w}_1,\dots,\mathbf{w}_m\} \quad \text{and} \quad \mathbf{w}_1,\dots,\mathbf{w}_m \text{ are in } \operatorname{Span}\{\mathbf{v}_1,\dots,\mathbf{v}_k\}.$$

Proof. If $\operatorname{Span}\{\mathbf{v}_1,\dots,\mathbf{v}_k\} = \operatorname{Span}\{\mathbf{w}_1,\dots,\mathbf{w}_m\}$, then obviously the vectors $\mathbf{v}_1,\dots,\mathbf{v}_k$ are in $\operatorname{Span}\{\mathbf{w}_1,\dots,\mathbf{w}_m\}$ and the vectors $\mathbf{w}_1,\dots,\mathbf{w}_m$ are in $\operatorname{Span}\{\mathbf{v}_1,\dots,\mathbf{v}_k\}$.

On the other hand, if the vectors $\mathbf{v}_1,\dots,\mathbf{v}_k$ are in $\operatorname{Span}\{\mathbf{w}_1,\dots,\mathbf{w}_m\}$ and the vectors $\mathbf{w}_1,\dots,\mathbf{w}_m$ are in $\operatorname{Span}\{\mathbf{v}_1,\dots,\mathbf{v}_k\}$, then

$$\operatorname{Span}\{\mathbf{v}_1,\dots,\mathbf{v}_k,\mathbf{w}_1,\dots,\mathbf{w}_m\} = \operatorname{Span}\{\mathbf{v}_1,\dots,\mathbf{v}_k\}$$

and

$$\operatorname{Span}\{\mathbf{v}_1,\dots,\mathbf{v}_k,\mathbf{w}_1,\dots,\mathbf{w}_m\} = \operatorname{Span}\{\mathbf{w}_1,\dots,\mathbf{w}_m\},$$

by Theorem 2.27. Hence $\operatorname{Span}\{\mathbf{v}_1,\dots,\mathbf{v}_k\} = \operatorname{Span}\{\mathbf{w}_1,\dots,\mathbf{w}_m\}$. □

In Example 2.26 we show that

$$\operatorname{Span}\left\{\begin{bmatrix}1\\3\\0\\2\end{bmatrix},\begin{bmatrix}1\\1\\1\\1\end{bmatrix},\begin{bmatrix}1\\9\\-3\\5\end{bmatrix}\right\} = \operatorname{Span}\left\{\begin{bmatrix}1\\3\\0\\2\end{bmatrix},\begin{bmatrix}1\\1\\1\\1\end{bmatrix}\right\},$$

so both $\operatorname{Span}\left\{\begin{bmatrix}1\\3\\0\\2\end{bmatrix},\begin{bmatrix}1\\1\\1\\1\end{bmatrix},\begin{bmatrix}1\\9\\-3\\5\end{bmatrix}\right\}$ and $\operatorname{Span}\left\{\begin{bmatrix}1\\3\\0\\2\end{bmatrix},\begin{bmatrix}1\\1\\1\\1\end{bmatrix}\right\}$ describe the same subspace of $\mathbb{R}^4$. The span using only two vectors seems to be a better, more efficient, way to describe that subspace. Clearly, it is not necessary to include the third vector. It does not provide any new information. On the other hand, it should be clear that two vectors are necessary. It is natural to ask whether there is always the least number of vectors necessary to describe a subspace and how to determine whether all vectors we are using to span a subspace are necessary. These questions are of fundamental importance in linear algebra. To answer these questions we proceed in a number of small steps.

Definition 2.29. Let $\mathscr{V}$ be a vector subspace of $\mathbb{R}^n$. A collection of vectors $\{\mathbf{v}_1,\dots,\mathbf{v}_k\}$ in $\mathscr{V}$ is a *basis* of $\mathscr{V}$ if the following two conditions are satisfied:

(a) the vectors $\mathbf{v}_1,\dots,\mathbf{v}_k$ are linearly independent;

(b) $\mathscr{V} = \operatorname{Span}\{\mathbf{v}_1,\dots,\mathbf{v}_k\}$.

In other words, by a basis of a subspace $\mathscr{V}$ we mean any linearly independent collection of vectors that span $\mathscr{V}$.

Example 2.30. Show that the set $\left\{\begin{bmatrix}1\\3\\0\\2\end{bmatrix}, \begin{bmatrix}1\\1\\1\\1\end{bmatrix}\right\}$ is a basis of the subspace

$$\operatorname{Span}\left\{\begin{bmatrix}1\\3\\0\\2\end{bmatrix}, \begin{bmatrix}1\\1\\1\\1\end{bmatrix}, \begin{bmatrix}1\\9\\-3\\5\end{bmatrix}\right\}.$$

Solution. In Example 2.26 we show that

$$\operatorname{Span}\left\{\begin{bmatrix}1\\3\\0\\2\end{bmatrix}, \begin{bmatrix}1\\1\\1\\1\end{bmatrix}, \begin{bmatrix}1\\9\\-3\\5\end{bmatrix}\right\} = \operatorname{Span}\left\{\begin{bmatrix}1\\3\\0\\2\end{bmatrix}, \begin{bmatrix}1\\1\\1\\1\end{bmatrix}\right\}.$$

It is easy to see that the vectors $\begin{bmatrix}1\\3\\0\\2\end{bmatrix}$ and $\begin{bmatrix}1\\1\\1\\1\end{bmatrix}$ are linearly independent. Therefore, the set $\left\{\begin{bmatrix}1\\3\\0\\2\end{bmatrix}, \begin{bmatrix}1\\1\\1\\1\end{bmatrix}\right\}$ is a basis of the subspace $\operatorname{Span}\left\{\begin{bmatrix}1\\3\\0\\2\end{bmatrix}, \begin{bmatrix}1\\1\\1\\1\end{bmatrix}, \begin{bmatrix}1\\9\\-3\\5\end{bmatrix}\right\}$.

Note that $\left\{\begin{bmatrix}1\\3\\0\\2\end{bmatrix}, \begin{bmatrix}1\\1\\1\\1\end{bmatrix}, \begin{bmatrix}1\\9\\-3\\5\end{bmatrix}\right\}$ is not a basis of $\operatorname{Span}\left\{\begin{bmatrix}1\\3\\0\\2\end{bmatrix}, \begin{bmatrix}1\\1\\1\\1\end{bmatrix}, \begin{bmatrix}1\\9\\-3\\5\end{bmatrix}\right\}$ because the vectors $\begin{bmatrix}1\\3\\0\\2\end{bmatrix}$, $\begin{bmatrix}1\\1\\1\\1\end{bmatrix}$, and $\begin{bmatrix}1\\9\\-3\\5\end{bmatrix}$ are not linearly independent since

$$\begin{bmatrix}1\\9\\-3\\5\end{bmatrix} = 4\begin{bmatrix}1\\3\\0\\2\end{bmatrix} - 3\begin{bmatrix}1\\1\\1\\1\end{bmatrix}.$$

□

If $\{\mathbf{v}_1, \dots, \mathbf{v}_k\}$ is a basis in $\mathscr{V}$, then every vector $\mathbf{x}$ in $\mathscr{V}$ can be written as a linear combination of the vectors $\mathbf{v}_1, \ldots, \mathbf{v}_k$, that is,

$$\mathbf{x} = c_1\mathbf{v}_1 + \cdots + c_k\mathbf{v}_k$$

for some real numbers $c_1, \ldots, c_k$. This representation is unique. In other words, if

$$\mathbf{x} = c_1\mathbf{v}_1 + \cdots + c_k\mathbf{v}_k = d_1\mathbf{v}_1 + \cdots + d_k\mathbf{v}_k, \tag{2.6}$$

then we must have

$$c_1 = d_1, c_2 = d_2, \ldots, c_k = d_k. \tag{2.7}$$

Indeed, if (2.6) holds, then

$$(c_1 - d_1)\mathbf{v}_1 + \cdots + (c_k - d_k)\mathbf{v}_k = \mathbf{0},$$

and thus

$$c_1 - d_1 = \cdots = c_k - d_k = 0,$$

by linear independence of the vectors $\mathbf{v}_1, \ldots, \mathbf{v}_k$. The unique numbers $c_1, \ldots, c_k$ such that $\mathbf{x} = c_1\mathbf{v}_1 + \cdots + c_k\mathbf{v}_k$ are called the coordinates of $\mathbf{x}$ in the basis $\{\mathbf{v}_1, \ldots, \mathbf{v}_k\}$.

The column space of a matrix

Definition 2.31. Let $A = [\mathbf{c}_1 \ \ldots \ \mathbf{c}_n]$ be an $m \times n$ matrix with columns $\mathbf{c}_1, \ldots, \mathbf{c}_n$. By the *column space* of A, denoted by $\mathbf{C}(A)$, we mean the subspace of $\mathbb{R}^m$ spanned by the columns of A, that is,

$$\mathbf{C}(A) = \text{Span}\{\mathbf{c}_1, \ldots, \mathbf{c}_n\}.$$

The next theorem gives us a practical way of finding a basis of the column space of a matrix.

Theorem 2.32. *The set of pivot columns of a matrix A is a basis of $\mathbf{C}(A)$.*

Proof of a particular case. We illustrate the proof using a 4×7 matrix A. Suppose we have

$$A = \begin{bmatrix} a_{11} & a_{12} & a_{13} & a_{14} & a_{15} & a_{16} & a_{17} \\ a_{21} & a_{22} & a_{23} & a_{24} & a_{25} & a_{26} & a_{27} \\ a_{31} & a_{32} & a_{33} & a_{34} & a_{35} & a_{36} & a_{37} \\ a_{41} & a_{42} & a_{43} & b_{44} & a_{45} & a_{46} & a_{47} \end{bmatrix} \sim \begin{bmatrix} 1 & p & 0 & q & s & 0 & u \\ 0 & 0 & 1 & r & t & 0 & v \\ 0 & 0 & 0 & 0 & 0 & 1 & w \\ 0 & 0 & 0 & 0 & 0 & 0 & 0 \end{bmatrix}.$$

First we show that the pivot columns are linearly independent. Indeed, since the equation

$$x\begin{bmatrix} a_{11} \\ a_{21} \\ a_{31} \\ a_{41} \end{bmatrix} + y\begin{bmatrix} a_{13} \\ a_{23} \\ a_{33} \\ a_{43} \end{bmatrix} + z\begin{bmatrix} a_{16} \\ a_{26} \\ a_{36} \\ a_{46} \end{bmatrix} = \begin{bmatrix} 0 \\ 0 \\ 0 \\ 0 \end{bmatrix}$$

is equivalent to the equation

$$x\begin{bmatrix}1\\0\\0\\0\end{bmatrix}+y\begin{bmatrix}0\\1\\0\\0\end{bmatrix}+z\begin{bmatrix}0\\0\\1\\0\end{bmatrix}=\begin{bmatrix}x\\y\\z\\0\end{bmatrix}=\begin{bmatrix}0\\0\\0\\0\end{bmatrix},$$

it has only the trivial solution $x=y=z=0$. This implies that the pivot columns $\begin{bmatrix}a_{11}\\a_{21}\\a_{31}\\a_{41}\end{bmatrix},\begin{bmatrix}a_{13}\\a_{23}\\a_{33}\\a_{43}\end{bmatrix},\begin{bmatrix}a_{16}\\a_{26}\\a_{36}\\a_{46}\end{bmatrix}$ are linearly independent.

Now, from Theorem 2.5, we get that there are real numbers p,q,r,s,t,u,v,w such that

$$\begin{bmatrix}a_{12}\\a_{22}\\a_{32}\\a_{42}\end{bmatrix}=p\begin{bmatrix}a_{11}\\a_{21}\\a_{31}\\a_{41}\end{bmatrix},$$

$$\begin{bmatrix}a_{14}\\a_{24}\\a_{34}\\a_{44}\end{bmatrix}=q\begin{bmatrix}a_{11}\\a_{21}\\a_{31}\\a_{41}\end{bmatrix}+r\begin{bmatrix}a_{13}\\a_{23}\\a_{33}\\a_{43}\end{bmatrix},$$

$$\begin{bmatrix}a_{15}\\a_{25}\\a_{35}\\a_{45}\end{bmatrix}=s\begin{bmatrix}a_{11}\\a_{21}\\a_{31}\\a_{41}\end{bmatrix}+t\begin{bmatrix}a_{13}\\a_{23}\\a_{33}\\a_{43}\end{bmatrix},$$

$$\begin{bmatrix}a_{17}\\a_{27}\\a_{37}\\a_{47}\end{bmatrix}=u\begin{bmatrix}a_{11}\\a_{21}\\a_{31}\\a_{41}\end{bmatrix}+v\begin{bmatrix}a_{13}\\a_{23}\\a_{33}\\a_{43}\end{bmatrix}+w\begin{bmatrix}a_{16}\\a_{26}\\a_{36}\\a_{46}\end{bmatrix}.$$

Hence, by Theorem 2.27, we have

$$\begin{aligned}&\operatorname{Span}\left\{\begin{bmatrix}a_{11}\\a_{21}\\a_{31}\\a_{41}\end{bmatrix},\begin{bmatrix}a_{12}\\a_{22}\\a_{32}\\a_{42}\end{bmatrix},\begin{bmatrix}a_{13}\\a_{23}\\a_{33}\\a_{43}\end{bmatrix},\begin{bmatrix}a_{14}\\a_{24}\\a_{34}\\a_{44}\end{bmatrix},\begin{bmatrix}a_{15}\\a_{25}\\a_{35}\\a_{45}\end{bmatrix},\begin{bmatrix}a_{16}\\a_{26}\\a_{36}\\a_{46}\end{bmatrix},\begin{bmatrix}a_{17}\\a_{27}\\a_{37}\\a_{47}\end{bmatrix}\right\}\\&=\operatorname{Span}\left\{\begin{bmatrix}a_{11}\\a_{21}\\a_{31}\\a_{41}\end{bmatrix},\begin{bmatrix}a_{13}\\a_{23}\\a_{33}\\a_{43}\end{bmatrix},\begin{bmatrix}a_{16}\\a_{26}\\a_{36}\\a_{46}\end{bmatrix}\right\}.\end{aligned}$$

The above, combined with the previously established linear independence of the pivot columns, proves that the set $\left\{\begin{bmatrix}a_{11}\\a_{21}\\a_{31}\\a_{41}\end{bmatrix},\begin{bmatrix}a_{13}\\a_{23}\\a_{33}\\a_{43}\end{bmatrix},\begin{bmatrix}a_{16}\\a_{26}\\a_{36}\\a_{46}\end{bmatrix}\right\}$ is a basis of $\mathbf{C}(A)$. □

Example 2.33. Show that the set $\left\{\begin{bmatrix}1\\1\\0\end{bmatrix}, \begin{bmatrix}1\\0\\1\end{bmatrix}\right\}$ is a basis of the subspace

$$\operatorname{Span}\left\{\begin{bmatrix}1\\3\\-2\end{bmatrix}, \begin{bmatrix}3\\1\\2\end{bmatrix}, \begin{bmatrix}1\\1\\0\end{bmatrix}, \begin{bmatrix}1\\0\\1\end{bmatrix}\right\}.$$

Solution. Since

$$\begin{bmatrix}1&1&1&3\\1&0&3&1\\0&1&-2&2\end{bmatrix} \sim \begin{bmatrix}1&1&1&3\\1&0&3&1\\0&1&-2&2\end{bmatrix} \sim \begin{bmatrix}1&0&3&1\\0&1&-2&2\\0&0&0&0\end{bmatrix},$$

the columns $\begin{bmatrix}1\\1\\0\end{bmatrix}$ and $\begin{bmatrix}1\\0\\1\end{bmatrix}$ are the pivot columns of the matrix $\begin{bmatrix}1&1&1&3\\1&0&3&1\\0&1&-2&2\end{bmatrix}$, the set $\left\{\begin{bmatrix}1\\1\\0\end{bmatrix}, \begin{bmatrix}1\\0\\1\end{bmatrix}\right\}$ is a basis of the subspace $\operatorname{Span}\left\{\begin{bmatrix}1\\3\\-2\end{bmatrix}, \begin{bmatrix}3\\1\\2\end{bmatrix}, \begin{bmatrix}1\\1\\0\end{bmatrix}, \begin{bmatrix}1\\0\\1\end{bmatrix}\right\}$. □

Example 2.34. In Example 2.33 we show that the set $\left\{\begin{bmatrix}1\\1\\0\end{bmatrix}, \begin{bmatrix}1\\0\\1\end{bmatrix}\right\}$ is a basis of the subspace $\operatorname{Span}\left\{\begin{bmatrix}1\\3\\-2\end{bmatrix}, \begin{bmatrix}3\\1\\2\end{bmatrix}, \begin{bmatrix}1\\1\\0\end{bmatrix}, \begin{bmatrix}1\\0\\1\end{bmatrix}\right\}$. If instead we were asked to find a basis for the same subspace, we might have proceeded as follows.

First we note that

$$\begin{bmatrix}1&3&1&1\\3&1&1&0\\-2&2&0&1\end{bmatrix} \sim \begin{bmatrix}1&0&\frac{1}{4}&-\frac{1}{8}\\0&1&\frac{1}{4}&\frac{3}{8}\\0&0&0&0\end{bmatrix}.$$

Since $\begin{bmatrix}1\\3\\-2\end{bmatrix}$ and $\begin{bmatrix}3\\1\\2\end{bmatrix}$ are the pivot columns of the matrix $\begin{bmatrix}1&3&1&1\\3&1&1&0\\-2&2&0&1\end{bmatrix}$, we conclude that the set $\left\{\begin{bmatrix}1\\3\\-2\end{bmatrix}, \begin{bmatrix}3\\1\\2\end{bmatrix}\right\}$ is a basis of $\operatorname{Span}\left\{\begin{bmatrix}1\\3\\-2\end{bmatrix}, \begin{bmatrix}3\\1\\2\end{bmatrix}, \begin{bmatrix}1\\1\\0\end{bmatrix}, \begin{bmatrix}1\\0\\1\end{bmatrix}\right\}$.

In order to find a basis of a column space of a matrix A we use row operations on the rows of the matrix A. It is important to remember that row operations change the column space of the matrix. In the above example we have

$$\begin{bmatrix} 1 & 3 & 1 & 1 \\ 3 & 1 & 1 & 0 \\ -2 & 2 & 0 & 1 \end{bmatrix} \sim \begin{bmatrix} 1 & 0 & \frac{1}{4} & -\frac{1}{8} \\ 0 & 1 & \frac{1}{4} & \frac{3}{8} \\ 0 & 0 & 0 & 0 \end{bmatrix},$$

but clearly

$$\text{Span}\left\{ \begin{bmatrix} 1 \\ 3 \\ -2 \end{bmatrix}, \begin{bmatrix} 3 \\ 1 \\ 2 \end{bmatrix}, \begin{bmatrix} 1 \\ 1 \\ 0 \end{bmatrix}, \begin{bmatrix} 1 \\ 0 \\ 1 \end{bmatrix} \right\} \neq \text{Span}\left\{ \begin{bmatrix} 1 \\ 0 \\ 0 \end{bmatrix}, \begin{bmatrix} 0 \\ 1 \\ 0 \end{bmatrix}, \begin{bmatrix} \frac{1}{4} \\ \frac{1}{4} \\ 0 \end{bmatrix}, \begin{bmatrix} -\frac{1}{8} \\ \frac{3}{8} \\ 0 \end{bmatrix} \right\}$$

and

$$\mathbf{C}\left(\begin{bmatrix} 1 & 3 & 1 & 1 \\ 3 & 1 & 1 & 0 \\ -2 & 2 & 0 & 1 \end{bmatrix} \right) \neq \mathbf{C}\left(\begin{bmatrix} 1 & 0 & \frac{1}{4} & -\frac{1}{8} \\ 0 & 1 & \frac{1}{4} & \frac{3}{8} \\ 0 & 0 & 0 & 0 \end{bmatrix} \right).$$

The only information from the matrix $\begin{bmatrix} 1 & 0 & \frac{1}{4} & -\frac{1}{8} \\ 0 & 1 & \frac{1}{4} & \frac{3}{8} \\ 0 & 0 & 0 & 0 \end{bmatrix}$ we use is the position of the pivot columns.

The use of row operations to find the column space of a matrix A may seem somewhat strange since it's all about the columns of A, not rows. Is it possible to apply elementary operations directly to columns of a matrix A without changing the column space of A? The answer is yes, as we explain below.

First we show that "elementary operations on columns" do not affect the column space of the matrix.

Lemma 2.35. *The order of vectors spanning a subspace does not matter, that is,*

$$\text{Span}\{\mathbf{v}_1, \dots, \mathbf{v}_j, \dots, \mathbf{v}_i, \dots, \mathbf{v}_m\} = \text{Span}\{\mathbf{v}_1, \dots, \mathbf{v}_i, \dots, \mathbf{v}_j, \dots, \mathbf{v}_m\}.$$

Proof. This is a direct consequence of the obvious equality

$$a_1\mathbf{v}_1 + \cdots + a_j\mathbf{v}_j + \cdots + a_i\mathbf{v}_i + \cdots + a_m\mathbf{v}_m = a_1\mathbf{v}_1 + \cdots + a_i\mathbf{v}_i + \cdots + a_j\mathbf{v}_j + \cdots + a_m\mathbf{v}_m.$$

□

Lemma 2.36. *If $c \neq 0$, then*

$$\text{Span}\{\mathbf{v}_1, \dots, c\mathbf{v}_j, \dots, \mathbf{v}_m\} = \text{Span}\{\mathbf{v}_1, \dots, \mathbf{v}_j, \dots, \mathbf{v}_m\}.$$

Proof. This is a direct consequence of the equality

$$a_1\mathbf{v}_1 + \cdots + a_j\mathbf{v}_j + \cdots + a_m\mathbf{v}_m = a_1\mathbf{v}_1 + \cdots + \frac{a_j}{c}(c\mathbf{v}_j) + \cdots + a_m\mathbf{v}_m.$$

□

Lemma 2.37. *For any real number c we have*

$$\operatorname{Span}\{\mathbf{v}_1, \ldots, \mathbf{v}_i + c\mathbf{v}_j, \ldots, \mathbf{v}_m\} = \operatorname{Span}\{\mathbf{v}_1, \ldots, \mathbf{v}_i \ldots, \mathbf{v}_m\}.$$

Proof. This is a direct consequence of the following equalities:

$$a_1\mathbf{v}_1 + \cdots + a_i(\mathbf{v}_i + c\mathbf{v}_j) + \cdots + a_j\mathbf{v}_j + \cdots + a_m\mathbf{v}_m = a_1\mathbf{v}_1 + \cdots + a_i\mathbf{v}_i + \cdots + (a_j + a_i c)\mathbf{v}_j + \cdots + a_m\mathbf{v}_m$$

and

$$a_1\mathbf{v}_1 + \cdots + a_i\mathbf{v}_i + \cdots + a_j\mathbf{v}_j + \cdots + a_m\mathbf{v}_m = a_1\mathbf{v}_1 + \cdots + a_i(\mathbf{v}_i + c\mathbf{v}_j) + \cdots + (a_j - a_i c)\mathbf{v}_j + \cdots + a_m\mathbf{v}_m.$$

□

Note that the operations on the columns of a matrix A described in the above three lemmas correspond to elementary row operations on the rows of the transpose of A. This point of view is often beneficial. We express this idea in the following theorem.

Theorem 2.38. *If $A^T \sim B$, then $\mathbf{C}(A) = \mathbf{C}(B^T)$.*

Proof. This is a direct consequence of Lemmas 2.35, 2.36, and 2.37. □

The following example illustrates how Theorem 2.38 can be used in practice to find a basis of the column space of a matrix. This method seems to be more natural than the method based on identifying pivot columns in the reduced row echelon form of the matrix.

Example 2.39. Consider the matrix

$$A = \begin{bmatrix} 1 & 2 & -3 & 1 \\ 1 & 5 & -2 & 2 \\ 1 & 8 & -1 & 3 \\ 0 & 3 & 1 & 1 \\ 2 & 13 & -3 & 5 \end{bmatrix}.$$

Find a basis of $\mathbf{C}(A)$ using row operations on the matrix A^T and then write the columns of the matrix A as linear combinations of this basis.

Solution. Since

$$A^T = \begin{bmatrix} 1 & 1 & 1 & 0 & 2 \\ 2 & 5 & 8 & 3 & 13 \\ -3 & -2 & -1 & 1 & -3 \\ 1 & 2 & 3 & 1 & 5 \end{bmatrix} \sim \begin{bmatrix} 1 & 0 & -1 & -1 & -1 \\ 0 & 1 & 2 & 1 & 3 \\ 0 & 0 & 0 & 0 & 0 \\ 0 & 0 & 0 & 0 & 0 \end{bmatrix}$$

and

$$\begin{bmatrix} 1 & 0 & -1 & -1 & -1 \\ 0 & 1 & 2 & 1 & 3 \\ 0 & 0 & 0 & 0 & 0 \\ 0 & 0 & 0 & 0 & 0 \end{bmatrix}^T = \begin{bmatrix} 1 & 0 & 0 & 0 \\ 0 & 1 & 0 & 0 \\ -1 & 2 & 0 & 0 \\ -1 & 1 & 0 & 0 \\ -1 & 3 & 0 & 0 \end{bmatrix},$$

the set $\left\{ \begin{bmatrix} 1 \\ 0 \\ -1 \\ -1 \\ -1 \end{bmatrix}, \begin{bmatrix} 0 \\ 1 \\ 2 \\ 1 \\ 3 \end{bmatrix} \right\}$ is a basis of $\operatorname{Span}\left\{ \begin{bmatrix} 1 \\ 1 \\ 1 \\ 0 \\ 2 \end{bmatrix}, \begin{bmatrix} 2 \\ 5 \\ 8 \\ 3 \\ 13 \end{bmatrix}, \begin{bmatrix} -3 \\ -2 \\ -1 \\ 1 \\ -3 \end{bmatrix}, \begin{bmatrix} 1 \\ 2 \\ 3 \\ 1 \\ 5 \end{bmatrix} \right\}$, by Theorem 2.38.

Now we find

$$\begin{bmatrix} 1 \\ 1 \\ 1 \\ 0 \\ 2 \end{bmatrix} = \begin{bmatrix} 1 \\ 0 \\ -1 \\ -1 \\ -1 \end{bmatrix} + \begin{bmatrix} 0 \\ 1 \\ 2 \\ 1 \\ 3 \end{bmatrix}, \qquad \begin{bmatrix} 2 \\ 5 \\ 8 \\ 3 \\ 13 \end{bmatrix} = 2\begin{bmatrix} 1 \\ 0 \\ -1 \\ -1 \\ -1 \end{bmatrix} + 5\begin{bmatrix} 0 \\ 1 \\ 2 \\ 1 \\ 3 \end{bmatrix},$$

$$\begin{bmatrix} -3 \\ -2 \\ -1 \\ 1 \\ -3 \end{bmatrix} = -3\begin{bmatrix} 1 \\ 0 \\ -1 \\ -1 \\ -1 \end{bmatrix} - 2\begin{bmatrix} 0 \\ 1 \\ 2 \\ 1 \\ 3 \end{bmatrix}, \qquad \begin{bmatrix} 1 \\ 2 \\ 3 \\ 1 \\ 5 \end{bmatrix} = \begin{bmatrix} 1 \\ 0 \\ -1 \\ -1 \\ -1 \end{bmatrix} + 2\begin{bmatrix} 0 \\ 1 \\ 2 \\ 1 \\ 3 \end{bmatrix}.$$

□

Changing A to the transpose of A is not really necessary. We could work instead with the columns of A as stated in Lemmas 2.35, 2.36, and 2.37. The reason for switching to A^T is that we use elementary row operations much more often, so it seems easier.

The null space of a matrix

Now we consider another subspace defined by a matrix, namely the null space of a matrix. We will see later that, in some sense, the null space of a matrix is complementary to the column space of that matrix. We start with an example that motivates the definitions that follow.

Example 2.40. Find the general solution of the equation

$$x_1 \begin{bmatrix} 1 \\ 2 \\ 1 \end{bmatrix} + x_2 \begin{bmatrix} 2 \\ 1 \\ -1 \end{bmatrix} + x_3 \begin{bmatrix} 3 \\ 1 \\ -2 \end{bmatrix} + x_4 \begin{bmatrix} 1 \\ 2 \\ 2 \end{bmatrix} + x_5 \begin{bmatrix} 2 \\ 1 \\ 1 \end{bmatrix} = \begin{bmatrix} 0 \\ 0 \\ 0 \end{bmatrix}. \tag{2.8}$$

Solution. The reduced row echelon form of the matrix

$$\begin{bmatrix} 1 & 2 & 3 & 1 & 2 & 0 \\ 2 & 1 & 1 & 2 & 1 & 0 \\ 1 & -1 & -2 & 2 & 1 & 0 \end{bmatrix}$$

is

$$\begin{bmatrix} 1 & 0 & -\frac{1}{3} & 0 & -2 & 0 \\ 0 & 1 & \frac{5}{3} & 0 & 1 & 0 \\ 0 & 0 & 0 & 1 & 2 & 0 \end{bmatrix}.$$

The corresponding system of equations is

$$\begin{cases} x_1 \phantom{{}+x_2} - \frac{1}{3}x_3 \phantom{{}+x_4} - 2x_5 = 0 \\ \phantom{x_1 +{}} x_2 + \frac{5}{3}x_3 \phantom{{}+x_4} + x_5 = 0 \\ \phantom{x_1 + x_2 + \frac{5}{3}x_3 +{}} x_4 + 2x_5 = 0 \end{cases},$$

where x_3 and x_5 are free variable. The general solution can be written as

$$\begin{bmatrix} x_1 \\ x_2 \\ x_3 \\ x_4 \\ x_5 \end{bmatrix} = \begin{bmatrix} \frac{1}{3}x_3 + 2x_5 \\ -\frac{5}{3}x_3 - x_5 \\ x_3 \\ -2x_5 \\ x_5 \end{bmatrix} = x_3 \begin{bmatrix} \frac{1}{3} \\ -\frac{5}{3} \\ 1 \\ 0 \\ 0 \end{bmatrix} + x_5 \begin{bmatrix} 2 \\ -1 \\ 0 \\ -2 \\ 1 \end{bmatrix}.$$

Note that $\begin{bmatrix} \frac{1}{3} \\ -\frac{5}{3} \\ 1 \\ 0 \\ 0 \end{bmatrix}$ is a particular solution obtained by choosing $x_3 = 1$ and $x_5 = 0$.

Similarly, $\begin{bmatrix} 2 \\ -1 \\ 0 \\ -2 \\ 1 \end{bmatrix}$ is a particular solution obtained by choosing $x_3 = 0$ and $x_5 = 1$. □

Definition 2.41. Let $\mathbf{c}_1, \dots, \mathbf{c}_n$ be column vectors in $\mathbb{R}^m$. A solution of the equation

$$x_1\mathbf{c}_1 + \dots + x_n\mathbf{c}_n = \mathbf{0}$$

is called a *special solution* if it is obtained from the general solution by letting one of the free variables be 1 and the remaining free variables be 0.

The equation (2.8), considered in Example 2.40, has two special solutions:

$$\begin{bmatrix} \frac{1}{3} \\ -\frac{5}{3} \\ 1 \\ 0 \\ 0 \end{bmatrix} \quad \text{and} \quad \begin{bmatrix} 2 \\ -1 \\ 0 \\ -2 \\ 1 \end{bmatrix}.$$

Note that these are the only special solutions of the equation (2.8).

Definition 2.42. Let A be an $m \times n$ matrix. By the *null space* of A, denoted by $\mathbf{N}(A)$, we mean the set of vectors $\mathbf{x}$ in $\mathbb{R}^n$ such that $A\mathbf{x} = \mathbf{0}$.

If $A = [\mathbf{c}_1 \ \dots \ \mathbf{c}_n]$, where $\mathbf{c}_1, \dots, \mathbf{c}_n$ are in $\mathbb{R}^m$ and are the columns of the matrix A, then the null space of A is the set of vectors $\mathbf{x} = \begin{bmatrix} x_1 \\ \vdots \\ x_n \end{bmatrix}$ which are solutions of the equation

$$x_1\mathbf{c}_1 + \dots + x_n\mathbf{c}_n = \mathbf{0}.$$

In Example 2.40 we show that the null space of the matrix

$$A = \begin{bmatrix} 1 & 2 & 3 & 1 & 2 \\ 2 & 1 & 1 & 2 & 1 \\ 1 & -1 & -2 & 2 & 1 \end{bmatrix},$$

is the set of all linear combinations of the vectors

$$\begin{bmatrix} \frac{1}{3} \\ -\frac{5}{3} \\ 1 \\ 0 \\ 0 \end{bmatrix} \quad \text{and} \quad \begin{bmatrix} 2 \\ -1 \\ 0 \\ -2 \\ 1 \end{bmatrix}.$$

Since these vectors are linearly independent, they form a basis of the null space of A. It turns out that this is not a special property of the matrix A, but rather a general rule.

Theorem 2.43. *Let A be $m \times n$ matrix. The null space $\mathbf{N}(A)$ is a subspace of $\mathbb{R}^n$ spanned by the special solutions of the equation $A\mathbf{x} = \mathbf{0}$ and the set of special solutions is a basis of $\mathbf{N}(A)$.*

Proof of a particular case. We consider the matrix

$$A = \begin{bmatrix} a_{11} & a_{12} & a_{13} & a_{14} & a_{15} \\ a_{21} & a_{22} & a_{23} & a_{24} & a_{25} \\ a_{31} & a_{32} & a_{33} & a_{34} & a_{35} \\ a_{41} & a_{42} & a_{43} & b_{44} & a_{45} \end{bmatrix}$$

and we suppose that

$$A = \begin{bmatrix} a_{11} & a_{12} & a_{13} & a_{14} & a_{15} \\ a_{21} & a_{22} & a_{23} & a_{24} & a_{25} \\ a_{31} & a_{32} & a_{33} & a_{34} & a_{35} \\ a_{41} & a_{42} & a_{43} & b_{44} & a_{45} \end{bmatrix} \sim \begin{bmatrix} 1 & 0 & a & 0 & c \\ 0 & 1 & b & 0 & d \\ 0 & 0 & 0 & 1 & e \\ 0 & 0 & 0 & 0 & 0 \end{bmatrix}.$$

The solutions of the equation

$$A \begin{bmatrix} x_1 \\ x_2 \\ x_3 \\ x_4 \\ x_5 \end{bmatrix} = \begin{bmatrix} 0 \\ 0 \\ 0 \\ 0 \end{bmatrix} \tag{2.9}$$

are the same as the solutions of the equation

$$\begin{bmatrix} 1 & 0 & a & 0 & c \\ 0 & 1 & b & 0 & d \\ 0 & 0 & 0 & 1 & e \\ 0 & 0 & 0 & 0 & 0 \end{bmatrix} \begin{bmatrix} x_1 \\ x_2 \\ x_3 \\ x_4 \\ x_5 \end{bmatrix} = \begin{bmatrix} 0 \\ 0 \\ 0 \\ 0 \end{bmatrix}.$$

Consequently,

$$x_1 + x_3 a + x_5 c = 0, \quad x_2 + x_3 b + x_5 d = 0, \quad \text{and} \quad x_4 + x_5 e = 0,$$

which means that the general solution of the equation (2.9) is

$$\begin{bmatrix} x_1 \\ x_2 \\ x_3 \\ x_4 \\ x_5 \end{bmatrix} = \begin{bmatrix} -x_3 a - x_5 c \\ -x_3 b - x_5 d \\ x_3 \\ -x_5 e \\ x_5 \end{bmatrix} = x_3 \begin{bmatrix} -a \\ -b \\ 1 \\ 0 \\ 0 \end{bmatrix} + x_5 \begin{bmatrix} -c \\ -d \\ 0 \\ -e \\ 1 \end{bmatrix}$$

where the vectors $\begin{bmatrix} -a \\ -b \\ 1 \\ 0 \\ 0 \end{bmatrix}$ and $\begin{bmatrix} -c \\ -d \\ 0 \\ -e \\ 1 \end{bmatrix}$ are the special solutions.

To complete the proof we have to show that the special solutions $\begin{bmatrix} -a \\ -b \\ 1 \\ 0 \\ 0 \end{bmatrix}$ and $\begin{bmatrix} -c \\ -d \\ 0 \\ -e \\ 1 \end{bmatrix}$ are linearly independent. Since the equation

$$x \begin{bmatrix} -a \\ -b \\ 1 \\ 0 \\ 0 \end{bmatrix} + y \begin{bmatrix} -c \\ -d \\ 0 \\ -e \\ 1 \end{bmatrix} = \begin{bmatrix} 0 \\ 0 \\ 0 \\ 0 \\ 0 \end{bmatrix}$$

has only the trivial solution $x = y = 0$, the special solutions are linearly independent and consequently the set

$$\left\{ \begin{bmatrix} -a \\ -b \\ 1 \\ 0 \\ 0 \end{bmatrix}, \begin{bmatrix} -c \\ -d \\ 0 \\ -e \\ 1 \end{bmatrix} \right\}$$

is a basis in $\mathbf{N}(A)$. □

Exercises 2.2

Determine the subspace spanned by the given vectors.

1. $\begin{bmatrix} 1 \\ 2 \\ 1 \\ 3 \end{bmatrix}, \begin{bmatrix} 2 \\ 1 \\ 4 \\ 5 \end{bmatrix}$

2. $\begin{bmatrix} 1 \\ 1 \\ 1 \\ 1 \end{bmatrix}, \begin{bmatrix} 1 \\ 2 \\ 1 \\ 0 \end{bmatrix}, \begin{bmatrix} 1 \\ 0 \\ 0 \\ 0 \end{bmatrix}$

3. Show that the set of all vectors of the form $\begin{bmatrix} a \\ b+5c \\ a-2b \\ a \end{bmatrix}$, where a, b, and c are arbitrary real numbers, is a subspace of $\mathbb{R}^4$.

4. Show that the set of all vectors of the form $\begin{bmatrix} a+b+c \\ c \\ b \\ a \end{bmatrix}$, where a, b, and c are arbitrary real numbers, is a subspace of $\mathbb{R}^4$.

5. Show that the set of all vectors of the form $\begin{bmatrix} a+b+1 \\ a \\ b \end{bmatrix}$, where a and b are arbitrary real numbers, is not a subspace of $\mathbb{R}^3$.

6. Show that the set of all vectors of the form $\begin{bmatrix} a+3 \\ a \\ a \\ 2a+b \end{bmatrix}$, where a and b are arbitrary real numbers, is not a subspace of $\mathbb{R}^4$.

7. Show that the set of all vectors of the form $\begin{bmatrix} a \\ a-b \\ 2 \\ b \end{bmatrix}$, where a and b are arbitrary real numbers, is not a subspace of $\mathbb{R}^4$.

8. Show that the set of all vectors of the form $\begin{bmatrix} 0 \\ 0 \\ a+b+7 \\ 0 \end{bmatrix}$, where a and b are arbitrary real numbers, is not a subspace of $\mathbb{R}^4$.

9. Show that $\begin{bmatrix} a \\ b \\ c \\ d \\ e \end{bmatrix}$ where a, b, c, d, e are real numbers such that $a-2b=0$, $c-3a=0$, $d-5a=0$, and $e-a-3b=0$ is a subspace of $\mathbb{R}^5$.

10. Show that $\begin{bmatrix} a \\ b \\ c \\ d \end{bmatrix}$ where a, b, c, d are real numbers such that $a+2d=0$ and $b+5c=0$ is a subspace of $\mathbb{R}^4$.

11. Show that $\begin{bmatrix} a \\ b \\ c \\ d \end{bmatrix}$ where a, b, c, d are real numbers such that $a + 2b + 2c + d = 0$ and $2a + b + c + d = 0$ is a subspace of $\mathbb{R}^4$.

12. Show that $\begin{bmatrix} a \\ b \\ c \\ d \end{bmatrix}$ where a, b, c, d are real numbers such that $a + b + c + d = 0$, $2a + 3b + 4c - 2d = 0$ is a subspace of $\mathbb{R}^4$.

Determine if the given vector $\mathbf{v}$ is in the subspace $\mathcal{V}$.

13. $\mathbf{v} = \begin{bmatrix} 5 \\ 4 \\ 5 \\ 3 \end{bmatrix}, \mathcal{V} = \text{Span}\left\{ \begin{bmatrix} 3 \\ 2 \\ 1 \\ 1 \end{bmatrix}, \begin{bmatrix} 1 \\ 1 \\ 2 \\ 1 \end{bmatrix} \right\}$

14. $\mathbf{v} = \begin{bmatrix} 1 \\ 2 \\ 1 \\ 5 \end{bmatrix}, \mathcal{V} = \text{Span}\left\{ \begin{bmatrix} 1 \\ 1 \\ 1 \\ 0 \end{bmatrix}, \begin{bmatrix} 3 \\ 1 \\ 3 \\ 4 \end{bmatrix} \right\}$

15. $\mathbf{v} = \begin{bmatrix} 0 \\ 0 \\ 0 \\ 1 \end{bmatrix}, \mathcal{V} = \text{Span}\left\{ \begin{bmatrix} 1 \\ 2 \\ 1 \\ 1 \end{bmatrix}, \begin{bmatrix} 1 \\ 1 \\ 0 \\ 1 \end{bmatrix}, \begin{bmatrix} 2 \\ 2 \\ 1 \\ 1 \end{bmatrix} \right\}$

16. $\mathbf{v} = \begin{bmatrix} 1 \\ 1 \\ 1 \\ 1 \end{bmatrix}, \mathcal{V} = \text{Span}\left\{ \begin{bmatrix} 1 \\ 2 \\ 1 \\ 0 \end{bmatrix}, \begin{bmatrix} 1 \\ 1 \\ 0 \\ -1 \end{bmatrix}, \begin{bmatrix} 2 \\ 1 \\ 3 \\ 5 \end{bmatrix} \right\}$

17. $\mathbf{v} = \begin{bmatrix} 1 \\ 1 \\ 1 \\ 1 \end{bmatrix}, \mathcal{V} = \text{Span}\left\{ \begin{bmatrix} 1 \\ 1 \\ 2 \\ 1 \end{bmatrix}, \begin{bmatrix} 1 \\ 2 \\ 1 \\ 1 \end{bmatrix} \right\}$

18. $\mathbf{v} = \begin{bmatrix} 1 \\ 0 \\ 0 \\ 0 \end{bmatrix}, \mathcal{V} = \text{Span}\left\{ \begin{bmatrix} 2 \\ 1 \\ 2 \\ 3 \end{bmatrix}, \begin{bmatrix} 1 \\ 1 \\ 1 \\ 1 \end{bmatrix} \right\}$

19. $\mathbf{v} = \begin{bmatrix} 1 \\ 0 \\ 1 \\ 0 \end{bmatrix}, \mathcal{V} = \text{Span}\left\{ \begin{bmatrix} 4 \\ 1 \\ 0 \\ 5 \end{bmatrix}, \begin{bmatrix} 1 \\ 3 \\ 0 \\ 1 \end{bmatrix}, \begin{bmatrix} 1 \\ 3 \\ 0 \\ 2 \end{bmatrix} \right\}$

20. $\mathbf{v} = \begin{bmatrix} 1 \\ 1 \\ 1 \\ 0 \end{bmatrix}, \mathcal{V} = \text{Span}\left\{ \begin{bmatrix} 1 \\ 0 \\ 0 \\ 1 \end{bmatrix}, \begin{bmatrix} 1 \\ 3 \\ 0 \\ 1 \end{bmatrix}, \begin{bmatrix} 1 \\ 3 \\ 0 \\ 2 \end{bmatrix} \right\}$

Determine a basis of the column space $\mathbf{C}(A)$ using row operations on the matrix A^T.

21. $A = \begin{bmatrix} 1 & 1 & 2 & 3 \\ 2 & 3 & 3 & 5 \\ 1 & 2 & 1 & 2 \end{bmatrix}$

22. $A = \begin{bmatrix} 3 & 1 & 1 & 1 \\ 1 & 3 & 1 & 2 \\ 2 & 3 & 1 & 1 \end{bmatrix}$

23. $A = \begin{bmatrix} 1 & 2 & 3 \\ 2 & 3 & 5 \\ 2 & 2 & 4 \\ 3 & 4 & 7 \end{bmatrix}$

24. $A = \begin{bmatrix} 2 & 1 & 1 \\ 2 & 3 & -1 \\ 2 & 2 & 0 \\ 5 & 4 & 1 \end{bmatrix}$

25. Show that $\operatorname{Span}\left\{\begin{bmatrix}1\\1\\0\\1\end{bmatrix}, \begin{bmatrix}1\\1\\2\\1\end{bmatrix}, \begin{bmatrix}1\\1\\1\\1\end{bmatrix}\right\} = \operatorname{Span}\left\{\begin{bmatrix}1\\1\\0\\1\end{bmatrix}, \begin{bmatrix}1\\1\\2\\1\end{bmatrix}\right\}$.

26. Show that $\operatorname{Span}\left\{\begin{bmatrix}1\\1\\1\\1\end{bmatrix}, \begin{bmatrix}2\\1\\4\\-1\end{bmatrix}, \begin{bmatrix}-1\\1\\-5\\5\end{bmatrix}\right\} = \operatorname{Span}\left\{\begin{bmatrix}1\\1\\1\\1\end{bmatrix}, \begin{bmatrix}2\\1\\4\\-1\end{bmatrix}\right\}$.

27. Find real numbers a and b such that the vector $\begin{bmatrix}12\\2\\a\\b\end{bmatrix}$ is in $\operatorname{Span}\left\{\begin{bmatrix}2\\2\\5\\3\end{bmatrix}, \begin{bmatrix}3\\2\\-1\\1\end{bmatrix}\right\}$.

28. Find real numbers a and b such that the vector $\begin{bmatrix}3\\a\\5\\b\end{bmatrix}$ is in $\operatorname{Span}\left\{\begin{bmatrix}1\\2\\1\\3\end{bmatrix}, \begin{bmatrix}1\\1\\2\\4\end{bmatrix}\right\}$.

Determine a basis of the column space $\mathbf{C}(A)$ for the given matrix A.

29. $A = \begin{bmatrix}1 & 2 & 1\\1 & 3 & 2\\2 & 3 & 1\\3 & 5 & 2\end{bmatrix}$

30. $A = \begin{bmatrix}1 & -2 & 1\\2 & -4 & 2\\-2 & 4 & 2\\4 & -8 & 1\end{bmatrix}$

Determine a basis of the null space $\mathbf{N}(A)$ for the given matrix A.

31. $A = \begin{bmatrix}1 & 2 & 1 & 1\\1 & -1 & 3 & 2\end{bmatrix}$

32. $A = \begin{bmatrix}3 & 4 & 2 & 1\\1 & 2 & 1 & 1\end{bmatrix}$

33. $A = \begin{bmatrix}2 & 1 & 1 & 1\\1 & -2 & 3 & 2\\3 & -1 & 4 & 1\end{bmatrix}$

34. $A = \begin{bmatrix}2 & 1 & 1 & 1 & 1\\1 & 1 & 1 & 2 & 1\end{bmatrix}$

35. $A = \begin{bmatrix}2 & 1 & 1 & 1\\1 & -2 & 3 & 2\\3 & -1 & 4 & 3\end{bmatrix}$

36. $A = \begin{bmatrix}4 & 3 & 5 & 1\\1 & 1 & 3 & 2\\3 & 2 & 2 & -1\end{bmatrix}$

37. $A = \begin{bmatrix}2 & 1 & 4 & 3 & 5\\1 & 2 & -1 & 0 & -2\\1 & 1 & 1 & 1 & 1\end{bmatrix}$

38. $A = \begin{bmatrix}1 & 3 & 1 & 1 & 1\\2 & 1 & 1 & 3 & 2\\1 & 1 & 4 & 2 & 1\end{bmatrix}$

2.3 The dimension of a subspace of $\mathbb{R}^n$

The dimension of a subspace $\mathcal{V}$ is defined as the number of vectors in any basis of $\mathcal{V}$. For such a definition to make sense we first have to show that all bases of a subspace have the same number of vectors. This is not obvious and will be accomplished in a number of smaller steps.

Theorem 2.44. *If a subspace* $\text{Span}\{\mathbf{v}_1,\dots,\mathbf{v}_k\}$ *of* $\mathbb{R}^n$ *contains* k *linearly independent vectors* $\mathbf{w}_1,\dots,\mathbf{w}_k$, *then the vectors* $\mathbf{v}_1,\dots,\mathbf{v}_k$ *are linearly independent and*

$$\text{Span}\{\mathbf{w}_1,\dots,\mathbf{w}_k\} = \text{Span}\{\mathbf{v}_1,\dots,\mathbf{v}_k\}.$$

In other words, every collection of k linearly independent vectors $\mathbf{w}_1,\dots,\mathbf{w}_k$ in a subspace spanned by k vectors $\mathbf{v}_1,\dots,\mathbf{v}_k$ is a basis of the subspace $\text{Span}\{\mathbf{v}_1,\dots,\mathbf{v}_k\}$ and the set $\{\mathbf{v}_1,\dots,\mathbf{v}_k\}$ is also a basis of that subspace.

Proof. Assume that $\mathbf{w}_1,\dots,\mathbf{w}_k$ are linearly independent vectors in $\text{Span}\{\mathbf{v}_1,\dots,\mathbf{v}_k\}$. Then every vector in $\{\mathbf{w}_1,\dots,\mathbf{w}_k\}$ is a linear combination of vectors $\mathbf{v}_1,\dots,\mathbf{v}_k$, that is, for every $j = 1,2,\dots,k$ there exist real numbers $a_{1j},\dots,a_{kj}$ such that

$$\mathbf{w}_j = a_{1j}\mathbf{v}_1 + \dots + a_{kj}\mathbf{v}_k.$$

If we let $\mathbf{a}_j = \begin{bmatrix} a_{1j} \\ \vdots \\ a_{kj} \end{bmatrix}$, then the above equations can be written as

$$\mathbf{w}_j = \begin{bmatrix} \mathbf{v}_1 & \dots & \mathbf{v}_k \end{bmatrix} \mathbf{a}_j,$$

or as a single matrix equation

$$\begin{bmatrix} \mathbf{w}_1 & \dots & \mathbf{w}_k \end{bmatrix} = \begin{bmatrix} \mathbf{v}_1 & \dots & \mathbf{v}_k \end{bmatrix} \begin{bmatrix} \mathbf{a}_1 & \dots & \mathbf{a}_k \end{bmatrix}. \tag{2.10}$$

Consequently, if $\mathbf{x} = \begin{bmatrix} x_1 \\ \vdots \\ x_k \end{bmatrix}$ is an arbitrary vector in $\mathbb{R}^k$, then

$$\begin{bmatrix} \mathbf{w}_1 & \dots & \mathbf{w}_k \end{bmatrix} \mathbf{x} = \begin{bmatrix} \mathbf{v}_1 & \dots & \mathbf{v}_k \end{bmatrix} \begin{bmatrix} \mathbf{a}_1 & \dots & \mathbf{a}_k \end{bmatrix} \mathbf{x}.$$

If $\begin{bmatrix} \mathbf{a}_1 & \dots & \mathbf{a}_k \end{bmatrix} \mathbf{x} = \mathbf{0}$, then $\begin{bmatrix} \mathbf{w}_1 & \dots & \mathbf{w}_k \end{bmatrix} \mathbf{x} = \mathbf{0}$ and, since the vectors $\mathbf{w}_1,\dots,\mathbf{w}_k$ are linearly independent, $\mathbf{x} = \mathbf{0}$. This implies that the $k \times k$ matrix $\begin{bmatrix} \mathbf{a}_1 & \dots & \mathbf{a}_k \end{bmatrix}$ is invertible. Hence, from (2.10) we get

$$\begin{bmatrix} \mathbf{w}_1 & \dots & \mathbf{w}_k \end{bmatrix} \begin{bmatrix} \mathbf{a}_1 & \dots & \mathbf{a}_k \end{bmatrix}^{-1} = \begin{bmatrix} \mathbf{v}_1 & \dots & \mathbf{v}_k \end{bmatrix},$$

which means that the vectors $\mathbf{v}_1,\dots,\mathbf{v}_k$ are elements of $\text{Span}\{\mathbf{w}_1,\dots,\mathbf{w}_k\}$ and thus

$$\text{Span}\{\mathbf{w}_1,\dots,\mathbf{w}_k\} = \text{Span}\{\mathbf{v}_1,\dots,\mathbf{v}_k\},$$

by Theorem 2.28.

To show that the vectors $\mathbf{v}_1,\dots,\mathbf{v}_k$ are linearly independent we assume that

$$\begin{bmatrix}\mathbf{v}_1 & \dots & \mathbf{v}_k\end{bmatrix}\begin{bmatrix}x_1\\ \vdots\\ x_k\end{bmatrix} = \mathbf{0}.$$

Since this equation is equivalent to the equation

$$\begin{bmatrix}\mathbf{w}_1 & \dots & \mathbf{w}_k\end{bmatrix}\begin{bmatrix}\mathbf{a}_1 & \dots & \mathbf{a}_k\end{bmatrix}^{-1}\begin{bmatrix}x_1\\ \vdots\\ x_k\end{bmatrix} = \mathbf{0},$$

we must have

$$\begin{bmatrix}\mathbf{a}_1 & \dots & \mathbf{a}_k\end{bmatrix}^{-1}\begin{bmatrix}x_1\\ \vdots\\ x_k\end{bmatrix} = \mathbf{0},$$

because the vectors $\mathbf{w}_1,\dots,\mathbf{w}_k$ are linearly independent. Now we multiply both sides of the above equation by $\begin{bmatrix}\mathbf{a}_1 & \dots & \mathbf{a}_k\end{bmatrix}$ and get

$$\begin{bmatrix}\mathbf{a}_1 & \dots & \mathbf{a}_k\end{bmatrix}\begin{bmatrix}\mathbf{a}_1 & \dots & \mathbf{a}_k\end{bmatrix}^{-1}\begin{bmatrix}x_1\\ \vdots\\ x_k\end{bmatrix} = \begin{bmatrix}x_1\\ \vdots\\ x_k\end{bmatrix} = \begin{bmatrix}\mathbf{a}_1 & \dots & \mathbf{a}_k\end{bmatrix}\mathbf{0} = \mathbf{0}.$$

Since $\begin{bmatrix}\mathbf{v}_1 & \dots & \mathbf{v}_k\end{bmatrix}\begin{bmatrix}x_1\\ \vdots\\ x_k\end{bmatrix} = \mathbf{0}$ implies $\begin{bmatrix}x_1\\ \vdots\\ x_k\end{bmatrix} = \mathbf{0}$, the vectors $\mathbf{v}_1,\dots,\mathbf{v}_k$ are linearly independent. □

Corollary 2.45.

(a) *Any set of n linearly independent vectors in $\mathbb{R}^n$ is a basis in $\mathbb{R}^n$.*

(b) *Any $n+1$ vectors in $\mathbb{R}^n$ are linearly dependent.*

Proof. To prove (a) we assume that $\mathbf{w}_1,\dots,\mathbf{w}_n$ are linearly independent vectors in $\mathbb{R}^n$. Since $\mathbb{R}^n = \operatorname{Span}\{\mathbf{e}_1,\dots,\mathbf{e}_n\}$, where $\mathbf{e}_1,\dots,\mathbf{e}_n$ are the columns of the unity matrix I_n, we have

$$\operatorname{Span}\{\mathbf{w}_1,\dots,\mathbf{w}_n\} = \mathbb{R}^n$$

by Theorem 2.44.

Now we prove (b). Let $\mathbf{w}_1,\mathbf{w}_2,\dots,\mathbf{w}_{n+1}$ be vectors in $\mathbb{R}^n$. If the vectors $\mathbf{w}_1,\dots,\mathbf{w}_n$ are linearly dependent, we are done. If the vectors $\mathbf{w}_1,\dots,\mathbf{w}_n$ are linearly independent, then we have $\operatorname{Span}\{\mathbf{w}_1,\dots,\mathbf{w}_n\} = \mathbb{R}^n$, by (a). But then the vector $\mathbf{w}_{n+1}$ is a linear combination of the vectors $\mathbf{w}_1,\dots,\mathbf{w}_n$, so the vectors $\mathbf{w}_1,\dots,\mathbf{w}_n,\mathbf{w}_{n+1}$ are linearly dependent. □

Using Theorem 2.44 we can prove a result announced earlier in this chapter, namely, that every subspace is a span. We present a proof for subspaces of $\mathbb{R}^4$. It should be clear how the presented argument generalizes to any dimension.

Theorem 2.46. *If $\mathcal{V}$ is a subspace of $\mathbb{R}^n$, then there are linearly independent vectors $\mathbf{v}_1, \ldots, \mathbf{v}_k$ in $\mathbb{R}^n$, for some $k \le n$, such that*

$$\mathcal{V} = \text{Span}\{\mathbf{v}_1, \ldots, \mathbf{v}_k\}.$$

Proof of a particular case. Suppose $\mathcal{V}$ is a subspace of $\mathbb{R}^4$.

If $\mathcal{V}$ contains a vector $\mathbf{v}_1 \neq \mathbf{0}$, then $\mathcal{V}$ must contain $\text{Span}\{\mathbf{v}_1\}$. If $\mathcal{V} = \text{Span}\{\mathbf{v}_1\}$, then we are done.

If $\mathcal{V} \neq \text{Span}\{\mathbf{v}_1\}$, then $\mathcal{V}$ contains a vector $\mathbf{v}_2$ which is not in $\text{Span}\{\mathbf{v}_1\}$. The subspace $\mathcal{V}$ must contain $\text{Span}\{\mathbf{v}_1, \mathbf{v}_2\}$ and the vectors $\mathbf{v}_1$ and $\mathbf{v}_2$ are linearly independent. If $\mathcal{V} = \text{Span}\{\mathbf{v}_1, \mathbf{v}_2\}$, then we are done.

If $\mathcal{V} \neq \text{Span}\{\mathbf{v}_1, \mathbf{v}_2\}$, then $\mathcal{V}$ contains a vector $\mathbf{v}_3$ which is not in $\text{Span}\{\mathbf{v}_1, \mathbf{v}_2\}$. The subspace $\mathcal{V}$ must contain $\text{Span}\{\mathbf{v}_1, \mathbf{v}_2, \mathbf{v}_3\}$ and the vectors $\mathbf{v}_1$, $\mathbf{v}_2$, and $\mathbf{v}_3$ are linearly independent. If $\mathcal{V} = \text{Span}\{\mathbf{v}_1, \mathbf{v}_2, \mathbf{v}_3\}$, then we are done.

If $\mathcal{V} \neq \text{Span}\{\mathbf{v}_1, \mathbf{v}_2, \mathbf{v}_3\}$, then the subspace $\mathcal{V}$ must contain a vector $\mathbf{v}_4$ which is not in $\text{Span}\{\mathbf{v}_1, \mathbf{v}_2, \mathbf{v}_3\}$ and $\mathcal{V}$ must contain $\text{Span}\{\mathbf{v}_1, \mathbf{v}_2, \mathbf{v}_3, \mathbf{v}_4\}$ and the vectors $\mathbf{v}_1$, $\mathbf{v}_2$, $\mathbf{v}_3$, and $\mathbf{v}_4$ are linearly independent. In this case since $\mathbb{R}^4 = \text{Span}\{\mathbf{e}_1, \mathbf{e}_2, \mathbf{e}_3, \mathbf{e}_4\}$, where $\mathbf{e}_1, \mathbf{e}_2, \mathbf{e}_3, \mathbf{e}_4$ are the columns of the unity matrix I_4, we have

$$\mathcal{V} = \text{Span}\{\mathbf{v}_1, \mathbf{v}_2, \mathbf{v}_3, \mathbf{v}_4\} = \mathbb{R}^4,$$

by Theorem 2.44. □

Finally we are in a position to prove the result stated at the beginning of this section.

Theorem 2.47. *All bases of a subspace of $\mathbb{R}^n$ have the same number of elements. More precisely, if $\{\mathbf{v}_1, \ldots, \mathbf{v}_k\}$ and $\{\mathbf{w}_1, \ldots, \mathbf{w}_m\}$ are bases of the same vector subspace $\mathcal{V}$ of $\mathbb{R}^n$, then $k = m$.*

Proof. Suppose, to the contrary, that $\{\mathbf{v}_1, \ldots, \mathbf{v}_k\}$ and $\{\mathbf{w}_1, \ldots, \mathbf{w}_m\}$ are bases of a vector subspace $\mathcal{V}$ of $\mathbb{R}^n$ and $k \neq m$. Without loss of generality we can assume that $m > k$. Since the vectors $\mathbf{w}_1, \ldots, \mathbf{w}_k$ are linearly independent we have

$$\text{Span}\{\mathbf{w}_1, \ldots, \mathbf{w}_k\} = \text{Span}\{\mathbf{v}_1, \ldots, \mathbf{v}_k\} = \mathcal{V},$$

by Theorem 2.44. But this contradicts the fact that the vectors $\mathbf{w}_1, \ldots, \mathbf{w}_m$ are linearly independent, since

$$\text{Span}\{\mathbf{w}_1, \ldots, \mathbf{w}_m\} = \mathcal{V} = \text{Span}\{\mathbf{w}_1, \ldots, \mathbf{w}_k\}.$$

□

The unique number of vectors in any basis of a given subspace is an essential attribute of that subspace.

Definition 2.48. If $\{\mathbf{v}_1, \ldots, \mathbf{v}_k\}$ is a basis of a subspace $\mathcal{V}$, then the number k is called the *dimension* of the subspace $\mathcal{V}$ and denoted by $\dim \mathcal{V}$.

The next result is often useful when proving that a set of vectors is a basis in a subspace.

Theorem 2.49. *Let $\mathcal{V}$ be a vector subspace of $\mathbb{R}^n$. If $\dim \mathcal{V} = k$, then for any collection of vectors $\{\mathbf{v}_1, \ldots, \mathbf{v}_k\}$ in $\mathcal{V}$ the following three conditions are equivalent:*

(a) *The set $\{\mathbf{v}_1, \ldots, \mathbf{v}_k\}$ is basis in $\mathcal{V}$;*

(b) *The vectors $\mathbf{v}_1, \ldots, \mathbf{v}_k$ are linearly independent;*

(c) $\mathcal{V} = \operatorname{Span}\{\mathbf{v}_1, \ldots, \mathbf{v}_k\}$.

Proof. First note that (a) implies (b) by the definition of a basis and (b) implies (c) by Theorem 2.44.

Now let $\{\mathbf{w}_1, \ldots, \mathbf{w}_k\}$ be a basis of $\mathcal{V}$. If $\mathcal{V} = \operatorname{Span}\{\mathbf{v}_1, \ldots, \mathbf{v}_k\}$, then the linearly independent vectors $\mathbf{w}_1, \ldots, \mathbf{w}_k$ are in $\operatorname{Span}\{\mathbf{v}_1, \ldots, \mathbf{v}_k\}$ and the vectors $\mathbf{v}_1, \ldots, \mathbf{v}_k$ are linearly independent, again by Theorem 2.44. This means that the set $\{\mathbf{v}_1, \ldots, \mathbf{v}_k\}$ is a basis in $\mathcal{V}$. Thus (c) implies (a). $\square$

Example 2.50. Show that

$$\dim \operatorname{Span}\left\{\begin{bmatrix}1\\2\\1\\2\end{bmatrix}, \begin{bmatrix}3\\4\\1\\3\end{bmatrix}, \begin{bmatrix}5\\8\\3\\7\end{bmatrix}, \begin{bmatrix}1\\1\\1\\3\end{bmatrix}\right\} = 3$$

and determine all subsets of 3 vectors from the set $\left\{\begin{bmatrix}1\\2\\1\\2\end{bmatrix}, \begin{bmatrix}3\\4\\1\\3\end{bmatrix}, \begin{bmatrix}5\\8\\3\\7\end{bmatrix}, \begin{bmatrix}1\\1\\1\\3\end{bmatrix}\right\}$ that are bases.

Solution. Since

$$\begin{bmatrix} 1 & 3 & 5 & 1 & 0 \\ 2 & 4 & 8 & 1 & 0 \\ 1 & 1 & 3 & 1 & 0 \\ 2 & 3 & 7 & 3 & 0 \end{bmatrix} \sim \begin{bmatrix} 1 & 0 & 2 & 0 & 0 \\ 0 & 1 & 1 & 0 & 0 \\ 0 & 0 & 0 & 1 & 0 \\ 0 & 0 & 0 & 0 & 0 \end{bmatrix}, \tag{2.11}$$

we have

$$\begin{bmatrix} 5 \\ 8 \\ 3 \\ 7 \end{bmatrix} = 2\begin{bmatrix} 1 \\ 2 \\ 1 \\ 2 \end{bmatrix} + \begin{bmatrix} 3 \\ 4 \\ 1 \\ 3 \end{bmatrix}$$

and thus

$$\text{Span}\left\{\begin{bmatrix} 1 \\ 2 \\ 1 \\ 2 \end{bmatrix}, \begin{bmatrix} 3 \\ 4 \\ 1 \\ 3 \end{bmatrix}, \begin{bmatrix} 5 \\ 8 \\ 3 \\ 7 \end{bmatrix}, \begin{bmatrix} 1 \\ 1 \\ 1 \\ 3 \end{bmatrix}\right\} = \text{Span}\left\{\begin{bmatrix} 1 \\ 2 \\ 1 \\ 2 \end{bmatrix}, \begin{bmatrix} 3 \\ 4 \\ 1 \\ 3 \end{bmatrix}, \begin{bmatrix} 1 \\ 1 \\ 1 \\ 3 \end{bmatrix}\right\}.$$

The vectors $\begin{bmatrix} 1 \\ 2 \\ 1 \\ 2 \end{bmatrix}, \begin{bmatrix} 3 \\ 4 \\ 1 \\ 3 \end{bmatrix}$ and $\begin{bmatrix} 1 \\ 1 \\ 1 \\ 3 \end{bmatrix}$ are linearly independent, because, in view of (2.11), the only solution of the equation

$$x_1\begin{bmatrix} 1 \\ 2 \\ 1 \\ 2 \end{bmatrix} + x_2\begin{bmatrix} 3 \\ 4 \\ 1 \\ 3 \end{bmatrix} + x_3\begin{bmatrix} 1 \\ 1 \\ 1 \\ 3 \end{bmatrix} = \begin{bmatrix} 0 \\ 0 \\ 0 \\ 0 \end{bmatrix}$$

is the trivial solution, that is, $x_1 = x_2 = x_3 = 0$. This means that the set of vectors

$$\left\{\begin{bmatrix} 1 \\ 2 \\ 1 \\ 2 \end{bmatrix}, \begin{bmatrix} 3 \\ 4 \\ 1 \\ 3 \end{bmatrix}, \begin{bmatrix} 1 \\ 1 \\ 1 \\ 3 \end{bmatrix}\right\}$$

is a basis of the subspace $\text{Span}\left\{\begin{bmatrix} 1 \\ 2 \\ 1 \\ 2 \end{bmatrix}, \begin{bmatrix} 3 \\ 4 \\ 1 \\ 3 \end{bmatrix}, \begin{bmatrix} 5 \\ 8 \\ 3 \\ 7 \end{bmatrix}, \begin{bmatrix} 1 \\ 1 \\ 1 \\ 3 \end{bmatrix}\right\}$.

Now we look for other choices for a basis from the vectors

$$\begin{bmatrix} 1 \\ 2 \\ 1 \\ 2 \end{bmatrix}, \begin{bmatrix} 3 \\ 4 \\ 1 \\ 3 \end{bmatrix}, \begin{bmatrix} 5 \\ 8 \\ 3 \\ 7 \end{bmatrix}, \begin{bmatrix} 1 \\ 1 \\ 1 \\ 3 \end{bmatrix}.$$

We know that we need to find three vectors that are linearly independent.

First we consider the vectors $\begin{bmatrix}1\\2\\1\\2\end{bmatrix}, \begin{bmatrix}5\\8\\3\\7\end{bmatrix}, \begin{bmatrix}1\\1\\1\\3\end{bmatrix}$. By (2.11), the equation

$$x_1\begin{bmatrix}1\\2\\1\\2\end{bmatrix}+x_2\begin{bmatrix}5\\8\\3\\7\end{bmatrix}+x_3\begin{bmatrix}1\\1\\1\\3\end{bmatrix}=\begin{bmatrix}0\\0\\0\\0\end{bmatrix}$$

is equivalent to the equation

$$x_1\begin{bmatrix}1\\0\\0\\0\end{bmatrix}+x_2\begin{bmatrix}2\\1\\0\\0\end{bmatrix}+x_3\begin{bmatrix}0\\0\\1\\0\end{bmatrix}=\begin{bmatrix}0\\0\\0\\0\end{bmatrix}$$

which only has the trivial solution $x_1 = x_2 = x_3 = 0$. Hence the set $\left\{\begin{bmatrix}1\\2\\1\\2\end{bmatrix}, \begin{bmatrix}5\\8\\3\\7\end{bmatrix}, \begin{bmatrix}1\\1\\1\\3\end{bmatrix}\right\}$ is a basis.

Using a similar argument we can show that the set $\left\{\begin{bmatrix}3\\4\\1\\3\end{bmatrix}, \begin{bmatrix}5\\8\\3\\7\end{bmatrix}, \begin{bmatrix}1\\1\\1\\3\end{bmatrix}\right\}$ is a basis.

The set $\left\{\begin{bmatrix}1\\2\\1\\2\end{bmatrix}, \begin{bmatrix}3\\4\\1\\3\end{bmatrix}, \begin{bmatrix}5\\8\\3\\7\end{bmatrix}, \right\}$ is not a basis, because the vectors of this set are linearly dependent. □

The following theorem will be used frequently in Chapter 7 as well as in Chapter 11 where we illustrate applications of a computer algebra system to solve practical linear algebra problems.

Theorem 2.51. *Let* $\mathbf{v}_1,\dots,\mathbf{v}_m$ *be linearly independent vectors in* $\mathbb{R}^n$ *for some* $m \le n$ *and let* $\mathbf{w}_1,\dots,\mathbf{w}_k$ *be linearly independent vectors in* $\operatorname{Span}\{\mathbf{v}_1,\dots,\mathbf{v}_m\}$ *for some* $k < m$. *There are vectors* $\mathbf{w}_{k+1},\dots,\mathbf{w}_m$ *in the set* $\{\mathbf{v}_1,\dots,\mathbf{v}_m\}$ *such that*

$$\{\mathbf{w}_1,\dots,\mathbf{w}_k,\mathbf{w}_{k+1},\dots,\mathbf{w}_m\}$$

is a basis in $\operatorname{Span}\{\mathbf{v}_1,\dots,\mathbf{v}_m\}$.

Proof. The pivot columns of the matrix

$$\begin{bmatrix} \mathbf{w}_1 & \dots & \mathbf{w}_k & \mathbf{v}_1 & \dots & \mathbf{v}_m \end{bmatrix}$$

determine a basis of the vector subspace $\text{Span}\{\mathbf{w}_1,\dots,\mathbf{w}_k,\mathbf{v}_1,\dots,\mathbf{v}_m\} = \text{Span}\{\mathbf{v}_1,\dots,\mathbf{v}_m\}$. This basis contains the vectors $\mathbf{w}_1,\dots,\mathbf{w}_k$ because being linearly independent these vectors are pivot columns of the matrix $\begin{bmatrix} \mathbf{w}_1 & \dots & \mathbf{w}_k & \mathbf{v}_1 & \dots & \mathbf{v}_m \end{bmatrix}$. □

Corollary 2.52. *If $\mathbf{v}_1,\dots,\mathbf{v}_k$ are linearly independent vectors in $\mathbb{R}^n$ for some $k < n$, then there are vectors $\mathbf{v}_{k+1},\dots,\mathbf{v}_n$ in $\mathbb{R}^n$ such that*

$$\{\mathbf{v}_1,\dots,\mathbf{v}_k,\mathbf{v}_{k+1},\dots,\mathbf{v}_n\}$$

is a basis of $\mathbb{R}^n$.

Note that the proof of Theorem 2.51 gives us a method of construction of an extended basis.

Example 2.53. We consider the vectors $\mathbf{v}_1 = \begin{bmatrix} 1 \\ 4 \\ -3 \end{bmatrix}$ and $\mathbf{v}_2 = \begin{bmatrix} 2 \\ -4 \\ 3 \end{bmatrix}$. Find a vector $\mathbf{v}_3$ in $\mathbb{R}^3$ such that $\{\mathbf{v}_1,\mathbf{v}_2,\mathbf{v}_3\}$ is a basis in $\mathbb{R}^3$.

Solution. The reduced row echelon form of the matrix

$$\begin{bmatrix} 1 & 2 & 1 & 0 & 0 \\ 4 & -4 & 0 & 1 & 0 \\ -3 & 3 & 0 & 0 & 1 \end{bmatrix}$$

is

$$\begin{bmatrix} 1 & 0 & \frac{1}{3} & 0 & -\frac{2}{9} \\ 0 & 1 & \frac{1}{3} & 0 & \frac{1}{9} \\ 0 & 0 & 0 & 1 & \frac{4}{3} \end{bmatrix}.$$

This means that we can take $\mathbf{v}_3 = \begin{bmatrix} 0 \\ 1 \\ 0 \end{bmatrix}$. □

The rank theorem and the rank nullity theorem

In this section we are interested in properties of the dimensions of the column space and the null space of a matrix. As we will see, these two numbers provide an important piece of information about the matrix.

Definition 2.54. The dimension of $\mathbf{C}(A)$ is called the *rank* of A and is denoted rank(A).

Example 2.55. In Example 2.33 we show that the set $\left\{\begin{bmatrix}1\\1\\0\end{bmatrix},\begin{bmatrix}1\\0\\1\end{bmatrix}\right\}$ is a basis of the subspace

$$\operatorname{Span}\left\{\begin{bmatrix}1\\3\\-2\end{bmatrix},\begin{bmatrix}3\\1\\2\end{bmatrix},\begin{bmatrix}1\\1\\0\end{bmatrix},\begin{bmatrix}1\\0\\1\end{bmatrix}\right\}.$$

Consequently,

$$\begin{aligned}\operatorname{rank}\left(\begin{bmatrix}1&3&1&1\\3&1&1&0\\-2&2&0&1\end{bmatrix}\right)&=\dim\mathbf{C}\left(\begin{bmatrix}1&3&1&1\\3&1&1&0\\-2&2&0&1\end{bmatrix}\right)\\&=\dim\operatorname{Span}\left\{\begin{bmatrix}1\\3\\-2\end{bmatrix},\begin{bmatrix}3\\1\\2\end{bmatrix},\begin{bmatrix}1\\1\\0\end{bmatrix},\begin{bmatrix}1\\0\\1\end{bmatrix}\right\}\\&=\dim\operatorname{Span}\left\{\begin{bmatrix}1\\1\\0\end{bmatrix},\begin{bmatrix}1\\0\\1\end{bmatrix}\right\}=2.\end{aligned}$$

Here is an important and somewhat unexpected property of the rank of a matrix.

Theorem 2.56 (The rank theorem). *For any matrix A we have*

$$\operatorname{rank}(A)=\operatorname{rank}(A^T).$$

Proof of a particular case. Suppose that

$$A=\begin{bmatrix}a_{11}&a_{12}&a_{13}&a_{14}&a_{15}\\a_{21}&a_{22}&a_{23}&a_{24}&a_{25}\\a_{31}&a_{32}&a_{33}&a_{34}&a_{35}\\a_{41}&a_{42}&a_{43}&b_{44}&a_{45}\end{bmatrix}\sim\begin{bmatrix}1&0&a&0&c\\0&1&b&0&d\\0&0&0&1&e\\0&0&0&0&0\end{bmatrix}.$$

Then, by Theorem 2.32, we have rank$(A)=\dim\mathbf{C}(A)=3$.

Since the vectors $\begin{bmatrix}1\\0\\a\\0\\c\end{bmatrix},\begin{bmatrix}0\\1\\b\\0\\d\end{bmatrix}$, and $\begin{bmatrix}0\\0\\0\\1\\e\end{bmatrix}$ are linearly independent, they form a basis of $\mathbf{C}(A^T)$, by Theorem 2.38 applied to the matrix A^T. Consequently, rank$(A^T)=\dim\mathbf{C}(A^T)=3$. □

Example 2.57. Verify the rank theorem for the matrix

$$A = \begin{bmatrix} 1 & 2 & 0 & 2 \\ 3 & 1 & 2 & 3 \\ 2 & -1 & 2 & 1 \\ 5 & 5 & 2 & 7 \\ 5 & 0 & 4 & 4 \end{bmatrix}.$$

Solution. The reduced row echelon form of the matrix A is

$$\begin{bmatrix} 1 & 0 & \frac{4}{5} & \frac{4}{5} \\ 0 & 1 & -\frac{2}{5} & \frac{3}{5} \\ 0 & 0 & 0 & 0 \\ 0 & 0 & 0 & 0 \\ 0 & 0 & 0 & 0 \end{bmatrix}.$$

Since the matrix has 2 pivot columns, we have $\operatorname{rank}(A) = 2$.

The reduced row echelon form of the matrix

$$A^T = \begin{bmatrix} 1 & 3 & 2 & 5 & 5 \\ 2 & 1 & -1 & 5 & 0 \\ 0 & 2 & 2 & 2 & 4 \\ 2 & 3 & 1 & 7 & 4 \end{bmatrix}$$

is

$$\begin{bmatrix} 1 & 0 & -1 & 2 & -1 \\ 0 & 1 & 1 & 1 & 2 \\ 0 & 0 & 0 & 0 & 0 \\ 0 & 0 & 0 & 0 & 0 \end{bmatrix}.$$

This matrix has 2 pivot columns and thus $\operatorname{rank}(A^T) = 2$. □

Definition 2.58. The dimension of the subspace $\mathbf{N}(A)$ is called the *nullity* of A and is denoted by $\operatorname{nullity}(A)$.

The following important theorem connects the rank and nullity of a matrix with its size.

Theorem 2.59 (The rank-nullity theorem)**.** *Let A be an $m \times n$ matrix. Then*

$$\operatorname{rank}(A) + \operatorname{nullity}(A) = n.$$

Proof. Let R be the reduced row echelon form of a matrix A. The equations $A\mathbf{x} = \mathbf{0}$ and $R\mathbf{x} = \mathbf{0}$ are equivalent, that is, a vector $\mathbf{x}$ in $\mathbb{R}^n$ satisfies the equation $A\mathbf{x} = \mathbf{0}$ if and only if it satisfies the equation $R\mathbf{x} = \mathbf{0}$. The number of special solutions the equation $R\mathbf{x} = \mathbf{0}$ has is the same as the number of free variables of the equation $R\mathbf{x} = \mathbf{0}$ and is the same as the number of nonpivot columns in R. Consequently,

$$\begin{aligned}\begin{pmatrix}\text{the number of special solutions}\\ \text{of the equation } A\mathbf{x} = \mathbf{0}\end{pmatrix} &= \begin{pmatrix}\text{the number of free variables}\\ \text{of the equation } A\mathbf{x} = \mathbf{0}\end{pmatrix}\\ &= \begin{pmatrix}\text{the number of nonpivot columns}\\ \text{of the matrix } A\end{pmatrix}\\ &= n - \begin{pmatrix}\text{the number of pivot columns}\\ \text{of the matrix } A\end{pmatrix}.\end{aligned}$$

□

The solutions of the equation $A\mathbf{x} = \mathbf{0}$ are not affected by elementary row operations performed on A. Consequently, if B is a matrix obtained from a matrix A by elementary row operations, then $\mathbf{N}(A) = \mathbf{N}(B)$. This observation is useful when we need to find $\mathbf{N}(A)$.

Example 2.60. Verify the rank-nullity theorem for the matrix

$$A = \begin{bmatrix} 1 & 0 & 2 & 3 & 1\\ 1 & 1 & 1 & 1 & 2\\ 2 & 1 & 3 & 4 & 3\\ 1 & 2 & 0 & -1 & 3\end{bmatrix}.$$

Solution. Since

$$\begin{bmatrix} 1 & 0 & 2 & 3 & 1 & 0\\ 1 & 1 & 1 & 1 & 2 & 0\\ 2 & 1 & 3 & 4 & 3 & 0\\ 1 & 2 & 0 & -1 & 3 & 0\end{bmatrix} \sim \begin{bmatrix} 1 & 0 & 2 & 3 & 1 & 0\\ 0 & 1 & -1 & -2 & 1 & 0\\ 0 & 0 & 0 & 0 & 0 & 0\\ 0 & 0 & 0 & 0 & 0 & 0\end{bmatrix},$$

an element $\begin{bmatrix} x_1\\ x_2\\ x_3\\ x_4\\ x_5\end{bmatrix}$ from $\mathbf{N}(A)$ is characterized by the system of equations

$$\begin{cases} x_1 + \phantom{x_2 -{}} 2x_3 + 3x_4 + x_5 = 0\\ \phantom{x_1 +{}} x_2 - x_3 - 2x_4 + x_5 = 0\end{cases}.$$

Consequently, every element of $\mathbf{N}(A)$ is of the form

$$\begin{bmatrix} x_1 \\ x_2 \\ x_3 \\ x_4 \\ x_5 \end{bmatrix} = \begin{bmatrix} -2x_3 - 3x_4 - x_5 \\ x_3 + 2x_4 - x_5 \\ x_3 \\ x_4 \\ x_5 \end{bmatrix} = x_3 \begin{bmatrix} -2 \\ 1 \\ 1 \\ 0 \\ 0 \end{bmatrix} + x_4 \begin{bmatrix} -3 \\ 2 \\ 0 \\ 1 \\ 0 \end{bmatrix} + x_5 \begin{bmatrix} -1 \\ -1 \\ 0 \\ 0 \\ 1 \end{bmatrix}.$$

The equation $A\mathbf{x} = \mathbf{0}$ has 3 special solutions:

$$\begin{bmatrix} -2 \\ 1 \\ 1 \\ 0 \\ 0 \end{bmatrix}, \begin{bmatrix} -3 \\ 2 \\ 0 \\ 1 \\ 0 \end{bmatrix}, \begin{bmatrix} -1 \\ -1 \\ 0 \\ 0 \\ 1 \end{bmatrix}.$$

Since the matrix A has 2 pivot columns, $\operatorname{rank}(A) = 2$ and $n = 5$, the equality in the rank-nullity theorem holds. □

Fundamental subspaces of a matrix

In order to answer questions about a subspace $\operatorname{Span}\{\mathbf{c}_1, \dots, \mathbf{c}_n\}$ we form the matrix $A = [\mathbf{c}_1 \ \dots \ \mathbf{c}_n]$ and then use tools developed for dealing with matrices. Sometimes the situation requires the opposite. We answer questions about a matrix

$$A = [\mathbf{c}_1 \ \dots \ \mathbf{c}_n]$$

by considering the subspace $\operatorname{Span}\{\mathbf{c}_1, \dots, \mathbf{c}_n\}$ and its properties.

We note that the rank theorem can be formulated in terms of the dimensions of column spaces.

Theorem 2.61. *For any matrix A we have*

$$\dim \mathbf{C}(A) = \dim \mathbf{C}(A^T).$$

Similarly, the rank-nullity theorem can be formulated in terms of the dimensions of the column space and the null space.

Theorem 2.62. *For any $m \times n$ matrix we have*

$$\dim \mathbf{C}(A) + \dim \mathbf{N}(A) = n.$$

Definition 2.63. Let A be an $m \times n$ matrix. The four subspaces

$$\mathbf{C}(A), \quad \mathbf{N}(A), \quad \mathbf{C}(A^T), \quad \text{and} \quad \mathbf{N}(A^T)$$

are called the *fundamental subspaces* of the matrix A.

The following observations are useful when working with the fundamental subspaces of a matrix:

Let A be a $m \times n$ matrix.

(a) $\mathbf{C}(A)$ is a subspace of $\mathbf{R}^m$.

(b) $\mathbf{N}(A)$ is a subspace of $\mathbf{R}^n$ and we have $\dim \mathbf{N}(A) = n - \dim \mathbf{C}(A)$.

(c) $\mathbf{C}(A^T)$ is a subspace of $\mathbf{R}^n$ and we have $\dim \mathbf{C}(A) = \dim \mathbf{C}(A^T)$.

(d) $\mathbf{N}(A^T)$ is a subspace of $\mathbf{R}^m$ and we have $\dim \mathbf{N}(A^T) = m - \dim \mathbf{C}(A^T)$.

Example 2.64. We consider the matrix

$$A = \begin{bmatrix} 1 & 1 & 1 & 3 \\ 1 & 0 & 3 & 1 \\ 0 & 1 & -2 & 2 \end{bmatrix}.$$

Find bases in the fundamental subspaces $\mathbf{C}(A)$, $\mathbf{N}(A)$, $\mathbf{C}(A^T)$, and $\mathbf{N}(A^T)$. Verify the rank theorem for the matrix A and the rank-nullity theorem for the matrices A and A^T.

Solution. Since

$$\begin{bmatrix} A & \mathbf{0} \end{bmatrix} = \begin{bmatrix} 1 & 1 & 1 & 3 & 0 \\ 1 & 0 & 3 & 1 & 0 \\ 0 & 1 & -2 & 2 & 0 \end{bmatrix} \sim \begin{bmatrix} 1 & 0 & 3 & 1 & 0 \\ 0 & 1 & -2 & 2 & 0 \\ 0 & 0 & 0 & 0 & 0 \end{bmatrix},$$

the general solution of the equation $A\mathbf{x} = \mathbf{0}$ is

$$\begin{bmatrix} x_1 \\ x_2 \\ x_3 \\ x_4 \end{bmatrix} = \begin{bmatrix} -3x_3 - x_4 \\ 2x_3 - 2x_4 \\ x_3 \\ x_4 \end{bmatrix} = x_3 \begin{bmatrix} -3 \\ 2 \\ 1 \\ 0 \end{bmatrix} + x_4 \begin{bmatrix} -1 \\ -2 \\ 0 \\ 1 \end{bmatrix}.$$

Hence

$$\mathbf{N}(A) = \text{Span}\left\{\begin{bmatrix}-3\\2\\1\\0\end{bmatrix}, \begin{bmatrix}-1\\-2\\0\\1\end{bmatrix}\right\}$$

and $\left\{\begin{bmatrix}-3\\2\\1\\0\end{bmatrix}, \begin{bmatrix}-1\\-2\\0\\1\end{bmatrix}\right\}$ is a basis in $\mathbf{N}(A)$.

Moreover,

$$\mathbf{C}(A) = \text{Span}\left\{\begin{bmatrix}1\\1\\0\end{bmatrix}, \begin{bmatrix}1\\0\\1\end{bmatrix}\right\}$$

and $\left\{\begin{bmatrix}1\\1\\0\end{bmatrix}, \begin{bmatrix}1\\0\\1\end{bmatrix}\right\}$ is a basis in $\mathbf{C}(A)$.

Now, since

$$\begin{bmatrix}A^T & \mathbf{0}\end{bmatrix} = \begin{bmatrix}1 & 1 & 0 & 0\\1 & 0 & 1 & 0\\1 & 3 & -2 & 0\\3 & 1 & 2 & 0\end{bmatrix} \sim \begin{bmatrix}1 & 0 & 1 & 0\\0 & 1 & -1 & 0\\0 & 0 & 0 & 0\\0 & 0 & 0 & 0\end{bmatrix},$$

a vector $\begin{bmatrix}y_1\\y_2\\y_3\end{bmatrix}$ is in $\mathbf{N}(A^T)$ if

$$\begin{bmatrix}y_1\\y_2\\y_3\end{bmatrix} = \begin{bmatrix}-y_3\\y_3\\y_3\end{bmatrix} = y_3\begin{bmatrix}-1\\1\\1\end{bmatrix}.$$

This means that

$$\mathbf{N}(A^T) = \text{Span}\left\{\begin{bmatrix}-1\\1\\1\end{bmatrix}\right\}$$

and $\left\{\begin{bmatrix}-1\\1\\1\end{bmatrix}\right\}$ is a basis in $\mathbf{N}(A^T)$.

Moreover

$$\mathbf{C}(A^T) = \text{Span}\left\{\begin{bmatrix}1\\1\\1\\3\end{bmatrix}, \begin{bmatrix}1\\0\\3\\1\end{bmatrix}\right\}$$

and $\left\{ \begin{bmatrix} 1 \\ 1 \\ 1 \\ 3 \end{bmatrix}, \begin{bmatrix} 1 \\ 0 \\ 3 \\ 1 \end{bmatrix} \right\}$ is a basis in $\mathbf{C}(A^T)$.

Finally, we note that

$$\dim \mathbf{C}(A) + \dim \mathbf{N}(A) = 2 + 2 = 4,$$

$$\dim \mathbf{C}(A^T) + \dim \mathbf{N}(A^T) = 2 + 1 = 3,$$

and

$$\dim \mathbf{C}(A) = \dim \mathbf{C}(A^T) = 2.$$

□

We close this section with a theorem that connects some observations made at different places of this chapter.

Theorem 2.65. *For an $n \times n$ matrix A the following conditions are equivalent:*

(a) *The rank of the matrix A is n;*

(b) *The columns of the matrix A are linearly independent;*

(c) *The columns of the matrix A^T are linearly independent;*

(d) $\mathbf{N}(A) = \{\mathbf{0}\}$;

(e) *The matrix A is invertible.*

Proof. Equivalence of (a), (b), and (c) follows from the rank theorem (Theorem 2.56) and Theorem 2.49. Equivalence of (b) and (d) follows from the definition of the linearly independent vectors. Finally, equivalence of (b) and (e) follows from Theorem 2.17. □

Change of basis

Some applications require working with different bases. It is then important to have a tool that will allow us to switch between different bases. More precisely, if we know the coefficients of a vector with respect to one basis, we want to find the coefficients of that vector with respect to another basis. This can be done with the aid of the transition matrix.

Theorem 2.66. *Let $\mathscr{B} = \{\mathbf{v}_1, \dots, \mathbf{v}_m\}$ and $\mathscr{C} = \{\mathbf{w}_1, \dots, \mathbf{w}_m\}$ be bases of the same vector subspace of $\mathbb{R}^n$. Then the equality*

$$x_1\mathbf{v}_1 + \cdots + x_m\mathbf{v}_m = y_1\mathbf{w}_1 + \cdots + y_m\mathbf{w}_m \tag{2.12}$$

is equivalent to the equality

$$\begin{bmatrix} y_1 \\ \vdots \\ y_m \end{bmatrix} = B \begin{bmatrix} x_1 \\ \vdots \\ x_m \end{bmatrix}, \tag{2.13}$$

where B is the $m \times m$ matrix whose j-th column $\begin{bmatrix} b_{1j} \\ \vdots \\ b_{mj} \end{bmatrix}$ is defined by the equation

$$\mathbf{v}_j = b_{1j}\mathbf{w}_1 + \cdots + b_{mj}\mathbf{w}_m = \begin{bmatrix} \mathbf{w}_1 & \dots & \mathbf{w}_m \end{bmatrix} \begin{bmatrix} b_{1j} \\ \vdots \\ b_{mj} \end{bmatrix}.$$

Proof for $m = 3$. The equations

$$\begin{aligned} \mathbf{v}_1 &= b_{11}\mathbf{w}_1 + b_{21}\mathbf{w}_2 + b_{31}\mathbf{w}_3 \\ \mathbf{v}_2 &= b_{12}\mathbf{w}_1 + b_{22}\mathbf{w}_2 + b_{32}\mathbf{w}_3 \\ \mathbf{v}_3 &= b_{13}\mathbf{w}_1 + b_{23}\mathbf{w}_2 + b_{33}\mathbf{w}_3 \end{aligned}$$

are equivalent to the matrix equation

$$\begin{bmatrix} \mathbf{v}_1 & \mathbf{v}_2 & \mathbf{v}_3 \end{bmatrix} = \begin{bmatrix} \mathbf{w}_1 & \mathbf{w}_2 & \mathbf{w}_3 \end{bmatrix} \begin{bmatrix} b_{11} & b_{12} & b_{13} \\ b_{21} & b_{22} & b_{23} \\ b_{31} & b_{32} & b_{33} \end{bmatrix}.$$

Hence

$$\begin{aligned} x_1\mathbf{v}_1 + x_2\mathbf{v}_2 + x_3\mathbf{v}_3 &= \begin{bmatrix} \mathbf{v}_1 & \mathbf{v}_2 & \mathbf{v}_3 \end{bmatrix} \begin{bmatrix} x_1 \\ x_2 \\ x_3 \end{bmatrix} \\ &= \begin{bmatrix} \mathbf{w}_1 & \mathbf{w}_2 & \mathbf{w}_3 \end{bmatrix} \begin{bmatrix} b_{11} & b_{12} & b_{13} \\ b_{21} & b_{22} & b_{23} \\ b_{31} & b_{32} & b_{33} \end{bmatrix} \begin{bmatrix} x_1 \\ x_2 \\ x_3 \end{bmatrix} \\ &= \begin{bmatrix} \mathbf{w}_1 & \mathbf{w}_2 & \mathbf{w}_3 \end{bmatrix} \begin{bmatrix} y_1 \\ y_2 \\ y_3 \end{bmatrix} \\ &= y_1\mathbf{w}_1 + y_2\mathbf{w}_2 + y_3\mathbf{w}_3. \end{aligned}$$

□

Definition 2.67. Let $\{\mathbf{v}_1,\dots,\mathbf{v}_m\}$ and $\{\mathbf{w}_1,\dots,\mathbf{w}_m\}$ be bases of the same subspace of $\mathbb{R}^n$. The $m \times m$ matrix B from Theorem 2.66 is called the *transition matrix* from the basis $\{\mathbf{v}_1,\dots,\mathbf{v}_m\}$ to the basis $\{\mathbf{w}_1,\dots,\mathbf{w}_m\}$.

Theorem 2.68. *Let $\{\mathbf{v}_1,\dots,\mathbf{v}_m\}$ and $\{\mathbf{w}_1,\dots,\mathbf{w}_m\}$ be bases of the same subspace of $\mathbb{R}^n$. The transition matrix B from the basis $\{\mathbf{v}_1,\dots,\mathbf{v}_m\}$ to the basis $\{\mathbf{w}_1,\dots,\mathbf{w}_m\}$ is invertible and B^{-1} is the transition matrix from the basis $\{\mathbf{w}_1,\dots,\mathbf{w}_m\}$ to the basis $\{\mathbf{v}_1,\dots,\mathbf{v}_m\}$.*

Proof. The proof is obtained by applying Theorem 2.66 twice. □

Example 2.69. Show that $\mathscr{B} = \left\{ \begin{bmatrix} 2 \\ 1 \\ 1 \end{bmatrix}, \begin{bmatrix} 1 \\ 2 \\ -1 \end{bmatrix} \right\}$ and $\mathscr{C} = \left\{ \begin{bmatrix} 3 \\ 1 \\ 2 \end{bmatrix}, \begin{bmatrix} 1 \\ 4 \\ -3 \end{bmatrix} \right\}$ are bases of the same subspace of $\mathbb{R}^3$ and calculate the transition matrix from $\mathscr{B}$ to $\mathscr{C}$ and the transition matrix from $\mathscr{C}$ to $\mathscr{B}$.

Solution. Since

$$\begin{bmatrix} 3 & 1 & 2 & 1 \\ 1 & 4 & 1 & 2 \\ 2 & -3 & 1 & -1 \end{bmatrix} \sim \begin{bmatrix} 1 & 0 & \frac{7}{11} & \frac{2}{11} \\ 0 & 1 & \frac{1}{11} & \frac{5}{11} \\ 0 & 0 & 0 & 0 \end{bmatrix},$$

the transition matrix from $\mathscr{B}$ to $\mathscr{C}$ is

$$\begin{bmatrix} \frac{7}{11} & \frac{2}{11} \\ \frac{1}{11} & \frac{5}{11} \end{bmatrix}.$$

On the other hand, since

$$\begin{bmatrix} 2 & 1 & 3 & 1 \\ 1 & 2 & 1 & 4 \\ 1 & -1 & 2 & -3 \end{bmatrix} \sim \begin{bmatrix} 1 & 0 & \frac{5}{3} & -\frac{2}{3} \\ 0 & 1 & -\frac{1}{3} & \frac{7}{3} \\ 0 & 0 & 0 & 0 \end{bmatrix},$$

the transition matrix from $\mathscr{C}$ to $\mathscr{B}$ is

$$\begin{bmatrix} \frac{5}{3} & -\frac{2}{3} \\ -\frac{1}{3} & \frac{7}{3} \end{bmatrix}.$$

Note that we have

$$\begin{bmatrix} \frac{5}{3} & -\frac{2}{3} \\ -\frac{1}{3} & \frac{7}{3} \end{bmatrix} = \begin{bmatrix} \frac{7}{11} & \frac{2}{11} \\ \frac{1}{11} & \frac{5}{11} \end{bmatrix}^{-1}.$$

□

The next theorem tells us how the transition matrix between two bases in $\mathbb{R}^n$ can be calculated.

Theorem 2.70. *The transition matrix from a basis* $\{\mathbf{v}_1,\dots,\mathbf{v}_n\}$ *to a basis* $\{\mathbf{w}_1,\dots,\mathbf{w}_n\}$ *in* $\mathbb{R}^n$ *is the matrix*

$$B = \begin{bmatrix}\mathbf{w}_1 & \dots & \mathbf{w}_n\end{bmatrix}^{-1}\begin{bmatrix}\mathbf{v}_1 & \dots & \mathbf{v}_n\end{bmatrix}.$$

Proof. It suffices to note that

$$x_1\mathbf{v}_1 + \dots + x_n\mathbf{v}_n = y_1\mathbf{w}_1 + \dots + y_n\mathbf{w}_n$$

is equivalent to

$$\begin{bmatrix}\mathbf{v}_1 & \dots & \mathbf{v}_n\end{bmatrix}\begin{bmatrix}x_1 \\ \vdots \\ x_n\end{bmatrix} = \begin{bmatrix}\mathbf{w}_1 & \dots & \mathbf{w}_n\end{bmatrix}\begin{bmatrix}y_1 \\ \vdots \\ y_n\end{bmatrix}.$$

Since the vectors $\mathbf{w}_1,\dots,\mathbf{w}_n$ are linearly independent, the matrix $\begin{bmatrix}\mathbf{w}_1 & \dots & \mathbf{w}_n\end{bmatrix}$ is invertible and thus we have

$$\begin{bmatrix}\mathbf{w}_1 & \dots & \mathbf{w}_n\end{bmatrix}^{-1}\begin{bmatrix}\mathbf{v}_1 & \dots & \mathbf{v}_n\end{bmatrix}\begin{bmatrix}x_1 \\ \vdots \\ x_n\end{bmatrix} = \begin{bmatrix}y_1 \\ \vdots \\ y_n\end{bmatrix}.$$

□

Note that the above argument shows that the transition matrix between two bases in $\mathbb{R}^n$ is unique. In the following corollary we give two special cases of Theorem 2.70.

Corollary 2.71.

(a) *The transition matrix from an arbitrary basis* $\{\mathbf{v}_1, \dots, \mathbf{v}_n\}$ *to the standard basis* $\{\mathbf{e}_1, \dots, \mathbf{e}_n\}$ *is the matrix*

$$\begin{bmatrix} \mathbf{v}_1 & \dots & \mathbf{v}_n \end{bmatrix}.$$

(b) *The transition matrix from the standard basis* $\{\mathbf{e}_1, \dots, \mathbf{e}_n\}$ *to an arbitrary basis* $\{\mathbf{w}_1, \dots, \mathbf{w}_n\}$ *is the matrix*

$$\begin{bmatrix} \mathbf{w}_1 & \dots & \mathbf{w}_n \end{bmatrix}^{-1}.$$

Example 2.72. Find the transition matrix from the basis $\left\{ \begin{bmatrix} 2 \\ 1 \\ 1 \end{bmatrix}, \begin{bmatrix} 1 \\ 2 \\ 3 \end{bmatrix}, \begin{bmatrix} 1 \\ 0 \\ 1 \end{bmatrix} \right\}$ to the basis $\left\{ \begin{bmatrix} 1 \\ 1 \\ 1 \end{bmatrix}, \begin{bmatrix} 1 \\ 2 \\ 1 \end{bmatrix}, \begin{bmatrix} 1 \\ 1 \\ 2 \end{bmatrix} \right\}$.

Solution. We need to calculate the matrix

$$\begin{bmatrix} 1 & 1 & 1 \\ 1 & 2 & 1 \\ 1 & 1 & 2 \end{bmatrix}^{-1} \begin{bmatrix} 2 & 1 & 1 \\ 1 & 2 & 0 \\ 1 & 3 & 1 \end{bmatrix}.$$

Since

$$\begin{bmatrix} 1 & 1 & 1 & 2 & 1 & 1 \\ 1 & 2 & 1 & 1 & 2 & 0 \\ 1 & 1 & 2 & 1 & 3 & 1 \end{bmatrix} \sim \begin{bmatrix} 1 & 0 & 0 & 4 & -2 & 2 \\ 0 & 1 & 0 & -1 & 1 & -1 \\ 0 & 0 & 1 & -1 & 2 & 0 \end{bmatrix},$$

the transition matrix is

$$\begin{bmatrix} 4 & -2 & 2 \\ -1 & 1 & -1 \\ -1 & 2 & 0 \end{bmatrix}.$$

□

Exercises 2.3

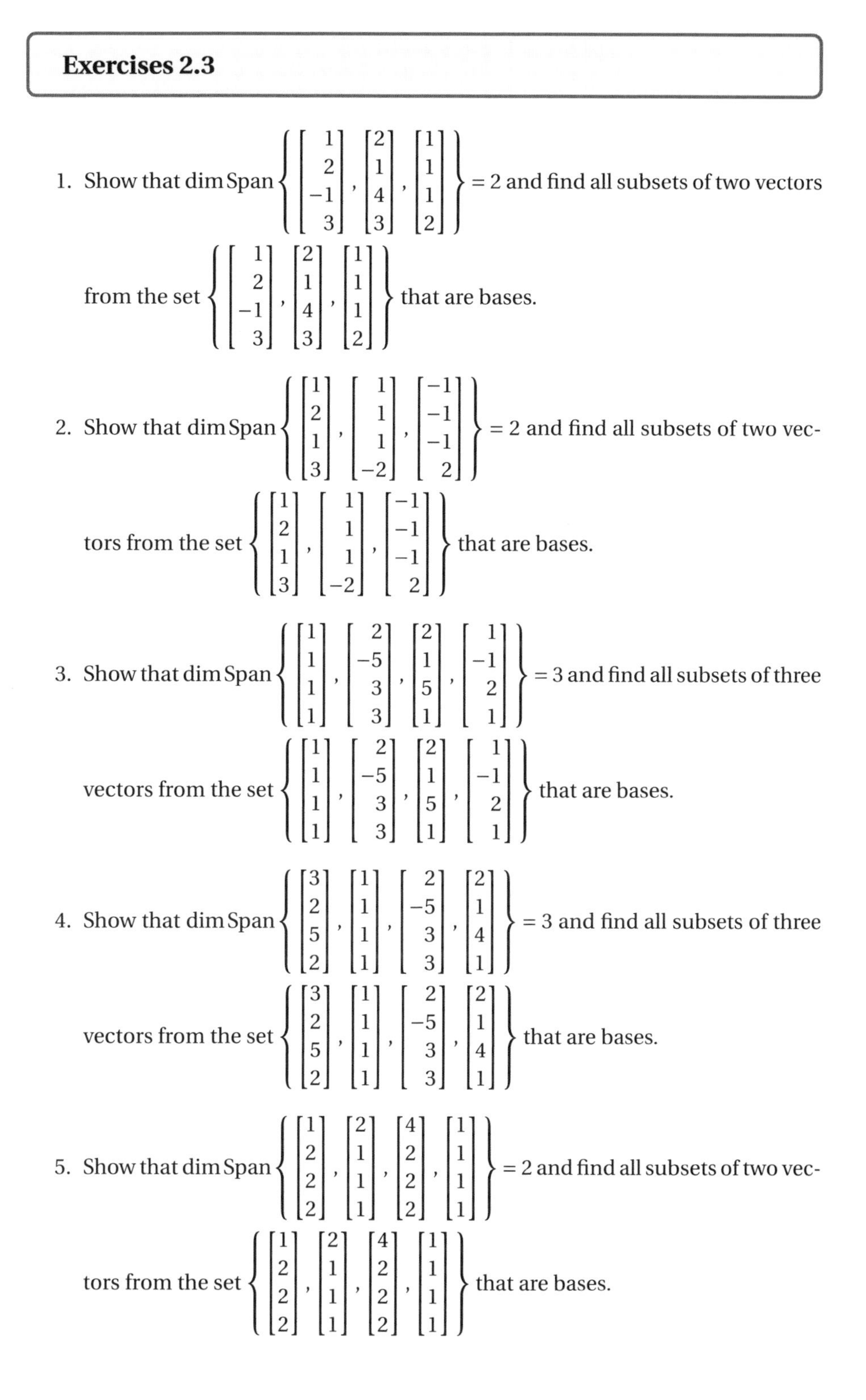

1. Show that $\dim \operatorname{Span}\left\{\begin{bmatrix}1\\2\\-1\\3\end{bmatrix},\begin{bmatrix}2\\1\\4\\3\end{bmatrix},\begin{bmatrix}1\\1\\1\\2\end{bmatrix}\right\}=2$ and find all subsets of two vectors from the set $\left\{\begin{bmatrix}1\\2\\-1\\3\end{bmatrix},\begin{bmatrix}2\\1\\4\\3\end{bmatrix},\begin{bmatrix}1\\1\\1\\2\end{bmatrix}\right\}$ that are bases.

2. Show that $\dim \operatorname{Span}\left\{\begin{bmatrix}1\\2\\1\\3\end{bmatrix},\begin{bmatrix}1\\1\\1\\-2\end{bmatrix},\begin{bmatrix}-1\\-1\\-1\\2\end{bmatrix}\right\}=2$ and find all subsets of two vectors from the set $\left\{\begin{bmatrix}1\\2\\1\\3\end{bmatrix},\begin{bmatrix}1\\1\\1\\-2\end{bmatrix},\begin{bmatrix}-1\\-1\\-1\\2\end{bmatrix}\right\}$ that are bases.

3. Show that $\dim \operatorname{Span}\left\{\begin{bmatrix}1\\1\\1\\1\end{bmatrix},\begin{bmatrix}2\\-5\\3\\3\end{bmatrix},\begin{bmatrix}2\\1\\5\\1\end{bmatrix},\begin{bmatrix}1\\-1\\2\\1\end{bmatrix}\right\}=3$ and find all subsets of three vectors from the set $\left\{\begin{bmatrix}1\\1\\1\\1\end{bmatrix},\begin{bmatrix}2\\-5\\3\\3\end{bmatrix},\begin{bmatrix}2\\1\\5\\1\end{bmatrix},\begin{bmatrix}1\\-1\\2\\1\end{bmatrix}\right\}$ that are bases.

4. Show that $\dim \operatorname{Span}\left\{\begin{bmatrix}3\\2\\5\\2\end{bmatrix},\begin{bmatrix}1\\1\\1\\1\end{bmatrix},\begin{bmatrix}2\\-5\\3\\3\end{bmatrix},\begin{bmatrix}2\\1\\4\\1\end{bmatrix}\right\}=3$ and find all subsets of three vectors from the set $\left\{\begin{bmatrix}3\\2\\5\\2\end{bmatrix},\begin{bmatrix}1\\1\\1\\1\end{bmatrix},\begin{bmatrix}2\\-5\\3\\3\end{bmatrix},\begin{bmatrix}2\\1\\4\\1\end{bmatrix}\right\}$ that are bases.

5. Show that $\dim \operatorname{Span}\left\{\begin{bmatrix}1\\2\\2\\2\end{bmatrix},\begin{bmatrix}2\\1\\1\\1\end{bmatrix},\begin{bmatrix}4\\2\\2\\2\end{bmatrix},\begin{bmatrix}1\\1\\1\\1\end{bmatrix}\right\}=2$ and find all subsets of two vectors from the set $\left\{\begin{bmatrix}1\\2\\2\\2\end{bmatrix},\begin{bmatrix}2\\1\\1\\1\end{bmatrix},\begin{bmatrix}4\\2\\2\\2\end{bmatrix},\begin{bmatrix}1\\1\\1\\1\end{bmatrix}\right\}$ that are bases.

6. Show that $\dim \operatorname{Span}\left\{\begin{bmatrix}1\\1\\1\\1\end{bmatrix}, \begin{bmatrix}3\\1\\3\\5\end{bmatrix}, \begin{bmatrix}2\\1\\2\\3\end{bmatrix}, \begin{bmatrix}1\\3\\1\\-1\end{bmatrix}\right\} = 2$ and find all subsets of two vectors from the set $\left\{\begin{bmatrix}1\\1\\1\\1\end{bmatrix}, \begin{bmatrix}3\\1\\3\\5\end{bmatrix}, \begin{bmatrix}2\\1\\2\\3\end{bmatrix}, \begin{bmatrix}1\\3\\1\\-1\end{bmatrix}\right\}$ that are bases.

7. Consider the matrix $A = \begin{bmatrix}\mathbf{c}_1 & \mathbf{c}_2 & \mathbf{c}_3 & \mathbf{c}_4 & \mathbf{c}_5\end{bmatrix} = \begin{bmatrix}1 & 2 & 1 & 4 & 3\\1 & 2 & 2 & 3 & 2\\2 & 4 & 1 & 9 & 7\end{bmatrix}$.

   (a) Show that $\dim \mathbf{C}(A) = 2$.

   (b) Determine all pairs of columns from the set $\{\mathbf{c}_1, \mathbf{c}_2, \mathbf{c}_3, \mathbf{c}_4, \mathbf{c}_5\}$ that are bases of $\mathbf{C}(A) = 2$.

8. Consider the matrix $A = \begin{bmatrix}\mathbf{c}_1 & \mathbf{c}_2 & \mathbf{c}_3 & \mathbf{c}_4\end{bmatrix} = \begin{bmatrix}1 & 2 & 1 & 1\\2 & 1 & 3 & 1\\4 & -1 & 1 & 1\end{bmatrix}$.

   (a) Show that $\dim \mathbf{C}(A) = 3$.

   (b) Determine all sets of three columns from the set $\{\mathbf{c}_1, \mathbf{c}_2, \mathbf{c}_3, \mathbf{c}_4\}$ that are bases of $\mathbf{C}(A) = 2$.

9. Consider the matrix $A = \begin{bmatrix}\mathbf{c}_1 & \mathbf{c}_2 & \mathbf{c}_3 & \mathbf{c}_4\end{bmatrix} = \begin{bmatrix}1 & 1 & 1 & 1\\2 & 1 & 1 & 1\\1 & 1 & 2 & 3\\1 & 2 & 1 & 0\end{bmatrix}$. Determine $\dim \mathbf{C}(A)$ and all subsets of $\{\mathbf{c}_1, \mathbf{c}_2, \mathbf{c}_3, \mathbf{c}_4\}$ that are bases of $\mathbf{C}(A) = 2$.

10. Consider the matrix $A = \begin{bmatrix}\mathbf{c}_1 & \mathbf{c}_2 & \mathbf{c}_3 & \mathbf{c}_4\end{bmatrix} = \begin{bmatrix}3 & 1 & 1 & 1\\2 & 2 & 1 & 1\\1 & 3 & 1 & 3\\4 & 0 & 1 & 0\end{bmatrix}$. Determine $\dim \mathbf{C}(A)$ and all subsets of $\{\mathbf{c}_1, \mathbf{c}_2, \mathbf{c}_3, \mathbf{c}_4\}$ that are bases of $\mathbf{C}(A) = 2$.

Verify the rank theorem for the given matrix.

11. $\begin{bmatrix} 1 & 2 & 3 & 2 \\ 3 & 1 & 2 & 3 \\ 5 & 5 & 8 & 7 \end{bmatrix}$

12. $\begin{bmatrix} 1 & 2 & 5 \\ 3 & 1 & 5 \\ 1 & 4 & 9 \\ 2 & 4 & 10 \end{bmatrix}$

13. $\begin{bmatrix} 1 & 2 & 1 \\ 2 & 1 & 3 \\ 1 & -1 & 2 \\ 3 & 3 & 4 \\ 5 & 4 & 7 \end{bmatrix}$

14. $\begin{bmatrix} 1 & 1 & 3 & 1 & 2 \\ 1 & 3 & 1 & 2 & 1 \\ 2 & 4 & 4 & 3 & 3 \\ 3 & 5 & 3 & 8 & 5 \\ 1 & 1 & 1 & 3 & 2 \end{bmatrix}$

The reduced row echelon form of the matrix A is given. Verify the rank-nullity theorem for the matrix A.

15. $A \sim \begin{bmatrix} 1 & 0 & 4 & 0 \\ 0 & 1 & 2 & 0 \\ 0 & 0 & 0 & 1 \end{bmatrix}$

16. $A \sim \begin{bmatrix} 1 & 3 & 0 & 2 \\ 0 & 0 & 1 & 5 \\ 0 & 0 & 0 & 0 \end{bmatrix}$

17. $A \sim \begin{bmatrix} 1 & 0 & 3 & 0 & 2 \\ 0 & 1 & 7 & 0 & 4 \\ 0 & 0 & 0 & 1 & 5 \\ 0 & 0 & 0 & 0 & 0 \end{bmatrix}$

18. $A \sim \begin{bmatrix} 1 & 3 & 0 \\ 0 & 0 & 1 \\ 0 & 0 & 0 \\ 0 & 0 & 0 \end{bmatrix}$

The reduced row echelon form of the matrix A is given. Determine bases in $\mathbf{C}(A)$ and $\mathbf{N}(A)$ and verify the rank-nullity theorem.

19. $A \sim \begin{bmatrix} 1 & a & 0 & b & 0 \\ 0 & 0 & 1 & c & 0 \\ 0 & 0 & 0 & 0 & 1 \\ 0 & 0 & 0 & 0 & 0 \end{bmatrix}$

20. $A \sim \begin{bmatrix} 1 & 0 & 0 & a & 0 \\ 0 & 1 & 0 & b & 0 \\ 0 & 0 & 1 & c & 0 \\ 0 & 0 & 0 & 0 & 1 \end{bmatrix}$

21. $A \sim \begin{bmatrix} 1 & 0 & a & 0 \\ 0 & 1 & b & 0 \\ 0 & 0 & 0 & 1 \\ 0 & 0 & 0 & 0 \\ 0 & 0 & 0 & 0 \end{bmatrix}$

22. $A \sim \begin{bmatrix} 1 & 0 & a & c \\ 0 & 1 & b & d \\ 0 & 0 & 0 & 0 \\ 0 & 0 & 0 & 0 \\ 0 & 0 & 0 & 0 \end{bmatrix}$

The reduced row echelon form of the matrix A is given. Determine the dimensions of the fundamental spaces of A.

23. $A \sim \begin{bmatrix} 1 & 0 & a & 0 & c \\ 0 & 1 & b & 0 & d \\ 0 & 0 & 0 & 1 & e \\ 0 & 0 & 0 & 0 & 0 \\ 0 & 0 & 0 & 0 & 0 \end{bmatrix}$

24. $A \sim \begin{bmatrix} 1 & a & 0 & 0 & 0 \\ 0 & 0 & 1 & 0 & 0 \\ 0 & 0 & 0 & 1 & 0 \\ 0 & 0 & 0 & 0 & 1 \\ 0 & 0 & 0 & 0 & 0 \end{bmatrix}$

25. $A \sim \begin{bmatrix} 1 & 0 & 0 & a & 0 & e & 0 \\ 0 & 1 & 0 & b & 0 & f & 0 \\ 0 & 0 & 1 & c & 0 & g & 0 \\ 0 & 0 & 0 & 0 & 1 & h & 0 \\ 0 & 0 & 0 & 0 & 0 & 0 & 1 \end{bmatrix}$

26. $A \sim \begin{bmatrix} 1 & 0 & a & 0 & 0 & 0 & c \\ 0 & 1 & b & 0 & 0 & 0 & d \\ 0 & 0 & 0 & 1 & 0 & 0 & e \\ 0 & 0 & 0 & 0 & 1 & 0 & f \\ 0 & 0 & 0 & 0 & 0 & 1 & g \end{bmatrix}$

Find bases in the fundamental subspaces of the given matrix.

27. $\begin{bmatrix} 3 & 1 & 1 \\ 1 & 1 & 2 \\ 1 & 3 & 7 \end{bmatrix}$

28. $\begin{bmatrix} 3 & 1 & 1 & 5 \\ 1 & 1 & 2 & 4 \\ 1 & 3 & 7 & 11 \end{bmatrix}$

29. $\begin{bmatrix} 2 & 1 & 1 & 3 \\ 1 & 2 & 3 & 1 \end{bmatrix}$

30. $A = \begin{bmatrix} 3 & 1 & 1 \\ 1 & 1 & 2 \\ 1 & 3 & 7 \\ 1 & 1 & 1 \end{bmatrix}$

31. Let A be $m \times n$ matrix. Show that the vector $\begin{bmatrix} x_1 \\ \vdots \\ x_m \end{bmatrix}$ is in $\mathbf{N}(A^T)$ if and only if $\begin{bmatrix} x_1 & \dots & x_m \end{bmatrix} A = \mathbf{0}$.

32. Suppose that the vector subspace $\mathscr{V}$ has basis $\{\mathbf{v}_1, \dots, \mathbf{v}_k\}$ and that $\mathbf{x} \neq \mathbf{0}$ is a vector in $\mathscr{V}$. Show that there exists a basis $\{\mathbf{x}, \mathbf{u}_1, \dots, \mathbf{u}_{k-1}\}$ where the vectors $\mathbf{u}_1, \dots, \mathbf{u}_{k-1}$ are in the basis $\{\mathbf{v}_1, \dots, \mathbf{v}_k\}$.

Show that $\mathscr{B}$ and $\mathscr{C}$ are bases of the same subspace of $\mathbb{R}^3$ and calculate the transition matrices from $\mathscr{B}$ to $\mathscr{C}$ and from $\mathscr{C}$ to $\mathscr{B}$.

33. $\mathscr{B} = \left\{ \begin{bmatrix} 3 \\ 1 \\ 7 \end{bmatrix}, \begin{bmatrix} 1 \\ 3 \\ 5 \end{bmatrix} \right\}, \mathscr{C} = \left\{ \begin{bmatrix} 1 \\ 1 \\ 3 \end{bmatrix}, \begin{bmatrix} 1 \\ 2 \\ 4 \end{bmatrix} \right\}$

34. $\mathscr{B} = \left\{ \begin{bmatrix} 2 \\ 1 \\ 4 \end{bmatrix}, \begin{bmatrix} 1 \\ 2 \\ 5 \end{bmatrix} \right\}, \mathscr{C} = \left\{ \begin{bmatrix} 3 \\ 1 \\ 5 \end{bmatrix}, \begin{bmatrix} 1 \\ 4 \\ 9 \end{bmatrix} \right\}$

Find the transition matrix from the basis $\mathscr{B}$ to the basis $\mathscr{C}$.

35. $\mathscr{B} = \left\{ \begin{bmatrix} 1 \\ 1 \end{bmatrix}, \begin{bmatrix} 2 \\ 3 \end{bmatrix} \right\}, \mathscr{C} = \left\{ \begin{bmatrix} 2 \\ 1 \end{bmatrix}, \begin{bmatrix} -1 \\ 1 \end{bmatrix} \right\}$

36. $\mathscr{B} = \left\{ \begin{bmatrix} 1 \\ 2 \end{bmatrix}, \begin{bmatrix} -2 \\ 1 \end{bmatrix} \right\}, \mathscr{C} = \left\{ \begin{bmatrix} 1 \\ 0 \end{bmatrix}, \begin{bmatrix} 0 \\ 1 \end{bmatrix} \right\}$

37. $\mathscr{B} = \left\{ \begin{bmatrix} 1 \\ 0 \end{bmatrix}, \begin{bmatrix} 0 \\ 1 \end{bmatrix} \right\}, \mathscr{C} = \left\{ \begin{bmatrix} 3 \\ 2 \end{bmatrix}, \begin{bmatrix} 4 \\ 5 \end{bmatrix} \right\}$

38. $\mathscr{B} = \left\{ \begin{bmatrix} 1 \\ 0 \end{bmatrix}, \begin{bmatrix} 0 \\ 1 \end{bmatrix} \right\}, \mathscr{C} = \left\{ \begin{bmatrix} 5 \\ 1 \end{bmatrix}, \begin{bmatrix} 7 \\ 3 \end{bmatrix} \right\}$

39. $\mathscr{B} = \left\{ \begin{bmatrix} 1 \\ 0 \\ 0 \end{bmatrix}, \begin{bmatrix} 0 \\ 1 \\ 0 \end{bmatrix}, \begin{bmatrix} 0 \\ 0 \\ 1 \end{bmatrix} \right\}, \mathscr{C} = \left\{ \begin{bmatrix} 1 \\ 1 \\ 1 \end{bmatrix}, \begin{bmatrix} 1 \\ 1 \\ 0 \end{bmatrix}, \begin{bmatrix} 0 \\ 1 \\ 1 \end{bmatrix} \right\}$

40. $\mathscr{B} = \left\{ \begin{bmatrix} 1 \\ 0 \\ 0 \end{bmatrix}, \begin{bmatrix} 0 \\ 1 \\ 0 \end{bmatrix}, \begin{bmatrix} 0 \\ 0 \\ 1 \end{bmatrix} \right\}, \mathscr{C} = \left\{ \begin{bmatrix} 2 \\ 1 \\ 1 \end{bmatrix}, \begin{bmatrix} 1 \\ 2 \\ 1 \end{bmatrix}, \begin{bmatrix} 1 \\ 1 \\ 2 \end{bmatrix} \right\}$

41. $\mathscr{B} = \left\{ \begin{bmatrix} 1 \\ 0 \\ 0 \end{bmatrix}, \begin{bmatrix} 1 \\ 1 \\ 0 \end{bmatrix}, \begin{bmatrix} 1 \\ 1 \\ 1 \end{bmatrix} \right\}, \mathscr{C} = \left\{ \begin{bmatrix} 0 \\ 1 \\ 0 \end{bmatrix}, \begin{bmatrix} 0 \\ 0 \\ 1 \end{bmatrix}, \begin{bmatrix} 2 \\ 1 \\ 1 \end{bmatrix} \right\}$

42. $\mathscr{B} = \left\{ \begin{bmatrix} 1 \\ 1 \\ 1 \end{bmatrix}, \begin{bmatrix} 1 \\ 2 \\ 2 \end{bmatrix}, \begin{bmatrix} 1 \\ 2 \\ 1 \end{bmatrix} \right\}, \mathscr{C} = \left\{ \begin{bmatrix} 1 \\ 0 \\ 1 \end{bmatrix}, \begin{bmatrix} 1 \\ 0 \\ 0 \end{bmatrix}, \begin{bmatrix} 1 \\ 1 \\ 0 \end{bmatrix} \right\}$

Chapter 3

Orthogonality in $\mathbb{R}^n$

In this chapter we discuss some geometric aspects of linear algebra. As we will see, geometric interpretations of algebraic concepts provide useful insights that help us solve algebraic problems. On the other hand, algebraic methods often give us powerful tools for dealing with problems that are geometric in nature.

3.1 The scalar product, norm, and distance in $\mathbb{R}^n$

We start by defining the scalar product of vectors in $\mathbb{R}^n$. This is a good example of a purely algebraic definition that turns out to have a clear geometric interpretation.

Definition 3.1. By the *scalar product* (or *dot product*) of vectors

$$\mathbf{u} = \begin{bmatrix} u_1 \\ u_2 \\ \vdots \\ u_n \end{bmatrix} \quad \text{and} \quad \mathbf{v} = \begin{bmatrix} v_1 \\ v_2 \\ \vdots \\ v_n \end{bmatrix},$$

denoted by $\mathbf{u} \cdot \mathbf{v}$, we mean the number

$$\mathbf{u} \cdot \mathbf{v} = u_1 v_1 + u_2 v_2 + \cdots + u_n v_n.$$

Example 3.2.

$$\begin{bmatrix} 1 \\ 3 \\ 7 \\ 4 \end{bmatrix} \cdot \begin{bmatrix} 2 \\ 5 \\ -2 \\ 3 \end{bmatrix} = 1 \cdot 2 + 3 \cdot 5 + 7 \cdot (-2) + 4 \cdot 3 = 15.$$

The following properties of the scalar product are direct consequences of the definition:

Theorem 3.3.

$$\mathbf{u}\cdot\mathbf{v} = \mathbf{v}\cdot\mathbf{u},$$

$$\mathbf{u}\cdot(\mathbf{v}+\mathbf{w}) = \mathbf{u}\cdot\mathbf{v}+\mathbf{u}\cdot\mathbf{w},$$

and

$$t(\mathbf{u}\cdot\mathbf{v}) = (t\mathbf{u})\cdot\mathbf{v} = \mathbf{u}\cdot(t\mathbf{v}),$$

where t is an arbitrary real number.

Note that the scalar product can be interpreted as matrix multiplication:

$$\mathbf{u}\cdot\mathbf{v} = \mathbf{u}^T\mathbf{v} = \mathbf{v}^T\mathbf{u}.$$

The scalar product of any vector with itself is always a nonnegative number. Indeed, we have

$$\begin{bmatrix} u_1 \\ u_2 \\ \vdots \\ u_n \end{bmatrix}\cdot\begin{bmatrix} u_1 \\ u_2 \\ \vdots \\ u_n \end{bmatrix} = u_1^2 + u_2^2 + \cdots + u_n^2 \ge 0.$$

The above also shows that $\mathbf{u}\cdot\mathbf{u} = 0$ only if $\mathbf{u} = \mathbf{0}$.

The scalar product is closely related to the norm of a vector.

Definition 3.4. By the *norm* (or *magnitude*) of a vector $\mathbf{u} = \begin{bmatrix} u_1 \\ u_2 \\ \vdots \\ u_n \end{bmatrix}$, denoted by $\|\mathbf{u}\|$, we mean the number

$$\|\mathbf{u}\| = \sqrt{\mathbf{u}\cdot\mathbf{u}} = \sqrt{u_1^2 + u_2^2 + \cdots + u_n^2}.$$

Example 3.5.

$$\left\|\begin{bmatrix} 1 \\ 2 \\ -2 \\ 4 \\ 0 \end{bmatrix}\right\| = \sqrt{1^2 + 2^2 + (-2)^2 + 4^2 + 0^2} = 5.$$

The expression in the definition of the norm should look familiar. If we think of a vector $\mathbf{u}$ as a point in $\mathbb{R}^n$, then the norm $\|\mathbf{u}\|$ is the distance of that point from the origin.

Definition 3.6. By the *distance* between vectors $\mathbf{u}$ and $\mathbf{v}$ we mean the number

$$d(\mathbf{u},\mathbf{v}) = \|\mathbf{u}-\mathbf{v}\|.$$

Note that

$$d(\mathbf{u},\mathbf{v}) = \|\mathbf{u}-\mathbf{v}\| = \|\mathbf{v}-\mathbf{u}\| = d(\mathbf{v},\mathbf{u}).$$

If

$$\mathbf{u} = \begin{bmatrix} u_1 \\ u_2 \\ \vdots \\ u_n \end{bmatrix} \quad \text{and} \quad \mathbf{v} = \begin{bmatrix} v_1 \\ v_2 \\ \vdots \\ v_n \end{bmatrix},$$

then

$$d(\mathbf{u},\mathbf{v}) = \sqrt{(u_1-v_1)^2 + (u_2-v_2)^2 + \cdots + (u_n-v_n)^2}.$$

Example 3.7.

$$d\left(\begin{bmatrix} 0 \\ 1 \\ 2 \\ 3 \\ 4 \end{bmatrix}, \begin{bmatrix} 4 \\ 3 \\ 2 \\ 1 \\ 0 \end{bmatrix}\right) = \sqrt{(0-4)^2 + (1-3)^2 + (2-2)^2 + (3-1)^2 + (4-0)^2} = 2\sqrt{10}.$$

If $\|\mathbf{u}\| = 1$, then we say that $\mathbf{u}$ is a *unit vector.* If $\mathbf{u}$ is a nonzero vector, then the vector $\frac{1}{\|\mathbf{u}\|}\mathbf{u}$ is a unit vector. If we multiply a nonzero vector $\mathbf{u}$ by $\frac{1}{\|\mathbf{u}\|}$, we say that we *normalize* the vector.

One of the most important consequences of having the scalar product in $\mathbb{R}^n$ is that we can define orthogonality of vectors. The definition of orthogonality is motivated by its geometric interpretation illustrated in Figure 3.1.

Definition 3.8. Two vectors $\mathbf{u}$ and $\mathbf{v}$ are called *orthogonal* if

$$\|\mathbf{u}+\mathbf{v}\| = \|\mathbf{u}-\mathbf{v}\|.$$

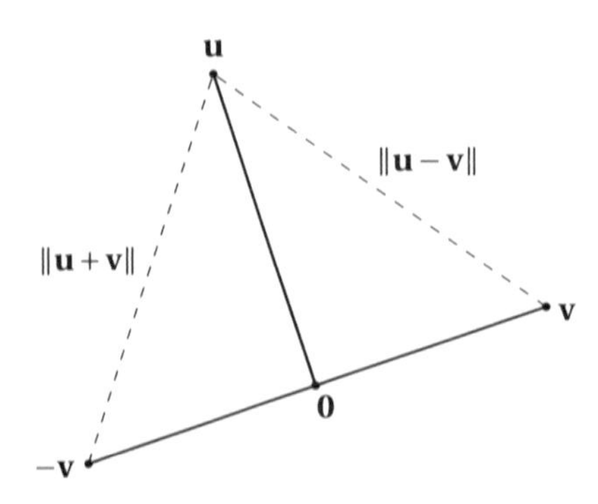

Figure 3.1: Vectors **u** and **v** are orthogonal.

Example 3.9. Since

$$\left\| \begin{bmatrix} 1 \\ 2 \\ -2 \\ 3 \\ 0 \end{bmatrix} + \begin{bmatrix} -1 \\ -4 \\ 0 \\ 3 \\ 5 \end{bmatrix} \right\| = \left\| \begin{bmatrix} 0 \\ -2 \\ -2 \\ 6 \\ 5 \end{bmatrix} \right\| = \sqrt{0^2 + (-2)^2 + (-2)^2 + 6^2 + 5^2} = \sqrt{69}$$

and

$$\left\| \begin{bmatrix} 1 \\ 2 \\ -2 \\ 3 \\ 0 \end{bmatrix} - \begin{bmatrix} -1 \\ -4 \\ 0 \\ 3 \\ 5 \end{bmatrix} \right\| = \left\| \begin{bmatrix} 2 \\ 6 \\ -2 \\ 0 \\ -5 \end{bmatrix} \right\| = \sqrt{2^2 + 6^2 + (-2)^2 + 0^2 + (-5)^2} = \sqrt{69},$$

the vectors $\begin{bmatrix} 1 \\ 2 \\ -2 \\ 3 \\ 0 \end{bmatrix}$ and $\begin{bmatrix} -1 \\ -4 \\ 0 \\ 3 \\ 5 \end{bmatrix}$ are orthogonal.

From the definition of orthogonality it is not obvious that there is an important connection between orthogonality and the scalar product.

Theorem 3.10. *Vectors* $\mathbf{u}$ *and* $\mathbf{v}$ *are orthogonal if and only if* $\mathbf{u} \cdot \mathbf{v} = 0$.

Proof. We need to show that vectors $\mathbf{u} = \begin{bmatrix} u_1 \\ u_2 \\ \vdots \\ u_n \end{bmatrix}$ and $\mathbf{v} = \begin{bmatrix} v_1 \\ v_2 \\ \vdots \\ v_n \end{bmatrix}$ are orthogonal if and only if $u_1v_1 + u_2v_2 + \cdots + u_nv_n = 0$.

First we find that

$$\begin{aligned}
\|\mathbf{u}-\mathbf{v}\|^2 &= (u_1 - v_1)^2 + (u_2 - v_2)^2 + \cdots + (u_n - v_n)^2 \\
&= u_1^2 + v_1^2 - 2u_1v_1 + u_2^2 + v_2^2 - 2u_2v_2 + \cdots + u_n^2 + v_n^2 - 2u_nv_n \\
&= u_1^2 + u_2^2 + \cdots + u_n^2 + v_1^2 + v_2^2 + \cdots + v_n^2 - 2u_1v_1 - 2u_2v_2 - \cdots - 2u_nv_n \\
&= \|\mathbf{u}\|^2 + \|\mathbf{v}\|^2 - 2(u_1v_1 + u_2v_2 + \cdots + u_nv_n).
\end{aligned}$$

Similarly, we find that

$$\|\mathbf{u}+\mathbf{v}\|^2 = \|\mathbf{u}\|^2 + \|\mathbf{v}\|^2 + 2(u_1v_1 + u_2v_2 + \cdots + u_nv_n).$$

Thus

$$\|\mathbf{u}+\mathbf{v}\|^2 = \|\mathbf{u}-\mathbf{v}\|^2$$

if and only if

$$u_1v_1 + u_2v_2 + \cdots + u_nv_n = 0.$$

This proves the theorem, since the norm is always nonnegative. □

Example 3.11. In Example 3.9 we use the definition of orthogonality to show that the vectors $\begin{bmatrix} 1 \\ 2 \\ -2 \\ 3 \\ 0 \end{bmatrix}$ and $\begin{bmatrix} -1 \\ -4 \\ 0 \\ 3 \\ 5 \end{bmatrix}$ are orthogonal. The calculations become simpler if we use the above theorem instead:

$$\begin{bmatrix} 1 \\ 2 \\ -2 \\ 3 \\ 0 \end{bmatrix} \cdot \begin{bmatrix} -1 \\ -4 \\ 0 \\ 3 \\ 5 \end{bmatrix} = 1\cdot(-1) + 2\cdot(-4) + (-2)\cdot 0 + 3\cdot 3 + 0\cdot 5 = 0.$$

We close this section with a theorem on orthogonal vectors in $\mathbb{R}^n$ that is a general form of the familiar Pythagorean Theorem.

Theorem 3.12 (The Pythagorean Theorem). *If the vectors* $\mathbf{u}$ *and* $\mathbf{v}$ *are orthogonal, then*

$$\|\mathbf{u}+\mathbf{v}\|^2 = \|\mathbf{u}\|^2 + \|\mathbf{v}\|^2.$$

Proof. If $\mathbf{u}$ and $\mathbf{v}$ are orthogonal, then $\mathbf{u}\cdot\mathbf{v}=0$ and thus

$$\|\mathbf{u}+\mathbf{v}\|^2=(\mathbf{u}+\mathbf{v})\cdot(\mathbf{u}+\mathbf{v})=\mathbf{u}\cdot\mathbf{u}+2(\mathbf{u}\cdot\mathbf{v})+\mathbf{v}\cdot\mathbf{v}=\|\mathbf{u}\|^2+\|\mathbf{v}\|^2.$$

□

Exercises 3.1

Find the norm of the given vector.

1. $\begin{bmatrix}1\\-1\\-1\\1\end{bmatrix}$ 2. $\begin{bmatrix}1\\1\\1\\3\end{bmatrix}$ 3. $\begin{bmatrix}3\\0\\1\\1\\-2\end{bmatrix}$ 4. $\begin{bmatrix}1\\2\\-1\\1\\-2\end{bmatrix}$

Calculate the distance $\|\mathbf{u}-\mathbf{v}\|$.

5. $\mathbf{u}=\begin{bmatrix}1\\1\\1\\3\end{bmatrix}$ and $\mathbf{v}=\begin{bmatrix}5\\4\\3\\2\end{bmatrix}$ 6. $\mathbf{u}=\begin{bmatrix}1\\2\\-1\\0\end{bmatrix}$ and $\mathbf{v}=\begin{bmatrix}1\\4\\0\\-2\end{bmatrix}$

Calculate the scalar product.

7. $\begin{bmatrix}1\\1\\1\\3\end{bmatrix}\cdot\begin{bmatrix}5\\4\\3\\-3\end{bmatrix}$ 8. $\begin{bmatrix}3\\-5\\1\\-2\\3\end{bmatrix}\cdot\begin{bmatrix}2\\4\\1\\1\\7\end{bmatrix}$

Normalize the given vector.

9. $\begin{bmatrix}2\\2\\1\end{bmatrix}$ 10. $\begin{bmatrix}1\\3\\2\end{bmatrix}$ 11. $\begin{bmatrix}1\\3\\1\\2\end{bmatrix}$ 12. $\begin{bmatrix}1\\2\\-1\\1\end{bmatrix}$

13. Find a real number a such that the vectors $\mathbf{u}=\begin{bmatrix}1\\2\\1\\a\end{bmatrix}$ and $\mathbf{v}=\begin{bmatrix}5\\1\\2a+3\\-3\end{bmatrix}$ are orthogonal.

3.2 Best approximation and projection in $\mathbb{R}^n$

The scalar product gives us a convenient and effective tool to deal with best approximation and projection problems.

Definition 3.13. Let $\mathbf{u}_1, \dots, \mathbf{u}_k$ and $\mathbf{b}$ be vectors in $\mathbb{R}^n$. A vector $\mathbf{p}$ from the subspace $\operatorname{Span}\{\mathbf{u}_1, \dots, \mathbf{u}_k\}$ is called the *best approximation* to the vector $\mathbf{b}$ by vectors from the subspace $\operatorname{Span}\{\mathbf{u}_1, \dots, \mathbf{u}_k\}$ if

$$\|\mathbf{b} - \mathbf{p}\| < \|\mathbf{b} - \mathbf{q}\|$$

for every $\mathbf{q}$ in $\operatorname{Span}\{\mathbf{u}_1, \dots, \mathbf{u}_k\}$ such that $\mathbf{q} \neq \mathbf{p}$.

It is clear from the definition that, if the best approximation exists, then it is unique. We need to address two questions. First, does the best approximation always exist? Second, how do we find it? It turns out that the scalar product will help us answer both questions. It is done through orthogonal projections.

Definition 3.14. Let $\mathbf{u}_1, \dots, \mathbf{u}_k$ and $\mathbf{b}$ be vectors in $\mathbb{R}^n$. A vector $\mathbf{p}$ in $\operatorname{Span}\{\mathbf{u}_1, \dots, \mathbf{u}_k\}$ is called an *orthogonal projection* of the vector $\mathbf{b}$ on the subspace $\operatorname{Span}\{\mathbf{u}_1, \dots, \mathbf{u}_k\}$ if

$$(\mathbf{b} - \mathbf{p}) \cdot \mathbf{v} = 0$$

for every vector $\mathbf{v}$ in $\operatorname{Span}\{\mathbf{u}_1, \dots, \mathbf{u}_k\}$.

The next theorem shows, as expected, that the best approximation to the vector $\mathbf{b}$ by vectors from a subspace $\mathcal{V}$ coincides with the orthogonal projection of the vector $\mathbf{b}$ on the subspace $\mathcal{V}$.

Theorem 3.15. *Let $\mathbf{u}_1, \dots, \mathbf{u}_k$ and $\mathbf{b}$ be vectors in $\mathbb{R}^n$, let $\mathcal{V} = \operatorname{Span}\{\mathbf{u}_1, \dots, \mathbf{u}_k\}$, and let $\mathbf{p}$ be a vector in $\mathcal{V}$. Then the following conditions are equivalent:*

(a) *The vector $\mathbf{p}$ is an orthogonal projection of the vector $\mathbf{b}$ on the subspace $\mathcal{V}$;*

(b) *The vector $\mathbf{p}$ is the best approximation to the vector $\mathbf{b}$ by vectors from the subspace $\mathcal{V}$.*

Proof. Let $\mathbf{p} = x_1\mathbf{u}_1 + \dots + x_k\mathbf{u}_k$ be an orthogonal projection of the vector $\mathbf{b}$ on the subspace $\mathcal{V} = \operatorname{Span}\{\mathbf{u}_1, \dots, \mathbf{u}_k\}$. Then $(\mathbf{b} - \mathbf{p}) \cdot \mathbf{v} = 0$ for every vector $\mathbf{v}$ in $\mathcal{V}$. Let

$$\mathbf{q} = y_1\mathbf{u}_1 + \dots + y_k\mathbf{u}_k$$

be an arbitrary vector from $\mathcal{V}$. Then

$$\|\mathbf{b}-\mathbf{q}\|^2 = \|\mathbf{b}-\mathbf{p}+\mathbf{p}-\mathbf{q}\|^2 = \|\mathbf{b}-\mathbf{p}+\mathbf{v}\|^2,$$

where

$$\mathbf{v} = \mathbf{p}-\mathbf{q} = (x_1-y_1)\mathbf{u}_1 + \cdots + (x_k-y_k)\mathbf{u}_k$$

is a vector from $\mathcal{V}$. Consequently,

$$\begin{aligned}
\|\mathbf{b}-\mathbf{q}\|^2 &= \|\mathbf{b}-\mathbf{p}+\mathbf{v}\|^2 \\
&= (\mathbf{b}-\mathbf{p}+\mathbf{v})\cdot(\mathbf{b}-\mathbf{p}+\mathbf{v}) \\
&= (\mathbf{b}-\mathbf{p})\cdot(\mathbf{b}-\mathbf{p}) + 2(\mathbf{b}-\mathbf{p})\cdot\mathbf{v} + \mathbf{v}\cdot\mathbf{v} \\
&= (\mathbf{b}-\mathbf{p})\cdot(\mathbf{b}-\mathbf{p}) + \mathbf{v}\cdot\mathbf{v} \\
&= \|\mathbf{b}-\mathbf{p}\|^2 + \|\mathbf{v}\|^2 \\
&= \|\mathbf{b}-\mathbf{p}\|^2 + \|\mathbf{p}-\mathbf{q}\|^2.
\end{aligned}$$

Hence the vector $\mathbf{p}$ is the best approximation to the vector $\mathbf{b}$ by vectors from $\mathcal{V}$ since $\|\mathbf{p}-\mathbf{q}\|^2$ is always nonnegative. This shows that (a) implies (b).

Now assume that the vector $\mathbf{p}$ is is the best approximation to the vector $\mathbf{b}$ by vectors from $\mathcal{V}$. Let $\mathbf{v}$ be an arbitrary nonzero vector in $\mathcal{V}$ and let t be a real number. Then

$$\|\mathbf{b}-\mathbf{p}+t\mathbf{v}\|^2 \geq \|\mathbf{b}-\mathbf{p}\|^2$$

and consequently

$$t^2\|\mathbf{v}\|^2 + 2t(\mathbf{b}-\mathbf{p})\cdot\mathbf{v} \geq 0.$$

If we take $t = -\frac{(\mathbf{b}-\mathbf{p})\cdot\mathbf{v}}{\|\mathbf{v}\|^2}$, then we get

$$-\frac{((\mathbf{b}-\mathbf{p})\cdot\mathbf{v})^2}{\|\mathbf{v}\|^2} \geq 0.$$

This implies that $(\mathbf{b}-\mathbf{p})\cdot\mathbf{v} = 0$ and thus $\mathbf{p}$ is an orthogonal projection of the vector $\mathbf{b}$ on $\mathcal{V}$. Therefore (b) implies (a). □

The above result shows that an orthogonal projection is unique for every vector $\mathbf{b}$ and every subspace $\operatorname{Span}\{\mathbf{u}_1,\ldots,\mathbf{u}_k\}$.

The question of existence of orthogonal projections is a consequence of the next theorem which characterizes a projection on a subspace as a solution of a system of linear equations and also gives us an effective way of finding orthogonal projections of vectors.

Theorem 3.16. *Let* $\mathbf{u}_1, \ldots, \mathbf{u}_k$ *and* $\mathbf{b}$ *be vectors in* $\mathbb{R}^n$*. The vector*

$$x_1\mathbf{u}_1 + \cdots + x_k\mathbf{u}_k$$

is the projection of the vector $\mathbf{b}$ *on the subspace* $\operatorname{Span}\{\mathbf{u}_1, \ldots, \mathbf{u}_k\}$ *if and only if the following equalities hold*

$$\begin{array}{ccccccc} x_1\mathbf{u}_1\cdot\mathbf{u}_1 & + & \cdots & + & x_k\mathbf{u}_k\cdot\mathbf{u}_1 & = & \mathbf{b}\cdot\mathbf{u}_1 \\ x_1\mathbf{u}_1\cdot\mathbf{u}_2 & + & \cdots & + & x_k\mathbf{u}_k\cdot\mathbf{u}_2 & = & \mathbf{b}\cdot\mathbf{u}_2 \\ & \vdots & & \vdots & & \vdots & \\ x_1\mathbf{u}_1\cdot\mathbf{u}_k & + & \cdots & + & x_k\mathbf{u}_k\cdot\mathbf{u}_k & = & \mathbf{b}\cdot\mathbf{u}_k. \end{array}$$

Proof of a particular case. We give the proof for $k = 3$. It illustrates the method of the proof well and generalizing it to higher dimensions is straightforward.

The vector $\mathbf{p} = x_1\mathbf{u}_1 + x_2\mathbf{u}_2 + x_3\mathbf{u}_3$ is the projection of a vector $\mathbf{b}$ on the subspace $\operatorname{Span}\{\mathbf{u}_1, \mathbf{u}_2, \mathbf{u}_3\}$ if and only if

$$(\mathbf{b} - \mathbf{p})\cdot\mathbf{v} = 0$$

for every $\mathbf{v}$ in $\operatorname{Span}\{\mathbf{u}_1, \mathbf{u}_2, \mathbf{u}_3\}$. Since every vector $\mathbf{v}$ in $\operatorname{Span}\{\mathbf{u}_1, \mathbf{u}_2, \mathbf{u}_3\}$ is of the form

$$\mathbf{v} = y_1\mathbf{u}_1 + y_2\mathbf{u}_2 + y_3\mathbf{u}_3,$$

the equation $(\mathbf{b} - \mathbf{p})\cdot\mathbf{v} = 0$ is equivalent to the following three equations

$$\begin{aligned} (\mathbf{b} - \mathbf{p})\cdot\mathbf{u}_1 &= 0 \\ (\mathbf{b} - \mathbf{p})\cdot\mathbf{u}_2 &= 0 \\ (\mathbf{b} - \mathbf{p})\cdot\mathbf{u}_3 &= 0 \end{aligned}$$

or

$$\begin{aligned} (\mathbf{b} - (x_1\mathbf{u}_1 + x_2\mathbf{u}_2 + x_3\mathbf{u}_3))\cdot\mathbf{u}_1 &= 0 \\ (\mathbf{b} - (x_1\mathbf{u}_1 + x_2\mathbf{u}_2 + x_3\mathbf{u}_3))\cdot\mathbf{u}_2 &= 0 \\ (\mathbf{b} - (x_1\mathbf{u}_1 + x_2\mathbf{u}_2 + x_3\mathbf{u}_3))\cdot\mathbf{u}_3 &= 0 \end{aligned}$$

which can also be written as

$$\begin{aligned} x_1\mathbf{u}_1\cdot\mathbf{u}_1 + x_2\mathbf{u}_2\cdot\mathbf{u}_1 + x_3\mathbf{u}_3\cdot\mathbf{u}_1 &= \mathbf{b}\cdot\mathbf{u}_1\\ x_1\mathbf{u}_1\cdot\mathbf{u}_2 + x_2\mathbf{u}_2\cdot\mathbf{u}_2 + x_3\mathbf{u}_3\cdot\mathbf{u}_2 &= \mathbf{b}\cdot\mathbf{u}_2\\ x_1\mathbf{u}_1\cdot\mathbf{u}_3 + x_2\mathbf{u}_2\cdot\mathbf{u}_3 + x_3\mathbf{u}_3\cdot\mathbf{u}_3 &= \mathbf{b}\cdot\mathbf{u}_3. \end{aligned}$$

□

Corollary 3.17. *Let $\mathbf{u}_1,\dots,\mathbf{u}_k$ and $\mathbf{b}$ be vectors in $\mathbb{R}^n$. The numbers $x_1,\dots,x_k$ minimize the norm*

$$\|\mathbf{b} - x_1\mathbf{u}_1 - \cdots - x_k\mathbf{u}_k\|$$

if and only if the following equalities hold

$$\begin{array}{ccccccc} x_1\mathbf{u}_1\cdot\mathbf{u}_1 & + & \cdots & + & x_k\mathbf{u}_k\cdot\mathbf{u}_1 & = & \mathbf{b}\cdot\mathbf{u}_1\\ x_1\mathbf{u}_1\cdot\mathbf{u}_2 & + & \cdots & + & x_k\mathbf{u}_k\cdot\mathbf{u}_2 & = & \mathbf{b}\cdot\mathbf{u}_2\\ & \vdots & & \vdots & & \vdots & \\ x_1\mathbf{u}_1\cdot\mathbf{u}_k & + & \cdots & + & x_k\mathbf{u}_k\cdot\mathbf{u}_k & = & \mathbf{b}\cdot\mathbf{u}_k. \end{array}$$

Corollary 3.18. *Let $\mathbf{u}_1,\mathbf{u}_2,\dots,\mathbf{u}_k$ and $\mathbf{b}$ be arbitrary vectors in $\mathbb{R}^n$. Then the orthogonal projection of the vector $\mathbf{b}$ on the subspace* $\operatorname{Span}\{\mathbf{u}_1,\dots,\mathbf{u}_k\}$ *exists.*

Proof for $k = 3$. Without loss of generality we can suppose, by Theorem 2.46, that the vectors $\mathbf{u}_1$, $\mathbf{u}_2$, and $\mathbf{u}_3$ are linearly independent. It is enough to prove that the system in Theorem 3.16 has a unique solution. To do this, according to Theorem 1.75, it is enough to prove that the matrix

$$\begin{bmatrix} \mathbf{u}_1\cdot\mathbf{u}_1 & \mathbf{u}_2\cdot\mathbf{u}_1 & \mathbf{u}_3\cdot\mathbf{u}_1\\ \mathbf{u}_1\cdot\mathbf{u}_2 & \mathbf{u}_2\cdot\mathbf{u}_2 & \mathbf{u}_3\cdot\mathbf{u}_2\\ \mathbf{u}_1\cdot\mathbf{u}_3 & \mathbf{u}_2\cdot\mathbf{u}_3 & \mathbf{u}_3\cdot\mathbf{u}_3 \end{bmatrix}$$

is invertible. Note that

$$\begin{bmatrix} \mathbf{u}_1\cdot\mathbf{u}_1 & \mathbf{u}_2\cdot\mathbf{u}_1 & \mathbf{u}_3\cdot\mathbf{u}_1\\ \mathbf{u}_1\cdot\mathbf{u}_2 & \mathbf{u}_2\cdot\mathbf{u}_2 & \mathbf{u}_3\cdot\mathbf{u}_2\\ \mathbf{u}_1\cdot\mathbf{u}_3 & \mathbf{u}_2\cdot\mathbf{u}_3 & \mathbf{u}_3\cdot\mathbf{u}_3 \end{bmatrix} = A^T A,$$

where A is the matrix $\begin{bmatrix}\mathbf{u}_1 & \mathbf{u}_2 & \mathbf{u}_3\end{bmatrix}$. Since the vectors $\mathbf{u}_1,\mathbf{u}_2,\mathbf{u}_3$ are linearly independent, the matrix A is invertible, by Theorem 2.17. Consequently, the matrix A^T invertible. Now the matrix $A^T A$ is invertible as a product of two invertible matrices. (Note that a different proof can be obtained from Lemma 3.55.) □

Before we consider some examples illustrating how to find projections on subspaces we introduce a convenient notation for projections and prove a simple theorem that is often useful when dealing with orthogonal projections.

Definition 3.19. The orthogonal projection of a vector $\mathbf{b}$ on a subspace $\mathcal{V}$ is denoted by

$$\text{proj}_{\mathcal{V}}\mathbf{b}.$$

Theorem 3.20. *Let $\mathcal{V}$ be a subspace in $\mathbb{R}^n$. For any vector $\mathbf{b}$ in $\mathbb{R}^n$, the following conditions are equivalent:*

(a) $\mathbf{b}$ *is in $\mathcal{V}$;*

(b) $\mathbf{b} = \text{proj}_{\mathcal{V}}\mathbf{b}$.

Proof. By Theorem 3.15, $\text{proj}_{\mathcal{V}}\mathbf{b}$ is the best approximation of $\mathbf{b}$ by vectors in $\mathcal{V}$. Consequently, if $\mathbf{b} = \text{proj}_{\mathcal{V}}\mathbf{b}$, then $\mathbf{b}$ must be a vector in $\mathcal{V}$.

Conversely, if $\mathbf{b}$ is in $\mathcal{V}$, then clearly the best approximation of $\mathbf{b}$ by vectors in $\mathcal{V}$ is $\mathbf{b}$. □

Example 3.21. Consider the vectors

$$\mathbf{u}_1 = \begin{bmatrix} 1 \\ 1 \\ 1 \\ 0 \\ 1 \end{bmatrix}, \ \mathbf{u}_2 = \begin{bmatrix} 1 \\ 0 \\ 1 \\ 1 \\ 1 \end{bmatrix}, \ \text{and } \mathbf{b} = \begin{bmatrix} 1 \\ 0 \\ 1 \\ 0 \\ 0 \end{bmatrix}.$$

Find the projection $\text{proj}_{\text{Span}\{\mathbf{u}_1,\mathbf{u}_2\}}\mathbf{b}$ and verify the result.

Solution. According to Theorem 3.16, the vector $\mathbf{p} = x_1\mathbf{u}_1 + x_2\mathbf{u}_2$ is the projection of the vector $\mathbf{b}$ on $\text{Span}\{\mathbf{u}_1, \mathbf{u}_2\}$ if and only if the numbers x_1 and x_2 are the solutions of the system

$$\begin{cases} x_1\mathbf{u}_1 \cdot \mathbf{u}_1 + x_2\mathbf{u}_2 \cdot \mathbf{u}_1 = \mathbf{b} \cdot \mathbf{u}_1 \\ x_1\mathbf{u}_1 \cdot \mathbf{u}_2 + x_2\mathbf{u}_2 \cdot \mathbf{u}_2 = \mathbf{b} \cdot \mathbf{u}_2 \end{cases}.$$

After calculating the dot products we obtain the system

$$\begin{cases} 4x_1 + 3x_2 = 2 \\ 3x_1 + 4x_2 = 2 \end{cases}.$$

We use Gaussian elimination to solve the system. Since

$$\begin{bmatrix} 4 & 3 & 2 \\ 3 & 4 & 2 \end{bmatrix} \sim \begin{bmatrix} 1 & 0 & \frac{2}{7} \\ 0 & 1 & \frac{2}{7} \end{bmatrix},$$

the solution is $x_1 = x_2 = \frac{2}{7}$. This means that

$$\text{proj}_{\text{Span}\{\mathbf{u}_1,\mathbf{u}_2\}}\mathbf{b} = x_1\mathbf{u}_1 + x_2\mathbf{u}_2 = x_1\begin{bmatrix}1\\1\\1\\0\\1\end{bmatrix} + x_2\begin{bmatrix}1\\0\\1\\1\\1\end{bmatrix} = \frac{2}{7}\begin{bmatrix}1\\1\\1\\0\\1\end{bmatrix} + \frac{2}{7}\begin{bmatrix}1\\0\\1\\1\\1\end{bmatrix} = \frac{2}{7}\begin{bmatrix}2\\1\\2\\1\\2\end{bmatrix}.$$

This also means, according to Corollary 3.17, that $x_1 = x_2 = \frac{2}{7}$ minimizes

$$\left\| \begin{bmatrix}1\\0\\1\\0\\0\end{bmatrix} - x_1\begin{bmatrix}1\\1\\1\\0\\1\end{bmatrix} - x_2\begin{bmatrix}1\\0\\1\\1\\1\end{bmatrix} \right\|^2 = (1-x_1-x_2)^2 + x_1^2 + (1-x_1-x_2)^2 + x_2^2 + (x_1+x_2)^2.$$

To verify that $\frac{2}{7}\begin{bmatrix}2\\1\\2\\1\\2\end{bmatrix}$ is the projection of $\mathbf{b}$ on $\text{Span}\{\mathbf{u}_1,\mathbf{u}_2\}$ we have to check that $(\mathbf{b}-\mathbf{p})\cdot\mathbf{u}_1 = 0$ and $(\mathbf{b}-\mathbf{p})\cdot\mathbf{u}_2 = 0$. Indeed, we have

$$\left(\begin{bmatrix}1\\0\\1\\0\\0\end{bmatrix} - \frac{2}{7}\begin{bmatrix}2\\1\\2\\1\\2\end{bmatrix}\right)\cdot\begin{bmatrix}1\\1\\1\\0\\1\end{bmatrix} = \frac{1}{7}\begin{bmatrix}3\\-2\\3\\-2\\-4\end{bmatrix}\cdot\begin{bmatrix}1\\1\\1\\0\\1\end{bmatrix} = 0$$

and

$$\left(\begin{bmatrix}1\\0\\1\\0\\0\end{bmatrix} - \frac{2}{7}\begin{bmatrix}2\\1\\2\\1\\2\end{bmatrix}\right)\cdot\begin{bmatrix}1\\0\\1\\1\\1\end{bmatrix} = \frac{1}{7}\begin{bmatrix}3\\-2\\3\\-2\\-4\end{bmatrix}\cdot\begin{bmatrix}1\\0\\1\\1\\1\end{bmatrix} = 0.$$

□

Example 3.22. Consider the vectors

$$\mathbf{u}_1 = \begin{bmatrix} 1 \\ 1 \\ -1 \\ 0 \end{bmatrix}, \mathbf{u}_2 = \begin{bmatrix} 1 \\ -1 \\ 1 \\ 1 \end{bmatrix}, \mathbf{u}_3 = \begin{bmatrix} 2 \\ 0 \\ 0 \\ 1 \end{bmatrix}, \text{ and } \mathbf{b} = \begin{bmatrix} 1 \\ 1 \\ 1 \\ 1 \end{bmatrix}.$$

Find $\text{proj}_{\mathcal{V}}\mathbf{b}$ where $\mathcal{V} = \text{Span}\{\mathbf{u}_1, \mathbf{u}_2, \mathbf{u}_3\}$.

Solution. According to Theorem 3.16 we need to solve the system

$$\begin{cases} 3x_1 - x_2 + 2x_3 = 1 \\ -x_1 + 4x_2 + 3x_3 = 2 \\ 2x_1 + 3x_2 + 5x_3 = 3 \end{cases}.$$

The Gaussian elimination method gives us

$$\begin{bmatrix} 3 & -1 & 2 & 1 \\ -1 & 4 & 3 & 2 \\ 2 & 3 & 5 & 3 \end{bmatrix} \sim \begin{bmatrix} 1 & 0 & 1 & \frac{6}{11} \\ 0 & 1 & 1 & \frac{7}{11} \\ 0 & 0 & 0 & 0 \end{bmatrix}.$$

Hence $x_1 = -x_3 + \frac{6}{11}$ and $x_2 = -x_3 + \frac{7}{11}$, where x_3 is a free variable. It may look like the problem has infinitely many solutions contrary to what we have learned. However, after calculating the projection we end up with a unique solution, as expected:

$$\text{proj}_{\mathcal{V}}\mathbf{b} = \left(-x_3 + \frac{6}{11}\right)\begin{bmatrix} 1 \\ 1 \\ -1 \\ 0 \end{bmatrix} + \left(-x_3 + \frac{7}{11}\right)\begin{bmatrix} 1 \\ -1 \\ 1 \\ 1 \end{bmatrix} + x_3\begin{bmatrix} 2 \\ 0 \\ 0 \\ 1 \end{bmatrix} = \begin{bmatrix} \frac{13}{11} \\ -\frac{1}{11} \\ \frac{1}{11} \\ \frac{7}{11} \end{bmatrix}.$$

A closer inspection of the problem reveals that the vectors $\mathbf{u}_1$, $\mathbf{u}_2$, and $\mathbf{u}_3$ are not linearly independent, so the unique projection of $\mathbf{b}$ on $\text{Span}\{\mathbf{u}_1, \mathbf{u}_2, \mathbf{u}_3\}$ can be represented as a linear combination of the vectors $\mathbf{u}_1$, $\mathbf{u}_2$, and $\mathbf{u}_3$ in infinitely many ways. □

Example 3.23. Find the real numbers x and y that minimize the sum

$$(1 - x - y)^2 + (1 - x + y)^2 + (1 + x)^2 + (1 + x - y)^2.$$

Solution. Note that

$$(1-x-y)^2+(1-x+y)^2+(1+x)^2+(1+x-y)^2=\left\|\begin{bmatrix}1\\1\\1\\1\end{bmatrix}-x\begin{bmatrix}1\\1\\-1\\-1\end{bmatrix}-y\begin{bmatrix}1\\-1\\0\\1\end{bmatrix}\right\|^2.$$

By Theorem 3.15, the real numbers x and y minimize the sum when the vector

$$x\begin{bmatrix}1\\1\\-1\\-1\end{bmatrix}+y\begin{bmatrix}1\\-1\\0\\1\end{bmatrix}$$

is the projection of the vector $\begin{bmatrix}1\\1\\1\\1\end{bmatrix}$ on the subspace $\operatorname{Span}\left\{\begin{bmatrix}1\\1\\-1\\-1\end{bmatrix},\begin{bmatrix}1\\-1\\0\\1\end{bmatrix}\right\}$. Now, according to Theorem 3.16, to find the numbers x and y we need to solve the system

$$\begin{cases} x\begin{bmatrix}1\\1\\-1\\-1\end{bmatrix}\cdot\begin{bmatrix}1\\1\\-1\\-1\end{bmatrix}+y\begin{bmatrix}1\\-1\\0\\1\end{bmatrix}\cdot\begin{bmatrix}1\\1\\-1\\-1\end{bmatrix}=\begin{bmatrix}1\\1\\1\\1\end{bmatrix}\cdot\begin{bmatrix}1\\1\\-1\\-1\end{bmatrix}\\ x\begin{bmatrix}1\\1\\-1\\-1\end{bmatrix}\cdot\begin{bmatrix}1\\-1\\0\\1\end{bmatrix}+y\begin{bmatrix}1\\-1\\0\\1\end{bmatrix}\cdot\begin{bmatrix}1\\-1\\0\\1\end{bmatrix}=\begin{bmatrix}1\\1\\1\\1\end{bmatrix}\cdot\begin{bmatrix}1\\-1\\0\\1\end{bmatrix}\end{cases}$$

or, after calculating the scalar products, the system

$$\begin{cases}-4x+\ \ y=\ \ 0\\ \ \ \ x-3y=-1\end{cases}.$$

Since the solution is

$$x=\frac{1}{11}\quad\text{and}\quad y=\frac{4}{11},$$

these are the numbers that minimize the sum

$$(1-x-y)^2+(1-x+y)^2+(1+x)^2+(1+x-y)^2.$$

□

Calculating the orthogonal projection of a vector on a subspace $\operatorname{Span}\{\mathbf{u}_1,\dots,\mathbf{u}_k\}$ is much simpler if the vectors $\mathbf{u}_1,\dots,\mathbf{u}_k$ are orthogonal.

Definition 3.24. Let $\mathbf{u}_1,\dots,\mathbf{u}_k$ be vectors in $\mathbb{R}^n$. We say that the set $\{\mathbf{u}_1,\dots,\mathbf{u}_k\}$ is an *orthogonal set* if $\mathbf{u}_i \cdot \mathbf{u}_j = 0$ whenever $i \neq j$.

Example 3.25. Since

$$\begin{bmatrix}1\\1\\1\\1\end{bmatrix}\cdot\begin{bmatrix}-2\\0\\2\\0\end{bmatrix}=0,\quad \begin{bmatrix}1\\1\\1\\1\end{bmatrix}\cdot\begin{bmatrix}0\\3\\0\\-3\end{bmatrix}=0,\quad \text{and}\quad \begin{bmatrix}-2\\0\\2\\0\end{bmatrix}\cdot\begin{bmatrix}0\\3\\0\\-3\end{bmatrix}=0,$$

the set

$$\left\{\begin{bmatrix}1\\1\\1\\1\end{bmatrix},\begin{bmatrix}-2\\0\\2\\0\end{bmatrix},\begin{bmatrix}0\\3\\0\\-3\end{bmatrix}\right\}$$

is an orthogonal set.

Theorem 3.26. *If $\{\mathbf{u}_1,\dots,\mathbf{u}_k\}$ is an orthogonal set of nonzero vectors in $\mathbb{R}^n$, then the vectors $\mathbf{u}_1,\dots,\mathbf{u}_k$ are linearly independent.*

Proof. Assume $\{\mathbf{u}_1,\dots,\mathbf{u}_k\}$ is an orthogonal set of nonzero vectors and

$$x_1\mathbf{u}_1+\cdots+x_k\mathbf{u}_k=\mathbf{0}$$

for some real numbers $x_1,\dots,x_k$. Then

$$\begin{aligned}(x_1\mathbf{u}_1+x_2\mathbf{u}_2+\cdots+x_k\mathbf{u}_k)\cdot\mathbf{u}_1 &= x_1\mathbf{u}_1\cdot\mathbf{u}_1+x_2\mathbf{u}_2\cdot\mathbf{u}_1+\cdots+x_k\mathbf{u}_k\cdot\mathbf{u}_1\\ &= x_1\mathbf{u}_1\cdot\mathbf{u}_1\\ &= x_1\|\mathbf{u}_1\|^2.\end{aligned}$$

On the other hand,

$$(x_1\mathbf{u}_1+x_2\mathbf{u}_2+\cdots+x_k\mathbf{u}_k)\cdot\mathbf{u}_1=\mathbf{0}\cdot\mathbf{u}_1=0.$$

Since $x_1\|\mathbf{u}_1\|^2=0$ and $\|\mathbf{u}_1\|\neq 0$, we must have $x_1=0$. Similarly, by considering the scalar product with $\mathbf{u}_2,\mathbf{u}_3,\dots,\mathbf{u}_k$ we obtain

$$x_2=x_3=\cdots=x_k=0.$$

Consequently, the vectors $\mathbf{u}_1,\dots,\mathbf{u}_k$ are linearly independent. □

Now we prove a theorem that gives us a simple way of finding the orthogonal projection on a subspace $\operatorname{Span}\{\mathbf{u}_1,\dots,\mathbf{u}_k\}$ if the vectors $\mathbf{u}_1,\dots,\mathbf{u}_k$ are orthogonal. Note that this result proves the existence of orthogonal projections for these kinds of subspaces.

Theorem 3.27. *Let $\{\mathbf{u}_1,\dots,\mathbf{u}_k\}$ be an orthogonal set of nonzero vectors in $\mathbb{R}^n$ and let $\mathcal{V} = \operatorname{Span}\{\mathbf{u}_1,\dots,\mathbf{u}_k\}$. Then*

$$\operatorname{proj}_{\mathcal{V}}\mathbf{b} = \frac{\mathbf{b}\cdot\mathbf{u}_1}{\mathbf{u}_1\cdot\mathbf{u}_1}\mathbf{u}_1 + \cdots + \frac{\mathbf{b}\cdot\mathbf{u}_k}{\mathbf{u}_k\cdot\mathbf{u}_k}\mathbf{u}_k$$

for every vector $\mathbf{b}$ in $\mathbb{R}^n$.

Proof of a particular case. We give a proof for $k = 4$. Let

$$\mathbf{p} = \frac{\mathbf{b}\cdot\mathbf{u}_1}{\mathbf{u}_1\cdot\mathbf{u}_1}\mathbf{u}_1 + \frac{\mathbf{b}\cdot\mathbf{u}_2}{\mathbf{u}_2\cdot\mathbf{u}_2}\mathbf{u}_2 + \frac{\mathbf{b}\cdot\mathbf{u}_3}{\mathbf{u}_3\cdot\mathbf{u}_3}\mathbf{u}_3 + \frac{\mathbf{b}\cdot\mathbf{u}_4}{\mathbf{u}_4\cdot\mathbf{u}_4}\mathbf{u}_4.$$

We have

$$\begin{aligned}
(\mathbf{b}-\mathbf{p})\cdot\mathbf{u}_1 &= \left(\mathbf{b} - \frac{\mathbf{b}\cdot\mathbf{u}_1}{\mathbf{u}_1\cdot\mathbf{u}_1}\mathbf{u}_1 - \frac{\mathbf{b}\cdot\mathbf{u}_2}{\mathbf{u}_2\cdot\mathbf{u}_2}\mathbf{u}_2 - \frac{\mathbf{b}\cdot\mathbf{u}_3}{\mathbf{u}_3\cdot\mathbf{u}_3}\mathbf{u}_3 - \frac{\mathbf{b}\cdot\mathbf{u}_4}{\mathbf{u}_4\cdot\mathbf{u}_4}\mathbf{u}_4\right)\cdot\mathbf{u}_1 \\
&= \mathbf{b}\cdot\mathbf{u}_1 - \frac{\mathbf{b}\cdot\mathbf{u}_1}{\mathbf{u}_1\cdot\mathbf{u}_1}\mathbf{u}_1\cdot\mathbf{u}_1 - \frac{\mathbf{b}\cdot\mathbf{u}_2}{\mathbf{u}_2\cdot\mathbf{u}_2}\mathbf{u}_2\cdot\mathbf{u}_1 - \frac{\mathbf{b}\cdot\mathbf{u}_3}{\mathbf{u}_3\cdot\mathbf{u}_3}\mathbf{u}_3\cdot\mathbf{u}_1 - \frac{\mathbf{b}\cdot\mathbf{u}_4}{\mathbf{u}_4\cdot\mathbf{u}_4}\mathbf{u}_4\cdot\mathbf{u}_1 \\
&= \mathbf{b}\cdot\mathbf{u}_1 - \frac{\mathbf{b}\cdot\mathbf{u}_1}{\mathbf{u}_1\cdot\mathbf{u}_1}\mathbf{u}_1\cdot\mathbf{u}_1 = 0.
\end{aligned}$$

Similarly we obtain

$$(\mathbf{b}-\mathbf{p})\cdot\mathbf{u}_2 = 0, \quad (\mathbf{b}-\mathbf{p})\cdot\mathbf{u}_3 = 0, \quad \text{and} \quad (\mathbf{b}-\mathbf{p})\cdot\mathbf{u}_4 = 0.$$

Since every vector $\mathbf{v}$ in $\operatorname{Span}\{\mathbf{u}_1,\mathbf{u}_2,\mathbf{u}_3,\mathbf{u}_4\}$ is of the form

$$\mathbf{v} = x_1\mathbf{u}_1 + x_2\mathbf{u}_2 + x_3\mathbf{u}_3 + x_4\mathbf{u}_4,$$

we have

$$(\mathbf{b}-\mathbf{p})\cdot\mathbf{v} = 0$$

for every $\mathbf{v}$ in $\operatorname{Span}\{\mathbf{u}_1,\mathbf{u}_2,\mathbf{u}_3,\mathbf{u}_4\}$, which means that $\mathbf{p} = \operatorname{proj}_{\mathcal{V}}\mathbf{b}$, by Definition 3.14. □

Note that Theorem 3.27 could be obtained as a consequence of Theorem 3.16.

Example 3.28. Let $\mathbf{u}_1 = \begin{bmatrix}1\\1\\1\\0\\1\end{bmatrix}$, $\mathbf{u}_2 = \begin{bmatrix}1\\-3\\1\\4\\1\end{bmatrix}$, and $\mathbf{b} = \begin{bmatrix}1\\1\\1\\1\\1\end{bmatrix}$. Find $\operatorname{proj}_{\operatorname{Span}\{\mathbf{u}_1,\mathbf{u}_2\}}\mathbf{b}$ and verify the result.

Solution. Since $\mathbf{u}_1 \cdot \mathbf{u}_2 = 0$, we can use Theorem 3.27:

$$\operatorname{proj}_{\operatorname{Span}\{\mathbf{u}_1,\mathbf{u}_2\}}\mathbf{b} = \frac{\mathbf{b}\cdot\mathbf{u}_1}{\mathbf{u}_1\cdot\mathbf{u}_1}\mathbf{u}_1 + \frac{\mathbf{b}\cdot\mathbf{u}_2}{\mathbf{u}_2\cdot\mathbf{u}_2}\mathbf{u}_2 = \frac{4}{4}\begin{bmatrix}1\\1\\1\\0\\1\end{bmatrix} + \frac{4}{28}\begin{bmatrix}1\\-3\\1\\4\\1\end{bmatrix}$$

$$= \begin{bmatrix}1\\1\\1\\0\\1\end{bmatrix} + \frac{1}{7}\begin{bmatrix}1\\-3\\1\\4\\1\end{bmatrix} = \frac{1}{7}\begin{bmatrix}8\\4\\8\\4\\8\end{bmatrix} = \frac{4}{7}\begin{bmatrix}2\\1\\2\\1\\2\end{bmatrix}.$$

To verify we calculate

$$\left(\begin{bmatrix}1\\1\\1\\1\\1\end{bmatrix} - \frac{4}{7}\begin{bmatrix}2\\1\\2\\1\\2\end{bmatrix}\right)\cdot\begin{bmatrix}1\\1\\1\\0\\1\end{bmatrix} = \frac{1}{7}\begin{bmatrix}-1\\3\\-1\\3\\-1\end{bmatrix}\cdot\begin{bmatrix}1\\1\\1\\0\\1\end{bmatrix} = 0$$

and

$$\left(\begin{bmatrix}1\\1\\1\\1\\1\end{bmatrix} - \frac{4}{7}\begin{bmatrix}2\\1\\2\\1\\2\end{bmatrix}\right)\cdot\begin{bmatrix}1\\-3\\1\\4\\1\end{bmatrix} = \frac{1}{7}\begin{bmatrix}-1\\3\\-1\\3\\-1\end{bmatrix}\cdot\begin{bmatrix}1\\-3\\1\\4\\1\end{bmatrix} = 0.$$

□

In the next result we show that the projection on a subspace $\operatorname{Span}\{\mathbf{u}_1,\dots,\mathbf{u}_k\}$, where $\{\mathbf{u}_1,\dots,\mathbf{u}_k\}$ is an orthogonal set of nonzero vectors, can be obtained by multiplication by a matrix. This is an important idea. The matrix is completely determined by the vectors $\mathbf{u}_1,\dots,\mathbf{u}_k$, so once we find the matrix for the subspace $\operatorname{Span}\{\mathbf{u}_1, \dots,\mathbf{u}_k\}$ it is easy to find the projection for any element of $\mathbb{R}^n$.

Theorem 3.29. *Let $\{\mathbf{u}_1, \dots, \mathbf{u}_k\}$ be an orthogonal set of nonzero vectors in $\mathbb{R}^n$ and let $\mathcal{V} = \operatorname{Span}\{\mathbf{u}_1, \dots, \mathbf{u}_k\}$. Then for any vector $\mathbf{x}$ in $\mathbb{R}^n$ we have*

$$\operatorname{proj}_{\mathcal{V}} \mathbf{x} = A\mathbf{x}$$

where

$$A = \frac{1}{\mathbf{u}_1 \cdot \mathbf{u}_1}(\mathbf{u}_1 \mathbf{u}_1^T) + \cdots + \frac{1}{\mathbf{u}_k \cdot \mathbf{u}_k}(\mathbf{u}_k \mathbf{u}_k^T).$$

Moreover, the matrix A is the unique matrix with this property.

Proof of a particular case. We give the proof for $m = 3$.

If $\{\mathbf{u}_1, \mathbf{u}_2, \mathbf{u}_3\}$ is an orthogonal set of nonzero vectors and $\mathcal{V} = \operatorname{Span}\{\mathbf{u}_1, \mathbf{u}_2, \mathbf{u}_3\}$, then by Theorem 3.27 we have

$$\begin{aligned}
\operatorname{proj}_{\mathcal{V}} \mathbf{x} &= \frac{\mathbf{x} \cdot \mathbf{u}_1}{\mathbf{u}_1 \cdot \mathbf{u}_1}\mathbf{u}_1 + \frac{\mathbf{x} \cdot \mathbf{u}_2}{\mathbf{u}_2 \cdot \mathbf{u}_2}\mathbf{u}_2 + \frac{\mathbf{x} \cdot \mathbf{u}_3}{\mathbf{u}_3 \cdot \mathbf{u}_3}\mathbf{u}_3 \\
&= \frac{\mathbf{u}_1^T \mathbf{x}}{\mathbf{u}_1 \cdot \mathbf{u}_1}\mathbf{u}_1 + \frac{\mathbf{u}_2^T \mathbf{x}}{\mathbf{u}_2 \cdot \mathbf{u}_2}\mathbf{u}_2 + \frac{\mathbf{u}_3^T \mathbf{x}}{\mathbf{u}_3 \cdot \mathbf{u}_3}\mathbf{u}_3 \\
&= \frac{1}{\mathbf{u}_1 \cdot \mathbf{u}_1}\mathbf{u}_1(\mathbf{u}_1^T \mathbf{x}) + \frac{1}{\mathbf{u}_2 \cdot \mathbf{u}_2}\mathbf{u}_2(\mathbf{u}_2^T \mathbf{x}) + \frac{1}{\mathbf{u}_3 \cdot \mathbf{u}_3}\mathbf{u}_3(\mathbf{u}_3^T \mathbf{x}) \\
&= \frac{1}{\mathbf{u}_1 \cdot \mathbf{u}_1}(\mathbf{u}_1 \mathbf{u}_1^T)\mathbf{x} + \frac{1}{\mathbf{u}_2 \cdot \mathbf{u}_2}(\mathbf{u}_2 \mathbf{u}_2^T)\mathbf{x} + \frac{1}{\mathbf{u}_3 \cdot \mathbf{u}_3}(\mathbf{u}_3 \mathbf{u}_3^T)\mathbf{x} \\
&= \left(\frac{1}{\mathbf{u}_1 \cdot \mathbf{u}_1}(\mathbf{u}_1 \mathbf{u}_1^T) + \frac{1}{\mathbf{u}_2 \cdot \mathbf{u}_2}(\mathbf{u}_2 \mathbf{u}_2^T) + \frac{1}{\mathbf{u}_3 \cdot \mathbf{u}_3}(\mathbf{u}_3 \mathbf{u}_3^T)\right)\mathbf{x}.
\end{aligned}$$

The uniqueness of the matrix results from Theorem 1.18. □

Definition 3.30. The unique matrix A such that $\operatorname{proj}_{\mathcal{V}} \mathbf{x} = A\mathbf{x}$ is called the *projection matrix* on the subspace $\mathcal{V}$.

Example 3.31. Find the projection matrix on the subspace

$$\mathcal{V} = \operatorname{Span}\left\{\begin{bmatrix} 1 \\ 1 \\ -1 \\ 1 \end{bmatrix}, \begin{bmatrix} 1 \\ 1 \\ 3 \\ 1 \end{bmatrix}, \begin{bmatrix} 2 \\ -1 \\ 0 \\ -1 \end{bmatrix}\right\}.$$

Solution. Since the set $\left\{\begin{bmatrix} 1 \\ 1 \\ -1 \\ 1 \end{bmatrix}, \begin{bmatrix} 1 \\ 1 \\ 3 \\ 1 \end{bmatrix}, \begin{bmatrix} 2 \\ -1 \\ 0 \\ -1 \end{bmatrix}\right\}$ is orthogonal, the projection matrix on $\mathcal{V}$ is

$$
\begin{aligned}
A &= \frac{1}{4}\begin{bmatrix} 1\\ 1\\ -1\\ 1\end{bmatrix}\begin{bmatrix} 1 & 1 & -1 & 1\end{bmatrix} + \frac{1}{12}\begin{bmatrix} 1\\ 1\\ 3\\ 1\end{bmatrix}\begin{bmatrix} 1 & 1 & 3 & 1\end{bmatrix} + \frac{1}{6}\begin{bmatrix} 2\\ -1\\ 0\\ -1\end{bmatrix}\begin{bmatrix} 2 & -1 & 0 & -1\end{bmatrix}\\
&= \frac{1}{4}\begin{bmatrix} 1 & 1 & -1 & 1\\ 1 & 1 & -1 & 1\\ -1 & -1 & 1 & -1\\ 1 & 1 & -1 & 1\end{bmatrix} + \frac{1}{12}\begin{bmatrix} 1 & 1 & 3 & 1\\ 1 & 1 & 3 & 1\\ 3 & 3 & 9 & 3\\ 1 & 1 & 3 & 1\end{bmatrix} + \frac{1}{6}\begin{bmatrix} 4 & -2 & 0 & -2\\ -2 & 1 & 0 & 1\\ 0 & 0 & 0 & 0\\ -2 & 1 & 0 & 1\end{bmatrix}\\
&= \frac{1}{2}\begin{bmatrix} 2 & 0 & 0 & 0\\ 0 & 1 & 0 & 1\\ 0 & 0 & 2 & 0\\ 0 & 1 & 0 & 1\end{bmatrix}.
\end{aligned}
$$

□

Definition 3.32. Let $\mathbf{u}_1, \dots, \mathbf{u}_k$ be vectors in $\mathbb{R}^n$. We say that the set $\{\mathbf{u}_1, \dots, \mathbf{u}_k\}$ is an *orthonormal set* if it is orthogonal and $\|\mathbf{u}_1\| = \cdots = \|\mathbf{u}_k\| = 1$.

Example 3.33. The set $\left\{\begin{bmatrix} 1\\ 1\\ -1\\ 1\end{bmatrix}, \begin{bmatrix} 1\\ 1\\ 3\\ 1\end{bmatrix}, \begin{bmatrix} 2\\ -1\\ 0\\ -1\end{bmatrix}\right\}$ is orthogonal, but not orthonormal because

$$
\left\|\begin{bmatrix} 1\\ 1\\ -1\\ 1\end{bmatrix}\right\| = 2, \quad \left\|\begin{bmatrix} 1\\ 1\\ 3\\ 1\end{bmatrix}\right\| = 2\sqrt{3}, \quad \text{and} \quad \left\|\begin{bmatrix} 2\\ -1\\ 0\\ -1\end{bmatrix}\right\| = \sqrt{6}.
$$

If we normalize vectors in an orthogonal set, we obtain an orthonormal set spanning the same subspace. In our case we obtain the orthonormal set

$$
\left\{\begin{bmatrix} \frac{1}{2}\\ \frac{1}{2}\\ -\frac{1}{2}\\ \frac{1}{2}\end{bmatrix}, \begin{bmatrix} \frac{1}{2\sqrt{3}}\\ \frac{1}{2\sqrt{3}}\\ \frac{3}{2\sqrt{3}}\\ \frac{1}{2\sqrt{3}}\end{bmatrix}, \begin{bmatrix} \frac{2}{\sqrt{6}}\\ -\frac{1}{\sqrt{6}}\\ 0\\ -\frac{1}{\sqrt{6}}\end{bmatrix}\right\}
$$

and we have

$$\operatorname{Span}\left\{\begin{bmatrix}1\\1\\-1\\1\end{bmatrix},\begin{bmatrix}1\\1\\3\\1\end{bmatrix},\begin{bmatrix}2\\-1\\0\\-1\end{bmatrix}\right\}=\operatorname{Span}\left\{\begin{bmatrix}\frac{1}{2}\\\frac{1}{2}\\-\frac{1}{2}\\\frac{1}{2}\end{bmatrix},\begin{bmatrix}\frac{1}{2\sqrt{3}}\\\frac{1}{2\sqrt{3}}\\\frac{3}{2\sqrt{3}}\\\frac{1}{2\sqrt{3}}\end{bmatrix},\begin{bmatrix}\frac{2}{\sqrt{6}}\\-\frac{1}{\sqrt{6}}\\0\\-\frac{1}{\sqrt{6}}\end{bmatrix}\right\}.$$

The formula in Theorem 3.27 is simpler for orthonormal sets since $\mathbf{u}_1\cdot\mathbf{u}_1=\cdots=\mathbf{u}_k\cdot\mathbf{u}_k=1$.

Corollary 3.34. *Let* $\{\mathbf{u}_1,\ldots,\mathbf{u}_k\}$ *be an orthonormal set of nonzero vectors in* $\mathbb{R}^n$ *and let* $\mathcal{V}=\operatorname{Span}\{\mathbf{u}_1,\ldots,\mathbf{u}_k\}$. *Then*

$$\operatorname{proj}_{\mathcal{V}}\mathbf{b}=(\mathbf{b}\cdot\mathbf{u}_1)\mathbf{u}_1+\cdots+(\mathbf{b}\cdot\mathbf{u}_k)\mathbf{u}_k$$

for every vector $\mathbf{b}$ *in* $\mathbb{R}^n$.

As expected, the expression for the projection matrix for an orthonormal set is also simpler.

Corollary 3.35. *Let* $\{\mathbf{u}_1,\ldots,\mathbf{u}_k\}$ *be an orthonormal set of nonzero vectors in* $\mathbb{R}^n$ *and let* $\mathcal{V}=\operatorname{Span}\{\mathbf{u}_1,\ldots,\mathbf{u}_k\}$. *Then for any vector* $\mathbf{x}$ *in* $\mathbb{R}^n$ *we have*

$$\operatorname{proj}_{\mathcal{V}}\mathbf{x}=A\mathbf{x}$$

where

$$A=\mathbf{u}_1\mathbf{u}_1^T+\mathbf{u}_2\mathbf{u}_2^T+\cdots+\mathbf{u}_k\mathbf{u}_k^T.$$

Exercises 3.2

Find the projection of $\mathbf{b}$ on the given subspace.

1. $\mathbf{b}=\begin{bmatrix}1\\1\\0\\1\end{bmatrix}$, $\operatorname{Span}\left\{\begin{bmatrix}1\\2\\2\\2\end{bmatrix}\right\}$

2. $\mathbf{b}=\begin{bmatrix}1\\1\\1\\1\end{bmatrix}$, $\operatorname{Span}\left\{\begin{bmatrix}1\\2\\-1\\1\end{bmatrix}\right\}$

3. $\mathbf{b}=\begin{bmatrix}1\\1\\0\\1\\1\end{bmatrix}$, $\operatorname{Span}\left\{\begin{bmatrix}1\\1\\-1\\1\\2\end{bmatrix}\right\}$

4. $\mathbf{b} = \begin{bmatrix}1\\0\\0\\1\\0\end{bmatrix}$, $\text{Span}\left\{\begin{bmatrix}1\\1\\2\\2\\1\end{bmatrix}\right\}$

5. $\mathbf{b} = \begin{bmatrix}1\\1\\1\\1\end{bmatrix}$, $\text{Span}\left\{\begin{bmatrix}0\\1\\-1\\0\end{bmatrix}, \begin{bmatrix}1\\0\\2\\1\end{bmatrix}\right\}$

6. $\mathbf{b} = \begin{bmatrix}1\\0\\0\\0\end{bmatrix}$, $\text{Span}\left\{\begin{bmatrix}3\\1\\1\\1\end{bmatrix}, \begin{bmatrix}2\\1\\1\\1\end{bmatrix}\right\}$

7. $\mathbf{b} = \begin{bmatrix}1\\0\\0\\0\\0\end{bmatrix}$, $\text{Span}\left\{\begin{bmatrix}1\\0\\1\\1\\1\end{bmatrix}, \begin{bmatrix}1\\1\\1\\1\\0\end{bmatrix}\right\}$

8. $\mathbf{b} = \begin{bmatrix}1\\0\\0\\1\\0\end{bmatrix}$, $\text{Span}\left\{\begin{bmatrix}1\\1\\1\\1\\1\end{bmatrix}, \begin{bmatrix}0\\0\\0\\1\\0\end{bmatrix}\right\}$

Find the real number x that minimizes the given sum.

9. $(1-2x)^2+(1+x)^2+(2-x)^2+(1-2x)^2$

10. $(2-x)^2+(2+x)^2+(1-2x)^2+(1+2x)^2$

Find the real numbers x and y that minimize the given sum.

11. $(1-x)^2 + y^2 + (1-x-y)^2 + (1-x+y)^2$
12. $(2+y)^2 + (1-2x)^2 + (1-x-y)^2 + (1+x+y)^2$
13. $(1-x-2y)^2 + (1-x-y)^2 + (1+x-y)^2 + (1-x+y)^2$
14. $(2-x+y)^2 + (1-x)^2 + (1-y)^2 + (1-x-y)^2$
15. $(1-x+y)^2 + (1-x)^2 + (1-y)^2 + (1-x-y)^2 + (1+x-y)^2$
16. $(1-x-y)^2 + (1-2x)^2 + (1+y)^2 + (1-3y)^2 + (2-x+y)^2$

Find the real numbers x, y, and z that minimize the given sum.

17. $(1-x)^2 + (1-y)^2 + (1-z)^2 + (1-x-y-z)^2$
18. $(1-x-y)^2 + (1-y-z)^2 + (1-x-z)^2 + (1-x)^2$

Find the projection of $\mathbf{b}$ on the given subspace. Use orthogonality of vectors, if possible.

19. $\mathbf{b} = \begin{bmatrix}1\\0\\0\\1\end{bmatrix}$, $\text{Span}\left\{\begin{bmatrix}0\\-1\\1\\1\end{bmatrix}, \begin{bmatrix}1\\1\\1\\0\end{bmatrix}\right\}$

20. $\mathbf{b} = \begin{bmatrix}1\\0\\0\\0\end{bmatrix}$, $\text{Span}\left\{\begin{bmatrix}2\\1\\-1\\1\end{bmatrix}, \begin{bmatrix}1\\2\\2\\-2\end{bmatrix}\right\}$

21. $\mathbf{b} = \begin{bmatrix} 1 \\ 0 \\ 0 \\ 0 \end{bmatrix}$, Span$\left\{ \begin{bmatrix} 0 \\ -1 \\ 1 \\ 1 \end{bmatrix}, \begin{bmatrix} 1 \\ 1 \\ 1 \\ 0 \end{bmatrix}, \begin{bmatrix} 1 \\ 0 \\ -1 \\ 1 \end{bmatrix} \right\}$

22. $\mathbf{b} = \begin{bmatrix} 0 \\ 0 \\ 1 \\ 0 \end{bmatrix}$, Span$\left\{ \begin{bmatrix} 1 \\ 1 \\ 1 \\ 1 \end{bmatrix}, \begin{bmatrix} 1 \\ -1 \\ -1 \\ 1 \end{bmatrix}, \begin{bmatrix} 1 \\ 0 \\ 0 \\ -1 \end{bmatrix} \right\}$

Find the projection matrix on the given subspace using Theorem 3.29.

23. Span$\left\{ \begin{bmatrix} 1 \\ 2 \\ 2 \\ 2 \end{bmatrix} \right\}$

24. Span$\left\{ \begin{bmatrix} 2 \\ 1 \\ -1 \\ 1 \end{bmatrix} \right\}$

25. Span$\left\{ \begin{bmatrix} 0 \\ -1 \\ 1 \\ 1 \end{bmatrix}, \begin{bmatrix} 1 \\ 1 \\ 1 \\ 0 \end{bmatrix} \right\}$

26. Span$\left\{ \begin{bmatrix} 2 \\ 1 \\ -1 \\ 1 \end{bmatrix}, \begin{bmatrix} 1 \\ 2 \\ 2 \\ -2 \end{bmatrix} \right\}$

27. Span$\left\{ \begin{bmatrix} 0 \\ -1 \\ 1 \\ 1 \end{bmatrix}, \begin{bmatrix} 1 \\ 1 \\ 1 \\ 0 \end{bmatrix}, \begin{bmatrix} 1 \\ 0 \\ -1 \\ 1 \end{bmatrix} \right\}$

28. Span$\left\{ \begin{bmatrix} 1 \\ 1 \\ 1 \\ 1 \end{bmatrix}, \begin{bmatrix} 1 \\ -1 \\ -1 \\ 1 \end{bmatrix}, \begin{bmatrix} 1 \\ 0 \\ 0 \\ -1 \end{bmatrix} \right\}$

3.3 Orthogonal bases in subspaces of $\mathbb{R}^n$

While calculating an orthogonal projection of a vector on a subspace Span$\{\mathbf{u}_1,\dots,\mathbf{u}_k\}$ is much simpler if the vectors $\mathbf{u}_1,\dots,\mathbf{u}_k$ are orthogonal, in practice we often have to find projections on subspaces Span$\{\mathbf{u}_1,\dots,\mathbf{u}_k\}$ where the vectors $\mathbf{u}_1,\dots,\mathbf{u}_k$ are not orthogonal. In this section we describe a process that will allow us to replace the original vectors $\mathbf{u}_1,\dots,\mathbf{u}_k$ with orthogonal vectors $\mathbf{v}_1,\dots,\mathbf{v}_m$ that define the same subspace, that is, such that Span$\{\mathbf{u}_1,\dots,\mathbf{u}_k\}$ = Span$\{\mathbf{v}_1,\dots,\mathbf{v}_m\}$. The following theorem is the basis for that process.

Theorem 3.36. *Let* $\{\mathbf{u}_1,\dots,\mathbf{u}_k\}$ *be an orthogonal set of nonzero vectors in* $\mathbb{R}^n$.

(a) *The vector* $\mathbf{v}$ *is in* $\operatorname{Span}\{\mathbf{u}_1,\dots,\mathbf{u}_k\}$ *if and only if*

$$\mathbf{v} = \operatorname{proj}_{\operatorname{Span}\{\mathbf{u}_1,\dots,\mathbf{u}_k\}}\mathbf{v} = \frac{\mathbf{u}_1\cdot\mathbf{v}}{\mathbf{u}_1\cdot\mathbf{u}_1}\mathbf{u}_1 + \frac{\mathbf{u}_2\cdot\mathbf{v}}{\mathbf{u}_2\cdot\mathbf{u}_2}\mathbf{u}_2 + \cdots + \frac{\mathbf{u}_k\cdot\mathbf{v}}{\mathbf{u}_k\cdot\mathbf{u}_k}\mathbf{u}_k.$$

In this case we have

$$\operatorname{Span}\{\mathbf{u}_1,\dots,\mathbf{u}_k,\mathbf{v}\} = \operatorname{Span}\{\mathbf{u}_1,\dots,\mathbf{u}_k\}.$$

(b) *The vector* $\mathbf{v}$ *is not in* $\operatorname{Span}\{\mathbf{u}_1,\dots,\mathbf{u}_k\}$ *if and only if the vector*

$$\mathbf{u}_{k+1} = \mathbf{v} - \frac{\mathbf{u}_1\cdot\mathbf{v}}{\mathbf{u}_1\cdot\mathbf{u}_1}\mathbf{u}_1 - \frac{\mathbf{u}_2\cdot\mathbf{v}}{\mathbf{u}_2\cdot\mathbf{u}_2}\mathbf{u}_2 - \cdots - \frac{\mathbf{u}_k\cdot\mathbf{v}}{\mathbf{u}_k\cdot\mathbf{u}_k}\mathbf{u}_k$$

is a nonzero vector. In this case we have

$$\mathbf{u}_{k+1}\cdot\mathbf{u}_1 = \mathbf{u}_{k+1}\cdot\mathbf{u}_2 = \cdots = \mathbf{u}_{k+1}\cdot\mathbf{u}_k = 0$$

and

$$\operatorname{Span}\{\mathbf{u}_1,\dots,\mathbf{u}_k,\mathbf{v}\} = \operatorname{Span}\{\mathbf{u}_1,\dots,\mathbf{u}_k,\mathbf{u}_{k+1}\}.$$

Proof of a particular case. We prove the theorem for $k = 3$.

Let $\{\mathbf{u}_1,\mathbf{u}_2,\mathbf{u}_3\}$ be an orthogonal set of nonzero vectors in $\mathbb{R}^n$ and let $\mathbf{v}$ be an arbitrary vector in $\mathbb{R}^n$. Then the vector

$$\mathbf{p} = \frac{\mathbf{u}_1\cdot\mathbf{v}}{\mathbf{u}_1\cdot\mathbf{u}_1}\mathbf{u}_1 + \frac{\mathbf{u}_2\cdot\mathbf{v}}{\mathbf{u}_2\cdot\mathbf{u}_2}\mathbf{u}_2 + \frac{\mathbf{u}_3\cdot\mathbf{v}}{\mathbf{u}_3\cdot\mathbf{u}_3}\mathbf{u}_3$$

is the orthogonal projection of $\mathbf{v}$ on $\operatorname{Span}\{\mathbf{u}_1,\mathbf{u}_2,\mathbf{u}_3\}$, by Theorem 3.27.

Since $\mathbf{v} = \mathbf{p}$ if and only if $\mathbf{v}$ is in $\operatorname{Span}\{\mathbf{u}_1,\mathbf{u}_2,\mathbf{u}_3\}$, part (a) follows.

To prove part (b) note that $\mathbf{v}$ is not in $\operatorname{Span}\{\mathbf{u}_1,\mathbf{u}_2,\mathbf{u}_3\}$ if and only if $\mathbf{v} \neq \mathbf{p}$, that is, if and only if

$$\mathbf{u}_4 = \mathbf{v} - \frac{\mathbf{u}_1\cdot\mathbf{v}}{\mathbf{u}_1\cdot\mathbf{u}_1}\mathbf{u}_1 - \frac{\mathbf{u}_2\cdot\mathbf{v}}{\mathbf{u}_2\cdot\mathbf{u}_2}\mathbf{u}_2 - \frac{\mathbf{u}_3\cdot\mathbf{v}}{\mathbf{u}_3\cdot\mathbf{u}_3}\mathbf{u}_3 \neq \mathbf{0}.$$

Since $\mathbf{u}_4$ is in $\operatorname{Span}\{\mathbf{u}_1,\mathbf{u}_2,\mathbf{u}_3,\mathbf{v}\}$ and $\mathbf{v}$ is in $\operatorname{Span}\{\mathbf{u}_1,\mathbf{u}_2,\mathbf{u}_3,\mathbf{u}_4\}$, we have

$$\operatorname{Span}\{\mathbf{u}_1,\mathbf{u}_2,\mathbf{u}_3,\mathbf{v}\} = \operatorname{Span}\{\mathbf{u}_1,\mathbf{u}_2,\mathbf{u}_3,\mathbf{u}_4,\mathbf{v}\} = \operatorname{Span}\{\mathbf{u}_1,\mathbf{u}_2,\mathbf{u}_3,\mathbf{u}_4\},$$

by Theorem 2.27. Moreover, since $\mathbf{p}$ is the projection of $\mathbf{v}$ on $\operatorname{Span}\{\mathbf{u}_1,\mathbf{u}_2,\mathbf{u}_3\}$, we have

$$\mathbf{u}_4\cdot\mathbf{u}_1 = (\mathbf{v}-\mathbf{p})\cdot\mathbf{u}_1 = 0, \quad \mathbf{u}_4\cdot\mathbf{u}_2 = (\mathbf{v}-\mathbf{p})\cdot\mathbf{u}_2 = 0, \quad \text{and} \quad \mathbf{u}_4\cdot\mathbf{u}_3 = (\mathbf{v}-\mathbf{p})\cdot\mathbf{u}_3 = 0.$$

□

Example 3.37. Apply Theorem 3.36 to the vectors $\mathbf{u}_1 = \begin{bmatrix} 1\\1\\0\\1\\1 \end{bmatrix}$ and $\mathbf{v} = \begin{bmatrix} 1\\0\\1\\1\\1 \end{bmatrix}$ in order to find a vector $\mathbf{u}_2$ such that $\operatorname{Span}\{\mathbf{u}_1, \mathbf{v}\} = \operatorname{Span}\{\mathbf{u}_1, \mathbf{u}_2\}$ and $\mathbf{u}_2 \cdot \mathbf{u}_1 = 0$.

Solution. The projection of the vector $\mathbf{v}$ on $\operatorname{Span}\{\mathbf{u}_1\}$ is

$$\frac{\mathbf{u}_1 \cdot \mathbf{v}}{\mathbf{u}_1 \cdot \mathbf{u}_1}\mathbf{u}_1 = \frac{3}{4}\begin{bmatrix} 1\\1\\0\\1\\1 \end{bmatrix}.$$

Since

$$\mathbf{v} - \frac{\mathbf{u}_1 \cdot \mathbf{v}}{\mathbf{u}_1 \cdot \mathbf{u}_1}\mathbf{u}_1 = \mathbf{v} - \frac{3}{4}\begin{bmatrix} 1\\1\\0\\1\\1 \end{bmatrix} = \begin{bmatrix} 1\\0\\1\\1\\1 \end{bmatrix} - \frac{3}{4}\begin{bmatrix} 1\\1\\0\\1\\1 \end{bmatrix} = \frac{1}{4}\begin{bmatrix} 1\\-3\\4\\1\\1 \end{bmatrix},$$

we can take

$$\mathbf{u}_2 = \begin{bmatrix} 1\\-3\\4\\1\\1 \end{bmatrix}.$$

□

If we normalize the orthogonal vectors $\begin{bmatrix} 1\\1\\0\\1\\1 \end{bmatrix}$ and $\begin{bmatrix} 1\\-3\\4\\1\\1 \end{bmatrix}$, then we get an orthonormal basis in $\operatorname{Span}\{\mathbf{u}_1, \mathbf{v}\} = \operatorname{Span}\{\mathbf{u}_1, \mathbf{u}_2\}$:

$$\left\{ \begin{bmatrix} \frac{1}{2}\\ \frac{1}{2}\\ 0\\ \frac{1}{2}\\ \frac{1}{2} \end{bmatrix}, \begin{bmatrix} \frac{1}{2\sqrt{7}}\\ -\frac{3}{2\sqrt{7}}\\ \frac{4}{2\sqrt{7}}\\ \frac{1}{2\sqrt{7}}\\ \frac{1}{2\sqrt{7}} \end{bmatrix} \right\}.$$

Example 3.38. For the orthogonal vectors $\mathbf{u}_1 = \begin{bmatrix}1\\1\\0\\1\\1\end{bmatrix}$ and $\mathbf{u}_2 = \begin{bmatrix}1\\-3\\4\\1\\1\end{bmatrix}$ and the vector $\mathbf{v} = \begin{bmatrix}1\\1\\1\\1\\1\end{bmatrix}$ find a vector $\mathbf{u}_3$ such that $\operatorname{Span}\{\mathbf{u}_1,\mathbf{u}_2,\mathbf{v}\} = \operatorname{Span}\{\mathbf{u}_1,\mathbf{u}_2,\mathbf{u}_3\}$ and $\mathbf{u}_3\cdot\mathbf{u}_1 = \mathbf{u}_3\cdot\mathbf{u}_2 = 0$.

Solution. Again we use Theorem 3.36. The projection of $\mathbf{v}$ on $\operatorname{Span}\{\mathbf{u}_1,\mathbf{u}_2\}$ is

$$\frac{\mathbf{u}_1\cdot\mathbf{v}}{\mathbf{u}_1\cdot\mathbf{u}_1}\mathbf{u}_1 + \frac{\mathbf{u}_2\cdot\mathbf{v}}{\mathbf{u}_2\cdot\mathbf{u}_2}\mathbf{u}_2 = \frac{4}{4}\begin{bmatrix}1\\1\\0\\1\\1\end{bmatrix} + \frac{4}{28}\begin{bmatrix}1\\-3\\4\\1\\1\end{bmatrix} = \frac{1}{7}\begin{bmatrix}8\\4\\4\\8\\8\end{bmatrix}.$$

Since

$$\mathbf{v} - \frac{\mathbf{u}_1\cdot\mathbf{v}}{\mathbf{u}_1\cdot\mathbf{u}_1}\mathbf{u}_1 + \frac{\mathbf{u}_2\cdot\mathbf{v}}{\mathbf{u}_2\cdot\mathbf{u}_2}\mathbf{u}_2 = \mathbf{v} - \frac{1}{7}\begin{bmatrix}8\\4\\4\\8\\8\end{bmatrix} = \begin{bmatrix}1\\1\\1\\1\\1\end{bmatrix} - \frac{1}{7}\begin{bmatrix}8\\4\\4\\8\\8\end{bmatrix} = \frac{1}{7}\begin{bmatrix}-1\\3\\3\\-1\\-1\end{bmatrix},$$

we can take

$$\mathbf{u}_3 = \begin{bmatrix}-1\\3\\3\\-1\\-1\end{bmatrix}.$$

□

Example 3.39. For the orthogonal vectors $\mathbf{u}_1 = \begin{bmatrix}1\\1\\0\\1\\1\end{bmatrix}$, $\mathbf{u}_2 = \begin{bmatrix}1\\-3\\4\\1\\1\end{bmatrix}$, $\mathbf{u}_3 = \begin{bmatrix}-1\\3\\3\\-1\\-1\end{bmatrix}$, and the vector $\mathbf{v} = \begin{bmatrix} 1 \\ 0 \\ 1 \\ 0 \\ 1 \end{bmatrix}$ find a vector $\mathbf{u}_4$ such that $\operatorname{Span}\{\mathbf{u}_1, \mathbf{u}_2, \mathbf{u}_3, \mathbf{v}\} = \operatorname{Span}\{\mathbf{u}_1, \mathbf{u}_2, \mathbf{u}_3, \mathbf{u}_4\}$ and $\mathbf{u}_4 \cdot \mathbf{u}_1 = \mathbf{u}_4 \cdot \mathbf{u}_2 = \mathbf{u}_4 \cdot \mathbf{u}_3 = 0$.

Solution. This is another application of Theorem 3.36. Since the projection of $\mathbf{v}$ on $\operatorname{Span}\{\mathbf{u}_1, \mathbf{u}_2, \mathbf{u}_3\}$ is

$$\frac{\mathbf{u}_1 \cdot \mathbf{v}}{\mathbf{u}_1 \cdot \mathbf{u}_1}\mathbf{u}_1 + \frac{\mathbf{u}_2 \cdot \mathbf{v}}{\mathbf{u}_2 \cdot \mathbf{u}_2}\mathbf{u}_2 + \frac{\mathbf{u}_3 \cdot \mathbf{v}}{\mathbf{u}_3 \cdot \mathbf{u}_3}\mathbf{u}_3 = \frac{1}{2}\begin{bmatrix} 1 \\ 1 \\ 0 \\ 1 \\ 1 \end{bmatrix} + \frac{3}{14}\begin{bmatrix} 1 \\ -3 \\ 4 \\ 1 \\ 1 \end{bmatrix} + \frac{1}{21}\begin{bmatrix} -1 \\ 3 \\ 3 \\ -1 \\ -1 \end{bmatrix} = \frac{1}{42}\begin{bmatrix} 28 \\ 0 \\ 42 \\ 28 \\ 28 \end{bmatrix} = \frac{1}{3}\begin{bmatrix} 2 \\ 0 \\ 3 \\ 2 \\ 2 \end{bmatrix}$$

and

$$\mathbf{v} - \frac{\mathbf{u}_1 \cdot \mathbf{v}}{\mathbf{u}_1 \cdot \mathbf{u}_1}\mathbf{u}_1 - \frac{\mathbf{u}_2 \cdot \mathbf{v}}{\mathbf{u}_2 \cdot \mathbf{u}_2}\mathbf{u}_2 + - \frac{\mathbf{u}_3 \cdot \mathbf{v}}{\mathbf{u}_3 \cdot \mathbf{u}_3}\mathbf{u}_3 = \mathbf{v} - \frac{1}{3}\begin{bmatrix} 2 \\ 0 \\ 3 \\ 2 \\ 2 \end{bmatrix} = \frac{1}{2}\begin{bmatrix} 0 \\ -1 \\ 1 \\ 0 \end{bmatrix} = \begin{bmatrix} 1 \\ 0 \\ 1 \\ 0 \\ 1 \end{bmatrix} - \frac{1}{3}\begin{bmatrix} 2 \\ 0 \\ 3 \\ 2 \\ 2 \end{bmatrix} = \frac{1}{3}\begin{bmatrix} 1 \\ 0 \\ 0 \\ -2 \\ 1 \end{bmatrix},$$

we can take

$$\mathbf{u}_4 = \begin{bmatrix} 1 \\ 0 \\ 0 \\ -2 \\ 1 \end{bmatrix}.$$

□

Definition 3.40. We say that $\{\mathbf{u}_1, \dots, \mathbf{u}_k\}$ is an *orthogonal basis* of a subspace $\mathcal{V}$ of $\mathbb{R}^n$ if $\{\mathbf{u}_1, \dots, \mathbf{u}_k\}$ is a basis of $\mathcal{V}$ and the set $\{\mathbf{u}_1, \dots, \mathbf{u}_k\}$ is orthogonal.

Theorem 3.41. *The set $\{\mathbf{u}_1, \dots, \mathbf{u}_k\}$ is an orthogonal basis of a subspace $\mathcal{V}$ of $\mathbb{R}^n$ if and only if $\{\mathbf{u}_1, \dots, \mathbf{u}_k\}$ is an orthogonal set of nonzero vectors and $\mathcal{V} =$ $\operatorname{Span}\{\mathbf{u}_1, \dots, \mathbf{u}_k\}$.*

Proof. This is an immediate consequence of Theorem 3.26. □

Example 3.42. Find an orthogonal basis $\{\mathbf{u}_1, \mathbf{u}_2, \mathbf{u}_3, \mathbf{u}_4\}$ of the subspace

$$\operatorname{Span}\left\{\begin{bmatrix}1\\1\\0\\1\\1\end{bmatrix}, \begin{bmatrix}1\\0\\1\\1\\1\end{bmatrix}, \begin{bmatrix}1\\1\\1\\1\\1\end{bmatrix}, \begin{bmatrix}1\\0\\1\\0\\1\end{bmatrix}\right\}$$

such that

$$\operatorname{Span}\left\{\begin{bmatrix}1\\1\\0\\1\\1\end{bmatrix}, \begin{bmatrix}1\\0\\1\\1\\1\end{bmatrix}\right\} = \operatorname{Span}\{\mathbf{u}_1, \mathbf{u}_2\} \text{ and } \operatorname{Span}\left\{\begin{bmatrix}1\\1\\0\\1\\1\end{bmatrix}, \begin{bmatrix}1\\0\\1\\1\\1\end{bmatrix}, \begin{bmatrix}1\\1\\1\\1\\1\end{bmatrix}\right\} = \operatorname{Span}\{\mathbf{u}_1, \mathbf{u}_2, \mathbf{u}_3\}.$$

Solution. By using the results obtained in the Examples 3.37, 3.38, and 3.39 we have

$$\operatorname{Span}\left\{\begin{bmatrix}1\\1\\0\\1\\1\end{bmatrix}, \begin{bmatrix}1\\0\\1\\1\\1\end{bmatrix}\right\} = \operatorname{Span}\left\{\begin{bmatrix}1\\1\\0\\1\\1\end{bmatrix}, \begin{bmatrix}1\\-3\\4\\1\\1\end{bmatrix}\right\},$$

$$\operatorname{Span}\left\{\begin{bmatrix}1\\1\\0\\1\\1\end{bmatrix}, \begin{bmatrix}1\\0\\1\\1\\1\end{bmatrix}, \begin{bmatrix}1\\1\\1\\1\\1\end{bmatrix}\right\} = \operatorname{Span}\left\{\begin{bmatrix}1\\1\\0\\1\\1\end{bmatrix}, \begin{bmatrix}1\\-3\\4\\1\\1\end{bmatrix}, \begin{bmatrix}1\\1\\1\\1\\1\end{bmatrix}\right\} = \operatorname{Span}\left\{\begin{bmatrix}1\\1\\0\\1\\1\end{bmatrix}, \begin{bmatrix}1\\-3\\4\\1\\1\end{bmatrix}, \begin{bmatrix}-1\\3\\3\\-1\\-1\end{bmatrix}\right\},$$

and

$$\operatorname{Span}\left\{\begin{bmatrix}1\\1\\0\\1\\1\end{bmatrix}, \begin{bmatrix}1\\0\\1\\1\\1\end{bmatrix}, \begin{bmatrix}1\\1\\1\\1\\1\end{bmatrix}, \begin{bmatrix}1\\0\\1\\0\\1\end{bmatrix}\right\} = \operatorname{Span}\left\{\begin{bmatrix}1\\1\\0\\1\\1\end{bmatrix}, \begin{bmatrix}1\\-3\\4\\1\\1\end{bmatrix}, \begin{bmatrix}-1\\3\\3\\-1\\-1\end{bmatrix}, \begin{bmatrix}1\\0\\1\\0\\1\end{bmatrix}\right\}$$

$$= \operatorname{Span}\left\{\begin{bmatrix}1\\1\\0\\1\\1\end{bmatrix}, \begin{bmatrix}1\\-3\\4\\1\\1\end{bmatrix}, \begin{bmatrix}-1\\3\\3\\-1\\-1\end{bmatrix}, \begin{bmatrix}1\\0\\0\\-2\\1\end{bmatrix}\right\}.$$

□

The process of transforming a set of vectors to an orthogonal set using Theorem 3.36, as described in the above example, is called the *Gram-Schmidt orthogonalization process* or simply the *Gram-Schmidt process.*

Example 3.43. For the vectors $\mathbf{u}_1 = \begin{bmatrix} 1 \\ 1 \\ 1 \\ 1 \end{bmatrix}$, $\mathbf{v}_2 = \begin{bmatrix} 1 \\ -1 \\ 0 \\ 1 \end{bmatrix}$, and $\mathbf{v}_3 = \begin{bmatrix} 1 \\ 0 \\ 1 \\ 1 \end{bmatrix}$, find vectors $\mathbf{u}_2$ and $\mathbf{u}_3$ such that $\mathbf{u}_1 \cdot \mathbf{u}_2 = \mathbf{u}_1 \cdot \mathbf{u}_3 = \mathbf{u}_2 \cdot \mathbf{u}_3 = 0$, $\operatorname{Span}\{\mathbf{u}_1, \mathbf{u}_2\} = \operatorname{Span}\{\mathbf{u}_1, \mathbf{v}_2\}$ and $\operatorname{Span}\{\mathbf{u}_1, \mathbf{u}_2, \mathbf{u}_3\} = \operatorname{Span}\{\mathbf{u}_1, \mathbf{v}_2, \mathbf{v}_3\}$.

Proof. Since

$$\mathbf{v}_2 - \frac{\mathbf{u}_1 \cdot \mathbf{v}_2}{\mathbf{u}_1 \cdot \mathbf{u}_1}\mathbf{u}_1 = \begin{bmatrix} 1 \\ -1 \\ 0 \\ 1 \end{bmatrix} - \frac{1}{4}\begin{bmatrix} 1 \\ 1 \\ 1 \\ 1 \end{bmatrix} = \frac{1}{4}\begin{bmatrix} 3 \\ -5 \\ -1 \\ 3 \end{bmatrix},$$

we can take $\mathbf{u}_2 = \begin{bmatrix} 3 \\ -5 \\ -1 \\ 3 \end{bmatrix}$. Then $\mathbf{u}_1 \cdot \mathbf{u}_2 = 0$ and $\operatorname{Span}\{\mathbf{u}_1, \mathbf{u}_2\} = \operatorname{Span}\{\mathbf{u}_1, \mathbf{v}_2\}$.

Now we calculate

$$\mathbf{v}_3 - \frac{\mathbf{u}_1 \cdot \mathbf{v}_3}{\mathbf{u}_1 \cdot \mathbf{u}_1}\mathbf{u}_1 - \frac{\mathbf{u}_2 \cdot \mathbf{v}_3}{\mathbf{u}_2 \cdot \mathbf{u}_2}\mathbf{u}_2 = \begin{bmatrix} 1 \\ 0 \\ 1 \\ 1 \end{bmatrix} - \frac{3}{4}\begin{bmatrix} 1 \\ 1 \\ 1 \\ 1 \end{bmatrix} - \frac{5}{44}\begin{bmatrix} 3 \\ -5 \\ -1 \\ 3 \end{bmatrix} = \frac{-4}{44}\begin{bmatrix} 1 \\ 2 \\ -4 \\ 1 \end{bmatrix}.$$

If we take $\mathbf{u}_3 = \begin{bmatrix} 1 \\ 2 \\ -4 \\ 1 \end{bmatrix}$, then we have $\mathbf{u}_1 \cdot \mathbf{u}_3 = \mathbf{u}_2 \cdot \mathbf{u}_3 = 0$ and

$$\operatorname{Span}\{\mathbf{u}_1, \mathbf{u}_2, \mathbf{u}_3\} = \operatorname{Span}\{\mathbf{u}_1, \mathbf{u}_2, \mathbf{v}_3\} = \operatorname{Span}\{\mathbf{u}_1, \mathbf{v}_2, \mathbf{v}_3\}.$$

□

Example 3.44. Let $\mathbf{u}_1 = \begin{bmatrix} 1 \\ 0 \\ -2 \\ 0 \end{bmatrix}$, $\mathbf{u}_2 = \begin{bmatrix} 2 \\ 1 \\ 1 \\ 1 \end{bmatrix}$, $\mathbf{u}_3 = \begin{bmatrix} 2 \\ -5 \\ 1 \\ 0 \end{bmatrix}$, and $\mathbf{v} = \begin{bmatrix} 5 \\ -4 \\ 0 \\ 1 \end{bmatrix}$. Show that

$$\operatorname{Span}\left\{\begin{bmatrix} 1 \\ 0 \\ -2 \\ 0 \end{bmatrix}, \begin{bmatrix} 2 \\ 1 \\ 1 \\ 1 \end{bmatrix}, \begin{bmatrix} 2 \\ -5 \\ 1 \\ 0 \end{bmatrix}, \begin{bmatrix} 5 \\ -4 \\ 0 \\ 1 \end{bmatrix}\right\} = \operatorname{Span}\left\{\begin{bmatrix} 1 \\ 0 \\ -2 \\ 0 \end{bmatrix}, \begin{bmatrix} 2 \\ 1 \\ 1 \\ 1 \end{bmatrix}, \begin{bmatrix} 2 \\ -5 \\ 1 \\ 0 \end{bmatrix}\right\}.$$

Solution. Since the vectors $\mathbf{u}_1$, $\mathbf{u}_2$, and $\mathbf{u}_3$ are orthogonal, the projection of the vector $\mathbf{v}$ on $\operatorname{Span}\{\mathbf{u}_1,\mathbf{u}_2,\mathbf{u}_3\}$ is

$$\frac{\mathbf{u}_1\cdot\mathbf{v}}{\mathbf{u}_1\cdot\mathbf{u}_1}\mathbf{u}_1+\frac{\mathbf{u}_2\cdot\mathbf{v}}{\mathbf{u}_2\cdot\mathbf{u}_2}\mathbf{u}_2+\frac{\mathbf{u}_3\cdot\mathbf{v}}{\mathbf{u}_3\cdot\mathbf{u}_3}\mathbf{u}_3=\frac{1}{1}\begin{bmatrix}1\\0\\-2\\0\end{bmatrix}+\frac{1}{1}\begin{bmatrix}2\\1\\1\\1\end{bmatrix}+\frac{1}{1}\begin{bmatrix}2\\-5\\1\\0\end{bmatrix}=\begin{bmatrix}5\\-4\\0\\1\end{bmatrix}=\mathbf{v}.$$

Therefore $\begin{bmatrix}5\\-4\\0\\1\end{bmatrix}$ is in $\operatorname{Span}\left\{\begin{bmatrix}1\\0\\-2\\0\end{bmatrix},\begin{bmatrix}2\\1\\1\\1\end{bmatrix},\begin{bmatrix}2\\-5\\1\\0\end{bmatrix}\right\}$ and consequently

$$\operatorname{Span}\left\{\begin{bmatrix}1\\0\\-2\\0\end{bmatrix},\begin{bmatrix}2\\1\\1\\1\end{bmatrix},\begin{bmatrix}2\\-5\\1\\0\end{bmatrix},\begin{bmatrix}5\\-4\\0\\1\end{bmatrix}\right\}=\operatorname{Span}\left\{\begin{bmatrix}1\\0\\-2\\0\end{bmatrix},\begin{bmatrix}2\\1\\1\\1\end{bmatrix},\begin{bmatrix}2\\-5\\1\\0\end{bmatrix}\right\}.$$

□

The following theorem captures some of the important ideas of this section. It is a direct consequence of Theorem 3.36.

Theorem 3.45.

(a) *For arbitrary vectors* $\mathbf{u}_1,\dots,\mathbf{u}_k$ *in* $\mathbb{R}^n$ *there are orthogonal vectors* $\mathbf{v}_1,\dots,\mathbf{v}_m$, *with* $m \le k$, *such that*

$$\operatorname{Span}\{\mathbf{u}_1,\dots,\mathbf{u}_k\}=\operatorname{Span}\{\mathbf{v}_1,\dots,\mathbf{v}_m\}.$$

(b) *For linearly independent vectors* $\mathbf{u}_1,\dots,\mathbf{u}_k$ *in* $\mathbb{R}^n$ *there are orthogonal vectors* $\mathbf{v}_1,\dots,\mathbf{v}_k$ *such that*

$$\operatorname{Span}\{\mathbf{u}_1,\dots,\mathbf{u}_k\}=\operatorname{Span}\{\mathbf{v}_1,\dots,\mathbf{v}_k\}.$$

We close this section with a result that describes the so-called *QR-decomposition* of a matrix. QR-decomposition can be interpreted as the matrix version of the Gram-Schmidt orthogonalization process.

A square matrix is called *upper triangular* if all entries below the main diagonal are zero.

Theorem 3.46. *If an $m \times n$ matrix*

$$A = \begin{bmatrix} \mathbf{c}_1 & \dots & \mathbf{c}_n \end{bmatrix}$$

has linearly independent columns, then A can be represented in the form

$$A = QR$$

where $Q = \begin{bmatrix} \mathbf{u}_1 & \dots & \mathbf{u}_n \end{bmatrix}$ for some orthonormal vectors $\mathbf{u}_1, \dots, \mathbf{u}_n$ in $\mathbb{R}^m$ and R is an upper triangular $n \times n$ matrix R with positive numbers on the diagonal.

Proof. We prove the result for $n = 3$. Using the Gram-Schmidt process we get nonzero orthogonal vectors $\mathbf{v}_1, \mathbf{v}_2, \mathbf{v}_3$ such that

$$\mathbf{v}_1 = \mathbf{c}_1, \quad \text{Span}\{\mathbf{c}_1, \mathbf{c}_2\} = \text{Span}\{\mathbf{v}_1, \mathbf{v}_2\}, \quad \text{and} \quad \text{Span}\{\mathbf{c}_1, \mathbf{c}_2, \mathbf{c}_3\} = \text{Span}\{\mathbf{v}_1, \mathbf{v}_2, \mathbf{v}_3\}.$$

Since $\mathbf{c}_2$ is in $\text{Span}\{\mathbf{v}_1, \mathbf{v}_2\}$ and $\mathbf{c}_3$ is in $\text{Span}\{\mathbf{v}_1, \mathbf{v}_2, \mathbf{v}_3\}$, we get

$$\mathbf{c}_2 = \frac{\mathbf{c}_2 \cdot \mathbf{v}_1}{\mathbf{v}_1 \cdot \mathbf{v}_1}\mathbf{v}_1 + \frac{\mathbf{c}_2 \cdot \mathbf{v}_2}{\mathbf{v}_2 \cdot \mathbf{v}_2}\mathbf{v}_2 = s_{1,2}\mathbf{v}_1 + s_{2,2}\mathbf{v}_2$$

and

$$\mathbf{c}_3 = \frac{\mathbf{c}_3 \cdot \mathbf{v}_1}{\mathbf{v}_1 \cdot \mathbf{v}_1}\mathbf{v}_1 + \frac{\mathbf{c}_3 \cdot \mathbf{v}_2}{\mathbf{v}_2 \cdot \mathbf{v}_2}\mathbf{v}_2 + \frac{\mathbf{c}_3 \cdot \mathbf{v}_3}{\mathbf{v}_3 \cdot \mathbf{v}_3}\mathbf{v}_3 = s_{1,3}\mathbf{v}_1 + s_{2,3}\mathbf{v}_2 + s_{3,3}\mathbf{v}_3.$$

Because the vectors $\mathbf{c}_2$ and $\mathbf{v}_1$ are independent, we have $s_{2,2} \neq 0$. Moreover, since the vector $\mathbf{c}_3$ is not in $\text{Span}\{\mathbf{v}_1, \mathbf{v}_2\}$, we have $s_{3,3} \neq 0$. We can always suppose that $s_{2,2} > 0$, because if not, then we can use $-\mathbf{v}_2$ instead of $\mathbf{v}_2$. Similarly, we can suppose that $s_{3,3} > 0$.

Now we normalize the vectors $\mathbf{v}_1, \mathbf{v}_2, \mathbf{v}_3$:

$$\mathbf{u}_1 = \frac{1}{\|\mathbf{v}_1\|}\mathbf{v}_1, \ \mathbf{u}_2 = \frac{1}{\|\mathbf{v}_2\|}\mathbf{v}_2, \ \mathbf{u}_3 = \frac{1}{\|\mathbf{v}_3\|}\mathbf{v}_3.$$

Next we let

$$r_{1,1} = \|\mathbf{v}_1\|, \ r_{1,2} = \|\mathbf{v}_1\| s_{1,2}, \ r_{1,3} = \|\mathbf{v}_1\| s_{1,3},$$

$$r_{2,2} = \|\mathbf{v}_2\| s_{2,2}, \ r_{2,3} = \|\mathbf{v}_2\| s_{2,3},$$

$$r_{3,3} = \|\mathbf{v}_3\| s_{3,3}.$$

Then we have

$$\mathbf{c}_1 = r_{1,1}\mathbf{u}_1, \quad \mathbf{c}_2 = r_{1,2}\mathbf{u}_1 + r_{2,2}\mathbf{u}_2, \quad \text{and} \quad \mathbf{c}_3 = r_{1,3}\mathbf{u}_1 + r_{2,3}\mathbf{u}_2 + r_{3,3}\mathbf{u}_3,$$

and consequently

$$\begin{bmatrix} \mathbf{c}_1 & \mathbf{c}_2 & \mathbf{c}_3 \end{bmatrix} = \begin{bmatrix} \mathbf{u}_1 & \mathbf{u}_2 & \mathbf{u}_3 \end{bmatrix} \begin{bmatrix} r_{11} & r_{12} & r_{13} \\ 0 & r_{22} & r_{23} \\ 0 & 0 & r_{33} \end{bmatrix}.$$

Note that $r_{1,1} > 0$, $r_{2,2} > 0$, and $r_{3,3} > 0$. □

Note that, if $\mathbf{u}_1, \dots, \mathbf{u}_n$ are orthonormal vectors in $\mathbb{R}^m$, then

$$\begin{bmatrix} \mathbf{u}_1 & \dots & \mathbf{u}_n \end{bmatrix}^T \begin{bmatrix} \mathbf{u}_1 & \dots & \mathbf{u}_n \end{bmatrix} = I_n.$$

Consequently, if $A = \begin{bmatrix} \mathbf{u}_1 & \dots & \mathbf{u}_n \end{bmatrix} R$, then

$$R = \begin{bmatrix} \mathbf{u}_1 & \dots & \mathbf{u}_n \end{bmatrix}^T A.$$

Example 3.47. Determine the QR decomposition of the matrix $A = \begin{bmatrix} 1 & 1 \\ 1 & 0 \\ 1 & -1 \\ -1 & 1 \end{bmatrix}$.

Solution. We apply the Gram-Schmidt process to the vectors $\begin{bmatrix} 1 \\ 1 \\ 1 \\ -1 \end{bmatrix}$ and $\begin{bmatrix} 1 \\ 0 \\ -1 \\ 1 \end{bmatrix}$.

Since

$$\begin{bmatrix} 1 \\ 0 \\ -1 \\ 1 \end{bmatrix} - \frac{\begin{bmatrix} 1 \\ 0 \\ -1 \\ 1 \end{bmatrix} \cdot \begin{bmatrix} 1 \\ 1 \\ 1 \\ -1 \end{bmatrix}}{\begin{bmatrix} 1 \\ 1 \\ 1 \\ -1 \end{bmatrix} \cdot \begin{bmatrix} 1 \\ 1 \\ 1 \\ -1 \end{bmatrix}} \begin{bmatrix} 1 \\ 1 \\ 1 \\ -1 \end{bmatrix} = \begin{bmatrix} 1 \\ 0 \\ -1 \\ 1 \end{bmatrix} + \frac{1}{4} \begin{bmatrix} 1 \\ 1 \\ 1 \\ -1 \end{bmatrix} = \frac{1}{4} \begin{bmatrix} 5 \\ 1 \\ -3 \\ 3 \end{bmatrix},$$

we have

$$\operatorname{Span}\left\{ \begin{bmatrix} 1 \\ 1 \\ 1 \\ -1 \end{bmatrix}, \begin{bmatrix} 1 \\ 0 \\ -1 \\ 1 \end{bmatrix} \right\} = \operatorname{Span}\left\{ \begin{bmatrix} 1 \\ 1 \\ 1 \\ -1 \end{bmatrix}, \begin{bmatrix} 5 \\ 1 \\ -3 \\ 3 \end{bmatrix} \right\}.$$

Now we calculate the norms

$$\left\| \begin{bmatrix} 1 \\ 1 \\ 1 \\ -1 \end{bmatrix} \right\| = 2 \quad \text{and} \quad \left\| \begin{bmatrix} 5 \\ 1 \\ -3 \\ 3 \end{bmatrix} \right\| = 2\sqrt{11}$$

and let

$$\mathbf{u}_1 = \frac{1}{2} \begin{bmatrix} 1 \\ 1 \\ 1 \\ -1 \end{bmatrix} \quad \text{and} \quad \mathbf{u}_2 = \frac{1}{2\sqrt{11}} \begin{bmatrix} 5 \\ 1 \\ -3 \\ 3 \end{bmatrix}.$$

Consequently

$$\begin{bmatrix} 1 \\ 1 \\ 1 \\ -1 \end{bmatrix} = 2\left(\frac{1}{2}\begin{bmatrix} 1 \\ 1 \\ 1 \\ -1 \end{bmatrix}\right) = 2\mathbf{u}_1.$$

Next we use part (a) of Theorem 3.36:

$$\begin{aligned}
\begin{bmatrix} 1 \\ 0 \\ -1 \\ 1 \end{bmatrix} &= \frac{\begin{bmatrix} 1 \\ 0 \\ -1 \\ 1 \end{bmatrix} \cdot \begin{bmatrix} 1 \\ 1 \\ 1 \\ -1 \end{bmatrix}}{\begin{bmatrix} 1 \\ 1 \\ 1 \\ -1 \end{bmatrix} \cdot \begin{bmatrix} 1 \\ 1 \\ 1 \\ -1 \end{bmatrix}} \begin{bmatrix} 1 \\ 1 \\ 1 \\ -1 \end{bmatrix} + \frac{\begin{bmatrix} 1 \\ 0 \\ -1 \\ 1 \end{bmatrix} \cdot \begin{bmatrix} 5 \\ 1 \\ -3 \\ 3 \end{bmatrix}}{\begin{bmatrix} 5 \\ 1 \\ -3 \\ 3 \end{bmatrix} \cdot \begin{bmatrix} 5 \\ 1 \\ -3 \\ 3 \end{bmatrix}} \begin{bmatrix} 5 \\ 1 \\ -3 \\ 3 \end{bmatrix} \\
&= -\frac{1}{4}\begin{bmatrix} 1 \\ 1 \\ 1 \\ -1 \end{bmatrix} + \frac{1}{4}\begin{bmatrix} 5 \\ 1 \\ -3 \\ 3 \end{bmatrix} \\
&= -\frac{1}{2}\left(\frac{1}{2}\begin{bmatrix} 1 \\ 1 \\ 1 \\ -1 \end{bmatrix}\right) + \frac{\sqrt{11}}{2}\left(\frac{1}{2\sqrt{11}}\begin{bmatrix} 5 \\ 1 \\ -3 \\ 3 \end{bmatrix}\right) \\
&= -\frac{1}{2}\mathbf{u}_1 + \frac{\sqrt{11}}{2}\mathbf{u}_2.
\end{aligned}$$

Now it is easy to obtain the QR decomposition of the matrix A:

$$A = \begin{bmatrix} 1 & 1 \\ 1 & 0 \\ 1 & -1 \\ -1 & 1 \end{bmatrix} = \begin{bmatrix} \mathbf{u}_1 & \mathbf{u}_2 \end{bmatrix} \begin{bmatrix} 2 & -\frac{1}{2} \\ 0 & \frac{\sqrt{11}}{2} \end{bmatrix}.$$

As remarked before the example, we also have

$$R = \begin{bmatrix} 2 & -\frac{1}{2} \\ 0 & \frac{\sqrt{11}}{2} \end{bmatrix} = Q^T A = \begin{bmatrix} \mathbf{u}_1^T \\ \mathbf{u}_2^T \end{bmatrix} \begin{bmatrix} 1 & 1 \\ 1 & 0 \\ 1 & -1 \\ -1 & 1 \end{bmatrix}.$$

□

Example 3.48. Determine the QR decomposition of the matrix $A = \begin{bmatrix} 1 & 1 & -2 \\ 1 & 1 & 1 \\ 0 & 1 & -1 \end{bmatrix}$.

Solution. We apply the Gram-Schmidt process to the vectors $\begin{bmatrix} 1 \\ 1 \\ 0 \end{bmatrix}$, $\begin{bmatrix} 1 \\ 1 \\ 1 \end{bmatrix}$ and $\begin{bmatrix} -2 \\ 1 \\ -1 \end{bmatrix}$.

Since

$$\begin{bmatrix} 1 \\ 1 \\ 1 \end{bmatrix} - \frac{\begin{bmatrix} 1 \\ 1 \\ 1 \end{bmatrix} \cdot \begin{bmatrix} 1 \\ 1 \\ 0 \end{bmatrix}}{\begin{bmatrix} 1 \\ 1 \\ 0 \end{bmatrix} \cdot \begin{bmatrix} 1 \\ 1 \\ 0 \end{bmatrix}} \begin{bmatrix} 1 \\ 1 \\ 0 \end{bmatrix} = \begin{bmatrix} 0 \\ 0 \\ 1 \end{bmatrix},$$

we have

$$\operatorname{Span}\left\{ \begin{bmatrix} 1 \\ 1 \\ 0 \end{bmatrix}, \begin{bmatrix} 1 \\ 1 \\ 1 \end{bmatrix} \right\} = \operatorname{Span}\left\{ \begin{bmatrix} 1 \\ 1 \\ 0 \end{bmatrix}, \begin{bmatrix} 0 \\ 0 \\ 1 \end{bmatrix} \right\}.$$

Next, since

$$\begin{bmatrix} -2 \\ 1 \\ -1 \end{bmatrix} - \frac{\begin{bmatrix} -2 \\ 1 \\ -1 \end{bmatrix} \cdot \begin{bmatrix} 1 \\ 1 \\ 0 \end{bmatrix}}{\begin{bmatrix} 1 \\ 1 \\ 0 \end{bmatrix} \cdot \begin{bmatrix} 1 \\ 1 \\ 0 \end{bmatrix}} \begin{bmatrix} 1 \\ 1 \\ 0 \end{bmatrix} - \frac{\begin{bmatrix} -2 \\ 1 \\ -1 \end{bmatrix} \cdot \begin{bmatrix} 0 \\ 0 \\ 1 \end{bmatrix}}{\begin{bmatrix} 0 \\ 0 \\ 1 \end{bmatrix} \cdot \begin{bmatrix} 0 \\ 0 \\ 1 \end{bmatrix}} \begin{bmatrix} 0 \\ 0 \\ 1 \end{bmatrix} = \begin{bmatrix} -2 \\ 1 \\ -1 \end{bmatrix} + \frac{1}{2} \begin{bmatrix} 1 \\ 1 \\ 0 \end{bmatrix} + \begin{bmatrix} 0 \\ 0 \\ 1 \end{bmatrix} = \frac{3}{2} \begin{bmatrix} -1 \\ 1 \\ 0 \end{bmatrix},$$

we have

$$\operatorname{Span}\left\{ \begin{bmatrix} 1 \\ 1 \\ 0 \end{bmatrix}, \begin{bmatrix} 1 \\ 1 \\ 1 \end{bmatrix}, \begin{bmatrix} -2 \\ 1 \\ -1 \end{bmatrix} \right\} = \operatorname{Span}\left\{ \begin{bmatrix} 1 \\ 1 \\ 0 \end{bmatrix}, \begin{bmatrix} 0 \\ 0 \\ 1 \end{bmatrix}, \begin{bmatrix} -1 \\ 1 \\ 0 \end{bmatrix} \right\}.$$

Now we normalize the vectors. First we calculate the norms

$$\left\| \begin{bmatrix} 1 \\ 1 \\ 0 \end{bmatrix} \right\| = \sqrt{2}, \quad \left\| \begin{bmatrix} 0 \\ 0 \\ 1 \end{bmatrix} \right\| = 1, \quad \text{and} \quad \left\| \begin{bmatrix} -1 \\ 1 \\ 0 \end{bmatrix} \right\| = \sqrt{2}$$

and then let $\mathbf{u}_1 = \frac{1}{\sqrt{2}}\begin{bmatrix}1\\1\\0\end{bmatrix}$, $\mathbf{u}_2 = \begin{bmatrix}0\\0\\1\end{bmatrix}$, and $\mathbf{u}_3 = \frac{1}{\sqrt{2}}\begin{bmatrix}-1\\1\\0\end{bmatrix}$. Consequently,

$$\begin{bmatrix}1\\1\\0\end{bmatrix} = \sqrt{2}\left(\frac{1}{\sqrt{2}}\begin{bmatrix}1\\1\\0\end{bmatrix}\right) = \sqrt{2}\mathbf{u}_1.$$

Next we use part (a) of Theorem 3.36 to get

$$\begin{bmatrix}1\\1\\1\end{bmatrix} = \frac{\begin{bmatrix}1\\1\\1\end{bmatrix}\cdot\begin{bmatrix}1\\1\\0\end{bmatrix}}{\begin{bmatrix}1\\1\\0\end{bmatrix}\cdot\begin{bmatrix}1\\1\\0\end{bmatrix}}\begin{bmatrix}1\\1\\0\end{bmatrix} + \frac{\begin{bmatrix}1\\1\\1\end{bmatrix}\cdot\begin{bmatrix}0\\0\\1\end{bmatrix}}{\begin{bmatrix}0\\0\\1\end{bmatrix}\cdot\begin{bmatrix}0\\0\\1\end{bmatrix}}\begin{bmatrix}0\\0\\1\end{bmatrix} = \begin{bmatrix}1\\1\\0\end{bmatrix} + \begin{bmatrix}0\\0\\1\end{bmatrix} = \sqrt{2}\mathbf{u}_1 + \mathbf{u}_2$$

and

$$\begin{aligned}\begin{bmatrix}-2\\1\\-1\end{bmatrix} &= \frac{\begin{bmatrix}-2\\1\\-1\end{bmatrix}\cdot\begin{bmatrix}1\\1\\0\end{bmatrix}}{\begin{bmatrix}1\\1\\0\end{bmatrix}\cdot\begin{bmatrix}1\\1\\0\end{bmatrix}}\begin{bmatrix}1\\1\\0\end{bmatrix} + \frac{\begin{bmatrix}-2\\1\\-1\end{bmatrix}\cdot\begin{bmatrix}0\\0\\1\end{bmatrix}}{\begin{bmatrix}0\\0\\1\end{bmatrix}\cdot\begin{bmatrix}0\\0\\1\end{bmatrix}}\begin{bmatrix}0\\0\\1\end{bmatrix} + \frac{\begin{bmatrix}-2\\1\\-1\end{bmatrix}\cdot\begin{bmatrix}-1\\1\\0\end{bmatrix}}{\begin{bmatrix}-1\\1\\0\end{bmatrix}\cdot\begin{bmatrix}-1\\1\\0\end{bmatrix}}\begin{bmatrix}-1\\1\\0\end{bmatrix}\\ &= -\frac{1}{\sqrt{2}}\begin{bmatrix}1\\1\\0\end{bmatrix} - \begin{bmatrix}0\\0\\1\end{bmatrix} + \frac{3}{2}\begin{bmatrix}-1\\1\\0\end{bmatrix}\\ &= -\frac{1}{2}\mathbf{u}_1 - \mathbf{u}_2 + \frac{3}{\sqrt{2}}\mathbf{u}_3.\end{aligned}$$

Now it is easy to obtain the QR decomposition of the matrix A:

$$A = \begin{bmatrix}1 & 1 & -2\\1 & 1 & 1\\0 & 1 & -1\end{bmatrix} = \begin{bmatrix}\mathbf{u}_1 & \mathbf{u}_2 & \mathbf{u}_3\end{bmatrix}\begin{bmatrix}\sqrt{2} & \sqrt{2} & -\frac{1}{\sqrt{2}}\\0 & 1 & -1\\0 & 0 & \frac{3}{\sqrt{2}}\end{bmatrix}.$$

We also get

$$R = \begin{bmatrix}\sqrt{2} & \sqrt{2} & -\frac{1}{\sqrt{2}}\\0 & 1 & -1\\0 & 0 & \frac{3}{\sqrt{2}}\end{bmatrix} = \begin{bmatrix}\mathbf{u}_1^T\\\mathbf{u}_2^T\\\mathbf{u}_3^T\end{bmatrix}\begin{bmatrix}1 & 1 & -2\\1 & 1 & 1\\0 & 1 & -1\end{bmatrix}.$$

□

Exercises 3.3

1. For $\mathbf{u}_1 = \begin{bmatrix} 1 \\ -3 \\ 3 \\ -1 \end{bmatrix}$, $\mathbf{u}_2 = \begin{bmatrix} 1 \\ 1 \\ 1 \\ 1 \end{bmatrix}$, and $\mathbf{v} = \begin{bmatrix} 3 \\ -1 \\ 5 \\ 1 \end{bmatrix}$ show that $\operatorname{Span}\{\mathbf{u}_1, \mathbf{u}_2, \mathbf{v}\} = \operatorname{Span}\{\mathbf{u}_1, \mathbf{u}_2\}$.

2. For $\mathbf{u}_1 = \begin{bmatrix} 2 \\ 1 \\ 0 \\ 2 \end{bmatrix}$, $\mathbf{u}_2 = \begin{bmatrix} 1 \\ 2 \\ 3 \\ -2 \end{bmatrix}$, and $\mathbf{v} = \begin{bmatrix} 5 \\ 3 \\ 1 \\ 4 \end{bmatrix}$ show that $\operatorname{Span}\{\mathbf{u}_1, \mathbf{u}_2, \mathbf{v}\} = \operatorname{Span}\{\mathbf{u}_1, \mathbf{u}_2\}$.

Determine an orthogonal basis for of the subspace $\operatorname{Span}\{\mathbf{u}_1, \mathbf{u}_2\}$ for the given vectors $\mathbf{u}_1$ and $\mathbf{u}_2$.

3. $\mathbf{u}_1 = \begin{bmatrix} 1 \\ 1 \\ 1 \\ 1 \end{bmatrix}$, $\mathbf{u}_2 = \begin{bmatrix} 1 \\ 0 \\ 0 \\ 0 \end{bmatrix}$

4. $\mathbf{u}_1 = \begin{bmatrix} 1 \\ 1 \\ 1 \\ 1 \end{bmatrix}$, $\mathbf{v} = \begin{bmatrix} 1 \\ -1 \\ 0 \\ 1 \end{bmatrix}$

5. $\mathbf{u}_1 = \begin{bmatrix} 1 \\ 1 \\ 0 \\ 1 \\ 0 \end{bmatrix}$, $\mathbf{u}_2 = \begin{bmatrix} 1 \\ 0 \\ 1 \\ 0 \\ 1 \end{bmatrix}$

6. $\mathbf{u}_1 = \begin{bmatrix} 1 \\ 1 \\ 1 \\ 1 \\ 1 \end{bmatrix}$, $\mathbf{u}_2 = \begin{bmatrix} 0 \\ 1 \\ 0 \\ 0 \\ 0 \end{bmatrix}$

7. For the orthogonal vectors $\mathbf{u}_1 = \begin{bmatrix} 1 \\ 1 \\ 1 \\ 0 \end{bmatrix}$ and $\mathbf{u}_2 = \begin{bmatrix} 1 \\ -1 \\ 0 \\ 1 \end{bmatrix}$ and the vector $\mathbf{v} = \begin{bmatrix} 1 \\ 0 \\ 0 \\ 0 \end{bmatrix}$ find a vector $\mathbf{u}_3$ such that $\operatorname{Span}\{\mathbf{u}_1, \mathbf{u}_2, \mathbf{v}\} = \operatorname{Span}\{\mathbf{u}_1, \mathbf{u}_2, \mathbf{u}_3\}$ and $\mathbf{u}_3 \cdot \mathbf{u}_1 = \mathbf{u}_3 \cdot \mathbf{u}_2 = 0$.

8. For the orthogonal vectors $\mathbf{u}_1 = \begin{bmatrix} 1 \\ 0 \\ 1 \\ 0 \end{bmatrix}$ and $\mathbf{u}_2 = \begin{bmatrix} 0 \\ 1 \\ 0 \\ 1 \end{bmatrix}$ and the vector $\mathbf{v} = \begin{bmatrix} 1 \\ 1 \\ 0 \\ 0 \end{bmatrix}$ find a vector $\mathbf{u}_3$ such that $\operatorname{Span}\{\mathbf{u}_1, \mathbf{u}_2, \mathbf{v}\} = \operatorname{Span}\{\mathbf{u}_1, \mathbf{u}_2, \mathbf{u}_3\}$ and $\mathbf{u}_3 \cdot \mathbf{u}_1 = \mathbf{u}_3 \cdot \mathbf{u}_2 = 0$.

9. For the vectors $\mathbf{u}_1 = \begin{bmatrix} 1 \\ 0 \\ 1 \\ 0 \end{bmatrix}$, $\mathbf{v}_2 = \begin{bmatrix} 1 \\ 1 \\ 0 \\ 0 \end{bmatrix}$, and $\mathbf{v}_3 = \begin{bmatrix} 1 \\ 0 \\ 0 \\ 1 \end{bmatrix}$ find vectors $\mathbf{u}_2$ and $\mathbf{u}_3$ such that $\mathbf{u}_1 \cdot \mathbf{u}_2 = \mathbf{u}_1 \cdot \mathbf{u}_3 = \mathbf{u}_2 \cdot \mathbf{u}_3 = 0$, $\operatorname{Span}\{\mathbf{u}_1, \mathbf{u}_2\} = \operatorname{Span}\{\mathbf{u}_1, \mathbf{v}_2\}$, and $\operatorname{Span}\{\mathbf{u}_1, \mathbf{u}_2, \mathbf{u}_3\} = \operatorname{Span}\{\mathbf{u}_1, \mathbf{v}_2, \mathbf{v}_3\}$.

10. For the vectors $\mathbf{u}_1 = \begin{bmatrix} 1 \\ 1 \\ 1 \\ 1 \end{bmatrix}$, $\mathbf{v}_2 = \begin{bmatrix} 1 \\ 1 \\ 1 \\ 0 \end{bmatrix}$, and $\mathbf{v}_3 = \begin{bmatrix} 1 \\ 1 \\ 0 \\ 0 \end{bmatrix}$ find vectors $\mathbf{u}_2$ and $\mathbf{u}_3$ such that $\mathbf{u}_1 \cdot \mathbf{u}_2 = \mathbf{u}_1 \cdot \mathbf{u}_3 = \mathbf{u}_2 \cdot \mathbf{u}_3 = 0$, $\text{Span}\{\mathbf{u}_1, \mathbf{u}_2\} = \text{Span}\{\mathbf{u}_1, \mathbf{v}_2\}$ and $\text{Span}\{\mathbf{u}_1, \mathbf{u}_2, \mathbf{u}_3\} = \text{Span}\{\mathbf{u}_1, \mathbf{v}_2, \mathbf{v}_3\}$.

Determine the QR decomposition of the given matrix.

11. $\begin{bmatrix} 1 & 2 \\ 1 & 1 \end{bmatrix}$

12. $\begin{bmatrix} 3 & 0 \\ 1 & 3 \end{bmatrix}$

13. $\begin{bmatrix} 1 & 1 \\ -1 & 1 \\ 1 & 3 \end{bmatrix}$

14. $\begin{bmatrix} 1 & 1 \\ 1 & 2 \\ 1 & 3 \end{bmatrix}$

15. $\begin{bmatrix} 1 & -1 \\ 0 & 1 \\ 1 & 1 \\ -1 & 1 \end{bmatrix}$

16. $\begin{bmatrix} 1 & 0 \\ 0 & 1 \\ 1 & 1 \\ 1 & 1 \end{bmatrix}$

17. $\begin{bmatrix} 1 & 1 & 0 \\ 0 & 0 & 1 \\ 1 & 0 & 1 \end{bmatrix}$

18. $\begin{bmatrix} 1 & 1 & -1 \\ 0 & 0 & 1 \\ 0 & 1 & 0 \end{bmatrix}$

19. $\begin{bmatrix} 1 & 0 & 1 \\ 1 & 1 & 0 \\ 1 & 1 & 1 \\ 0 & 1 & 1 \end{bmatrix}$

20. $A = \begin{bmatrix} 0 & 1 & 1 \\ 1 & 0 & 0 \\ 0 & 0 & 1 \\ 1 & 1 & 0 \end{bmatrix}$

3.4 Least squares approximation

Methods for solving systems of linear equations motivate a lot of ideas in linear algebra. Systems of linear equations are usually interpreted as matrix equations $A\mathbf{x} = \mathbf{b}$. We know that such equations may have exactly one solution, infinitely many solutions, or no solution. In this section we consider the following question. If the equation $A\mathbf{x} = \mathbf{b}$ has no solution, can we find a vector $\mathbf{x}$ such that $\|\mathbf{b} - A\mathbf{x}\|$ is minimized? We can interpret the number $\|\mathbf{b} - A\mathbf{x}\|$ as an error that we are trying to minimize. Linear algebra gives us an effective tool to deal with such problems. The method is based on the following theorem.

Theorem 3.49. *Let $\mathbf{u}_1,\dots,\mathbf{u}_k$ and $\mathbf{b}$ be vectors in $\mathbb{R}^n$ and let A be the $n \times k$ matrix with columns $\mathbf{u}_1,\dots,\mathbf{u}_k$, that is, $A = \begin{bmatrix}\mathbf{u}_1 & \dots & \mathbf{u}_k\end{bmatrix}$. The real numbers $x_1,\dots,x_k$ minimize the norm*

$$\|\mathbf{b} - x_1\mathbf{u}_1 - \dots - x_k\mathbf{u}_k\| = \|\mathbf{b} - A\mathbf{x}\|$$

if and only if the vector $\mathbf{x} = \begin{bmatrix} x_1 \\ \vdots \\ x_k \end{bmatrix}$ *satisfies the equation*

$$A^T A\mathbf{x} = A^T\mathbf{b}.$$

Proof. First note that for any numbers $x_1,\dots,x_k$ the vector $x_1\mathbf{u}_1 + \dots + x_k\mathbf{u}_k$ is in the subspace $\operatorname{Span}\{\mathbf{u}_1,\dots,\mathbf{u}_k\}$. Consequently, the distance

$$\|\mathbf{b} - (x_1\mathbf{u}_1 + \dots + x_k\mathbf{u}_k)\| = \|\mathbf{b} - x_1\mathbf{u}_1 - \dots - x_k\mathbf{u}_k\|$$

is minimized if the vector $x_1\mathbf{u}_1 + \dots + x_k\mathbf{u}_k$ is the orthogonal projection of $\mathbf{b}$ on the subspace $\operatorname{Span}\{\mathbf{u}_1,\dots,\mathbf{u}_k\}$. By Theorem 3.16, the projection is characterized by the equations

$$\begin{array}{ccccccc}
x_1\mathbf{u}_1\cdot\mathbf{u}_1 & + & \cdots & + & x_k\mathbf{u}_k\cdot\mathbf{u}_1 & = & \mathbf{b}\cdot\mathbf{u}_1 \\
x_1\mathbf{u}_1\cdot\mathbf{u}_2 & + & \cdots & + & x_k\mathbf{u}_k\cdot\mathbf{u}_2 & = & \mathbf{b}\cdot\mathbf{u}_2 \\
 & \vdots & & \vdots & & \vdots & \\
x_1\mathbf{u}_1\cdot\mathbf{u}_k & + & \cdots & + & x_k\mathbf{u}_k\cdot\mathbf{u}_k & = & \mathbf{b}\cdot\mathbf{u}_k.
\end{array}$$

We will show, in the case when $k = 3$, how these equations lead to the matrix equation $A^T A\mathbf{x} = A^T\mathbf{b}$.

The system

$$\begin{array}{ccccccc}
x_1\mathbf{u}_1\cdot\mathbf{u}_1 & + & x_2\mathbf{u}_2\cdot\mathbf{u}_1 & + & x_3\mathbf{u}_3\cdot\mathbf{u}_1 & = & \mathbf{b}\cdot\mathbf{u}_1 \\
x_1\mathbf{u}_1\cdot\mathbf{u}_2 & + & x_2\mathbf{u}_2\cdot\mathbf{u}_2 & + & x_3\mathbf{u}_3\cdot\mathbf{u}_2 & = & \mathbf{b}\cdot\mathbf{u}_2 \\
x_1\mathbf{u}_1\cdot\mathbf{u}_3 & + & x_2\mathbf{u}_2\cdot\mathbf{u}_3 & + & x_3\mathbf{u}_3\cdot\mathbf{u}_3 & = & \mathbf{b}\cdot\mathbf{u}_3
\end{array}$$

can be written as

$$\begin{array}{ccccccc}
x_1\mathbf{u}_1^T\mathbf{u}_1 & + & x_2\mathbf{u}_2^T\mathbf{u}_1 & + & x_3\mathbf{u}_3^T\mathbf{u}_1 & = & \mathbf{u}_1^T\mathbf{b} \\
x_1\mathbf{u}_1^T\mathbf{u}_2 & + & x_2\mathbf{u}_2^T\mathbf{u}_2 & + & x_3\mathbf{u}_3^T\mathbf{u}_2 & = & \mathbf{u}_2^T\mathbf{b} \\
x_1\mathbf{u}_1^T\mathbf{u}_3 & + & x_2\mathbf{u}_2^T\mathbf{u}_3 & + & x_3\mathbf{u}_3^T\mathbf{u}_3 & = & \mathbf{u}_3^T\mathbf{b}
\end{array}$$

or as the matrix equation

$$\begin{bmatrix}
\mathbf{u}_1^T\mathbf{u}_1 & \mathbf{u}_2^T\mathbf{u}_1 & \mathbf{u}_3^T\mathbf{u}_1 \\
\mathbf{u}_1^T\mathbf{u}_2 & \mathbf{u}_2^T\mathbf{u}_2 & \mathbf{u}_3^T\mathbf{u}_2 \\
\mathbf{u}_1^T\mathbf{u}_3 & \mathbf{u}_2^T\mathbf{u}_3 & \mathbf{u}_3^T\mathbf{u}_3
\end{bmatrix}
\begin{bmatrix} x_1 \\ x_2 \\ x_3 \end{bmatrix}
=
\begin{bmatrix} \mathbf{u}_1^T\mathbf{b} \\ \mathbf{u}_2^T\mathbf{b} \\ \mathbf{u}_3^T\mathbf{b} \end{bmatrix}. \tag{3.1}$$

Since

$$\begin{bmatrix} \mathbf{u}_1^T\mathbf{u}_1 & \mathbf{u}_2^T\mathbf{u}_1 & \mathbf{u}_3^T\mathbf{u}_1 \\ \mathbf{u}_1^T\mathbf{u}_2 & \mathbf{u}_2^T\mathbf{u}_2 & \mathbf{u}_3^T\mathbf{u}_2 \\ \mathbf{u}_1^T\mathbf{u}_3 & \mathbf{u}_2^T\mathbf{u}_3 & \mathbf{u}_3^T\mathbf{u}_3 \end{bmatrix} = \begin{bmatrix} \mathbf{u}_1^T \\ \mathbf{u}_2^T \\ \mathbf{u}_3^T \end{bmatrix} \begin{bmatrix} \mathbf{u}_1 & \mathbf{u}_2 & \mathbf{u}_3 \end{bmatrix} \quad \text{and} \quad \begin{bmatrix} \mathbf{u}_1^T\mathbf{b} \\ \mathbf{u}_2^T\mathbf{b} \\ \mathbf{u}_3^T\mathbf{b} \end{bmatrix} = \begin{bmatrix} \mathbf{u}_1^T \\ \mathbf{u}_2^T \\ \mathbf{u}_3^T \end{bmatrix} \mathbf{b},$$

equation (3.1) can be written as

$$A^T A\mathbf{x} = A^T\mathbf{b}$$

where

$$A = \begin{bmatrix} \mathbf{u}_1 & \mathbf{u}_2 & \mathbf{u}_3 \end{bmatrix} \quad \text{and} \quad \mathbf{x} = \begin{bmatrix} x_1 \\ x_2 \\ x_3 \end{bmatrix}.$$

□

Definition 3.50. The matrix equation

$$A^T A\mathbf{x} = A^T\mathbf{b}$$

is called the *normal equation* corresponding to the equation $A\mathbf{x} = \mathbf{b}$.

Definition 3.51. Let A be an $m \times n$ matrix and let $\mathbf{b}$ be a vector in $\mathbb{R}^m$. A vector $\hat{\mathbf{x}}$ in $\mathbb{R}^n$ is called a *least squares solution* of the equation $A\mathbf{x} = \mathbf{b}$ if $\hat{\mathbf{x}}$ is a solution of the normal equation, that is,

$$A^T A\hat{\mathbf{x}} = A^T\mathbf{b}.$$

Note that if a vector $\hat{\mathbf{x}}$ is a least squares solution of the equation $A\mathbf{x} = \mathbf{b}$, then

$$\|\mathbf{b} - A\hat{\mathbf{x}}\| \le \|\mathbf{b} - A\mathbf{x}\|$$

and $A\hat{\mathbf{x}}$ is the projection of the vector $\mathbf{b}$ on the subspace $\mathbf{C}(A)$, that is, the subspace of the columns of A.

Example 3.52. Let

$$A = \begin{bmatrix} 1 & 1 & 2 \\ 1 & -1 & 1 \\ 2 & 1 & 1 \\ 1 & 1 & -1 \end{bmatrix} \quad \text{and} \quad \mathbf{b} = \begin{bmatrix} 1 \\ 3 \\ 1 \\ 1 \end{bmatrix}.$$

Find the normal equation corresponding to the equation $A\mathbf{x} = \mathbf{b}$.

Solution. Since

$$A^T = \begin{bmatrix} 1 & 1 & 2 & 1 \\ 1 & -1 & 1 & 1 \\ 2 & 1 & 1 & -1 \end{bmatrix},$$

$$A^T A = \begin{bmatrix} 1 & 1 & 2 & 1 \\ 1 & -1 & 1 & 1 \\ 2 & 1 & 1 & -1 \end{bmatrix} \begin{bmatrix} 1 & 1 & 2 \\ 1 & -1 & 1 \\ 2 & 1 & 1 \\ 1 & 1 & -1 \end{bmatrix} = \begin{bmatrix} 7 & 3 & 4 \\ 3 & 4 & 1 \\ 4 & 1 & 7 \end{bmatrix},$$

and

$$A^T\mathbf{b} = \begin{bmatrix} 1 & 1 & 2 & 1 \\ 1 & -1 & 1 & 1 \\ 2 & 1 & 1 & -1 \end{bmatrix} \begin{bmatrix} 1 \\ 3 \\ 1 \\ 1 \end{bmatrix} = \begin{bmatrix} 7 \\ 0 \\ 5 \end{bmatrix},$$

the normal equation corresponding to the equation

$$\begin{bmatrix} 1 & 1 & 2 \\ 1 & -1 & 1 \\ 2 & 1 & 1 \\ 1 & 1 & -1 \end{bmatrix} \begin{bmatrix} x_1 \\ x_2 \\ x_3 \end{bmatrix} = \begin{bmatrix} 1 \\ 3 \\ 1 \\ 1 \end{bmatrix}$$

is the equation

$$\begin{bmatrix} 7 & 3 & 4 \\ 3 & 4 & 1 \\ 4 & 1 & 7 \end{bmatrix} \begin{bmatrix} x_1 \\ x_2 \\ x_3 \end{bmatrix} = \begin{bmatrix} 7 \\ 0 \\ 5 \end{bmatrix}.$$

□

Example 3.53. Find numbers x_1, x_2, and x_3 that minimize the sum

$$(1 - x_1 - x_2 - 2x_3)^2 + (3 - x_1 + x_2 - x_3)^2 + (1 - 2x_1 - x_2 - x_3)^2 + (1 - x_1 - x_2 + x_3)^2.$$

Solution. Since

$$(1 - x_1 - x_2 - 2x_3)^2 + (3 - x_1 + x_2 - x_3)^2 + (1 - 2x_1 - x_2 - x_3)^2 + (1 - x_1 - x_2 + x_3)^2$$

$$= \left\| \begin{bmatrix} 1 \\ 3 \\ 1 \\ 1 \end{bmatrix} - x_1 \begin{bmatrix} 1 \\ 1 \\ 2 \\ 1 \end{bmatrix} - x_2 \begin{bmatrix} 1 \\ -1 \\ 1 \\ 1 \end{bmatrix} - x_3 \begin{bmatrix} 2 \\ 1 \\ 1 \\ -1 \end{bmatrix} \right\|^2,$$

we can use Theorem 3.49 with

$$\mathbf{u}_1 = \begin{bmatrix} 1 \\ 1 \\ 2 \\ 1 \end{bmatrix}, \mathbf{u}_2 = \begin{bmatrix} 1 \\ -1 \\ 1 \\ 1 \end{bmatrix}, \mathbf{u}_3 = \begin{bmatrix} 2 \\ 1 \\ 1 \\ -1 \end{bmatrix}, \quad \text{and} \quad \mathbf{b} = \begin{bmatrix} 1 \\ 3 \\ 1 \\ 1 \end{bmatrix}.$$

We have

$$A = \begin{bmatrix} \mathbf{u}_1 & \mathbf{u}_2 & \mathbf{u}_3 \end{bmatrix} = \begin{bmatrix} 1 & 1 & 2 \\ 1 & -1 & 1 \\ 2 & 1 & 1 \\ 1 & 1 & -1 \end{bmatrix}$$

and the equation $A^T A\mathbf{x} = A^T\mathbf{b}$ is

$$\begin{bmatrix} 1 & 1 & 2 & 1 \\ 1 & -1 & 1 & 1 \\ 2 & 1 & 1 & -1 \end{bmatrix} \begin{bmatrix} 1 & 1 & 2 \\ 1 & -1 & 1 \\ 2 & 1 & 1 \\ 1 & 1 & -1 \end{bmatrix} \begin{bmatrix} x_1 \\ x_2 \\ x_3 \end{bmatrix} = \begin{bmatrix} 1 & 1 & 2 & 1 \\ 1 & -1 & 1 & 1 \\ 2 & 1 & 1 & -1 \end{bmatrix} \begin{bmatrix} 1 \\ 3 \\ 1 \\ 1 \end{bmatrix}.$$

In view of the calculations in Example 3.52, this equation reduces to

$$\begin{bmatrix} 7 & 3 & 4 \\ 3 & 4 & 1 \\ 4 & 1 & 7 \end{bmatrix} \begin{bmatrix} x_1 \\ x_2 \\ x_3 \end{bmatrix} = \begin{bmatrix} 7 \\ 0 \\ 5 \end{bmatrix}.$$

Since

$$\begin{bmatrix} 7 & 3 & 4 & 7 \\ 3 & 4 & 1 & 0 \\ 4 & 1 & 7 & 5 \end{bmatrix} \sim \begin{bmatrix} 1 & 0 & 0 & \frac{62}{43} \\ 0 & 1 & 0 & -\frac{47}{43} \\ 0 & 0 & 1 & \frac{2}{43} \end{bmatrix},$$

the solution is

$$x_1 = \frac{62}{43},\ x_2 = -\frac{47}{43},\ x_3 = \frac{2}{43}.$$

These numbers minimize the sum

$$(1 - x_1 - x_2 - 2x_3)^2 + (3 - x_1 + x_2 - x_3)^2 + (1 - 2x_1 - x_2 - x_3)^2 + (1 - x_1 - x_2 + x_3)^2.$$

□

Example 3.54. Find a least squares solution of the system

$$\begin{cases} 3x + \ \ y + 2z = 0 \\ \ \ x + 3y + 2z = 1 \\ \ \ x + \ \ y + \ \ z = 1 \end{cases}.$$

Solution. The reduced row echelon form of the augmented matrix of the system is

$$\begin{bmatrix} 1 & 0 & 1/2 & 0 \\ 0 & 1 & 1/2 & 0 \\ 0 & 0 & 0 & 1 \end{bmatrix}$$

which shows that the system is inconsistent, that is, has no solutions. We let

$$A = \begin{bmatrix} 3 & 1 & 2 \\ 1 & 3 & 2 \\ 1 & 1 & 1 \end{bmatrix} \quad \text{and} \quad \mathbf{b} = \begin{bmatrix} 0 \\ 1 \\ 1 \end{bmatrix}.$$

Then

$$A^T A = \begin{bmatrix} 3 & 1 & 1 \\ 1 & 3 & 1 \\ 2 & 2 & 1 \end{bmatrix} \begin{bmatrix} 3 & 1 & 2 \\ 1 & 3 & 2 \\ 1 & 1 & 1 \end{bmatrix} = \begin{bmatrix} 11 & 7 & 9 \\ 7 & 11 & 9 \\ 9 & 9 & 9 \end{bmatrix} \quad \text{and} \quad A^T \mathbf{b} = \begin{bmatrix} 2 \\ 4 \\ 3 \end{bmatrix}.$$

We have to solve the normal equation $A^T A\mathbf{x} = A^T \mathbf{b}$ which is equivalent to the system

$$\begin{cases} 11x_1 + \ 7x_2 + 9x_3 = 2 \\ \ 7x_1 + 11x_2 + 9x_3 = 4 \\ \ 9x_1 + \ 9x_2 + 9x_2 = 3 \end{cases}.$$

Since the reduced row echelon form of the augmented matrix

$$\begin{bmatrix} 11 & 7 & 9 & 2 \\ 7 & 11 & 9 & 4 \\ 9 & 9 & 9 & 3 \end{bmatrix}$$

is

$$\begin{bmatrix} 1 & 0 & \frac{1}{2} & -\frac{1}{12} \\ 0 & 1 & \frac{1}{2} & \frac{5}{12} \\ 0 & 0 & 0 & 0 \end{bmatrix},$$

the general solution is

$$x_1 = -\frac{1}{2}x_3 - \frac{1}{12} \quad \text{and} \quad x_2 = -\frac{1}{2}x_3 + \frac{5}{12},$$

where x_3 is a free variable.

Note that a least squares solution is not unique in this case.

We also note that the projection of the vector $\mathbf{b}$ on $\mathbf{C}(A)$ is

$$A \begin{bmatrix} x_1 \\ x_2 \\ x_3 \end{bmatrix} = A \begin{bmatrix} -\frac{1}{2}x_3 - \frac{1}{12} \\ -\frac{1}{2}x_3 + \frac{5}{12} \\ x_3 \end{bmatrix} = \frac{1}{6} \begin{bmatrix} 1 \\ 7 \\ 2 \end{bmatrix}.$$

□

In the described method for finding a least squares solution of a matrix equation $A\mathbf{x} = \mathbf{b}$ we do not assume any special properties of the matrix A. Consequently, the method works for an arbitrary matrix A. Now we are going to consider a special case when the columns of the matrix A are linearly independent. Under this additional assumption a least squares solution of a matrix equation $A\mathbf{x} = \mathbf{b}$ can be found by multiplying $\mathbf{b}$ by a matrix. This is especially useful if it is necessary to find least squares solutions of a matrix equation $A\mathbf{x} = \mathbf{b}$ for a fixed matrix A and different choices of the vector $\mathbf{b}$.

We start with the following important observation.

Lemma 3.55. *Let $\mathbf{u}_1, \dots, \mathbf{u}_k$ be vectors in $\mathbb{R}^n$ and let A be the $n \times k$ matrix with columns $\mathbf{u}_1, \dots, \mathbf{u}_k$, that is, $A = \begin{bmatrix} \mathbf{u}_1 & \mathbf{u}_2 & \dots & \mathbf{u}_k \end{bmatrix}$. Then the vectors $\mathbf{u}_1, \dots, \mathbf{u}_k$ are linearly independent if and only if the square matrix $A^T A$ is invertible.*

Proof of a particular case. We prove the lemma when $k = 3$ and $n = 4$. Let

$$A = \begin{bmatrix} \mathbf{u}_1 & \mathbf{u}_2 & \mathbf{u}_3 \end{bmatrix} = \begin{bmatrix} a_1 & b_1 & c_1 \\ a_2 & b_2 & c_2 \\ a_3 & b_3 & c_3 \\ a_4 & b_4 & c_4 \end{bmatrix}.$$

First assume that the vectors $\mathbf{u}_1$, $\mathbf{u}_2$, and $\mathbf{u}_3$ are linearly independent. We can show that $A^T A$ invertible by showing that the only solution of the equation

$$A^T A \begin{bmatrix} x \\ y \\ z \end{bmatrix} = \begin{bmatrix} 0 \\ 0 \\ 0 \end{bmatrix} \tag{3.2}$$

is the trivial solution, that is, $x = y = z = 0$.

If (3.2) holds for some vector $\begin{bmatrix} x \\ y \\ z \end{bmatrix}$, then we have

$$\left\| A \begin{bmatrix} x \\ y \\ z \end{bmatrix} \right\|^2 = \begin{bmatrix} x & y & z \end{bmatrix} A^T A \begin{bmatrix} x \\ y \\ z \end{bmatrix} = \begin{bmatrix} x & y & z \end{bmatrix} \begin{bmatrix} 0 \\ 0 \\ 0 \end{bmatrix} = 0.$$

Hence

$$A \begin{bmatrix} x \\ y \\ z \end{bmatrix} = x \begin{bmatrix} a_1 \\ a_2 \\ a_3 \\ a_4 \end{bmatrix} + y \begin{bmatrix} b_1 \\ b_2 \\ b_3 \\ b_4 \end{bmatrix} + z \begin{bmatrix} c_1 \\ c_2 \\ c_3 \\ c_4 \end{bmatrix} = \begin{bmatrix} 0 \\ 0 \\ 0 \\ 0 \end{bmatrix},$$

which gives us $x = y = z = 0$, because the vectors $\begin{bmatrix} a_1 \\ a_2 \\ a_3 \\ a_4 \end{bmatrix}$, $\begin{bmatrix} b_1 \\ b_2 \\ b_3 \\ b_4 \end{bmatrix}$, $\begin{bmatrix} c_1 \\ c_2 \\ c_3 \\ c_4 \end{bmatrix}$ are linearly independent.

Now assume that the matrix $A^T A$ is invertible. If

$$A\begin{bmatrix}x\\y\\z\end{bmatrix} = x\begin{bmatrix}a_1\\a_2\\a_3\\a_4\end{bmatrix} + y\begin{bmatrix}b_1\\b_2\\b_3\\b_4\end{bmatrix} + z\begin{bmatrix}c_1\\c_2\\c_3\\c_4\end{bmatrix} = \begin{bmatrix}0\\0\\0\\0\end{bmatrix},$$

then

$$A^T A\begin{bmatrix}x\\y\\z\end{bmatrix} = A^T\begin{bmatrix}0\\0\\0\\0\end{bmatrix} = \begin{bmatrix}0\\0\\0\end{bmatrix}.$$

Since the matrix $A^T A$ is invertible, we get

$$\begin{bmatrix}x\\y\\z\end{bmatrix} = (A^T A)^{-1}\begin{bmatrix}0\\0\\0\end{bmatrix} = \begin{bmatrix}0\\0\\0\end{bmatrix}$$

and thus $x = y = z = 0$. This shows that the vectors $\begin{bmatrix}a_1\\a_2\\a_3\\a_4\end{bmatrix}, \begin{bmatrix}b_1\\b_2\\b_3\\b_4\end{bmatrix}, \begin{bmatrix}c_1\\c_2\\c_3\\c_4\end{bmatrix}$ are linearly independent. □

Theorem 3.56. *Let* $\mathbf{u}_1, \dots, \mathbf{u}_k$ *be linearly independent vectors in* $\mathbb{R}^n$ *and let* A *be the matrix with columns* $\mathbf{u}_1, \dots, \mathbf{u}_k$, *that is,* $A = \begin{bmatrix}\mathbf{u}_1 & \mathbf{u}_2 & \dots & \mathbf{u}_k\end{bmatrix}$.

(a) *For every vector* $\mathbf{b}$ *in* $\mathbb{R}^n$ *the vector*

$$\hat{\mathbf{x}} = (A^T A)^{-1} A^T \mathbf{b}$$

is the unique least squares solution of the equation $A\mathbf{x} = \mathbf{b}$.

(b) *The projection matrix on the subspace* $\mathbf{C}(A) = \operatorname{Span}\{\mathbf{u}_1, \dots, \mathbf{u}_k\}$ *is* $A(A^T A)^{-1} A^T$.

Proof. The least squares solution of the equation

$$A\mathbf{x} = \mathbf{b}$$

is the solution of the normal equation

$$A^T A\mathbf{x} = A^T\mathbf{b},$$

which is

$$\hat{\mathbf{x}} = (A^T A)^{-1} A^T \mathbf{b},$$

because the matrix $A^T A$ is invertible.

Now, since the projection of the vector $\mathbf{b}$ on the subspace Span$\{\mathbf{u}_1,\dots,\mathbf{u}_k\}$ is $A\hat{\mathbf{x}}$ and

$$A\hat{\mathbf{x}} = A((A^T A)^{-1}A^T\mathbf{b}) = (A(A^T A)^{-1}A^T)\mathbf{b},$$

the projection matrix on the subspace Span$\{\mathbf{u}_1,\dots,\mathbf{u}_k\}$ is $A(A^T A)^{-1}A^T$. □

Example 3.57. Find the least squares solution of the system

$$\begin{cases} x+y = 1 \\ x+ z = 0 \\ y+z = 0 \\ x+y+z = 0 \end{cases}$$

and then the projection matrix on the subspace

$$\text{Span}\left\{\begin{bmatrix}1\\1\\0\\1\end{bmatrix}, \begin{bmatrix}1\\0\\1\\1\end{bmatrix}, \begin{bmatrix}0\\1\\1\\1\end{bmatrix}\right\}.$$

Solution. To find the least squares solution of the system

$$\begin{cases} x+y = 1 \\ x+ z = 0 \\ y+z = 0 \\ x+y+z = 0 \end{cases}$$

we first note the system can be written as

$$\begin{bmatrix}1&1&0\\1&0&1\\0&1&1\\1&1&1\end{bmatrix}\begin{bmatrix}x\\y\\z\end{bmatrix} = \begin{bmatrix}1\\0\\0\\0\end{bmatrix}.$$

We denote $A = \begin{bmatrix}1&1&0\\1&0&1\\0&1&1\\1&1&1\end{bmatrix}$ and $\mathbf{b} = \begin{bmatrix}1\\0\\0\\0\end{bmatrix}$. Since

$$A = \begin{bmatrix}1&1&0\\1&0&1\\0&1&1\\1&1&1\end{bmatrix} \sim \begin{bmatrix}1&0&0\\0&1&0\\0&0&1\\0&0&0\end{bmatrix},$$

the vectors $\begin{bmatrix}1\\1\\0\\1\end{bmatrix}, \begin{bmatrix}1\\0\\1\\1\end{bmatrix}, \begin{bmatrix}0\\1\\1\\1\end{bmatrix}$ are linearly independent.

Now we find

$$A^T = \begin{bmatrix}1&1&0&1\\1&0&1&1\\0&1&1&1\end{bmatrix},$$

then

$$A^T A = \begin{bmatrix}3&2&2\\2&3&2\\2&2&3\end{bmatrix},$$

and finally

$$(A^T A)^{-1} = \frac{1}{7}\begin{bmatrix}5&-2&-2\\-2&5&-2\\-2&-2&5\end{bmatrix}.$$

Consequently, the least squares solution is

$$\begin{bmatrix}x\\y\\z\end{bmatrix} = (A^T A)^{-1}A^T\mathbf{b} = \frac{1}{7}\begin{bmatrix}5&-2&-2\\-2&5&-2\\-2&-2&5\end{bmatrix}\begin{bmatrix}1&1&0&1\\1&0&1&1\\0&1&1&1\end{bmatrix}\begin{bmatrix}1\\0\\0\\0\end{bmatrix} = \begin{bmatrix}\frac{3}{7}\\-\frac{3}{7}\\-\frac{4}{7}\end{bmatrix}$$

and the projection matrix is

$$\begin{aligned}A(A^T A)^{-1}A^T &= \begin{bmatrix}1&1&0\\1&0&1\\0&1&1\\1&1&1\end{bmatrix}\begin{bmatrix}3&2&2\\2&3&2\\2&2&3\end{bmatrix}^{-1}\begin{bmatrix}1&1&0&1\\1&0&1&1\\0&1&1&1\end{bmatrix}\\ &= \begin{bmatrix}3/7&3/7&-4/7\\3/7&-4/7&3/7\\-4/7&3/7&3/7\\1/7&1/7&1/7\end{bmatrix}\begin{bmatrix}1&1&0&1\\1&0&1&1\\0&1&1&1\end{bmatrix}\\ &= \begin{bmatrix}6/7&-1/7&-1/7&2/7\\-1/7&6/7&-1/7&2/7\\-1/7&-1/7&6/7&2/7\\2/7&2/7&2/7&3/7\end{bmatrix}.\end{aligned}$$

□

Exercises 3.4

1. For $A = \begin{bmatrix} 1 & 2 & 2 \\ 1 & 2 & 2 \end{bmatrix}$ and $\mathbf{b} = \begin{bmatrix} 3 \\ 1 \end{bmatrix}$ find the normal equation corresponding to the equation $A \begin{bmatrix} x \\ y \\ z \end{bmatrix} = \mathbf{b}$.

2. For $A = \begin{bmatrix} 1 & 2 \\ 2 & 3 \\ 1 & 0 \\ 0 & 2 \end{bmatrix}$ and $\mathbf{b} = \begin{bmatrix} 1 \\ 1 \\ 1 \\ 1 \end{bmatrix}$ find the normal equation corresponding to the equation $A \begin{bmatrix} x \\ y \end{bmatrix} = \mathbf{b}$.

Use the method from Example 3.53 to find real numbers x and y that minimize the given sum.

3. $(1+2x-y)^2+(1-2x+y)^2+(1-4x+2y)^2+(2-4x+2y)^2$
4. $(1+x-2y)^2+(1-x)^2+(1+y)^2+(1-x-y)^2$

Find the least squares solution of the given system of equations.

5. $\begin{cases} x+y=1 \\ x-y=1 \\ -x+y=1 \end{cases}$

6. $\begin{cases} 2x+y=4 \\ 2x+y=1 \\ x+y=1 \end{cases}$

7. $\begin{cases} x+y=1 \\ x+y=2 \\ x+y=3 \\ x+y=4 \end{cases}$

8. $\begin{cases} 2x+y=2 \\ x+2y=1 \\ x+y=2 \\ x-y=1 \end{cases}$

9. $\begin{cases} x+y+z=1 \\ x+y+z=2 \end{cases}$

10. $\begin{cases} 2x+2y+3z=2 \\ 2x+2y+3z=5 \end{cases}$

11. $\begin{cases} x+2y+z=1 \\ 2x+y+z=1 \\ 4x-y+z=1 \\ 5x-2y+z=1 \end{cases}$

12. $\begin{cases} x+y+2z=1 \\ x+y+2z=0 \\ x+y+2z=2 \\ x+y+2z=-1 \end{cases}$

Find real numbers x and y that minimize the given sum using Theorem 3.56.

13. $(1-x-y)^2+(1+x-y)^2+(1-x+y)^2+(1-x-2y)^2$
14. $(1+x-2y)^2+(1+2x-y)^2+(1+y)^2+(1-x-y)^2$
15. $(1-x-y)^2+(1-y)^2+(1-x)^2+(1-x+y)^2+(1+x+y)^2$
16. $(1+x+2y)^2+(1+x-y)^2+(1+y)^2+(1-x-y)^2+(1+x)^2$

Find the projection of $\mathbf{b}$ on $\text{Span}\{\mathbf{u}_1, \mathbf{u}_2\}$ using Theorem 3.56.

17. $\mathbf{b} = \begin{bmatrix} 1 \\ 1 \\ 1 \\ 1 \end{bmatrix}, \mathbf{u}_1 = \begin{bmatrix} 1 \\ 1 \\ 1 \\ 0 \end{bmatrix}, \mathbf{u}_2 = \begin{bmatrix} 1 \\ -1 \\ 0 \\ 1 \end{bmatrix}$

18. $\mathbf{b} = \begin{bmatrix} 1 \\ 0 \\ 1 \\ 1 \end{bmatrix}, \mathbf{u}_1 = \begin{bmatrix} 1 \\ 0 \\ 1 \\ 0 \end{bmatrix}, \mathbf{u}_2 = \begin{bmatrix} 0 \\ 1 \\ 0 \\ 1 \end{bmatrix}$

19. $\mathbf{b} = \begin{bmatrix} 1 \\ 0 \\ 0 \\ 0 \\ 0 \end{bmatrix}, \mathbf{u}_1 = \begin{bmatrix} 1 \\ 1 \\ 1 \\ 1 \\ 0 \end{bmatrix}, \mathbf{u}_2 = \begin{bmatrix} 1 \\ 1 \\ 1 \\ 0 \\ 1 \end{bmatrix}$

20. $\mathbf{b} = \begin{bmatrix} 1 \\ 0 \\ 0 \\ 1 \\ 0 \end{bmatrix}, \mathbf{u}_1 = \begin{bmatrix} 1 \\ 0 \\ 0 \\ 0 \\ 0 \end{bmatrix}, \mathbf{u}_2 = \begin{bmatrix} 1 \\ 1 \\ 1 \\ 1 \\ 1 \end{bmatrix}$

Find the projection matrix on the given subspace.

21. $\operatorname{Span}\left\{ \begin{bmatrix} 1 \\ 0 \\ 0 \\ 1 \end{bmatrix}, \begin{bmatrix} 0 \\ 0 \\ 1 \\ 0 \end{bmatrix} \right\}$

22. $\operatorname{Span}\left\{ \begin{bmatrix} 1 \\ 1 \\ 1 \\ 1 \end{bmatrix}, \begin{bmatrix} 0 \\ 0 \\ 0 \\ 1 \end{bmatrix} \right\}$

23. $\operatorname{Span}\left\{ \begin{bmatrix} 1 \\ 1 \\ 0 \\ 0 \end{bmatrix}, \begin{bmatrix} 1 \\ 0 \\ 1 \\ 1 \end{bmatrix} \right\}$

24. $\operatorname{Span}\left\{ \begin{bmatrix} 1 \\ 1 \\ 0 \\ 1 \end{bmatrix}, \begin{bmatrix} 0 \\ 1 \\ 1 \\ 1 \end{bmatrix} \right\}$

3.5 The orthogonal complement of a subspace of $\mathbb{R}^n$

The scalar product allows us to decompose $\mathbb{R}^n$ into orthogonal components. This has important theoretical and practical consequences.

Definition 3.58. Let $\mathcal{V}$ be a subspace of $\mathbb{R}^n$. The set of all vectors in $\mathbb{R}^n$ orthogonal to every vector in $\mathcal{V}$ is called the *orthogonal complement* of $\mathcal{V}$ and is denoted by $\mathcal{V}^\perp$:

$$\mathcal{V}^\perp = \{\mathbf{x} : \mathbf{x} \text{ is in } \mathbb{R}^n \text{ and } \mathbf{x} \cdot \mathbf{v} = 0 \text{ for every } \mathbf{v} \text{ in } \mathcal{V}\}.$$

It is easy to show that the orthogonal complement of a subspace of $\mathbb{R}^n$ is also a subspace of $\mathbb{R}^n$. The fact that the orthogonal complement of a subspace is a subspace is also an immediate consequence of the following theorem.

Theorem 3.59. *For any matrix A we have*

$$(\mathbf{C}(A))^\perp = \mathbf{N}(A^T).$$

Proof for $n = 3$. If $A = \begin{bmatrix} \mathbf{c}_1 & \mathbf{c}_2 & \mathbf{c}_3 \end{bmatrix}$, then a vector $\mathbf{x}$ is in $(\mathbf{C}(A))^\perp$ if and only if

$$\mathbf{c}_1 \cdot \mathbf{x} = \mathbf{c}_2 \cdot \mathbf{x} = \mathbf{c}_3 \cdot \mathbf{x} = 0$$

or, equivalently,

$$\mathbf{c}_1^T \mathbf{x} = \mathbf{c}_2^T \mathbf{x} = \mathbf{c}_2^T \mathbf{x} = 0.$$

These equalities can be written as a single matrix equation

$$\begin{bmatrix} \mathbf{c}_1^T \\ \mathbf{c}_2^T \\ \mathbf{c}_3^T \end{bmatrix} \mathbf{x} = A^T \mathbf{x} = \begin{bmatrix} 0 \\ 0 \\ 0 \end{bmatrix},$$

which means that $\mathbf{x}$ is in $\mathbf{N}(A^T)$. □

The above theorem not only tells us that the orthogonal complement is a subspace but helps us find a basis of that subspace.

Example 3.60. Find a basis of the orthogonal complement of the subspace

$$\text{Span}\left\{ \begin{bmatrix} 2 \\ 1 \\ 1 \\ 1 \end{bmatrix}, \begin{bmatrix} 1 \\ 1 \\ 1 \\ 2 \end{bmatrix} \right\}.$$

Solution. The question is equivalent to finding a basis of the orthogonal complement of the subspace $\mathbf{C}(A)$, where A is the matrix

$$A = \begin{bmatrix} 2 & 1 \\ 1 & 1 \\ 1 & 1 \\ 1 & 2 \end{bmatrix}.$$

By Theorem 3.59,

$$(\mathbf{C}(A))^\perp = \mathbf{N}(A^T),$$

so we can solve the problem by finding a basis of

$$\mathbf{N}(A^T) = \mathbf{N}\left(\begin{bmatrix} 2 & 1 & 1 & 1 \\ 1 & 1 & 1 & 2 \end{bmatrix} \right).$$

Since

$$\begin{bmatrix} 2 & 1 & 1 & 1 \\ 1 & 1 & 1 & 2 \end{bmatrix} \sim \begin{bmatrix} 1 & 0 & 0 & -1 \\ 0 & 1 & 1 & 3 \end{bmatrix},$$

a vector $\begin{bmatrix} x_1 \\ x_2 \\ x_3 \\ x_4 \end{bmatrix}$ is in $\mathbf{N}(A^T)$ if it is a solution of the system

$$\begin{cases} x_1 \qquad\qquad - \ x_4 = 0 \\ \quad x_2 + x_3 + 3x_4 = 0 \end{cases}.$$

The general solution is

$$\begin{bmatrix} x_1 \\ x_2 \\ x_3 \\ x_4 \end{bmatrix} = \begin{bmatrix} x_4 \\ -x_3 - 3x_4 \\ x_3 \\ x_4 \end{bmatrix} = x_3 \begin{bmatrix} 0 \\ -1 \\ 1 \\ 0 \end{bmatrix} + x_4 \begin{bmatrix} 1 \\ -3 \\ 0 \\ 1 \end{bmatrix}$$

which means that

$$(\mathbf{C}(A))^\perp = \mathbf{N}(A^T) = \operatorname{Span}\left\{ \begin{bmatrix} 0 \\ -1 \\ 1 \\ 0 \end{bmatrix}, \begin{bmatrix} 1 \\ -3 \\ 0 \\ 1 \end{bmatrix} \right\}$$

and consequently $\left\{ \begin{bmatrix} 0 \\ -1 \\ 1 \\ 0 \end{bmatrix}, \begin{bmatrix} 1 \\ -3 \\ 0 \\ 1 \end{bmatrix} \right\}$ is a basis of the orthogonal complement of the subspace $\operatorname{Span}\left\{ \begin{bmatrix} 2 \\ 1 \\ 1 \\ 1 \end{bmatrix}, \begin{bmatrix} 1 \\ 1 \\ 1 \\ 2 \end{bmatrix} \right\}$. □

Theorem 3.61. *Let $\mathcal{V}$ be a subspace of $\mathbb{R}^n$ and let $\mathbf{b}$ be an arbitrary vector in $\mathbb{R}^n$. Then*

(a) $\mathbf{b} = \operatorname{proj}_{\mathcal{V}}(\mathbf{b}) + \operatorname{proj}_{\mathcal{V}^\perp}(\mathbf{b})$.

(b) *The decomposition of* $\mathbf{b}$ *in part* (a) is unique in the following sense: If

$$\mathbf{b} = \mathbf{v} + \mathbf{w} \text{ for some } \mathbf{v} \text{ in } \mathcal{V} \text{ and some } \mathbf{w} \text{ in } \mathcal{V}^\perp,$$

then

$$\mathbf{v} = \operatorname{proj}_{\mathcal{V}}(\mathbf{b}) \quad \text{and} \quad \mathbf{w} = \operatorname{proj}_{\mathcal{V}^\perp}(\mathbf{b}).$$

Proof. To prove part (a) we note that

$$\mathbf{b} = \operatorname{proj}_{\mathcal{V}}(\mathbf{b}) + \mathbf{b} - \operatorname{proj}_{\mathcal{V}}(\mathbf{b}).$$

Since $\mathbf{b} - \operatorname{proj}_{\mathcal{V}}(\mathbf{b})$ is a vector in the subspace $\mathcal{V}^{\perp}$ and

$$\left(\mathbf{b} - (\mathbf{b} - \operatorname{proj}_{\mathcal{V}}(\mathbf{b})\right) \cdot \mathbf{x} = \operatorname{proj}_{\mathcal{V}}(\mathbf{b}) \cdot \mathbf{x} = 0$$

for every $\mathbf{x}$ in $\mathcal{V}^{\perp}$, we have $\mathbf{b} - \operatorname{proj}_{\mathcal{V}}(\mathbf{b}) = \operatorname{proj}_{\mathcal{V}^{\perp}}(\mathbf{b})$ and thus

$$\mathbf{b} = \operatorname{proj}_{\mathcal{V}}(\mathbf{b}) + \operatorname{proj}_{\mathcal{V}^{\perp}}(\mathbf{b}).$$

Now, if $\mathbf{b} = \mathbf{v} + \mathbf{w}$, for some $\mathbf{v}$ in $\mathcal{V}$ and $\mathbf{w}$ in $\mathcal{V}^{\perp}$, then

$$(\mathbf{b} - \mathbf{v}) \cdot \mathbf{x} = \mathbf{w} \cdot \mathbf{x} = 0$$

for every $\mathbf{x}$ in $\mathcal{V}$. Hence $\mathbf{v} = \operatorname{proj}_{\mathcal{V}}(\mathbf{b})$. Moreover

$$(\mathbf{b} - \mathbf{w}) \cdot \mathbf{x} = \mathbf{v} \cdot \mathbf{x} = 0$$

for every $\mathbf{x}$ in $\mathcal{V}^{\perp}$. Consequently, $\mathbf{w} = \operatorname{proj}_{\mathcal{V}^{\perp}}(\mathbf{b})$. □

Example 3.62. Find $\mathbf{v}$ and $\mathbf{w}$ such that $\begin{bmatrix} 1 \\ 0 \\ 0 \\ 0 \end{bmatrix} = \mathbf{v} + \mathbf{w}$ where $\mathbf{v}$ is in

$$\operatorname{Span}\left\{ \begin{bmatrix} 2 \\ 1 \\ 1 \\ 1 \end{bmatrix}, \begin{bmatrix} 1 \\ 1 \\ 1 \\ 2 \end{bmatrix}, \begin{bmatrix} 3 \\ 1 \\ 2 \\ 1 \end{bmatrix} \right\}$$

and $\mathbf{v} \cdot \mathbf{w} = 0$.

Solution. Using one of the methods from this chapter we find that the projection of the vector $\begin{bmatrix} 1 \\ 0 \\ 0 \\ 0 \end{bmatrix}$ on the subspace $\operatorname{Span}\left\{ \begin{bmatrix} 2 \\ 1 \\ 1 \\ 1 \end{bmatrix}, \begin{bmatrix} 1 \\ 1 \\ 1 \\ 2 \end{bmatrix}, \begin{bmatrix} 3 \\ 1 \\ 2 \\ 1 \end{bmatrix} \right\}$ is $\frac{1}{7} \begin{bmatrix} 6 \\ 2 \\ 1 \\ -1 \end{bmatrix}$.

Now we have

$$\begin{bmatrix} 1 \\ 0 \\ 0 \\ 0 \end{bmatrix} = \frac{1}{7} \begin{bmatrix} 6 \\ 2 \\ 1 \\ -1 \end{bmatrix} + \frac{1}{7} \begin{bmatrix} 1 \\ -2 \\ -1 \\ 1 \end{bmatrix},$$

where $\begin{bmatrix} 6 \\ 2 \\ 1 \\ -1 \end{bmatrix}$ is in $\operatorname{Span}\left\{ \begin{bmatrix} 2 \\ 1 \\ 1 \\ 1 \end{bmatrix}, \begin{bmatrix} 1 \\ 1 \\ 1 \\ 2 \end{bmatrix}, \begin{bmatrix} 3 \\ 1 \\ 2 \\ 1 \end{bmatrix} \right\}$ and $\begin{bmatrix} 6 \\ 2 \\ 1 \\ -1 \end{bmatrix} \cdot \begin{bmatrix} 1 \\ -2 \\ -1 \\ 1 \end{bmatrix} = 0.$ □

From Theorem 3.61 we obtain the following useful result.

Corollary 3.63. *Let $\mathcal{V}$ be a subspace of $\mathbb{R}^n$.*

(a) *If $\{\mathbf{v}_1,\dots,\mathbf{v}_k\}$ be a basis of $\mathcal{V}$ and $\{\mathbf{w}_1,\dots,\mathbf{w}_l\}$ be a basis of $\mathcal{V}^\perp$, then $\{\mathbf{v}_1,\dots,\mathbf{v}_k,\mathbf{w}_1,\dots,\mathbf{w}_l\}$ is a basis of $\mathbb{R}^n$.*

(b) *If $\{\mathbf{v}_1,\dots,\mathbf{v}_k\}$ be an orthogonal basis of $\mathcal{V}$ and $\{\mathbf{w}_1,\dots,\mathbf{w}_l\}$ be an orthogonal basis of $\mathcal{V}^\perp$, then $\{\mathbf{v}_1,\dots,\mathbf{v}_k,\mathbf{w}_1,\dots,\mathbf{w}_l\}$ is an orthogonal basis of $\mathbb{R}^n$.*

Example 3.64. Find an orthogonal basis of the subspace Span$\left\{\begin{bmatrix}2\\1\\1\\1\end{bmatrix},\begin{bmatrix}1\\1\\1\\2\end{bmatrix}\right\}$. Extend that basis to an orthogonal basis of $\mathbb{R}^4$ and find the coordinates of the vector $\begin{bmatrix}0\\0\\1\\0\end{bmatrix}$ in that basis.

Solution. We find, using Gram-Schmidt process, that $\left\{\begin{bmatrix}2\\1\\1\\1\end{bmatrix},\begin{bmatrix}-5\\1\\1\\8\end{bmatrix}\right\}$ is an orthogonal basis of the subspace Span$\left\{\begin{bmatrix}2\\1\\1\\1\end{bmatrix},\begin{bmatrix}1\\1\\1\\2\end{bmatrix}\right\}$ and in Example 3.60 we found that $\left\{\begin{bmatrix}0\\-1\\1\\0\end{bmatrix},\begin{bmatrix}1\\-3\\0\\1\end{bmatrix}\right\}$ is a basis of the orthogonal complement of the subspace Span$\left\{\begin{bmatrix}2\\1\\1\\1\end{bmatrix},\begin{bmatrix}1\\1\\1\\2\end{bmatrix}\right\}$. The Gram-Schmidt process produces $\left\{\begin{bmatrix}0\\-1\\1\\0\end{bmatrix},\begin{bmatrix}2\\-3\\-3\\2\end{bmatrix}\right\}$ as an orthogonal basis of the orthogonal complement of the subspace Span$\left\{\begin{bmatrix}2\\1\\1\\1\end{bmatrix},\begin{bmatrix}1\\1\\1\\2\end{bmatrix}\right\}$

and consequently

$$\left\{\begin{bmatrix}2\\1\\1\\1\end{bmatrix},\begin{bmatrix}-5\\1\\1\\8\end{bmatrix},\begin{bmatrix}0\\-1\\1\\0\end{bmatrix},\begin{bmatrix}2\\-3\\-3\\2\end{bmatrix}\right\}$$

is an orthogonal basis of the space $\mathbb{R}^4$. Consequently, there are real numbers x, y, z, and w such that

$$\begin{bmatrix}0\\0\\1\\0\end{bmatrix}=x\begin{bmatrix}2\\1\\1\\1\end{bmatrix}+y\begin{bmatrix}-5\\1\\1\\8\end{bmatrix}+z\begin{bmatrix}0\\-1\\1\\0\end{bmatrix}+w\begin{bmatrix}2\\-3\\-3\\2\end{bmatrix}.$$

Since

$$\begin{bmatrix}0\\0\\1\\0\end{bmatrix}\cdot\begin{bmatrix}2\\1\\1\\1\end{bmatrix}=x\begin{bmatrix}2\\1\\1\\1\end{bmatrix}\cdot\begin{bmatrix}2\\1\\1\\1\end{bmatrix}+y\begin{bmatrix}-5\\1\\1\\8\end{bmatrix}\cdot\begin{bmatrix}2\\1\\1\\1\end{bmatrix}+z\begin{bmatrix}0\\-1\\1\\0\end{bmatrix}\cdot\begin{bmatrix}2\\1\\1\\1\end{bmatrix}+w\begin{bmatrix}2\\-3\\-3\\2\end{bmatrix}\cdot\begin{bmatrix}2\\1\\1\\1\end{bmatrix}$$

and

$$\begin{bmatrix}-5\\1\\1\\8\end{bmatrix}\cdot\begin{bmatrix}2\\1\\1\\1\end{bmatrix}=\begin{bmatrix}0\\-1\\1\\0\end{bmatrix}\cdot\begin{bmatrix}2\\1\\1\\1\end{bmatrix}=\begin{bmatrix}2\\-3\\-3\\2\end{bmatrix}\cdot\begin{bmatrix}2\\1\\1\\1\end{bmatrix}=0,$$

we have $\begin{bmatrix}0\\0\\1\\0\end{bmatrix}\cdot\begin{bmatrix}2\\1\\1\\1\end{bmatrix}=x\begin{bmatrix}2\\1\\1\\1\end{bmatrix}\cdot\begin{bmatrix}2\\1\\1\\1\end{bmatrix}$, and thus $x=\dfrac{\begin{bmatrix}0\\0\\1\\0\end{bmatrix}\cdot\begin{bmatrix}2\\1\\1\\1\end{bmatrix}}{\begin{bmatrix}2\\1\\1\\1\end{bmatrix}\cdot\begin{bmatrix}2\\1\\1\\1\end{bmatrix}}=\dfrac{1}{7}$. In a similar way we obtain

$$y=\frac{\begin{bmatrix}0\\0\\1\\0\end{bmatrix}\cdot\begin{bmatrix}-5\\1\\1\\8\end{bmatrix}}{\begin{bmatrix}-5\\1\\1\\8\end{bmatrix}\cdot\begin{bmatrix}-5\\1\\1\\8\end{bmatrix}}=\frac{1}{91},\quad z=\frac{\begin{bmatrix}0\\0\\1\\0\end{bmatrix}\cdot\begin{bmatrix}0\\-1\\1\\0\end{bmatrix}}{\begin{bmatrix}0\\-1\\1\\0\end{bmatrix}\cdot\begin{bmatrix}0\\-1\\1\\0\end{bmatrix}}=\frac{1}{2},\quad\text{and}\quad w=\frac{\begin{bmatrix}0\\0\\1\\0\end{bmatrix}\cdot\begin{bmatrix}2\\-3\\-3\\2\end{bmatrix}}{\begin{bmatrix}2\\-3\\-3\\2\end{bmatrix}\cdot\begin{bmatrix}2\\-3\\-3\\2\end{bmatrix}}=-\frac{3}{26}.$$

Consequently,

$$\begin{bmatrix}0\\0\\1\\0\end{bmatrix} = \frac{1}{7}\begin{bmatrix}2\\1\\1\\1\end{bmatrix} + \frac{1}{91}\begin{bmatrix}-5\\1\\1\\8\end{bmatrix} + \frac{1}{2}\begin{bmatrix}0\\-1\\1\\0\end{bmatrix} - \frac{3}{26}\begin{bmatrix}2\\-3\\-3\\2\end{bmatrix}.$$

□

Theorem 3.65.

(a) *For an arbitrary subspace $\mathcal{V}$ of $\mathbb{R}^n$ we have $(\mathcal{V}^\perp)^\perp = \mathcal{V}$.*

(b) *For an arbitrary matrix A we have $(\mathbf{N}(A))^\perp = \mathbf{C}(A^T)$.*

Proof. If $\mathbf{b}$ is in $(\mathcal{V}^\perp)^\perp$, then we have

$$\begin{aligned}0 &= \mathbf{b}\cdot \operatorname{proj}_{\mathcal{V}^\perp}(\mathbf{b})\\ &= (\operatorname{proj}_{\mathcal{V}}(\mathbf{b}) + \operatorname{proj}_{\mathcal{V}^\perp}(\mathbf{b}))\cdot \operatorname{proj}_{\mathcal{V}^\perp}(\mathbf{b})\\ &= \operatorname{proj}_{\mathcal{V}^\perp}(\mathbf{b})\cdot \operatorname{proj}_{\mathcal{V}^\perp}(\mathbf{b})\\ &= \|\operatorname{proj}_{\mathcal{V}^\perp}(\mathbf{b})\|^2,\end{aligned}$$

which implies that $\operatorname{proj}_{\mathcal{V}^\perp}(\mathbf{b}) = \mathbf{0}$. Consequently $\mathbf{b} = \operatorname{proj}_{\mathcal{V}}(\mathbf{b})$, which means that $\mathbf{b}$ is in the subspace $\mathcal{V}$. On the other hand, if $\mathbf{b}$ is in $\mathcal{V}$, then clearly $\mathbf{b}$ is orthogonal to every vector that is orthogonal to $\mathcal{V}$, which means that $\mathbf{b}$ is in $(\mathcal{V}^\perp)^\perp$. This concludes the proof of part (a).

Now we prove part (b). By Theorem 3.59 we have

$$(\mathbf{C}(A^T))^\perp = \mathbf{N}((A^T)^T) = \mathbf{N}(A),$$

because $(A^T)^T = A$. Hence, using part (a), we obtain

$$\mathbf{C}(A^T) = ((\mathbf{C}(A^T))^\perp)^\perp = (\mathbf{N}(A))^\perp.$$

□

Example 3.66. Verify that $(\mathbf{N}(A))^\perp = \mathbf{C}(A^T)$ for the matrix $A = \begin{bmatrix} 4 & 1 & 2 & -1 \\ 7 & 4 & 5 & 2 \\ 1 & 1 & 1 & 1 \end{bmatrix}$.

Solution. First we calculate $(\mathbf{N}(A))^\perp$. Since

$$\begin{bmatrix} 4 & 1 & 2 & -1 \\ 7 & 4 & 5 & 2 \\ 1 & 1 & 1 & 1 \end{bmatrix} \sim \begin{bmatrix} 1 & 0 & 1/3 & -2/3 \\ 0 & 1 & 2/3 & 5/3 \\ 0 & 0 & 0 & 0 \end{bmatrix},$$

we have

$$\mathbf{N}(A) = \mathbf{C}\left(\begin{bmatrix} -1/3 & 2/3 \\ -2/3 & -5/3 \\ 1 & 0 \\ 0 & 1 \end{bmatrix}\right) = \mathbf{C}\left(\begin{bmatrix} -1 & 2 \\ -2 & -5 \\ 3 & 0 \\ 0 & 3 \end{bmatrix}\right).$$

To find $(\mathbf{N}(A))^\perp = \left(\mathbf{C}\left(\begin{bmatrix} -1 & 2 \\ -2 & -5 \\ 3 & 0 \\ 0 & 3 \end{bmatrix}\right)\right)^\perp$ we use Theorem 3.59. The reduced row echelon form of the matrix $\begin{bmatrix} -1 & 2 \\ -2 & -5 \\ 3 & 0 \\ 0 & 3 \end{bmatrix}^T$ is $\begin{bmatrix} 1 & 0 & -5/3 & 2/3 \\ 0 & 1 & -2/3 & -1/3 \end{bmatrix}$ and consequently

$$(\mathbf{N}(A))^\perp = \mathbf{C}\left(\begin{bmatrix} 5 & -2 \\ 2 & 1 \\ 3 & 0 \\ 0 & 3 \end{bmatrix}\right).$$

Now, since $A^T = \begin{bmatrix} 4 & 7 & 1 \\ 1 & 4 & 1 \\ 2 & 5 & 1 \\ -1 & 2 & 1 \end{bmatrix}$ and

$$\begin{bmatrix} 5 & -2 & 4 & 7 & 1 \\ 2 & 1 & 1 & 4 & 1 \\ 3 & 0 & 2 & 5 & 1 \\ 0 & 3 & -1 & 2 & 1 \end{bmatrix} \sim \begin{bmatrix} 1 & 0 & 2/3 & 5/3 & 1/3 \\ 0 & 1 & -1/3 & 2/3 & 1/3 \\ 0 & 0 & 0 & 0 & 0 \\ 0 & 0 & 0 & 0 & 0 \end{bmatrix},$$

the columns of the matrix $\begin{bmatrix} 4 & 7 & 1 \\ 1 & 4 & 1 \\ 2 & 5 & 1 \\ -1 & 2 & 1 \end{bmatrix}$ are in $\mathbf{C}\left(\begin{bmatrix} 5 & -2 \\ 2 & 1 \\ 3 & 0 \\ 0 & 3 \end{bmatrix}\right) = (\mathbf{N}(A))^\perp$. On the other hand, since

$$\begin{bmatrix} 4 & 7 & 1 & 5 & -2 \\ 1 & 4 & 1 & 2 & 1 \\ 2 & 5 & 1 & 3 & 0 \\ -1 & 2 & 1 & 0 & 3 \end{bmatrix} \sim \begin{bmatrix} 1 & 0 & -1/3 & 2/3 & -5/3 \\ 0 & 1 & 1/3 & 1/3 & 2/3 \\ 0 & 0 & 0 & 0 & 0 \\ 0 & 0 & 0 & 0 & 0 \end{bmatrix},$$

the columns of the matrix $\begin{bmatrix} 5 & -2 \\ 2 & 1 \\ 3 & 0 \\ 0 & 3 \end{bmatrix}$ are in $\mathbf{C}\left(\begin{bmatrix} 4 & 7 & 1 \\ 1 & 4 & 1 \\ 2 & 5 & 1 \\ -1 & 2 & 1 \end{bmatrix}\right)$. We finish using Theorem 2.28. □

Theorem 3.67. *Let A be an arbitrary $m \times n$ matrix.*

(a) $\mathbf{C}(A)$ *and* $\mathbf{N}(A^T)$ *are subspaces of* $\mathbb{R}^m$ *and we have*

$$(\mathbf{C}(A))^\perp = \mathbf{N}(A^T),$$

(b) $\mathbf{C}(A^T)$ *and* $\mathbf{N}(A)$ *are subspaces of* $\mathbb{R}^n$ *and we have*

$$(\mathbf{N}(A))^\perp = \mathbf{C}(A^T).$$

Proof. The identities are consequences of Theorems 3.59 and 3.65. □

Exercises 3.5

Find the orthogonal complement of the given subspace.

1. $\operatorname{Span}\left\{\begin{bmatrix}1\\1\\2\\1\end{bmatrix}\right\}$

2. $\operatorname{Span}\left\{\begin{bmatrix}2\\1\\-1\\1\end{bmatrix}\right\}$

3. $\operatorname{Span}\left\{\begin{bmatrix}1\\2\\1\\1\end{bmatrix}, \begin{bmatrix}3\\1\\1\\2\end{bmatrix}\right\}$

4. $\operatorname{Span}\left\{\begin{bmatrix}1\\-1\\1\\1\end{bmatrix}, \begin{bmatrix}2\\3\\0\\1\end{bmatrix}\right\}$

5. $\operatorname{Span}\left\{\begin{bmatrix}1\\1\\2\\1\end{bmatrix}, \begin{bmatrix}1\\2\\1\\1\end{bmatrix}, \begin{bmatrix}3\\1\\1\\2\end{bmatrix}\right\}$

6. $\operatorname{Span}\left\{\begin{bmatrix}2\\1\\1\\1\end{bmatrix}, \begin{bmatrix}1\\1\\1\\2\end{bmatrix}, \begin{bmatrix}3\\1\\2\\1\end{bmatrix}\right\}$

7. $\text{Span}\left\{\begin{bmatrix}1\\1\\1\\1\end{bmatrix}, \begin{bmatrix}1\\0\\1\\0\end{bmatrix}, \begin{bmatrix}0\\1\\0\\1\end{bmatrix}\right\}$

8. $\text{Span}\left\{\begin{bmatrix}3\\2\\1\end{bmatrix}, \begin{bmatrix}1\\2\\3\end{bmatrix}, \begin{bmatrix}1\\1\\1\end{bmatrix}\right\}$

9. $\text{Span}\left\{\begin{bmatrix}1\\1\\1\\0\\1\end{bmatrix}, \begin{bmatrix}1\\1\\0\\1\\1\end{bmatrix}, \begin{bmatrix}1\\0\\1\\1\\1\end{bmatrix}\right\}$

10. $\text{Span}\left\{\begin{bmatrix}1\\-1\\0\\0\\0\end{bmatrix}, \begin{bmatrix}1\\1\\1\\1\\1\end{bmatrix}, \begin{bmatrix}1\\0\\0\\0\\-1\end{bmatrix}\right\}$

Find vectors $\mathbf{v}$ and $\mathbf{w}$ such that $\mathbf{x} = \mathbf{v} + \mathbf{w}$, where $\mathbf{v}$ is in $\mathcal{V}$ and $\mathbf{v} \cdot \mathbf{w} = 0$.

11. $\mathbf{x} = \begin{bmatrix}1\\0\\0\\1\end{bmatrix}, \mathcal{V} = \text{Span}\left\{\begin{bmatrix}1\\1\\1\\1\end{bmatrix}\right\}$

12. $\mathbf{x} = \begin{bmatrix}2\\1\\1\\0\end{bmatrix}, \mathcal{V} = \text{Span}\left\{\begin{bmatrix}1\\0\\0\\1\end{bmatrix}\right\}$

13. $\mathbf{x} = \begin{bmatrix}0\\0\\0\\1\end{bmatrix}, \mathcal{V} = \text{Span}\left\{\begin{bmatrix}1\\1\\2\\1\end{bmatrix}, \begin{bmatrix}1\\0\\1\\1\end{bmatrix}\right\}$

14. $\mathbf{x} = \begin{bmatrix}1\\1\\1\\1\end{bmatrix}, \mathcal{V} = \text{Span}\left\{\begin{bmatrix}1\\1\\-2\\1\end{bmatrix}, \begin{bmatrix}1\\0\\1\\1\end{bmatrix}\right\}$

Extend the given orthogonal set to an orthogonal basis of $\mathbb{R}^4$.

15. $\left\{\begin{bmatrix}2\\1\\1\\-1\end{bmatrix}, \begin{bmatrix}1\\1\\-1\\2\end{bmatrix}\right\}$

16. $\left\{\begin{bmatrix}1\\1\\1\\1\end{bmatrix}, \begin{bmatrix}1\\-3\\1\\1\end{bmatrix}\right\}$

Find the projection matrix on the given subspace.

17. $\text{Span}\left\{\begin{bmatrix}2\\1\\1\\-3\end{bmatrix}\right\}^{\perp}$

18. $\text{Span}\left\{\begin{bmatrix}3\\2\\-1\end{bmatrix}\right\}^{\perp}$

For the given matrix A verify that $(\mathbf{N}(A))^{\perp} = \mathbf{C}(A^T)$.

19. $A = \begin{bmatrix}2 & 1 & -1\\1 & 1 & 2\\3 & 2 & 1\\-1 & 0 & 3\end{bmatrix}$

20. $A = \begin{bmatrix}2 & 1 & -1 & 0 & 3\\1 & 2 & 4 & 3 & 0\\1 & 1 & 1 & 1 & 1\end{bmatrix}$

Chapter 4

Determinants

The determinant of a $n \times n$ matrix

The determinant is a number associated with an $n \times n$ matrix that gives just as useful information about the matrix. In this chapter we define determinants, study their properties, and give some examples of applications of determinants. To motivate the definition of the determinant of an $n \times n$ matrix we first consider determinants of 2×2 matrices and 3×3 matrices.

The determinant is one of the basic tools in linear algebra. It can be used to calculate solutions of systems of equations, to find the eigenvalues of a matrix, and to test a matrix for invertibility.

The determinant of a 2×2 matrix

Definition 4.1. By the *determinant* of a 2×2 matrix $\begin{bmatrix} a_1 & b_1 \\ a_2 & b_2 \end{bmatrix}$ we mean the number

$$\det \begin{bmatrix} a_1 & b_1 \\ a_2 & b_2 \end{bmatrix} = a_1 b_2 - a_2 b_1.$$

The following result is an easy consequence of the above definition.

Theorem 4.2. *Let A and B be two arbitrary* 2×2 *matrices.*

(a) *If B is obtained from A by multiplying one row of A by a real number t, then*

$$\det B = t \det A.$$

(b) *If B is obtained from A by interchanging the rows of A, then*

$$\det B = -\det A.$$

(c) *If B is obtained from A by multiplying one row of A by a real number t and adding it to the other row of A, then*

$$\det B = \det A.$$

(d)

$$\det \begin{bmatrix} 1 & 0 \\ 0 & 1 \end{bmatrix} = 1.$$

The determinant of a 3×3 matrix

Definition 4.3. By the *determinant* of a 3×3 matrix $\begin{bmatrix} a_1 & b_1 & c_1 \\ a_2 & b_2 & c_2 \\ a_3 & b_3 & c_3 \end{bmatrix}$ we mean the number

$$\det \begin{bmatrix} a_1 & b_1 & c_1 \\ a_2 & b_2 & c_2 \\ a_3 & b_3 & c_3 \end{bmatrix} = a_1 \det \begin{bmatrix} b_2 & c_2 \\ b_3 & c_3 \end{bmatrix} - a_2 \det \begin{bmatrix} b_1 & c_1 \\ b_3 & c_3 \end{bmatrix} + a_3 \det \begin{bmatrix} b_1 & c_1 \\ b_2 & c_2 \end{bmatrix}.$$

This definition connects determinants of 3×3 matrices with determinants of 2×2 matrices. It also gives us a practical way of calculating determinants of 3×3 matrices.

Properties of determinants of 2×2 matrices formulated in Theorem 4.2 are not specific to 2×2 matrices.

Theorem 4.4. *Let A and B be two arbitrary 3×3 matrices.*

(a) *If B is obtained from A by multiplying one row of A by a real number t, then*

$$\det B = t \det A.$$

(b) *If B is obtained from A by interchanging any two rows of A, then*

$$\det B = -\det A.$$

(c) *If B is obtained from A by multiplying one row of A by a real number t and adding it to another row of A, then*

$$\det B = \det A.$$

(d)

$$\det \begin{bmatrix} 1 & 0 & 0 \\ 0 & 1 & 0 \\ 0 & 0 & 1 \end{bmatrix} = 1.$$

Proof. Let $A = \begin{bmatrix} a_1 & b_1 & c_1 \\ a_2 & b_2 & c_2 \\ a_3 & b_3 & c_3 \end{bmatrix}$. We have to prove the following properties.

(a) For any real number t we have

$$\det \begin{bmatrix} ta_1 & tb_1 & tc_1 \\ a_2 & b_2 & c_2 \\ a_3 & b_3 & c_3 \end{bmatrix} = \det \begin{bmatrix} a_1 & b_1 & c_1 \\ ta_2 & tb_2 & tc_2 \\ a_3 & b_3 & c_3 \end{bmatrix} = \det \begin{bmatrix} a_1 & b_1 & c_1 \\ a_2 & b_2 & c_2 \\ ta_3 & tb_3 & tc_3 \end{bmatrix} = t \det \begin{bmatrix} a_1 & b_1 & c_1 \\ a_2 & b_2 & c_2 \\ a_3 & b_3 & c_3 \end{bmatrix}.$$

(b)

$$\det \begin{bmatrix} a_2 & b_2 & c_2 \\ a_1 & b_1 & c_1 \\ a_3 & b_3 & c_3 \end{bmatrix} = \det \begin{bmatrix} a_1 & b_1 & c_1 \\ a_3 & b_3 & c_3 \\ a_2 & b_2 & c_2 \end{bmatrix} = \det \begin{bmatrix} a_3 & b_3 & c_3 \\ a_2 & b_2 & c_2 \\ a_1 & b_1 & c_1 \end{bmatrix} = -\det \begin{bmatrix} a_1 & b_1 & c_1 \\ a_2 & b_2 & c_2 \\ a_3 & b_3 & c_3 \end{bmatrix}.$$

(c) For any real number t we have

$$\det\begin{bmatrix} a_1+ta_2 & b_1+tb_2 & c_1+tc_2 \\ a_2 & b_2 & c_2 \\ a_3 & b_3 & c_3 \end{bmatrix} = \det\begin{bmatrix} a_1+ta_3 & b_1+tb_3 & c_1+tc_3 \\ a_2 & b_2 & c_2 \\ a_3 & b_3 & c_3 \end{bmatrix}$$

$$= \det\begin{bmatrix} a_1 & b_1 & c_1 \\ a_2+ta_1 & b_2+tb_1 & c_2+tc_1 \\ a_3 & b_3 & c_3 \end{bmatrix} = \det\begin{bmatrix} a_1 & b_1 & c_1 \\ a_2+ta_3 & b_2+tb_3 & c_2+tc_3 \\ a_3 & b_3 & c_3 \end{bmatrix}$$

$$= \det\begin{bmatrix} a_1 & b_1 & c_1 \\ a_2 & b_2 & c_2 \\ a_3+ta_1 & b_3+tb_1 & c_3+tc_1 \end{bmatrix} = \det\begin{bmatrix} a_1 & b_1 & c_1 \\ a_2 & b_2 & c_2 \\ a_3+ta_2 & b_3+tb_2 & c_3+tc_2 \end{bmatrix}$$

$$= \det\begin{bmatrix} a_1 & b_1 & c_1 \\ a_2 & b_2 & c_2 \\ a_3 & b_3 & c_3 \end{bmatrix}.$$

(d) $\det\begin{bmatrix} 1 & 0 & 0 \\ 0 & 1 & 0 \\ 0 & 0 & 1 \end{bmatrix} = 1.$

We can verify these equalities by direct calculations. □

The determinant of an $n \times n$ matrix

Extending some concepts of linear algebra from $\mathbb{R}^2$ and $\mathbb{R}^3$ to higher dimensions is easy and natural. For example, the inner product in $\mathbb{R}^2$ and $\mathbb{R}^3$ is defined as

$$\begin{bmatrix} a_1 \\ a_2 \end{bmatrix} \cdot \begin{bmatrix} b_1 \\ b_2 \end{bmatrix} = a_1b_1 + a_2b_2 \quad \text{and} \quad \begin{bmatrix} a_1 \\ a_2 \\ a_3 \end{bmatrix} \cdot \begin{bmatrix} b_1 \\ b_2 \\ b_3 \end{bmatrix} = a_1b_1 + a_2b_2 + a_3b_3$$

and it is easy to guess that in $\mathbb{R}^n$ we have

$$\begin{bmatrix} a_1 \\ \vdots \\ a_n \end{bmatrix} \cdot \begin{bmatrix} b_1 \\ \vdots \\ b_n \end{bmatrix} = a_1b_1 + \cdots + a_nb_n.$$

The situation is somewhat different with the determinants. From the definition of the determinants for 2×2 matrices,

$$\det\begin{bmatrix} a_1 & b_1 \\ a_2 & b_2 \end{bmatrix} = a_1b_2 - a_2b_1,$$

it is not clear why we should define the determinants for 3×3 matrices as

$$\det\begin{bmatrix} a_1 & b_1 & c_1 \\ a_2 & b_2 & c_2 \\ a_3 & b_3 & c_3 \end{bmatrix} = a_1\det\begin{bmatrix} b_2 & c_2 \\ b_3 & c_3 \end{bmatrix} - a_2\det\begin{bmatrix} b_1 & c_1 \\ b_3 & c_3 \end{bmatrix} + a_3\det\begin{bmatrix} b_1 & c_1 \\ b_2 & c_2 \end{bmatrix}.$$

And how should the determinants for larger matrices be defined? It turns out that the properties of determinants of 2×2 matrices and 3×3 matrices in Theorems 4.2 and 4.4 help us answer this question, because they actually define determinants. This claim is formulated more precisely in the next theorem. The proof of this theorem requires more advanced methods and is beyond the scope of this book.

Theorem 4.5. *For every integer $n \geq 2$ there is a **unique** function* det *which assigns to every $n \times n$ matrix A a number* $\det A$ *such that the following conditions are satisfied.*

(a) *If B is obtained from A by multiplying one row of A by a real number t, then*

$$\det B = t \det A.$$

(b) *If B is obtained from A by interchanging any two rows of A, then*

$$\det B = -\det A.$$

(c) *If B is obtained from A by multiplying one row of A by a real number t and adding it to another row of A, then*

$$\det B = \det A.$$

(d)

$$\det I_n = 1.$$

In other words, the function det is determined by how it responds to elementary row operations on the matrix and the value of $\det I_n$.

Since elementary row operations correspond to multiplication by elementary matrices, properties (a), (b), and (c) in Theorem 4.5 can be formulated in terms of elementary matrices.

Theorem 4.6. *Let A be an arbitrary $n \times n$ matrix.*

(a) *If E is an elementary matrix obtained from the unit matrix I_n by multiplying one row by a real number t, then*

$$\det(EA) = t \det A.$$

(b) *If E is an elementary matrix obtained from the identity matrix I_n by interchanging any two rows, then*

$$\det(EA) = -\det A.$$

(c) *If E is an elementary matrix obtained from the identity matrix I_n by multiplying any row by a real number t and adding it to another row, then*

$$\det(EA) = \det A.$$

Example 4.7.

$$\det\left(\begin{bmatrix} 1 & 0 & 0 & 0 \\ 0 & 1 & 0 & 0 \\ 0 & 0 & t & 0 \\ 0 & 0 & 0 & 1 \end{bmatrix}\begin{bmatrix} a_1 & b_1 & c_1 & d_1 \\ a_2 & b_2 & c_2 & d_2 \\ a_3 & b_3 & c_3 & d_3 \\ a_4 & b_4 & c_4 & d_4 \end{bmatrix}\right) = t \det\begin{bmatrix} a_1 & b_1 & c_1 & d_1 \\ a_2 & b_2 & c_2 & d_2 \\ a_3 & b_3 & c_3 & d_3 \\ a_4 & b_4 & c_4 & d_4 \end{bmatrix}$$

$$\det\left(\begin{bmatrix} 0 & 0 & 0 & 1 \\ 0 & 1 & 0 & 0 \\ 0 & 0 & 1 & 0 \\ 1 & 0 & 0 & 0 \end{bmatrix}\begin{bmatrix} a_1 & b_1 & c_1 & d_1 \\ a_2 & b_2 & c_2 & d_2 \\ a_3 & b_3 & c_3 & d_3 \\ a_4 & b_4 & c_4 & d_4 \end{bmatrix}\right) = -\det\begin{bmatrix} a_1 & b_1 & c_1 & d_1 \\ a_2 & b_2 & c_2 & d_2 \\ a_3 & b_3 & c_3 & d_3 \\ a_4 & b_4 & c_4 & d_4 \end{bmatrix}$$

$$\det\left(\begin{bmatrix} 1 & 0 & t & 0 \\ 0 & 1 & 0 & 0 \\ 0 & 0 & 1 & 0 \\ 0 & 0 & 0 & 1 \end{bmatrix}\begin{bmatrix} a_1 & b_1 & c_1 & d_1 \\ a_2 & b_2 & c_2 & d_2 \\ a_3 & b_3 & c_3 & d_3 \\ a_4 & b_4 & c_4 & d_4 \end{bmatrix}\right) = \det\begin{bmatrix} a_1 & b_1 & c_1 & d_1 \\ a_2 & b_2 & c_2 & d_2 \\ a_3 & b_3 & c_3 & d_3 \\ a_4 & b_4 & c_4 & d_4 \end{bmatrix}.$$

As a direct consequence of Theorem 4.6 we obtain the following corollary.

Corollary 4.8.

(a) *If E is an elementary matrix obtained from the unit matrix I_n by multiplying one row by a real number t, then*

$$\det E = t.$$

(b) *If E is an elementary matrix obtained from the identity matrix I_n by interchanging any two rows, then*

$$\det E = -1.$$

(c) *If E is an elementary matrix obtained from the identity matrix I_n by multiplying any row by a real number t and adding it to another row, then*

$$\det E = 1.$$

One of the fundamental properties of determinants is that the determinant of a product of matrices is the product of the determinants of those matrices. We prove this property in Theorem 4.13. The following corollary to Theorem 4.6 is the first step in that direction.

Corollary 4.9. *If the $n \times n$ matrix A is a matrix such that*

$$A = E_1 \dots E_k B,$$

where $E_1, \dots, E_k$ are elementary matrices, then

$$\det A = \det E_1 \dots \det E_k \det B.$$

The following simple observation is often useful in proofs and in calculations.

Theorem 4.10. *If a square matrix A has a row with only 0 entries, then*

$$\det A = 0.$$

Proof. Suppose that the i-th row of the matrix A has only 0 entries. If the matrix B is obtained from the matrix A by multiplying the i-th row by 2, then $\det B = 2\det A$. But A and B are the same matrix, so $\det B = \det A$. This is only possible if $\det A = 0$. □

Determinants can be used to test the invertibility of a matrix. For 2×2 and 3×3 matrices it is often the quickest way.

Theorem 4.11. *Let A be an arbitrary square matrix. Then*

$$A \text{ is invertible if and only if } \det A \neq 0.$$

Proof. If the matrix A is invertible, then

$$A = E_1 \cdots E_k$$

for some elementary matrices $E_1, \dots, E_k$. From Corollary 4.9 we have

$$\det A = \det E_1 \cdots \det E_k.$$

Since the determinant of any elementary matrix is different from 0, by Corollary 4.8, we conclude

$$\det A = \det E_1 \cdots \det E_k \neq 0.$$

If the matrix A is not invertible, then

$$A = E_1 \cdots E_k N$$

for some elementary matrices $E_1, \dots, E_k$ and some matrix N with a row whose entries are all 0. Since $\det N = 0$, by Theorem 4.10, we have

$$\det A = \det E_1 \cdots \det E_k \det N = 0.$$

□

From Corollary 1.67 and Theorem 4.11 we get the following simple but useful observation.

Corollary 4.12. *If an $n \times n$ matrix A has a column whose all entries are* 0, *then* $\det A = 0$.

Now we are ready to prove the important result mentioned earlier.

Theorem 4.13. *Let A and B be arbitrary $n \times n$ matrices. Then*

$$\det(AB) = \det A \det B.$$

Proof. We consider three cases: both A and B are invertible, A is invertible and B is not invertible, and A is not invertible.

If A and B are invertible, then

$$A = E_1 \cdots E_k \quad \text{and} \quad B = F_1 \cdots F_l$$

for some elementary matrices $E_1,\dots,E_k,F_1,\dots,F_l$. Since

$$AB = E_1\cdots E_k F_1\cdots F_l,$$

we have

$$\det AB = \det E_1\cdots\det E_k \det F_1\cdots\det F_l = \det A\det B,$$

by Corollary 4.9.

If A is invertible and B is not invertible, then

$$A = E_1\cdots E_k \quad\text{and}\quad B = F_1\cdots F_l N,$$

where $E_1,\dots,E_k,F_1,\dots,F_l$ are elementary matrices and N is a matrix with a row whose entries are all 0. Then

$$AB = E_1\dots E_k F_1\dots F_l N$$

and, consequently,

$$\det AB = \det E_1\dots\det E_k\det F_1\dots\det F_l\det N = 0,$$

by Corollary 4.9 and Theorem 4.10. Since $\det B = 0$, by Theorem 4.11, we have $\det A\det B = 0$. Thus $\det(AB) = \det A\det B$ in this case.

If AB is invertible, then there is an $n\times n$ matrix C such that $ABC = I_n$, so the matrix A is invertible. Therefore, if A is not invertible, then the matrix AB is not invertible. Consequently, if A is not invertible, then $\det A = \det AB = 0$, by Theorem 4.11, and thus $\det AB = \det A\det B$. □

It is easy to verify that $\det A = \det A^T$ for 2×2 matrices and 3×3 matrices. Now we are in a position to prove this identity for square matrices of arbitrary size.

Theorem 4.14. *For an arbitrary square matrix A we have*

$$\det A^T = \det A.$$

Proof. First we note that $\det E^T = \det E$ for any elementary matrix E. Indeed, if E is an elementary matrix obtained from the identity matrix by multiplying one row by a real number or by interchanging any two rows, then $E^T = E$ and thus $\det E^T = \det E$. If E is the elementary matrix obtained from the identity matrix by multiplying the i-th row by a real number c and adding it to the j-th row, then E^T is the elementary matrix obtained from the identity matrix by multiplying the j-th row by c and adding it to the i-th row. For example,

$$\begin{bmatrix} 1 & 0 & 0 & 0\\ 0 & 1 & 0 & 0\\ 0 & 0 & 1 & 0\\ 0 & c & 0 & 1 \end{bmatrix}^T = \begin{bmatrix} 1 & 0 & 0 & 0\\ 0 & 1 & 0 & c\\ 0 & 0 & 1 & 0\\ 0 & 0 & 0 & 1 \end{bmatrix}.$$

Consequently, $\det E^T = \det E = 1$ in this case.

Now we proceed to prove the identity for an arbitrary square matrix A.

If A is invertible, then

$$A = E_1 \cdots E_k$$

for some elementary matrices $E_1, \dots, E_k$. Consequently,

$$A^T = (E_1 \cdots E_k)^T = E_k^T \cdots E_1^T$$

and

$$\det A^T = \det E_k^T \cdots \det E_1^T = \det E_k \cdots \det E_1 = \det A.$$

If A is not invertible, then

$$A = E_1 \cdots E_k N$$

where $E_1, \dots, E_k$ are elementary matrices and N is a matrix with a row whose entries are all 0. Since

$$A^T = (E_1 \cdots E_k N)^T = N^T E_k^T \cdots E_1^T,$$

we have

$$\det A^T = \det N^T \det(E_k^T \cdots E_1^T).$$

Since the matrix N^T has a column whose entries are all 0, we have $\det N^T = 0$, by Corollary 4.12. Therefore,

$$\det A^T = 0 = \det A.$$

□

We note that Theorem 4.14 allows us to replace row by column in Theorem 4.5.

While we discussed several properties of determinants, we have not addressed the question of how to calculate determinants beyond 2×2 and 3×3 matrices. The next theorem gives us a practical way of calculating determinants of larger matrices.

Theorem 4.15. *Let C be an $n \times n$ matrix, $\mathbf{b}$ a $1 \times n$ matrix, $\mathbf{0}$ the $n \times 1$ vector whose entries are all 0, and a a real number. Then for the $(n+1) \times (n+1)$ matrix $\begin{bmatrix} a & \mathbf{b} \\ \mathbf{0} & C \end{bmatrix}$ we have*

$$\det \begin{bmatrix} a & \mathbf{b} \\ \mathbf{0} & C \end{bmatrix} = a \det C.$$

Proof of a particular case. We prove the result for matrices of the type

$$\det \begin{bmatrix} a & b_1 & c_1 & d_1 & e_1 \\ 0 & b_2 & c_2 & d_2 & e_2 \\ 0 & b_3 & c_3 & d_3 & e_3 \\ 0 & b_4 & c_4 & d_4 & e_4 \\ 0 & b_4 & c_4 & d_5 & e_5 \end{bmatrix}.$$

If $a = 0$, then

$$\det\begin{bmatrix} a & b_1 & c_1 & d_1 & e_1 \\ 0 & b_2 & c_2 & d_2 & e_2 \\ 0 & b_3 & c_3 & d_3 & e_3 \\ 0 & b_4 & c_4 & d_4 & e_4 \\ 0 & b_4 & c_4 & d_5 & e_5 \end{bmatrix} = \det\begin{bmatrix} 0 & b_1 & c_1 & d_1 & e_1 \\ 0 & b_2 & c_2 & d_2 & e_2 \\ 0 & b_3 & c_3 & d_3 & e_3 \\ 0 & b_4 & c_4 & d_4 & e_4 \\ 0 & b_4 & c_4 & d_5 & e_5 \end{bmatrix} = 0,$$

because the matrix has no more than 4 pivots and thus the reduced row echelon form has a row whose entries are all 0.

If $a \neq 0$, then the function f defined by

$$f\left(\begin{bmatrix} b_2 & c_2 & d_2 & e_2 \\ b_3 & c_3 & d_3 & e_3 \\ b_4 & c_4 & d_4 & e_4 \\ b_5 & c_5 & d_5 & e_5 \end{bmatrix}\right) = \frac{1}{a}\det\begin{bmatrix} a & b_1 & c_1 & d_1 & e_1 \\ 0 & b_2 & c_2 & d_2 & e_2 \\ 0 & b_3 & c_3 & d_3 & e_3 \\ 0 & b_4 & c_4 & d_4 & e_4 \\ 0 & b_5 & c_5 & d_5 & e_5 \end{bmatrix}$$

satisfies the four properties from Theorem 4.5. We verify that

$$\frac{1}{a}\det\begin{bmatrix} a & b_1 & c_1 & d_1 & e_1 \\ 0 & 1 & 0 & 0 & 0 \\ 0 & 0 & 1 & 0 & 0 \\ 0 & 0 & 0 & 1 & 0 \\ 0 & 0 & 0 & 0 & 1 \end{bmatrix} = 1,$$

because the other conditions are easy to verify.

In the matrix

$$\begin{bmatrix} a & b_1 & c_1 & d_1 & e_1 \\ 0 & 1 & 0 & 0 & 0 \\ 0 & 0 & 1 & 0 & 0 \\ 0 & 0 & 0 & 1 & 0 \\ 0 & 0 & 0 & 0 & 1 \end{bmatrix}$$

we add to the first row the second row multiplied by $-b_1$, the third row multiplied by $-c_1$, the forth row multiplied by $-d_1$, and the fifth row multiplied by $-e_1$. These operations give us the matrix

$$\begin{bmatrix} a & 0 & 0 & 0 & 0 \\ 0 & 1 & 0 & 0 & 0 \\ 0 & 0 & 1 & 0 & 0 \\ 0 & 0 & 0 & 1 & 0 \\ 0 & 0 & 0 & 0 & 1 \end{bmatrix}.$$

Since

$$\frac{1}{a}\det\begin{bmatrix} a & b_1 & c_1 & d_1 & e_1 \\ 0 & b_2 & c_2 & d_2 & e_2 \\ 0 & b_3 & c_3 & d_3 & e_3 \\ 0 & b_4 & c_4 & d_4 & e_4 \\ 0 & b_5 & c_5 & d_5 & e_5 \end{bmatrix} = \frac{1}{a}\det\begin{bmatrix} a & 0 & 0 & 0 & 0 \\ 0 & 1 & 0 & 0 & 0 \\ 0 & 0 & 1 & 0 & 0 \\ 0 & 0 & 0 & 1 & 0 \\ 0 & 0 & 0 & 0 & 1 \end{bmatrix} = \frac{1}{a}\cdot a\det\begin{bmatrix} 1 & 0 & 0 & 0 & 0 \\ 0 & 1 & 0 & 0 & 0 \\ 0 & 0 & 1 & 0 & 0 \\ 0 & 0 & 0 & 1 & 0 \\ 0 & 0 & 0 & 0 & 1 \end{bmatrix} = 1,$$

we must have

$$\det\begin{bmatrix} b_2 & c_2 & d_2 & e_2 \\ b_3 & c_3 & d_3 & e_3 \\ b_4 & c_4 & d_4 & e_4 \\ b_5 & c_5 & d_5 & e_5 \end{bmatrix} = \frac{1}{a}\det\begin{bmatrix} a & b_1 & c_1 & d_1 & e_1 \\ 0 & b_2 & c_2 & d_2 & e_2 \\ 0 & b_3 & c_3 & d_3 & e_3 \\ 0 & b_4 & c_4 & d_4 & e_4 \\ 0 & b_5 & c_5 & d_5 & e_5 \end{bmatrix}.$$

□

Example 4.16. Calculate $\det\begin{bmatrix} 2 & 3 & 2 & 2 \\ 3 & 1 & 2 & 1 \\ 2 & 4 & 5 & 3 \\ 4 & 1 & 3 & 2 \end{bmatrix}$.

Solution.

$$\begin{aligned}
\det\begin{bmatrix} 2 & 3 & 2 & 2 \\ 3 & 1 & 2 & 1 \\ 2 & 4 & 5 & 3 \\ 4 & 1 & 3 & 2 \end{bmatrix} &= \det\begin{bmatrix} -7 & 0 & -4 & -1 \\ 3 & 1 & 2 & 1 \\ -10 & 0 & -3 & -1 \\ 1 & 0 & 1 & 1 \end{bmatrix} \\
&= -\det\begin{bmatrix} 3 & 1 & 2 & 1 \\ -7 & 0 & -4 & -1 \\ -10 & 0 & -3 & -1 \\ 1 & 0 & 1 & 1 \end{bmatrix} \\
&= -\left(-\det\begin{bmatrix} 1 & 3 & 2 & 1 \\ 0 & -7 & -4 & -1 \\ 0 & -10 & -3 & -1 \\ 0 & 1 & 1 & 1 \end{bmatrix}\right) \\
&= \det\begin{bmatrix} -7 & -4 & -1 \\ -10 & -3 & -1 \\ 1 & 1 & 1 \end{bmatrix} = -15.
\end{aligned}$$

□

Cramer's rule

Now we are going to discuss some applications of determinants. The first one is a method of solving systems of n linear equations with n variables known as Cramer's Rule. In order to use determinants, we write such a system as a matrix equation.

Theorem 4.17 (Cramer's rule). *Let A be an invertible $n \times n$ matrix and let* $\mathbf{y}$ *be a vector in* $\mathbb{R}^n$. *Then the solution of the equation*

$$A\mathbf{x} = \mathbf{y}$$

is the vector $\mathbf{x} = \begin{bmatrix} x_1 \\ \vdots \\ x_n \end{bmatrix}$ *where, for every* $j = 1, \dots, n$, *we have*

$$x_j = \frac{\det A_j}{\det A}$$

and A_j *is the matrix whose* j*-th column is the vector* $\mathbf{y}$ *and all the other columns are the same as in the matrix* A.

Proof of a particular case. We consider the equation

$$\begin{bmatrix} a_1 & b_1 & c_1 & d_1 \\ a_2 & b_2 & c_2 & d_2 \\ a_3 & b_3 & c_3 & d_3 \\ a_4 & b_4 & c_4 & d_4 \end{bmatrix} \begin{bmatrix} x_1 \\ x_2 \\ x_3 \\ x_4 \end{bmatrix} = \begin{bmatrix} y_1 \\ y_2 \\ y_3 \\ y_4 \end{bmatrix},$$

where the matrix $A = \begin{bmatrix} a_1 & b_1 & c_1 & d_1 \\ a_2 & b_2 & c_2 & d_2 \\ a_3 & b_3 & c_3 & d_3 \\ a_4 & b_4 & c_4 & d_4 \end{bmatrix}$ is invertible and consequently $\det A \neq 0$. We will calculate x_3.

If $\begin{bmatrix} x_1 \\ x_2 \\ x_3 \\ x_4 \end{bmatrix}$ is the solution of the equation, then

$$\begin{bmatrix} a_1 & b_1 & c_1 & d_1 \\ a_2 & b_2 & c_2 & d_2 \\ a_3 & b_3 & c_3 & d_3 \\ a_4 & b_4 & c_4 & d_4 \end{bmatrix} \begin{bmatrix} 1 & 0 & x_1 & 0 \\ 0 & 1 & x_2 & 0 \\ 0 & 0 & x_3 & 0 \\ 0 & 0 & x_4 & 1 \end{bmatrix} = \begin{bmatrix} a_1 & b_1 & y_1 & d_1 \\ a_2 & b_2 & y_2 & d_2 \\ a_3 & b_3 & y_3 & d_3 \\ a_4 & b_4 & y_4 & d_4 \end{bmatrix},$$

and consequently

$$\det\left(\begin{bmatrix} a_1 & b_1 & c_1 & d_1 \\ a_2 & b_2 & c_2 & d_2 \\ a_3 & b_3 & c_3 & d_3 \\ a_4 & b_4 & c_4 & d_4 \end{bmatrix} \begin{bmatrix} 1 & 0 & x_1 & 0 \\ 0 & 1 & x_2 & 0 \\ 0 & 0 & x_3 & 0 \\ 0 & 0 & x_4 & 1 \end{bmatrix} \right) = \det \begin{bmatrix} a_1 & b_1 & y_1 & d_1 \\ a_2 & b_2 & y_2 & d_2 \\ a_3 & b_3 & y_3 & d_3 \\ a_4 & b_4 & y_4 & d_4 \end{bmatrix},$$

which gives us

$$\det\begin{bmatrix} a_1 & b_1 & c_1 & d_1 \\ a_2 & b_2 & c_2 & d_2 \\ a_3 & b_3 & c_3 & d_3 \\ a_4 & b_4 & c_4 & d_4 \end{bmatrix} \det\begin{bmatrix} 1 & 0 & x_1 & 0 \\ 0 & 1 & x_2 & 0 \\ 0 & 0 & x_3 & 0 \\ 0 & 0 & x_4 & 1 \end{bmatrix} = \det\begin{bmatrix} a_1 & b_1 & y_1 & d_1 \\ a_2 & b_2 & y_2 & d_2 \\ a_3 & b_3 & y_3 & d_3 \\ a_4 & b_4 & y_4 & d_4 \end{bmatrix}.$$

Since $\det\begin{bmatrix} 1 & 0 & x_1 & 0 \\ 0 & 1 & x_2 & 0 \\ 0 & 0 & x_3 & 0 \\ 0 & 0 & x_4 & 1 \end{bmatrix} = x_3$, we get

$$x_3 = \frac{\det\begin{bmatrix} a_1 & b_1 & y_1 & d_1 \\ a_2 & b_2 & y_2 & d_2 \\ a_3 & b_3 & y_3 & d_3 \\ a_4 & b_4 & y_4 & d_4 \end{bmatrix}}{\det\begin{bmatrix} a_1 & b_1 & c_1 & d_1 \\ a_2 & b_2 & c_2 & d_2 \\ a_3 & b_3 & c_3 & d_3 \\ a_4 & b_4 & c_4 & d_4 \end{bmatrix}}.$$

□

Example 4.18. We consider the system

$$\begin{cases} 3x + 2y + z + w = 1 \\ 5x + 3y + 2z + 2w = 3 \\ 2x + y + z + 2w = 7 \\ x + 3y + 4z + w = 2 \end{cases}.$$

Use Theorem 4.17 to find y.

Solution.

$$y = \frac{\det\begin{bmatrix} 3 & 1 & 1 & 1 \\ 5 & 3 & 2 & 2 \\ 2 & 7 & 1 & 2 \\ 1 & 2 & 4 & 1 \end{bmatrix}}{\det\begin{bmatrix} 3 & 2 & 1 & 1 \\ 5 & 3 & 2 & 2 \\ 2 & 1 & 1 & 2 \\ 1 & 3 & 4 & 1 \end{bmatrix}} = \frac{2}{6} = \frac{1}{3}.$$

□

Calculating the inverse of a matrix using determinants

Calculating the inverse of a matrix is often necessary and can be time-consuming. In Chapter 1 we calculated inverse matrices using elementary matrices and Gaussian elimination. The following theorem gives us a formula for the entries of the inverse matrix in terms of determinants.

Theorem 4.19. *Let A be an invertible $n \times n$ matrix. The (i, j) entry of the inverse matrix A^{-1} is*

$$(-1)^{i+j}\frac{\det A_{ji}}{\det A},$$

where the matrix A_{ji} is the $(n-1)\times(n-1)$ matrix obtained from the matrix A by deleting the j-th row and the i-th column.

Proof of a particular case. Consider the matrix $A = \begin{bmatrix} a_1 & b_1 & c_1 & d_1 \\ a_2 & b_2 & c_2 & d_2 \\ a_3 & b_3 & c_3 & d_3 \\ a_4 & b_4 & c_4 & d_4 \end{bmatrix}$. We will find the entry of the inverse matrix A^{-1} in the third row and the second column. That entry is the number x_3 in the solution of the equation

$$\begin{bmatrix} a_1 & b_1 & c_1 & d_1 \\ a_2 & b_2 & c_2 & d_2 \\ a_3 & b_3 & c_3 & d_3 \\ a_4 & b_4 & c_4 & d_4 \end{bmatrix}\begin{bmatrix} x_1 \\ x_2 \\ x_3 \\ x_4 \end{bmatrix} = \begin{bmatrix} 0 \\ 1 \\ 0 \\ 0 \end{bmatrix}.$$

Using Theorem 4.17 we find

$$x_3 = \frac{\det\begin{bmatrix} a_1 & b_1 & 0 & d_1 \\ a_2 & b_2 & 1 & d_2 \\ a_3 & b_3 & 0 & d_3 \\ a_4 & b_4 & 0 & d_4 \end{bmatrix}}{\det\begin{bmatrix} a_1 & b_1 & c_1 & d_1 \\ a_2 & b_2 & c_2 & d_2 \\ a_3 & b_3 & c_3 & d_3 \\ a_4 & b_4 & c_4 & d_4 \end{bmatrix}} = \frac{(-1)^{3+2}\det\begin{bmatrix} a_1 & b_1 & d_1 \\ a_3 & b_3 & d_3 \\ a_4 & b_4 & d_4 \end{bmatrix}}{\det\begin{bmatrix} a_1 & b_1 & c_1 & d_1 \\ a_2 & b_2 & c_2 & d_2 \\ a_3 & b_3 & c_3 & d_3 \\ a_4 & b_4 & c_4 & d_4 \end{bmatrix}}.$$

□

Example 4.20. We consider the matrix

$$A = \begin{bmatrix} 2 & 1 & 1 & 1 \\ 3 & 1 & 2 & 3 \\ 2 & 1 & 3 & 4 \\ 1 & 5 & 4 & 1 \end{bmatrix}.$$

Find the entry of the inverse matrix A^{-1} in the third row and the second column.

Solution.

$$\frac{\det\begin{bmatrix} 2 & 1 & 0 & 1 \\ 3 & 1 & 1 & 3 \\ 2 & 1 & 0 & 4 \\ 1 & 5 & 0 & 1 \end{bmatrix}}{\det\begin{bmatrix} 2 & 1 & 1 & 1 \\ 3 & 1 & 2 & 3 \\ 2 & 1 & 3 & 4 \\ 1 & 5 & 4 & 1 \end{bmatrix}} = \frac{(-1)^{3+2}\det\begin{bmatrix} 2 & 1 & 1 \\ 2 & 1 & 4 \\ 1 & 5 & 1 \end{bmatrix}}{\det\begin{bmatrix} 2 & 1 & 1 & 1 \\ 3 & 1 & 2 & 3 \\ 2 & 1 & 3 & 4 \\ 1 & 5 & 4 & 1 \end{bmatrix}} = \frac{-(-27)}{-4} = -\frac{27}{4}.$$

□

Using the identities $AA^{-1} = I_n$ and $A^{-1}A = I_n$ and Theorem 4.19 we obtain the following result that can be used to calculate determinants.

Corollary 4.21. *Let A be an invertible $n \times n$ matrix. For every $i = 1, \dots, n$ we have*

$$(-1)^{i+1} a_{i1} \det A_{i1} + \cdots + (-1)^{i+n} a_{in} \det A_{in} = \det A$$

and for every $j = 1, \dots, n$ we have

$$(-1)^{1+j} a_{1j} \det A_{1j} + \cdots + (-1)^{n+j} a_{nj} \det A_{nj} = \det A.$$

Actually the equalities from the preceding corollary are true even if the matrix A is not invertible.

Exercises 4.1

Find the determinant of the given matrix if $\det\begin{bmatrix} a_1 & b_1 & c_1 & d_1 \\ a_2 & b_2 & c_2 & d_2 \\ a_3 & b_3 & c_3 & d_3 \\ a_4 & b_4 & c_4 & d_4 \end{bmatrix} = 7$.

1. $\det\begin{bmatrix} a_2 & b_2 & c_2 & d_2 \\ a_4 & b_4 & c_4 & d_4 \\ 2a_3 & 2b_3 & 2c_3 & 2d_3 \\ a_1 & b_1 & c_1 & d_1 \end{bmatrix}$

2. $\det\begin{bmatrix} a_1 & b_1+7c_1 & c_1 & 5d_1 \\ a_2 & b_2+7c_2 & c_2 & 5d_2 \\ a_3 & b_3+7c_3 & c_3 & 5d_3 \\ a_4 & b_4+7c_4 & c_4 & 5d_4 \end{bmatrix}$

3. $\det\begin{bmatrix} a_3 & b_3 & c_3 & d_3 \\ a_1+8a_4 & b_1+8b_4 & c_1+8c_4 & d_1+8d_4 \\ a_2 & b_2 & c_2 & d_2 \\ a_4 & b_4 & c_4 & d_4 \end{bmatrix}$

4. $\det\begin{bmatrix} a_1 & b_1 & c_1 & d_1 \\ a_2+5a_3 & b_2+5b_3 & c_2+5c_3 & d_2+5d_3 \\ 9a_3 & 9b_3 & 9c_3 & 9d_3 \\ a_4+a_1 & b_4+b_1 & c_4+c_1 & d_4+d_1 \end{bmatrix}$

Find the determinant of the given matrix.

5. $\begin{bmatrix} 2 & 1 & 1 & 2 \\ 0 & 4 & 0 & 0 \\ 3 & 2 & 2 & 1 \\ 2 & 3 & 1 & 4 \end{bmatrix}$

6. $\begin{bmatrix} 2 & 1 & 3 & 1 \\ 5 & 4 & 2 & 3 \\ 0 & 0 & 3 & 0 \\ 3 & 1 & 1 & 2 \end{bmatrix}$

7. $\begin{bmatrix} 2 & 1 & 0 & 1 \\ 3 & 1 & 5 & 2 \\ 3 & 2 & 0 & 1 \\ 2 & 3 & 0 & 4 \end{bmatrix}$

8. $\begin{bmatrix} 5 & 0 & 3 & 3 \\ 1 & 0 & 5 & 2 \\ 2 & 2 & 1 & 2 \\ 3 & 0 & 4 & 1 \end{bmatrix}$

9. $\begin{bmatrix} 3 & 4 & 1 & 2 \\ 2 & 3 & 2 & 1 \\ 0 & 2 & 1 & 0 \\ 2 & 2 & 3 & 2 \end{bmatrix}$

10. $\begin{bmatrix} 2 & 2 & 3 & 1 \\ 0 & 3 & 2 & 0 \\ 1 & 4 & 1 & 1 \\ 2 & 1 & 4 & 3 \end{bmatrix}$

11. $\begin{bmatrix} 2 & 4 & 3 & 2 \\ 1 & 5 & 1 & 3 \\ 1 & 1 & 1 & 2 \\ 3 & 2 & 3 & 4 \end{bmatrix}$

12. $\begin{bmatrix} 3 & 2 & 3 & 1 \\ 5 & 1 & 1 & 2 \\ 1 & 4 & 1 & 1 \\ 2 & 1 & 2 & 3 \end{bmatrix}$

13. $\begin{bmatrix} 2 & 1 & 1 & 1 \\ 1 & 2 & 1 & 1 \\ 1 & 1 & 2 & 1 \\ 1 & 1 & 1 & 2 \end{bmatrix}$

14. $\begin{bmatrix} 2 & 1 & 1 & 1 & 1 \\ 1 & 2 & 1 & 1 & 1 \\ 1 & 1 & 2 & 1 & 1 \\ 1 & 1 & 1 & 2 & 1 \\ 1 & 1 & 1 & 1 & 2 \end{bmatrix}$

15. Show that $\det \begin{bmatrix} a_1 & b_1 & c_1 & d_1 & e_1 \\ a_2 & b_2 & 0 & d_2 & e_2 \\ a_3 & b_3 & 0 & d_3 & e_3 \\ a_4 & b_4 & 0 & d_4 & e_4 \\ a_5 & b_5 & 0 & d_5 & e_5 \end{bmatrix} = -c_1 \det \begin{bmatrix} a_2 & b_2 & d_2 & e_2 \\ a_3 & b_3 & d_3 & e_3 \\ a_4 & b_4 & d_4 & e_4 \\ a_5 & b_5 & d_5 & e_5 \end{bmatrix}$.

16. Show that $\det \begin{bmatrix} a_1 & b_1 & 0 & d_1 & e_1 \\ a_2 & b_2 & c_2 & d_2 & e_2 \\ a_3 & b_3 & 0 & d_3 & e_3 \\ a_4 & b_4 & 0 & d_4 & e_4 \\ a_5 & b_5 & 0 & d_5 & e_5 \end{bmatrix} = -c_2 \det \begin{bmatrix} a_1 & b_1 & d_1 & e_1 \\ a_3 & b_3 & d_3 & e_3 \\ a_4 & b_4 & d_4 & e_4 \\ a_5 & b_5 & d_5 & e_5 \end{bmatrix}$.

17. Show that $\det \begin{bmatrix} a_1 & b_1 & c_1 & d_1 & e_1 \\ a_2 & b_2 & c_2 & d_2 & e_2 \\ 0 & 0 & 0 & d_3 & 0 \\ a_4 & b_4 & c_4 & d_4 & e_4 \\ a_5 & b_5 & c_5 & d_5 & e_5 \end{bmatrix} = -d_3 \det \begin{bmatrix} a_1 & b_1 & c_1 & e_1 \\ a_2 & b_2 & c_2 & e_2 \\ a_4 & b_4 & c_4 & e_4 \\ a_5 & b_5 & c_5 & e_5 \end{bmatrix}$.

18. Show that $\det\begin{bmatrix} a_1 & b_1 & c_1 & d_1 & e_1 \\ 0 & 0 & c_2 & 0 & 0 \\ a_3 & b_3 & c_3 & d_3 & e_3 \\ a_4 & b_4 & c_4 & d_4 & e_4 \\ a_5 & b_5 & c_5 & d_5 & e_5 \end{bmatrix} = -c_2 \det\begin{bmatrix} a_1 & b_1 & d_1 & e_1 \\ a_3 & b_3 & d_3 & e_3 \\ a_4 & b_4 & d_4 & e_4 \\ a_5 & b_5 & d_5 & e_5 \end{bmatrix}$.

19. Show that $\det\begin{bmatrix} a_1 & b_1 & c_1 & d_1 \\ a_2 & b_2 & c_2 & d_2 \\ 0 & b_3 & c_3 & d_3 \\ 0 & b_4 & c_4 & d_4 \end{bmatrix} = a_1 \det\begin{bmatrix} b_2 & c_2 & d_2 \\ b_3 & c_3 & d_3 \\ b_4 & c_4 & d_4 \end{bmatrix} - a_2 \det\begin{bmatrix} b_1 & c_1 & d_1 \\ b_3 & c_3 & d_3 \\ b_4 & c_4 & d_4 \end{bmatrix}$.

20. Show that $\det\begin{bmatrix} 0 & b_1 & c_1 & d_1 \\ a_2 & b_2 & c_2 & d_2 \\ 0 & b_3 & c_3 & d_3 \\ a_4 & b_4 & c_4 & d_4 \end{bmatrix} = -a_2 \det\begin{bmatrix} b_1 & c_1 & d_1 \\ b_3 & c_3 & d_3 \\ b_4 & c_4 & d_4 \end{bmatrix} - a_4 \det\begin{bmatrix} b_1 & c_1 & d_1 \\ b_2 & c_2 & d_2 \\ b_3 & c_3 & d_3 \end{bmatrix}$.

21. Show that $\det\begin{bmatrix} a_1 & b_1 & c_1 & d_1 \\ a_2 & b_2 & c_2 & d_2 \\ a_3 & b_3 & c_3 & d_3 \\ 0 & b_4 & c_4 & d_4 \end{bmatrix} = a_1 \det\begin{bmatrix} b_2 & c_2 & d_2 \\ b_3 & c_3 & d_3 \\ b_4 & c_4 & d_4 \end{bmatrix} - a_2 \det\begin{bmatrix} b_1 & c_1 & d_1 \\ b_3 & c_3 & d_3 \\ b_4 & c_4 & d_4 \end{bmatrix} + a_3 \det\begin{bmatrix} b_1 & c_1 & d_1 \\ b_2 & c_2 & d_2 \\ b_4 & c_4 & d_4 \end{bmatrix}$.

22. Show that $\det\begin{bmatrix} 0 & b_1 & c_1 & d_1 \\ a_2 & b_2 & c_2 & d_2 \\ a_3 & b_3 & c_3 & d_3 \\ a_4 & b_4 & c_4 & d_4 \end{bmatrix} = -a_2 \det\begin{bmatrix} b_1 & c_1 & d_1 \\ b_3 & c_3 & d_3 \\ b_4 & c_4 & d_4 \end{bmatrix} + a_3 \det\begin{bmatrix} b_1 & c_1 & d_1 \\ b_2 & c_2 & d_2 \\ b_4 & c_4 & d_4 \end{bmatrix} - a_4 \det\begin{bmatrix} b_1 & c_1 & d_1 \\ b_2 & c_2 & d_2 \\ b_3 & c_3 & d_3 \end{bmatrix}$.

23. Show that $\det\begin{bmatrix} a_1 & b_1 & c_1 & d_1 \\ a_2 & b_2 & c_2 & d_2 \\ a_3 & b_3 & c_3 & d_3 \\ a_4 & b_4 & c_4 & d_4 \end{bmatrix} = a_1 \det\begin{bmatrix} b_2 & c_2 & d_2 \\ b_3 & c_3 & d_3 \\ b_4 & c_4 & d_4 \end{bmatrix} - a_2 \det\begin{bmatrix} b_1 & c_1 & d_1 \\ b_3 & c_3 & d_3 \\ b_4 & c_4 & d_4 \end{bmatrix} + a_3 \det\begin{bmatrix} b_1 & c_1 & d_1 \\ b_2 & c_2 & d_2 \\ b_4 & c_4 & d_4 \end{bmatrix} - a_4 \det\begin{bmatrix} b_1 & c_1 & d_1 \\ b_2 & c_2 & d_2 \\ b_3 & c_3 & d_3 \end{bmatrix}$.

24. Show that $\det\begin{bmatrix} a_1 & b_1 & c_1 & d_1 \\ 0 & b_2 & c_2 & d_2 \\ 0 & 0 & c_3 & d_3 \\ 0 & 0 & 0 & d_4 \end{bmatrix} = a_1 b_2 c_3 d_4$.

25. If the matrix A is invertible, show that $\det A^{-1} = \frac{1}{\det A}$.

26. Let A and B be $n \times n$ matrices. If the matrix B is invertible, show that $\det\left(B^{-1}AB\right) = \det A$.

27. If C is an $n \times n$ matrix, $\mathbf{b}$ is a $n \times 1$ vector, $\mathbf{0}$ is the $1 \times n$ matrix whose entries are all 0, and a is a real number, show that $\det \begin{bmatrix} a & \mathbf{0} \\ \mathbf{b} & C \end{bmatrix} = a \det C$.

28. Show that $\det(AA^T) = (\det A)^2$.

29. Show that

$$\det \begin{bmatrix} a_1 & b_1 & c_1 & d_1 \\ a_2 & b_2 & c_2 & d_2 \\ a_3 & b_3 & c_3 & d_3 \\ a_4 & b_4 & c_4 & d_4 \end{bmatrix} = -b_1 \det \begin{bmatrix} a_2 & c_2 & d_2 \\ a_3 & c_3 & d_3 \\ a_4 & c_4 & d_4 \end{bmatrix} + b_2 \det \begin{bmatrix} a_1 & c_1 & d_1 \\ a_3 & c_3 & d_3 \\ a_4 & c_4 & d_4 \end{bmatrix}$$
$$- b_3 \det \begin{bmatrix} a_1 & c_1 & d_1 \\ a_2 & c_2 & d_2 \\ a_4 & c_4 & d_4 \end{bmatrix} + b_4 \det \begin{bmatrix} a_1 & c_1 & d_1 \\ a_2 & c_2 & d_2 \\ a_3 & c_3 & d_3 \end{bmatrix}.$$

30. Show that

$$\det \begin{bmatrix} a_1 & b_1 & c_1 & d_1 \\ a_2 & b_2 & c_2 & d_2 \\ a_3 & b_3 & c_3 & d_3 \\ a_4 & b_4 & c_4 & d_4 \end{bmatrix} = c_1 \det \begin{bmatrix} a_2 & b_2 & d_2 \\ a_3 & b_3 & d_3 \\ a_4 & b_4 & d_4 \end{bmatrix} - c_2 \det \begin{bmatrix} a_1 & b_1 & d_1 \\ a_3 & b_3 & d_3 \\ a_4 & b_4 & d_4 \end{bmatrix}$$
$$+ c_3 \det \begin{bmatrix} a_1 & b_1 & d_1 \\ a_2 & b_2 & d_2 \\ a_4 & b_4 & d_4 \end{bmatrix} - c_4 \det \begin{bmatrix} a_1 & b_1 & d_1 \\ a_2 & b_2 & d_2 \\ a_3 & b_3 & d_3 \end{bmatrix}.$$

31. Show that

$$\det \begin{bmatrix} a_1 & b_1 & c_1 & d_1 \\ a_2 & b_2 & c_2 & d_2 \\ a_3 & b_3 & c_3 & d_3 \\ a_4 & b_4 & c_4 & d_4 \end{bmatrix} = a_3 \det \begin{bmatrix} b_1 & c_1 & d_1 \\ b_2 & c_2 & d_2 \\ b_4 & c_4 & d_4 \end{bmatrix} - b_3 \det \begin{bmatrix} a_1 & c_1 & d_1 \\ a_2 & c_2 & d_2 \\ a_4 & c_4 & d_4 \end{bmatrix}$$
$$+ c_3 \det \begin{bmatrix} a_1 & b_1 & d_1 \\ a_2 & b_2 & d_2 \\ a_4 & b_4 & d_4 \end{bmatrix} - d_3 \det \begin{bmatrix} a_1 & b_1 & c_1 \\ a_2 & b_2 & c_2 \\ a_4 & b_4 & c_4 \end{bmatrix}.$$

32. Show that

$$\det \begin{bmatrix} a_1 & b_1 & c_1 & d_1 \\ a_2 & b_2 & c_2 & d_2 \\ a_3 & b_3 & c_3 & d_3 \\ a_4 & b_4 & c_4 & d_4 \end{bmatrix} = -a_4 \det \begin{bmatrix} b_1 & c_1 & d_1 \\ b_2 & c_2 & d_2 \\ b_3 & c_3 & d_3 \end{bmatrix} + b_4 \det \begin{bmatrix} a_1 & c_1 & d_1 \\ a_2 & c_2 & d_2 \\ a_3 & c_3 & d_3 \end{bmatrix}$$
$$- c_4 \det \begin{bmatrix} a_1 & b_1 & d_1 \\ a_2 & b_2 & d_2 \\ a_3 & b_3 & d_3 \end{bmatrix} + d_4 \det \begin{bmatrix} a_1 & b_1 & c_1 \\ a_2 & b_2 & c_2 \\ a_3 & b_3 & c_3 \end{bmatrix}.$$

33. Consider the system

$$\begin{cases} 3x + 2y + z = 3 \\ 4x + y + 2z = 0 \\ 5x + 2y + 2z = 1 \end{cases}.$$

Use Theorem 4.17 to determine y.

34. Consider the system

$$\begin{cases} x + 4y + 2z = 1 \\ 2x + y + z = 1 \\ 3x + 3y + 2z = 1 \end{cases}.$$

Use Theorem 4.17 to determine z.

35. Consider the system

$$\begin{cases} 2x + y + 3z + 2w = 0 \\ 4x - y + z + w = 1 \\ 3x + y - z - w = 0 \\ 2x + 2y + z + 3w = 1 \end{cases}.$$

Use Theorem 4.17 to determine x.

36. Consider the system

$$\begin{cases} x + y + z + w = 2 \\ x + y - z + w = 3 \\ x + y + z - w = 1 \\ x + 2y + z + w = 1 \end{cases}.$$

Use Theorem 4.17 to determine z.

37. Consider the matrix $A = \begin{bmatrix} 2 & 2 & 3 \\ 2 & 3 & 2 \\ 3 & 2 & 1 \end{bmatrix}$. Determine the entry of the inverse matrix A^{-1} which is in the third row and the first column.

38. Consider the matrix $A = \begin{bmatrix} 1 & 1 & 1 \\ 1 & 1 & 2 \\ 2 & 1 & 3 \end{bmatrix}$. Determine the entry of the inverse matrix A^{-1} which is in the second row and the third column.

39. Consider the matrix $A = \begin{bmatrix} 1 & 0 & 5 & 1 \\ 2 & 2 & 2 & 1 \\ 4 & 1 & 0 & 1 \\ 1 & 3 & 1 & 1 \end{bmatrix}$. Determine the entry of the inverse matrix A^{-1} which is in the third row and the fourth column.

40. Consider the matrix $A = \begin{bmatrix} 2 & 1 & 1 & 1 \\ 1 & 3 & 0 & 1 \\ 0 & 1 & 1 & 1 \\ 2 & 0 & 1 & 1 \end{bmatrix}$. Determine the entry of the inverse matrix A^{-1} which is in the fourth row and the third column.

41. If B is an $m \times p$ matrix, C is an $m \times q$ matrix, D is an $n \times p$ matrix, E is an $n \times q$ matrix, $\mathbf{0}_1$ is an $m \times 1$ vector, $\mathbf{0}_2$ is a $1 \times p$ matrix, $\mathbf{0}_3$ is a $1 \times q$ matrix, $\mathbf{0}_4$ is an $n \times 1$ matrix, all entries of $\mathbf{0}_1, \mathbf{0}_2, \mathbf{0}_3, \mathbf{0}_4$ are 0, a is a real number, and $m + n = p + q$, show that

$$\det \begin{bmatrix} B & \mathbf{0}_1 & C \\ \mathbf{0}_2 & a & \mathbf{0}_3 \\ D & \mathbf{0}_4 & E \end{bmatrix} = (-1)^{m+p} a \det \begin{bmatrix} B & C \\ D & E \end{bmatrix}.$$

Chapter 5

Eigenvalues and eigenvectors

Diagonal matrices play a crucial role in this chapter.

Definition 5.1. An $n \times n$ matrix A is called *diagonal* if there are real numbers $a_1, \dots, a_n$ such that

$$A = \begin{bmatrix} a_1 & 0 & \cdots & 0 \\ 0 & a_2 & \cdots & 0 \\ \vdots & \vdots & \ddots & \vdots \\ 0 & 0 & \cdots & a_n \end{bmatrix}.$$

In other words, a diagonal matrix is a square matrix whose entries outside the main diagonal are all 0.

Diagonal matrices have properties that make them easy to work with. For example,

$$\begin{bmatrix} a_1 & 0 & \cdots & 0 \\ 0 & a_2 & \cdots & 0 \\ \vdots & \vdots & \ddots & \vdots \\ 0 & 0 & \cdots & a_n \end{bmatrix} \begin{bmatrix} x_1 \\ x_2 \\ \vdots \\ x_n \end{bmatrix} = \begin{bmatrix} a_1 x_1 \\ a_2 x_2 \\ \vdots \\ a_n x_n \end{bmatrix}$$

and

$$\begin{bmatrix} a_1 & 0 & \cdots & 0 \\ 0 & a_2 & \cdots & 0 \\ \vdots & \vdots & \ddots & \vdots \\ 0 & 0 & \cdots & a_n \end{bmatrix} \begin{bmatrix} b_1 & 0 & \cdots & 0 \\ 0 & b_2 & \cdots & 0 \\ \vdots & \vdots & \ddots & \vdots \\ 0 & 0 & \cdots & b_n \end{bmatrix} = \begin{bmatrix} a_1 b_1 & 0 & \cdots & 0 \\ 0 & a_2 b_2 & \cdots & 0 \\ \vdots & \vdots & \ddots & \vdots \\ 0 & 0 & \cdots & a_n b_n \end{bmatrix}. \tag{5.1}$$

A diagonal matrix is invertible if and only if all entries on its diagonal are different from 0 and then

$$\begin{bmatrix} a_1 & 0 & \cdots & 0 \\ 0 & a_2 & \cdots & 0 \\ \vdots & \vdots & \ddots & \vdots \\ 0 & 0 & \cdots & a_n \end{bmatrix}^{-1} = \begin{bmatrix} a_1^{-1} & 0 & \cdots & 0 \\ 0 & a_2^{-1} & \cdots & 0 \\ \vdots & \vdots & \ddots & \vdots \\ 0 & 0 & \cdots & a_n^{-1} \end{bmatrix}.$$

From (5.1) we obtain the following nice and useful property of diagonal matrices

$$\underbrace{\begin{bmatrix} a_1 & 0 & \cdots & 0 \\ 0 & a_2 & \cdots & 0 \\ \vdots & \vdots & \ddots & \vdots \\ 0 & 0 & \dots & a_n \end{bmatrix} \cdots \begin{bmatrix} a_1 & 0 & \cdots & 0 \\ 0 & a_2 & \cdots & 0 \\ \vdots & \vdots & \ddots & \vdots \\ 0 & 0 & \dots & a_n \end{bmatrix}}_{k \text{ times}} = \begin{bmatrix} a_1 & 0 & \cdots & 0 \\ 0 & a_2 & \cdots & 0 \\ \vdots & \vdots & \ddots & \vdots \\ 0 & 0 & \dots & a_n \end{bmatrix}^k = \begin{bmatrix} a_1^k & 0 & \cdots & 0 \\ 0 & a_2^k & \cdots & 0 \\ \vdots & \vdots & \ddots & \vdots \\ 0 & 0 & \dots & a_n^k \end{bmatrix}.$$

This property can be extended to any matrix that can be written in the form PDP^{-1} where P is an invertible $n \times n$ matrix and D is a diagonal $n \times n$ matrix. Indeed, for such a matrix we have

$$(PDP^{-1})^k = \underbrace{PDP^{-1}PDP^{-1}\dots PDP^{-1}}_{k \text{ times}} = PD^kP^{-1},$$

which means that

$$\left(P \begin{bmatrix} a_1 & 0 & \cdots & 0 \\ 0 & a_2 & \cdots & 0 \\ \vdots & \vdots & \ddots & \vdots \\ 0 & 0 & \dots & a_n \end{bmatrix} P^{-1} \right)^k = P \begin{bmatrix} a_1^k & 0 & \cdots & 0 \\ 0 & a_2^k & \cdots & 0 \\ \vdots & \vdots & \ddots & \vdots \\ 0 & 0 & \dots & a_n^k \end{bmatrix} P^{-1}$$

for any numbers $a_1, \dots, a_n$ and any invertible $n \times n$ matrix P.

It is not clear whether every matrix can be written in the form PDP^{-1}. Moreover, even if we know that a matrix has such a representation, it is not obvious how to find P and D. In the next section we investigate these questions and we show how the properties of diagonal matrices can be useful when dealing with matrices that are not diagonal.

5.1 Eigenvalues and eigenvectors of $n \times n$ matrices

We start with the definition of an eigenvalue, arguably one of the most important ideas in linear algebra.

Definition 5.2. A real number λ is called an *eigenvalue* of an $n \times n$ matrix A if there is a nonzero vector $\mathbf{x}$ in $\mathbb{R}^n$ such that

$$A\mathbf{x} = \lambda\mathbf{x}.$$

When finding eigenvalues of a $n \times n$ matrix we use the following theorem.

Theorem 5.3. *Let A be a $n \times n$ matrix. The real number λ is an eigenvalue of the matrix A if and only if*

$$\det(A - \lambda I_n) = 0.$$

Proof. If for some real number λ the equation $A\mathbf{x} = \lambda\mathbf{x}$ has a solution $\mathbf{x} \neq \mathbf{0}$, then $\mathbf{x}$ is a nontrivial solution of the equation $(A - \lambda I_n)\mathbf{x} = \mathbf{0}$. But this means that the matrix $(A - \lambda I_n)$ is not invertible and thus $\det(A - \lambda I_n) = 0$, by Theorem 4.11. □

Example 5.4. The number $\lambda = 3$ is an eigenvalue of the matrix $\begin{bmatrix} 3 & 3 & 1 & 4 \\ 4 & 5 & 3 & 5 \\ 2 & 7 & 4 & 9 \\ 4 & 2 & 3 & 8 \end{bmatrix}$ because

$$\det \left[\begin{bmatrix} 3 & 3 & 1 & 4 \\ 4 & 5 & 3 & 5 \\ 2 & 7 & 4 & 9 \\ 4 & 2 & 3 & 8 \end{bmatrix} - 3 \begin{bmatrix} 1 & 0 & 0 & 0 \\ 0 & 1 & 0 & 0 \\ 0 & 0 & 1 & 0 \\ 0 & 0 & 0 & 1 \end{bmatrix} \right] = \det \begin{bmatrix} 0 & 3 & 1 & 4 \\ 4 & 2 & 3 & 5 \\ 2 & 7 & 1 & 9 \\ 4 & 2 & 3 & 5 \end{bmatrix} = 0.$$

Note that in this case it is not necessary to calculate the last determinant: it has to be 0 because it has two identical rows.

The determinant $\det(A - \lambda I_n)$ is a polynomial in λ.

Example 5.5. If $A = \begin{bmatrix} 4 & 1 & -2 \\ 3 & 2 & -2 \\ 4 & 2 & -2 \end{bmatrix}$, then

$$\begin{aligned} \det(A - \lambda I_3) &= \det \begin{bmatrix} 4-\lambda & 1 & -2 \\ 3 & 2-\lambda & -2 \\ 4 & 2 & -2-\lambda \end{bmatrix} \\ &= (4-\lambda)\det \begin{bmatrix} 2-\lambda & -2 \\ 2 & -2-\lambda \end{bmatrix} - \det \begin{bmatrix} 3 & -2 \\ 4 & -2-\lambda \end{bmatrix} - 2\det \begin{bmatrix} 3 & 2-\lambda \\ 4 & 2 \end{bmatrix} \\ &= (4-\lambda)\lambda^2 - (-3\lambda + 2) - 2(4\lambda - 2) \\ &= -\lambda^3 + 4\lambda^2 - 5\lambda + 2. \end{aligned}$$

Definition 5.6. Let A be a $n \times n$ matrix. The polynomial $\det(A - \lambda I_n)$ is called the *characteristic polynomial* and the equation $\det(A - \lambda I_n) = 0$ is called the *characteristic equation.*

In Example 5.5 we find that the characteristic polynomial of the matrix $A = \begin{bmatrix} 4 & 1 & -2 \\ 3 & 2 & -2 \\ 4 & 2 & -2 \end{bmatrix}$ is $-\lambda^3 + 4\lambda^2 - 5\lambda + 2$. To find the eigenvalues of the matrix A we need to solve the characteristic equation $-\lambda^3 + 4\lambda^2 - 5\lambda + 2 = 0$. In the next example we calculate eigenvalues of a 3×3 matrix using a somewhat different, possibly simpler, approach.

Example 5.7. Calculate the eigenvalues of the matrix

$$A = \begin{bmatrix} -77 & 100 & -220 \\ 20 & -22 & 55 \\ 40 & -50 & 113 \end{bmatrix}.$$

Solution. We have to solve the equation

$$\det \begin{bmatrix} -77-\lambda & 100 & -220 \\ 20 & -22-\lambda & 55 \\ 40 & -50 & 113-\lambda \end{bmatrix} = 0.$$

To simplify our calculations we use Theorem 4.5. First we multiply the second row by -2 and add to the third row:

$$\det \begin{bmatrix} -77-\lambda & 100 & -220 \\ 20 & -22-\lambda & 55 \\ 0 & 2\lambda-6 & 3-\lambda \end{bmatrix} = 0.$$

Then we factor $3 - \lambda$ from the third row:

$$(3-\lambda)\det \begin{bmatrix} -77-\lambda & 100 & -220 \\ 20 & -22-\lambda & 55 \\ 0 & -2 & 1 \end{bmatrix} = 0.$$

Next we multiply the second row by 4 and add to the first row:

$$(3-\lambda)\det \begin{bmatrix} 3-\lambda & 12-4\lambda & 0 \\ 20 & -22-\lambda & 55 \\ 0 & -2 & 1 \end{bmatrix} = 0.$$

Now we factor $3-\lambda$ from the first row:

$$(3-\lambda)^2 \det \begin{bmatrix} 1 & 4 & 0 \\ 20 & -22-\lambda & 55 \\ 0 & -2 & 1 \end{bmatrix} = 0.$$

Finally, in order to get one more 0 in the first column, we multiply the first row by -20 and add to the second row:

$$(3-\lambda)^2 \det \begin{bmatrix} 1 & 4 & 0 \\ 0 & -102-\lambda & 55 \\ 0 & -2 & 1 \end{bmatrix} = (3-\lambda)^2 \det \begin{bmatrix} -102-\lambda & 55 \\ -2 & 1 \end{bmatrix} = (3-\lambda)^2(8-\lambda) = 0.$$

Consequently, the matrix A has two eigenvalues: 3 and 8.

Instead of getting one more 0 in the first column we could get one more 0 in the first row by multiplying the first column by -4 and adding to the second column. In this case the equation becomes

$$(3-\lambda)^2 \det \begin{bmatrix} 1 & 0 & 0 \\ 20 & -102-\lambda & 55 \\ 0 & -2 & 1 \end{bmatrix} = (3-\lambda)^2 \det \begin{bmatrix} -102-\lambda & 55 \\ -2 & 1 \end{bmatrix} = 0.$$

□

Finding the eigenvalues of a 3×3 matrix can be algebraically challenging. Clearly, the problem becomes even more difficult for larger matrices. In practice computer algebra systems are often used to find eigenvalues of larger matrices.

Definition 5.8. Let λ be an eigenvalue of an $n \times n$ matrix A. A vector $\mathbf{x} \neq \mathbf{0}$ is called an *eigenvector* corresponding to the eigenvalue λ if $A\mathbf{x} = \lambda\mathbf{x}$.

Example 5.9. The vector $\begin{bmatrix} 7 \\ 34 \\ 18 \\ -30 \end{bmatrix}$ is an eigenvector corresponding to the eigenvalue

3 of the matrix $\begin{bmatrix} 3 & 3 & 1 & 4 \\ 4 & 5 & 3 & 5 \\ 2 & 7 & 4 & 9 \\ 4 & 2 & 3 & 8 \end{bmatrix}$ because

$$\begin{bmatrix} 3 & 3 & 1 & 4 \\ 4 & 5 & 3 & 5 \\ 2 & 7 & 4 & 9 \\ 4 & 2 & 3 & 8 \end{bmatrix} \begin{bmatrix} 7 \\ 34 \\ 18 \\ -30 \end{bmatrix} = 3 \begin{bmatrix} 7 \\ 34 \\ 18 \\ -30 \end{bmatrix}.$$

Definition 5.10. Let λ be an eigenvalue of a $n \times n$ matrix A. The set of all vectors $\mathbf{x}$ in $\mathbb{R}^n$ such that $A\mathbf{x} = \lambda \mathbf{x}$ is called the *eigenspace* of A corresponding to the eigenvalue λ.

Note that the eigenspace of a matrix A corresponding to an eigenvalue λ is the subspace $\mathbf{N}(A - \lambda I_n)$ and consequently, by Theorem 2.43, an eigenspace of an $n \times n$ matrix is a vector subspace of $\mathbb{R}^n$.

Example 5.11. Show that 1 is an eigenvalue of the matrix

$$A = \begin{bmatrix} 2 & 2 & 1 & 4 \\ 1 & 3 & 1 & 4 \\ 1 & 2 & 2 & 4 \\ 1 & 2 & 1 & 5 \end{bmatrix}$$

and then find a basis of the eigenspace of A corresponding to the eigenvalue 1.

Solution. It is easy to show that 1 is an eigenvalue because

$$\det \begin{bmatrix} 2-1 & 2 & 1 & 4 \\ 1 & 3-1 & 1 & 4 \\ 1 & 2 & 2-1 & 4 \\ 1 & 2 & 1 & 5-1 \end{bmatrix} = \det \begin{bmatrix} 1 & 2 & 1 & 4 \\ 1 & 2 & 1 & 4 \\ 1 & 2 & 1 & 4 \\ 1 & 2 & 1 & 4 \end{bmatrix} = 0.$$

A vector $\begin{bmatrix} x \\ y \\ z \\ w \end{bmatrix}$ is in the eigenspace of A corresponding to the eigenvalue 1 if it is a solution of the equation

$$\begin{bmatrix} 2 & 2 & 1 & 4 \\ 1 & 3 & 1 & 4 \\ 1 & 2 & 2 & 4 \\ 1 & 2 & 1 & 5 \end{bmatrix} \begin{bmatrix} x \\ y \\ z \\ w \end{bmatrix} = 1 \cdot \begin{bmatrix} x \\ y \\ z \\ w \end{bmatrix} = \begin{bmatrix} x \\ y \\ z \\ w \end{bmatrix},$$

or, equivalently, a solution of the equation

$$\begin{bmatrix} 1 & 2 & 1 & 4 \\ 1 & 2 & 1 & 4 \\ 1 & 2 & 1 & 4 \\ 1 & 2 & 1 & 4 \end{bmatrix} \begin{bmatrix} x \\ y \\ z \\ w \end{bmatrix} = \begin{bmatrix} 0 \\ 0 \\ 0 \\ 0 \end{bmatrix}.$$

To find all such vectors we need to solve the system of equations

$$\begin{cases} x + 2y + z + 4w = 0 \\ x + 2y + z + 4w = 0 \\ x + 2y + z + 4w = 0 \\ x + 2y + z + 4w = 0 \end{cases},$$

which is equivalent to the single equation $x = -2y - z - 4w$. Hence

$$\begin{bmatrix} x \\ y \\ z \\ w \end{bmatrix} = \begin{bmatrix} -2y - z - 4w \\ y \\ z \\ w \end{bmatrix} = y \begin{bmatrix} -2 \\ 1 \\ 0 \\ 0 \end{bmatrix} + z \begin{bmatrix} -1 \\ 0 \\ 1 \\ 0 \end{bmatrix} + w \begin{bmatrix} -4 \\ 0 \\ 0 \\ 1 \end{bmatrix},$$

so the eigenspace of A corresponding to the eigenvalue 1 is

$$\operatorname{Span}\left\{ \begin{bmatrix} -2 \\ 1 \\ 0 \\ 0 \end{bmatrix}, \begin{bmatrix} -1 \\ 0 \\ 1 \\ 0 \end{bmatrix}, \begin{bmatrix} -4 \\ 0 \\ 0 \\ 1 \end{bmatrix} \right\}.$$

□

Definition 5.12. A $n \times n$ matrix A is called *diagonalizable* if there is a diagonal $n \times n$ matrix D and an invertible $n \times n$ matrix P such that

$$A = PDP^{-1}.$$

A representation of a matrix A in the form $A = PDP^{-1}$ is called a *diagonalization* of A.

For example, a matrix

$$\begin{bmatrix} a_1 & b_1 & c_1 & d_1 \\ a_2 & b_2 & c_2 & d_2 \\ a_3 & b_3 & c_3 & d_3 \\ a_4 & b_4 & c_4 & d_4 \end{bmatrix}$$

is diagonalizable if there are real numbers α, β, γ and δ and an invertible matrix

$$\begin{bmatrix} q_1 & r_1 & s_1 & t_1 \\ q_2 & r_2 & s_2 & t_2 \\ q_3 & r_3 & s_3 & t_3 \\ q_4 & r_4 & s_4 & t_4 \end{bmatrix}$$

such that

$$\begin{bmatrix} a_1 & b_1 & c_1 & d_1 \\ a_2 & b_2 & c_2 & d_2 \\ a_3 & b_3 & c_3 & d_3 \\ a_4 & b_4 & c_4 & d_4 \end{bmatrix} = \begin{bmatrix} q_1 & r_1 & s_1 & t_1 \\ q_2 & r_2 & s_2 & t_2 \\ q_3 & r_3 & s_3 & t_3 \\ q_4 & r_4 & s_4 & t_4 \end{bmatrix} \begin{bmatrix} \alpha & 0 & 0 & 0 \\ 0 & \beta & 0 & 0 \\ 0 & 0 & \gamma & 0 \\ 0 & 0 & 0 & \delta \end{bmatrix} \begin{bmatrix} q_1 & r_1 & s_1 & t_1 \\ q_2 & r_2 & s_2 & t_2 \\ q_3 & r_3 & s_3 & t_3 \\ q_4 & r_4 & s_4 & t_4 \end{bmatrix}^{-1}.$$

Not every matrix is diagonalizable. Characterizing diagonalizable matrices is one of the main goals of this chapter.

Theorem 5.13. *An $n \times n$ matrix is diagonalizable if and only if it has n linearly independent eigenvectors.*

Proof for $n = 4$. We have to prove that a 4×4 matrix A is diagonalizable if and only if A has 4 linearly independent eigenvectors.

In other words, we need to show that A is diagonalizable if and only if there exist real numbers α, β, γ, and δ, not necessarily different, and linearly independent vectors $\mathbf{p}_1$, $\mathbf{p}_2$, $\mathbf{p}_3$, and $\mathbf{p}_4$ such that $A\mathbf{p}_1 = \alpha\mathbf{p}_1$, $A\mathbf{p}_2 = \beta\mathbf{p}_2$, $A\mathbf{p}_3 = \gamma\mathbf{p}_3$, and $A\mathbf{p}_4 = \delta\mathbf{p}_4$.

We first note that the above four equations can be written as a single matrix equation

$$A\begin{bmatrix}\mathbf{p}_1 & \mathbf{p}_2 & \mathbf{p}_3 & \mathbf{p}_4\end{bmatrix} = \begin{bmatrix}\mathbf{p}_1 & \mathbf{p}_2 & \mathbf{p}_3 & \mathbf{p}_4\end{bmatrix} \begin{bmatrix} \alpha & 0 & 0 & 0 \\ 0 & \beta & 0 & 0 \\ 0 & 0 & \gamma & 0 \\ 0 & 0 & 0 & \delta \end{bmatrix}.$$

By Theorem 2.17, the vectors $\mathbf{p}_1$, $\mathbf{p}_2$, $\mathbf{p}_3$, and $\mathbf{p}_4$ are linearly independent if and only if the matrix $P = \begin{bmatrix}\mathbf{p}_1 & \mathbf{p}_2 & \mathbf{p}_3 & \mathbf{p}_4\end{bmatrix}$ is invertible. Consequently, to prove the theorem it suffices to show that a matrix A is diagonalizable if and only if there exists an invertible matrix P and a diagonal matrix D such that $AP = PD$.

If A is diagonalizable, then there exists an invertible matrix P and a diagonal matrix D such that $A = PDP^{-1}$. Consequently,

$$AP = PDP^{-1}P = PD.$$

Now, if there exists an invertible matrix P and a diagonal matrix D such that $AP = PD$, then

$$PDP^{-1} = APP^{-1} = A,$$

so A is diagonalizable. □

Note that the proof of Theorem 5.13 gives us the following result that is used in diagonalization of matrices.

Theorem 5.14. *If $\mathbf{p}_1, \dots, \mathbf{p}_n$ are linearly independent eigenvectors of an $n \times n$ matrix A corresponding to the eigenvalues $\lambda_1, \dots, \lambda_n$, respectively, then*

$$A = \begin{bmatrix} \mathbf{p}_1 & \dots & \mathbf{p}_n \end{bmatrix} \begin{bmatrix} \lambda_1 & 0 & \cdots & 0 \\ 0 & \lambda_2 & \cdots & 0 \\ \vdots & \vdots & \ddots & \vdots \\ 0 & 0 & \cdots & \lambda_n \end{bmatrix} \begin{bmatrix} \mathbf{p}_1 & \dots & \mathbf{p}_n \end{bmatrix}^{-1}.$$

Example 5.15. The matrix

$$\begin{bmatrix} -77 & 100 & -220 \\ 20 & -22 & 55 \\ 40 & -50 & 113 \end{bmatrix}$$

has two eigenvalues: 3 and 8. The vectors $\begin{bmatrix} 1 \\ 3 \\ 1 \end{bmatrix}$ and $\begin{bmatrix} 4 \\ 1 \\ -1 \end{bmatrix}$ are eigenvectors corresponding to the eigenvalue 3 because we have

$$\begin{bmatrix} -77 & 100 & -220 \\ 20 & -22 & 55 \\ 40 & -50 & 113 \end{bmatrix} \begin{bmatrix} 1 \\ 3 \\ 1 \end{bmatrix} = \begin{bmatrix} 3 \\ 9 \\ 3 \end{bmatrix} = 3 \begin{bmatrix} 1 \\ 3 \\ 1 \end{bmatrix}$$

and

$$\begin{bmatrix} -77 & 100 & -220 \\ 20 & -22 & 55 \\ 40 & -50 & 113 \end{bmatrix} \begin{bmatrix} 4 \\ 1 \\ -1 \end{bmatrix} = \begin{bmatrix} 12 \\ 3 \\ -3 \end{bmatrix} = 3 \begin{bmatrix} 4 \\ 1 \\ -1 \end{bmatrix}.$$

The vector $\begin{bmatrix} -4 \\ 1 \\ 2 \end{bmatrix}$ is an eigenvector corresponding to the eigenvalue 8 because we have

$$\begin{bmatrix} -77 & 100 & -220 \\ 20 & -22 & 55 \\ 40 & -50 & 113 \end{bmatrix} \begin{bmatrix} -4 \\ 1 \\ 2 \end{bmatrix} = \begin{bmatrix} -32 \\ 8 \\ 16 \end{bmatrix} = 8 \begin{bmatrix} -4 \\ 1 \\ 2 \end{bmatrix}.$$

The vectors $\begin{bmatrix} 1 \\ 3 \\ 1 \end{bmatrix}$, $\begin{bmatrix} 4 \\ 1 \\ -1 \end{bmatrix}$, and $\begin{bmatrix} -4 \\ 1 \\ 2 \end{bmatrix}$ are linearly independent since the determinant of the matrix $\begin{bmatrix} 1 & 4 & -4 \\ 3 & 1 & 1 \\ 1 & -1 & 2 \end{bmatrix}$ is -1. Consequently, by Theorem 5.14, we can write

$$\begin{bmatrix} -77 & 100 & -220 \\ 20 & -22 & 55 \\ 40 & -50 & 113 \end{bmatrix} = \begin{bmatrix} 1 & 4 & -4 \\ 3 & 1 & 1 \\ 1 & -1 & 2 \end{bmatrix} \begin{bmatrix} 3 & 0 & 0 \\ 0 & 3 & 0 \\ 0 & 0 & 8 \end{bmatrix} \begin{bmatrix} 1 & 4 & -4 \\ 3 & 1 & 1 \\ 1 & -1 & 2 \end{bmatrix}^{-1}.$$

Linear independence of the eigenvectors $\mathbf{p}_1, \dots, \mathbf{p}_n$ in Theorem 5.14 is essential, because without it the matrix $\begin{bmatrix} \mathbf{p}_1 & \dots & \mathbf{p}_n \end{bmatrix}$ is not invertible. It turns out that the eigenvectors corresponding to distinct eigenvalues are always linearly independent.

Theorem 5.16. *If $\mathbf{p}_1, \dots, \mathbf{p}_j$ are eigenvectors of an $n \times n$ matrix A corresponding to different eigenvalues, then the vectors $\mathbf{p}_1, \dots, \mathbf{p}_j$ are linearly independent.*

Proof of a particular case. If $\mathbf{p}_1, \dots, \mathbf{p}_j$ are eigenvectors of an $n \times n$ matrix A corresponding to different eigenvalues, then

$$A\mathbf{p}_1 = \lambda_1 \mathbf{p}_1,\ A\mathbf{p}_2 = \lambda_2 \mathbf{p}_2,\ \dots,\ A\mathbf{p}_j = \lambda_j \mathbf{p}_j,$$

where $\lambda_1, \dots, \lambda_j$ are distinct real numbers.

Now suppose that the vectors $\mathbf{p}_1, \dots, \mathbf{p}_j$ are linearly dependent. Since the vectors $\mathbf{p}_1, \dots, \mathbf{p}_j$ are nonzero, there is a number k such that the vector $\mathbf{p}_k$, is a linear combination of $\mathbf{p}_1, \dots, \mathbf{p}_{k-1}$ and the vectors $\mathbf{p}_1, \dots, \mathbf{p}_{k-1}$ are linearly independent, by Corollary 2.12. Suppose $k = 5$. Then there are real numbers a_1, a_2, a_3, and a_4 such that

$$\mathbf{p}_5 = a_1 \mathbf{p}_1 + a_2 \mathbf{p}_2 + a_3 \mathbf{p}_3 + a_4 \mathbf{p}_4.$$

Hence

$$A\mathbf{p}_5 = a_1 A\mathbf{p}_1 + a_2 A\mathbf{p}_2 + a_3 A\mathbf{p}_3 + a_4 A\mathbf{p}_4$$

and, consequently,

$$\lambda_5 \mathbf{p}_5 = a_1 \lambda_1 \mathbf{p}_1 + a_2 \lambda_2 \mathbf{p}_2 + a_3 \lambda_3 \mathbf{p}_3 + a_4 \lambda_4 \mathbf{p}_4.$$

Since we also have

$$\lambda_5 \mathbf{p}_5 = a_1 \lambda_5 \mathbf{p}_1 + a_2 \lambda_5 \mathbf{p}_2 + a_3 \lambda_5 \mathbf{p}_3 + a_4 \lambda_5 \mathbf{p}_4,$$

we can conclude that

$$\mathbf{0} = a_1(\lambda_5 - \lambda_1)\mathbf{p}_1 + a_2(\lambda_5 - \lambda_2)\mathbf{p}_2 + a_3(\lambda_5 - \lambda_3)\mathbf{p}_3 + a_4(\lambda_5 - \lambda_4)\mathbf{p}_4.$$

Linear independence of the vectors $\mathbf{p}_1$, $\mathbf{p}_2$, $\mathbf{p}_3$, and $\mathbf{p}_4$ implies

$$a_1 = a_2 = a_3 = a_4 = 0,$$

which is not possible because $\mathbf{p}_5 \neq \mathbf{0}$, being an eigenvector. □

From Theorems 5.13 and 5.16 we obtain the following useful results.

Corollary 5.17. *If a $n \times n$ matrix A has n different eigenvalues, then A is diagonalizable.*

Example 5.18. Diagonalize the matrix $A = \begin{bmatrix} 3 & 3 & 4 \\ 1 & 5 & 4 \\ 1 & 1 & 8 \end{bmatrix}$ using the fact that it has eigenvalues 2, 4, and 10.

Solution. A vector $\begin{bmatrix} x \\ y \\ z \end{bmatrix}$ is an eigenvector corresponding to the eigenvalue 2 if

$$\begin{bmatrix} 3-2 & 3 & 4 \\ 1 & 5-2 & 4 \\ 1 & 1 & 8-2 \end{bmatrix} \begin{bmatrix} x \\ y \\ z \end{bmatrix} = \begin{bmatrix} 1 & 3 & 4 \\ 1 & 3 & 4 \\ 1 & 1 & 6 \end{bmatrix} \begin{bmatrix} x \\ y \\ z \end{bmatrix} = \begin{bmatrix} 0 \\ 0 \\ 0 \end{bmatrix}.$$

This is equivalent to the system of equations

$$\begin{cases} x + 3y + 4z = 0 \\ x + y + 6z = 0 \end{cases}.$$

We find that $x = 7, y = -1, z = -1$ is a solution of the system, so the vector $\begin{bmatrix} 7 \\ -1 \\ -1 \end{bmatrix}$ is an eigenvector of the matrix A corresponding to the eigenvalue 2.

In the same way we find that $\begin{bmatrix} 2 \\ 2 \\ -1 \end{bmatrix}$ is an eigenvector corresponding to the eigenvalue 4 and $\begin{bmatrix} 1 \\ 1 \\ 1 \end{bmatrix}$ is an eigenvector corresponding to the eigenvalue 10.

Consequently,

$$\begin{bmatrix} 3 & 3 & 4 \\ 1 & 5 & 4 \\ 1 & 1 & 8 \end{bmatrix} = \begin{bmatrix} 7 & 1 & 1 \\ -1 & 2 & 1 \\ -1 & -1 & 1 \end{bmatrix} \begin{bmatrix} 2 & 0 & 0 \\ 0 & 4 & 0 \\ 0 & 0 & 10 \end{bmatrix} \begin{bmatrix} 7 & 1 & 1 \\ -1 & 2 & 1 \\ -1 & -1 & 1 \end{bmatrix}^{-1}.$$

Note that the order of the eigenvalues in the matrix $\begin{bmatrix} 2 & 0 & 0 \\ 0 & 4 & 0 \\ 0 & 0 & 10 \end{bmatrix}$ corresponds to the order of the eigenvectors in the matrix $\begin{bmatrix} 7 & 1 & 1 \\ -1 & 2 & 1 \\ -1 & -1 & 1 \end{bmatrix}$. □

Not every matrix has real eigenvalues. For example, since the characteristic polynomial of the matrix $A = \begin{bmatrix} 1 & -1 \\ 1 & 1 \end{bmatrix}$ is

$$\det(A - \lambda I_2) = \det \begin{bmatrix} 1-\lambda & -1 \\ 1 & 1-\lambda \end{bmatrix} = (1-\lambda)^2 + 1,$$

the matrix does not have any real eigenvalues.

If $\lambda_1, \dots, \lambda_k$ are distinct real eigenvalues of an $n \times n$ matrix, then clearly $k \le n$ and the characteristic polynomial of A can be written as

$$\det(A - \lambda I_n) = (\lambda - \lambda_1)^{m_1} \cdots (\lambda - \lambda_k)^{m_k} p(\lambda),$$

where $m_1, \dots, m_k$ are nonnegative integers such that $m_1 + \cdots + m_k \le n$ and $p(\lambda)$ is a polynomial with no real zeros, possibly a constant. The number m_j is called the *algebraic multiplicity* of the eigenvalue λ_j. For example, the characteristic polynomial of the matrix

$$A = \begin{bmatrix} 1 & 0 & \frac{1}{2} & -\frac{1}{2} & 2 \\ \frac{1}{2} & 2 & \frac{1}{2} & -\frac{1}{2} & \frac{1}{2} \\ 2 & 0 & \frac{5}{2} & -\frac{7}{2} & 2 \\ 2 & 0 & \frac{1}{2} & -\frac{3}{2} & 2 \\ 2 & 0 & \frac{1}{2} & -\frac{1}{2} & 1 \end{bmatrix}$$

is

$$\det(A - \lambda I_5) = (\lambda + 1)^2 (\lambda - 2)^2 (\lambda - 3).$$

This means that A has three distinct eigenvalues: -1, 2, and 3. Eigenvalues -1 and 2 are of algebraic multiplicity 2. The algebraic multiplicity of the eigenvalue 3 is 1.

If an $n \times n$ matrix A has n distinct eigenvalues, then A has n linearly independent eigenvectors and thus the dimension of every eigenspace of A is 1. If the algebraic multiplicity of an eigenvalue is greater than 1, then the dimension of the corresponding eigenspace can be greater than 1.

Example 5.19. Find the eigenvalues and the corresponding eigenspaces of the matrix

$$A = \begin{bmatrix} 2 & 1 & 1 & 3 \\ 1 & 2 & 1 & 3 \\ 1 & 1 & 3 & 2 \\ 1 & 1 & 1 & 4 \end{bmatrix}.$$

Solution. The eigenvalues are the solutions of the equation

$$\det\begin{bmatrix} 2-\lambda & 1 & 1 & 3 \\ 1 & 2-\lambda & 1 & 3 \\ 1 & 1 & 3-\lambda & 2 \\ 1 & 1 & 1 & 4-\lambda \end{bmatrix} = 0.$$

In order to calculate the determinant we subtract the last row from all other rows and get

$$\det\begin{bmatrix} 1-\lambda & 0 & 0 & \lambda-1 \\ 0 & 1-\lambda & 1 & \lambda-1 \\ 0 & 0 & 2-\lambda & \lambda-2 \\ 1 & 1 & 1 & 4-\lambda \end{bmatrix} = (1-\lambda)^2(2-\lambda)\det\begin{bmatrix} 1 & 0 & 0 & -1 \\ 0 & 1 & 1 & -1 \\ 0 & 0 & 1 & -1 \\ 1 & 1 & 1 & 4-\lambda \end{bmatrix}.$$

Now we subtract the first row from the last row and get

$$\det\begin{bmatrix} 1 & 0 & 0 & -1 \\ 0 & 1 & 1 & -1 \\ 0 & 0 & 1 & -1 \\ 1 & 1 & 1 & 4-\lambda \end{bmatrix} = \det\begin{bmatrix} 1 & 0 & 0 & -1 \\ 0 & 1 & 1 & -1 \\ 0 & 0 & 1 & -1 \\ 0 & 1 & 1 & 5-\lambda \end{bmatrix} = \det\begin{bmatrix} 1 & 1 & -1 \\ 0 & 1 & -1 \\ 1 & 1 & 5-\lambda \end{bmatrix} = 7-\lambda.$$

Consequently,

$$\det\begin{bmatrix} 1-\lambda & 0 & 0 & \lambda-1 \\ 0 & 1-\lambda & 1 & \lambda-1 \\ 0 & 0 & 2-\lambda & \lambda-2 \\ 1 & 1 & 1 & 4-\lambda \end{bmatrix} = (1-\lambda)^2(2-\lambda)(7-\lambda)$$

and the eigenvalues are 1, 2, and 7. Eigenvalues 2 and 7 are of algebraic multiplicity 1. The algebraic multiplicity of the eigenvalue 1 is 2.

The eigenspace corresponding to the eigenvalue 1 is the null space of the matrix

$$\begin{bmatrix} 1 & 1 & 1 & 3 \\ 1 & 1 & 1 & 3 \\ 1 & 1 & 2 & 2 \\ 1 & 1 & 1 & 3 \end{bmatrix},$$

which we find to be $\operatorname{Span}\left\{\begin{bmatrix} -4 \\ 0 \\ 1 \\ 1 \end{bmatrix}, \begin{bmatrix} -1 \\ 1 \\ 0 \\ 0 \end{bmatrix}\right\}$. Since the vectors $\begin{bmatrix} -4 \\ 0 \\ 1 \\ 1 \end{bmatrix}$ and $\begin{bmatrix} -1 \\ 1 \\ 0 \\ 0 \end{bmatrix}$ are linearly independent, the dimension of the eigenspace corresponding to the eigenvalue 1 is 2, which is the same as the algebraic multiplicity of the eigenvalue 1.

Next we determine that the eigenspace corresponding to the eigenvalue 2 is $\operatorname{Span}\left\{\begin{bmatrix}1\\1\\-4\\1\end{bmatrix}\right\}$ and the eigenspace corresponding to the eigenvalue 7 is $\operatorname{Span}\left\{\begin{bmatrix}1\\1\\1\\1\end{bmatrix}\right\}$. Note that the algebraic multiplicities of the eigenvalues 2 and 7 are both 1 and the dimensions of the corresponding eigenspaces are also 1. □

The algebraic multiplicity of an eigenvalue can be greater than the dimension of the corresponding eigenspace. Indeed, consider the matrix

$$A = \begin{bmatrix}5 & 1 & 1\\0 & 5 & 0\\0 & 0 & 5\end{bmatrix}.$$

The characteristic polynomial of A is $(5-\lambda)^3$, so A has only one eigenvalue whose algebraic multiplicity is 3. On the other hand, since the dimension of the null space of the matrix

$$\begin{bmatrix}0 & 1 & 1\\0 & 0 & 0\\0 & 0 & 0\end{bmatrix}$$

is 2, the dimension of the eigenspace corresponding to the eigenvalue 5 is 2.

The dimension of the eigenspace of a matrix A corresponding to an eigenvalue λ is called the *geometric multiplicity* of λ. The geometric multiplicity of an eigenvalue is always less than or equal to the algebraic multiplicity of an eigenvalue. Moreover, if A is a $n \times n$ matrix with k distinct eigenvalues $\lambda_1, \dots, \lambda_k$ and m_j is the geometric multiplicity of λ_j for each $j = 1, \dots, k$, then $m_1 + \cdots + m_k \le n$.

The following result is useful in diagonalizing matrices with eigenvalues of geometric multiplicity greater than 1. The theorem complements Theorem 5.14 and is usually proven in a second course in Linear Algebra. If the abstract formulation of the theorem appears difficult to follow, it is probably a good idea to skip this result for now and come back to it after studying the examples.

Theorem 5.20. *Let A be an $n \times n$ matrix with k distinct real eigenvalues $\lambda_1, \dots, \lambda_k$ and let m_j be the geometric multiplicity of λ_j for each $j = 1, \dots, k$. If*

$$m_1 + \cdots + m_k = n,$$

then the matrix A is diagonalizable and, if for each $j = 1, \dots, k$ the set $\{\mathbf{u}_{1,j}, \dots, \mathbf{u}_{m_j,j}\}$ is a basis of the eigenspace corresponding to the eigenvalue λ_j, then

$$\{\mathbf{u}_{1,1}, \dots, \mathbf{u}_{m_1,1}, \mathbf{u}_{1,2}, \dots, \mathbf{u}_{m_2,2}, \dots, \mathbf{u}_{1,k}, \dots, \mathbf{u}_{m_k,k}\}$$

is a basis in $\mathbb{R}^n$ and $A = PDP^{-1}$, where

$$P = \begin{bmatrix} \mathbf{u}_{1,1} & \dots & \mathbf{u}_{m_1,1} & \mathbf{u}_{1,2} & \dots & \mathbf{u}_{m_2,2} & \dots & \mathbf{u}_{1,k} & \dots & \mathbf{u}_{m_k,k} \end{bmatrix}$$

and D is the diagonal matrix with

$$\underbrace{\lambda_1, \dots, \lambda_1}_{m_1 \text{ times}}, \underbrace{\lambda_2, \dots, \lambda_2}_{m_2 \text{ times}}, \dots, \underbrace{\lambda_k, \dots, \lambda_k}_{m_k \text{ times}}$$

along the main diagonal.
Conversely, if the matrix A is diagonalizable then

$$m_1 + \cdots + m_k = n.$$

Example 5.21. In Example 5.19 we found that the matrix

$$A = \begin{bmatrix} 2 & 1 & 1 & 3 \\ 1 & 2 & 1 & 3 \\ 1 & 1 & 3 & 2 \\ 1 & 1 & 1 & 4 \end{bmatrix}$$

has three eigenvalues, 1, 2 and 7, of geometric multiplicity 2, 1, and 1, respectively. Since $2+1+1 = 4$, the size of the matrix A, the matrix A is diagonalizable. Moreover, we found that

$$\left\{ \begin{bmatrix} -4 \\ 0 \\ 1 \\ 1 \end{bmatrix}, \begin{bmatrix} -1 \\ 1 \\ 0 \\ 0 \end{bmatrix} \right\}, \quad \left\{ \begin{bmatrix} 1 \\ 1 \\ -4 \\ 1 \end{bmatrix} \right\}, \quad \text{and} \quad \left\{ \begin{bmatrix} 1 \\ 1 \\ 1 \\ 1 \end{bmatrix} \right\}$$

are bases of the eigenspaces of the eigenvalues 1, 2 and 7, respectively. Consequently, by the above theorem,

$$\left\{\begin{bmatrix}-4\\0\\1\\1\end{bmatrix},\begin{bmatrix}-1\\1\\0\\0\end{bmatrix},\begin{bmatrix}1\\1\\-4\\1\end{bmatrix},\begin{bmatrix}1\\1\\1\\1\end{bmatrix}\right\}$$

is a basis in $\mathbb{R}^4$ and we have

$$A=\begin{bmatrix}-4&-1&1&1\\0&1&1&1\\1&0&-4&1\\1&0&1&1\end{bmatrix}\begin{bmatrix}1&0&0&0\\0&1&0&0\\0&0&2&0\\0&0&0&7\end{bmatrix}\begin{bmatrix}-4&-1&1&1\\0&1&1&1\\1&0&-4&1\\1&0&1&1\end{bmatrix}^{-1}.$$

Example 5.22. If possible, diagonalize the matrix $A=\begin{bmatrix}4&1&-2\\3&2&-2\\4&2&-2\end{bmatrix}$.

Solution. First we calculate the eigenvalues of A: 1 and 2. Next we determine that the eigenspace corresponding to the eigenvalue 1 is $\operatorname{Span}\left\{\begin{bmatrix}1\\1\\2\end{bmatrix}\right\}$ and the eigenspace corresponding to the eigenvalue 2 is $\operatorname{Span}\left\{\begin{bmatrix}2\\2\\3\end{bmatrix}\right\}$. Since the set

$$\left\{\begin{bmatrix}1\\1\\2\end{bmatrix},\begin{bmatrix}2\\2\\3\end{bmatrix}\right\}$$

is not a basis of $\mathbb{R}^3$ the matrix A cannot be diagonalized. □

Exercises 5.1

Determine the characteristic polynomial of the matrix A.

1. $A=\begin{bmatrix}2&1&1\\1&0&2\\1&1&1\end{bmatrix}$

2. $A=\begin{bmatrix}5&0&1\\1&1&0\\0&1&2\end{bmatrix}$

3. $A = \begin{bmatrix} 3 & 1 & 0 & 2 \\ 0 & 0 & 2 & 2 \\ 0 & 1 & 0 & 0 \\ 0 & 1 & 1 & 1 \end{bmatrix}$

4. $A = \begin{bmatrix} 2 & 0 & 0 & 0 \\ 3 & 0 & 2 & 0 \\ 5 & 1 & 1 & 0 \\ 1 & 0 & 1 & 2 \end{bmatrix}$

Determine an eigenvalue λ of the matrix A without calculating the characteristic polynomial.

5. $A = \begin{bmatrix} 2 & 5 & 9 \\ 4 & 5 & 1 \\ 4 & 2 & 4 \end{bmatrix}$

6. $A = \begin{bmatrix} 2 & 3 & 4 \\ 1 & 4 & 4 \\ 5 & 2 & 7 \end{bmatrix}$

7. $A = \begin{bmatrix} 3 & 2 & 5 & 1 \\ 1 & 4 & 5 & 1 \\ 1 & 1 & 0 & 4 \\ 2 & 1 & 7 & 3 \end{bmatrix}$

8. $A = \begin{bmatrix} 3 & 4 & 5 & 7 \\ 1 & 7 & 5 & 2 \\ 1 & 4 & 8 & 2 \\ 2 & 9 & 8 & 5 \end{bmatrix}$

Determine two eigenvalues of the matrix A without calculating the characteristic polynomial.

9. $A = \begin{bmatrix} 5 & 1 & 7 \\ 2 & 4 & 7 \\ 3 & 1 & 9 \end{bmatrix}$

10. $A = \begin{bmatrix} 7 & 2 & 5 \\ 3 & 3 & 8 \\ 3 & 2 & 9 \end{bmatrix}$

11. $A = \begin{bmatrix} 3 & 1 & 5 & 1 \\ 0 & 4 & 5 & 1 \\ 1 & 1 & 7 & 4 \\ 1 & 1 & 2 & 9 \end{bmatrix}$

12. $A = \begin{bmatrix} 7 & 4 & 8 & 9 \\ 3 & 3 & 5 & 2 \\ 3 & 1 & 7 & 2 \\ 3 & 1 & 0 & 9 \end{bmatrix}$

Determine three eigenvalues of the matrix A without calculating the characteristic polynomial.

13. $A = \begin{bmatrix} 2 & 1 & 1 & 1 \\ 1 & 2 & 1 & 1 \\ 1 & 0 & 3 & 1 \\ 1 & 2 & 1 & 1 \end{bmatrix}$

14. $A = \begin{bmatrix} 5 & 1 & 1 & 0 \\ 1 & 4 & 1 & 1 \\ 1 & 2 & 3 & 1 \\ 1 & 1 & 1 & 4 \end{bmatrix}$

Find a basis of the eigenspace corresponding to the given eigenvalue λ for the given matrix A.

15. $\lambda = 2$ and $A = \begin{bmatrix} 1 & 5 & 2 \\ 3 & 2 & 1 \\ 3 & 0 & 3 \end{bmatrix}$

16. $\lambda = 1$ and $A = \begin{bmatrix} 2 & 3 & 2 \\ 1 & 4 & 2 \\ 5 & 2 & 1 \end{bmatrix}$

17. $\lambda = 1$ and $A = \begin{bmatrix} 4 & 4 & 2 \\ 3 & 5 & 2 \\ 3 & 4 & 3 \end{bmatrix}$

18. $\lambda = 3$ and $A = \begin{bmatrix} 5 & 1 & 1 \\ 2 & 4 & 1 \\ 2 & 1 & 4 \end{bmatrix}$

19. $\lambda = 3$ and $A = \begin{bmatrix} 4 & 1 & 1 & 1 \\ 1 & 4 & 1 & 1 \\ 0 & 1 & 2 & 2 \\ 1 & 0 & 0 & 1 \end{bmatrix}$

20. $\lambda = 4$ and $A = \begin{bmatrix} 2 & 0 & 0 & 1 \\ 1 & 4 & 0 & 1 \\ 1 & 0 & 4 & 1 \\ 1 & 1 & 1 & 1 \end{bmatrix}$

21. $\lambda = 2$ and $A = \begin{bmatrix} 3 & 1 & 0 & 1 \\ 1 & 3 & 0 & 1 \\ 1 & 1 & 3 & 2 \\ 1 & 1 & 1 & 4 \end{bmatrix}$

22. $\lambda = 1$ and $A = \begin{bmatrix} 3 & 1 & 1 & 1 \\ 2 & 2 & 1 & 1 \\ 1 & 1 & 3 & 1 \\ 1 & 1 & 2 & 2 \end{bmatrix}$

23. $\lambda = 2$ and $A = \begin{bmatrix} 3 & 1 & 1 & 1 \\ 3 & 5 & 3 & 3 \\ 1 & 1 & 3 & 1 \\ 1 & 1 & 1 & 3 \end{bmatrix}$

24. $\lambda = 2$ and $A = \begin{bmatrix} 2 & 1 & 1 & 1 \\ 1 & 2 & 1 & 1 \\ 1 & 1 & 2 & 1 \\ 3 & 3 & 3 & 4 \end{bmatrix}$

Find the eigenvalues of the given matrix.

25. $\begin{bmatrix} 2 & 2 & 3 \\ 1 & 3 & 3 \\ 1 & 1 & 5 \end{bmatrix}$

26. $A = \begin{bmatrix} 2 & 1 & 2 \\ 0 & 3 & 2 \\ 0 & 2 & 3 \end{bmatrix}$

27. $A = \begin{bmatrix} 4 & 4 & 2 \\ 3 & 5 & 2 \\ 3 & 4 & 3 \end{bmatrix}$

28. $A = \begin{bmatrix} 5 & 2 & 1 \\ 3 & 4 & 1 \\ 3 & 2 & 3 \end{bmatrix}$

29. $A = \begin{bmatrix} 5 & 1 & 1 & 0 \\ 1 & 4 & 1 & 1 \\ 1 & 2 & 3 & 1 \\ 1 & 1 & 1 & 4 \end{bmatrix}$

30. $A = \begin{bmatrix} 2 & 1 & 1 & 1 \\ 1 & 2 & 1 & 1 \\ 1 & 0 & 3 & 1 \\ 1 & 2 & 1 & 1 \end{bmatrix}$

31. $A = \begin{bmatrix} 2 & 1 & 1 & 1 \\ 1 & 2 & 1 & 1 \\ 1 & 1 & 2 & 1 \\ 3 & 3 & 3 & 4 \end{bmatrix}$

32. $A = \begin{bmatrix} 3 & 1 & 1 & 1 \\ 3 & 5 & 3 & 3 \\ 1 & 1 & 3 & 1 \\ 1 & 1 & 1 & 3 \end{bmatrix}$

Write the given matrix A in the form $A = PDP^{-1}$ where P is an invertible matrix and D is a diagonal matrix.

33. $A = \begin{bmatrix} 4 & 1 \\ 5 & 8 \end{bmatrix}$

34. $A = \begin{bmatrix} 4 & 2 \\ 12 & 9 \end{bmatrix}$

35. $A = \begin{bmatrix} 1 & 2 & 2 \\ 2 & 4 & 4 \\ 2 & 3 & 5 \end{bmatrix}$

36. $A = \begin{bmatrix} 3 & 3 & 2 \\ 1 & 5 & 2 \\ 3 & 3 & 2 \end{bmatrix}$

37. $\begin{bmatrix} 4 & 2 & 2 \\ 1 & 5 & 2 \\ 1 & 2 & 5 \end{bmatrix}$

38. $\begin{bmatrix} 4 & 1 & 2 \\ 4 & 1 & 2 \\ 8 & 2 & 4 \end{bmatrix}$

39. $A = \begin{bmatrix} 2 & 1 & 1 & 1 \\ 1 & 2 & 1 & 1 \\ 1 & 1 & 2 & 1 \\ 3 & 3 & 3 & 4 \end{bmatrix}$

40. $A = \begin{bmatrix} 3 & 1 & 1 & 1 \\ 3 & 5 & 3 & 3 \\ 1 & 1 & 3 & 1 \\ 1 & 1 & 1 & 3 \end{bmatrix}$

Show that the matrix A cannot be diagonalized.

41. $A = \begin{bmatrix} 3 & -3 & 7 \\ 4 & -4 & 10 \\ 1 & -1 & 3 \end{bmatrix}$

42. $\begin{bmatrix} 11 & -5 & 9 \\ 5 & 1 & 9 \\ 4 & -4 & 10 \end{bmatrix}$

43. $A = \begin{bmatrix} 3 & -1 & 5 \\ 3 & -1 & 7 \\ 2 & -2 & 4 \end{bmatrix}$

44. $A = \begin{bmatrix} 5 & 1 & 2 & 4 \\ 0 & 3 & 5 & 2 \\ 0 & 0 & 3 & 9 \\ 0 & 0 & 0 & 3 \end{bmatrix}$

5.2 Symmetric $n \times n$ matrices

Recall that a matrix is called symmetric if $A^T = A$.

Definition 5.23. A $n \times n$ matrix P is called *orthogonal* if it is invertible and we have

$$P^T = P^{-1}.$$

Note that we can also say that an $n \times n$ matrix P is orthogonal if

$$P^T P = I_n.$$

Example 5.24. Since

$$\begin{bmatrix} \frac{1}{3} & -\frac{2}{3} & \frac{2}{3} \\ \frac{2}{3} & -\frac{1}{3} & -\frac{2}{3} \\ \frac{2}{3} & \frac{2}{3} & \frac{1}{3} \end{bmatrix}^T \begin{bmatrix} \frac{1}{3} & -\frac{2}{3} & \frac{2}{3} \\ \frac{2}{3} & -\frac{1}{3} & -\frac{2}{3} \\ \frac{2}{3} & \frac{2}{3} & \frac{1}{3} \end{bmatrix} = \begin{bmatrix} \frac{1}{3} & \frac{2}{3} & \frac{2}{3} \\ -\frac{2}{3} & -\frac{1}{3} & \frac{2}{3} \\ \frac{2}{3} & -\frac{2}{3} & \frac{1}{3} \end{bmatrix} \begin{bmatrix} \frac{1}{3} & -\frac{2}{3} & \frac{2}{3} \\ \frac{2}{3} & -\frac{1}{3} & -\frac{2}{3} \\ \frac{2}{3} & \frac{2}{3} & \frac{1}{3} \end{bmatrix} = \begin{bmatrix} 1 & 0 & 0 \\ 0 & 1 & 0 \\ 0 & 0 & 1 \end{bmatrix},$$

the matrix

$$\begin{bmatrix} \frac{1}{3} & -\frac{2}{3} & \frac{2}{3} \\ \frac{2}{3} & -\frac{1}{3} & -\frac{2}{3} \\ \frac{2}{3} & \frac{2}{3} & \frac{1}{3} \end{bmatrix}$$

is orthogonal.

The following theorem gives an indication why such matrices are called orthogonal.

Theorem 5.25. *If the columns of a $n \times n$ matrix P are pairwise orthogonal unit vectors, then P is an orthogonal matrix.*

Proof. Let $P = \begin{bmatrix} \mathbf{p}_1 & \dots & \mathbf{p}_n \end{bmatrix}$ where $\mathbf{p}_1, \dots, \mathbf{p}_n$ are vectors such that

$$\mathbf{p}_j \cdot \mathbf{p}_k = 0 \quad \text{and} \quad \mathbf{p}_j \cdot \mathbf{p}_j = 1,$$

for $j, k = 1, \dots, n$ with $j \neq k$. Then

$$P^T P = \begin{bmatrix} \mathbf{p}_1^T \\ \vdots \\ \mathbf{p}_n^T \end{bmatrix} \begin{bmatrix} \mathbf{p}_1 & \dots & \mathbf{p}_n \end{bmatrix} = \begin{bmatrix} \mathbf{p}_1^T \mathbf{p}_1 & \dots & \mathbf{p}_1^T \mathbf{p}_n \\ \vdots & \dots & \vdots \\ \mathbf{p}_n^T \mathbf{p}_1 & \dots & \mathbf{p}_n^T \mathbf{p}_n \end{bmatrix} = \begin{bmatrix} \mathbf{p}_1 \cdot \mathbf{p}_1 & \dots & \mathbf{p}_1 \cdot \mathbf{p}_n \\ \vdots & \dots & \vdots \\ \mathbf{p}_n \cdot \mathbf{p}_1 & \dots & \mathbf{p}_n \cdot \mathbf{p}_n \end{bmatrix} = I_n.$$

This means that $P^T = P^{-1}$, by Theorem 1.71, and thus P is an orthogonal matrix. □

Note that the columns of the matrix

$$\begin{bmatrix} \frac{1}{3} & -\frac{2}{3} & \frac{2}{3} \\ \frac{2}{3} & -\frac{1}{3} & -\frac{2}{3} \\ \frac{2}{3} & \frac{2}{3} & \frac{1}{3} \end{bmatrix},$$

considered in Example 5.24, are pairwise orthogonal unit vectors.

Definition 5.26. We say that a matrix A is *orthogonally diagonalizable* if there are an orthogonal matrix P and a diagonal matrix D such that $A = PDP^{-1} = PDP^T$.

Example 5.27. In the previous example we show that the matrix

$$\begin{bmatrix} \frac{1}{3} & -\frac{2}{3} & \frac{2}{3} \\ \frac{2}{3} & -\frac{1}{3} & -\frac{2}{3} \\ \frac{2}{3} & \frac{2}{3} & \frac{1}{3} \end{bmatrix}$$

is orthogonal. Consequently, the matrix

$$\begin{bmatrix} \frac{1}{3} & -\frac{2}{3} & \frac{2}{3} \\ \frac{2}{3} & -\frac{1}{3} & -\frac{2}{3} \\ \frac{2}{3} & \frac{2}{3} & \frac{1}{3} \end{bmatrix} \begin{bmatrix} 2 & 0 & 0 \\ 0 & -1 & 0 \\ 0 & 0 & 3 \end{bmatrix} \begin{bmatrix} \frac{1}{3} & \frac{2}{3} & \frac{2}{3} \\ -\frac{2}{3} & -\frac{1}{3} & \frac{2}{3} \\ \frac{2}{3} & -\frac{2}{3} & \frac{1}{3} \end{bmatrix} = \begin{bmatrix} \frac{2}{9} & -\frac{2}{9} & \frac{10}{9} \\ -\frac{2}{9} & \frac{11}{9} & \frac{8}{9} \\ \frac{10}{9} & \frac{8}{9} & \frac{5}{9} \end{bmatrix}$$

is orthogonally diagonalizable.

We know that the matrix

$$\begin{bmatrix} \frac{2}{9} & -\frac{2}{9} & \frac{10}{9} \\ -\frac{2}{9} & \frac{11}{9} & \frac{8}{9} \\ \frac{10}{9} & \frac{8}{9} & \frac{5}{9} \end{bmatrix}$$

in the example above is orthogonally diagonalizable because it was constructed as a product PDP^T where P is an orthogonal matrix P and D is a diagonal matrix, but how could we check if a matrix is orthogonally diagonalizable? The following theorem can help with such questions.

Theorem 5.28. *If a $n \times n$ matrix A has n orthogonal eigenvectors, then A is symmetric and orthogonally diagonalizable.*

Proof. Let $\mathbf{v}_1, \dots, \mathbf{v}_n$ be orthogonal eigenvectors of the matrix A corresponding to the eigenvalues $\alpha_1, \dots, \alpha_n$, respectively. This means that

$$A\mathbf{v}_j = \alpha_j \mathbf{v}_j, \quad \text{and} \quad \mathbf{v}_j \cdot \mathbf{v}_k = 0,$$

for $j, k = 1, \dots, n$ with $j \neq k$. If we let

$$\mathbf{p}_j = \frac{1}{\|\mathbf{v}_j\|} \mathbf{v}_j,$$

then we have

$$\mathbf{p}_j \cdot \mathbf{p}_k = 0 \quad \text{and} \quad \|\mathbf{p}_j\| = 1,$$

for $j, k = 1, \dots, n$ with $j \neq k$. Let P be the matrix whose columns are the vectors $\mathbf{p}_1, \dots, \mathbf{p}_n$, that is,

$$P = \begin{bmatrix} \mathbf{p}_1 & \dots & \mathbf{p}_n \end{bmatrix}.$$

Then P is an orthogonal matrix, by Theorem 5.25.

The vectors $\mathbf{p}_1, \dots, \mathbf{p}_n$ satisfy the equations

$$A\mathbf{p}_j = \alpha_j \mathbf{p}_j, \quad j = 1, \dots, n,$$

that can be written as a single matrix equation

$$A\begin{bmatrix}\mathbf{p}_1 & \dots & \mathbf{p}_n\end{bmatrix} = \begin{bmatrix}\mathbf{p}_1 & \dots & \mathbf{p}_n\end{bmatrix}\begin{bmatrix}\alpha_1 & 0 & \dots & 0\\ 0 & \alpha_2 & \dots & 0\\ \vdots & \vdots & \ddots & \vdots\\ 0 & 0 & \dots & \alpha_n\end{bmatrix}$$

which is

$$AP = P\begin{bmatrix}\alpha_1 & 0 & \dots & 0\\ 0 & \alpha_2 & \dots & 0\\ \vdots & \vdots & \ddots & \vdots\\ 0 & 0 & \dots & \alpha_n\end{bmatrix}.$$

Hence

$$A = AI_n = APP^{-1} = P\begin{bmatrix}\alpha_1 & 0 & \dots & 0\\ 0 & \alpha_2 & \dots & 0\\ \vdots & \vdots & \ddots & \vdots\\ 0 & 0 & \dots & \alpha_n\end{bmatrix}P^{-1} = P\begin{bmatrix}\alpha_1 & 0 & \dots & 0\\ 0 & \alpha_2 & \dots & 0\\ \vdots & \vdots & \ddots & \vdots\\ 0 & 0 & \dots & \alpha_n\end{bmatrix}P^T.$$

Since P is an orthogonal matrix, the matrix A is orthogonally diagonalizable. Moreover, since

$$\begin{aligned}A^T &= \left(P\begin{bmatrix}\alpha_1 & 0 & \dots & 0\\ 0 & \alpha_2 & \dots & 0\\ \vdots & \vdots & \ddots & \vdots\\ 0 & 0 & \dots & \alpha_n\end{bmatrix}P^T\right)^T = (P^T)^T\begin{bmatrix}\alpha_1 & 0 & \dots & 0\\ 0 & \alpha_2 & \dots & 0\\ \vdots & \vdots & \ddots & \vdots\\ 0 & 0 & \dots & \alpha_n\end{bmatrix}^T P^T\\ &= P\begin{bmatrix}\alpha_1 & 0 & \dots & 0\\ 0 & \alpha_2 & \dots & 0\\ \vdots & \vdots & \ddots & \vdots\\ 0 & 0 & \dots & \alpha_n\end{bmatrix}P^T = A,\end{aligned}$$

the matrix A is symmetric. □

From Theorem 5.28 and its proof we get the following corollary that is of practical importance in diagonalizing orthogonal matrices.

Corollary 5.29. *If A is an $n \times n$ matrix which has n orthogonal eigenvectors $\mathbf{v}_1, \dots, \mathbf{v}_n$ corresponding to the real eigenvalues $\alpha_1, \dots, \alpha_n$, that is,*

$$A\mathbf{v}_1 = \alpha_1 \mathbf{v}_1, \dots, A\mathbf{v}_n = \alpha_n \mathbf{v}_n,$$

and if $P = \begin{bmatrix} \mathbf{p}_1 & \dots & \mathbf{p}_n \end{bmatrix}$ is the orthogonal matrix with columns $\mathbf{p}_1 = \frac{1}{\|\mathbf{v}_1\|}\mathbf{v}_1, \dots, \mathbf{p}_n = \frac{1}{\|\mathbf{v}_n\|}\mathbf{v}_n$ and D the diagonal matrix with $\alpha_1, \dots, \alpha_n$ on the main diagonal then,

$$A = PDP^T.$$

Calculating the diagonal form of an $n \times n$ symmetric matrix

We first prove the following useful property of the dot product.

Theorem 5.30. *If A is an arbitrary $n \times n$ matrix, then*

$$A\mathbf{u} \cdot \mathbf{v} = \mathbf{u} \cdot A^T \mathbf{v}$$

for any vectors $\mathbf{u}$ and $\mathbf{v}$ in $\mathbb{R}^n$.

Proof. Using the connection between the dot product and the product of matrices, namely

$$\mathbf{x} \cdot \mathbf{y} = \mathbf{x}^T \mathbf{y},$$

associativity of matrix multiplication, and a property of the transpose operation, we obtain

$$A\mathbf{u} \cdot \mathbf{v} = (A\mathbf{u})^T \mathbf{v} = (\mathbf{u}^T A^T)\mathbf{v} = \mathbf{u}^T (A^T \mathbf{v}) = \mathbf{u} \cdot A^T \mathbf{v}.$$

□

Note that for symmetric matrices we have

$$A\mathbf{u} \cdot \mathbf{v} = \mathbf{u} \cdot A\mathbf{v}.$$

Theorem 5.31. *Eigenvectors corresponding to distinct eigenvalues of a symmetric matrix are orthogonal.*

Proof. If $A\mathbf{u} = \alpha \mathbf{u}$ and $A\mathbf{v} = \beta \mathbf{v}$, then

$$(\alpha - \beta)(\mathbf{u} \cdot \mathbf{v}) = \alpha(\mathbf{u} \cdot \mathbf{v}) - \beta(\mathbf{u} \cdot \mathbf{v}) = (\alpha \mathbf{u}) \cdot \mathbf{v} - \mathbf{u} \cdot \beta(\mathbf{v}) = A\mathbf{u} \cdot \mathbf{v} - \mathbf{u} \cdot A\mathbf{v}.$$

For a symmetric matrix A we have $A\mathbf{u} \cdot \mathbf{v} - \mathbf{u} \cdot A\mathbf{v} = 0$ and consequently $(\alpha - \beta)(\mathbf{u} \cdot \mathbf{v}) = 0$. If $\alpha \neq \beta$, we must have $\mathbf{u} \cdot \mathbf{v} = 0$. □

The following theorem summarizes what we have learned about symmetric matrices. The theorem is similar to Theorem 5.20, but there are some essential differences. For instance, the condition $m_1 + \cdots + m_k = n$ was an assumption in Theorem 5.20. The theorem complements Theorem 5.28 and is usually proven in a second course in Linear Algebra. As in the case of Theorem 5.20, it is a good idea to first read Example 5.34 carefully and then read the theorem.

Theorem 5.32. *Let A be an $n \times n$ symmetric matrix with k distinct eigenvalues $\lambda_1, \dots, \lambda_k$ and let m_j be the geometric multiplicity of λ_j for each $j = 1, \dots, k$. Then*

(a) $m_1 + \cdots + m_k = n$,

(b) *If the set $\{\mathbf{v}_{1,j}, \dots, \mathbf{v}_{m_j,j}\}$ is an orthogonal basis of the eigenspace corresponding to the eigenvalue λ_j for $j = 1, \dots, k$, then*

$$\{\mathbf{v}_{1,1}, \dots, \mathbf{v}_{m_1,1}, \mathbf{v}_{1,2}, \dots, \mathbf{v}_{m_2,2}, \dots, \mathbf{v}_{1,k}, \dots, \mathbf{v}_{m_k,k}\}$$

is an orthogonal basis in $\mathbb{R}^n$,

(c) *A is orthogonally diagonalizable: $A = PDP^T$, where*

$$P = \begin{bmatrix} \frac{\mathbf{v}_{1,1}}{\|\mathbf{v}_{1,1}\|} & \cdots & \frac{\mathbf{v}_{m_1,1}}{\|\mathbf{v}_{m_1,1}\|} & \frac{\mathbf{v}_{1,2}}{\|\mathbf{v}_{1,2}\|} & \cdots & \frac{\mathbf{v}_{m_2,2}}{\|\mathbf{v}_{m_2,2}\|} & \cdots & \frac{\mathbf{v}_{1,k}}{\|\mathbf{v}_{1,k}\|} & \cdots & \frac{\mathbf{v}_{m_k,k}}{\|\mathbf{v}_{m_k,k}\|} \end{bmatrix}$$

and D is the diagonal matrix with

$$\underbrace{\lambda_1, \dots, \lambda_1}_{m_1 \text{ times}}, \underbrace{\lambda_2, \dots, \lambda_2}_{m_2 \text{ times}}, \dots, \underbrace{\lambda_k, \dots, \lambda_k}_{m_k \text{ times}}$$

along the main diagonal.

Corollary 5.33. *Every symmetric matrix has an orthogonal basis of eigenvectors.*

Example 5.34. Orthogonally diagonalize the matrix

$$A = \begin{bmatrix} 2 & 1 & 1 & 1 \\ 1 & 2 & 1 & 1 \\ 1 & 1 & 2 & 1 \\ 1 & 1 & 1 & 2 \end{bmatrix}.$$

Solution. First we calculate

$$\det(A - \lambda I_4) = \det \begin{bmatrix} 2-\lambda & 1 & 1 & 1 \\ 1 & 2-\lambda & 1 & 1 \\ 1 & 1 & 2-\lambda & 1 \\ 1 & 1 & 1 & 2-\lambda \end{bmatrix}.$$

We add all the rows to the first one and get

$$\det(A - \lambda I_4) = \det \begin{bmatrix} 5-\lambda & 5-\lambda & 5-\lambda & 5-\lambda \\ 1 & 2-\lambda & 1 & 1 \\ 1 & 1 & 2-\lambda & 1 \\ 1 & 1 & 1 & 2-\lambda \end{bmatrix} = (5-\lambda)\det \begin{bmatrix} 1 & 1 & 1 & 1 \\ 1 & 2-\lambda & 1 & 1 \\ 1 & 1 & 2-\lambda & 1 \\ 1 & 1 & 1 & 2-\lambda \end{bmatrix}.$$

Then we subtract the first column from the remaining three columns and get

$$\begin{aligned} \det(A - \lambda I_4) &= (5-\lambda)\det \begin{bmatrix} 1 & 0 & 0 & 0 \\ 1 & 1-\lambda & 0 & 0 \\ 1 & 0 & 1-\lambda & 0 \\ 1 & 0 & 0 & 1-\lambda \end{bmatrix} \\ &= (5-\lambda)\det \begin{bmatrix} 1-\lambda & 0 & 0 \\ 0 & 1-\lambda & 0 \\ 0 & 0 & 1-\lambda \end{bmatrix} \\ &= (5-\lambda)(1-\lambda)^3. \end{aligned}$$

Thus the eigenvalues are 5 and 1.

The eigenspace corresponding to the eigenvalue 1 is given by the equation

$$\begin{bmatrix} 2 & 1 & 1 & 1 \\ 1 & 2 & 1 & 1 \\ 1 & 1 & 2 & 1 \\ 1 & 1 & 1 & 2 \end{bmatrix} \begin{bmatrix} x \\ y \\ z \\ w \end{bmatrix} = \begin{bmatrix} x \\ y \\ z \\ w \end{bmatrix}$$

which reduces to the equation

$$x + y + z + w = 0.$$

Since this equation is equivalent to the equation

$$w = -x - y - z,$$

the general solution is

$$\begin{bmatrix} x \\ y \\ z \\ w \end{bmatrix} = \begin{bmatrix} x \\ y \\ z \\ -x-y-z \end{bmatrix} = x\begin{bmatrix} 1 \\ 0 \\ 0 \\ -1 \end{bmatrix} + y\begin{bmatrix} 0 \\ 1 \\ 0 \\ -1 \end{bmatrix} + z\begin{bmatrix} 0 \\ 0 \\ 1 \\ -1 \end{bmatrix}.$$

This means that the eigenspace corresponding to the eigenvalue 1 is

$$\operatorname{Span}\left\{\begin{bmatrix} 1 \\ 0 \\ 0 \\ -1 \end{bmatrix}, \begin{bmatrix} 0 \\ 1 \\ 0 \\ -1 \end{bmatrix}, \begin{bmatrix} 0 \\ 0 \\ 1 \\ -1 \end{bmatrix}\right\}.$$

Using the Gram-Schmidt process we get

$$\operatorname{Span}\left\{\begin{bmatrix} 1 \\ 0 \\ 0 \\ -1 \end{bmatrix}, \begin{bmatrix} 0 \\ 1 \\ 0 \\ -1 \end{bmatrix}, \begin{bmatrix} 0 \\ 0 \\ 1 \\ -1 \end{bmatrix}\right\} = \operatorname{Span}\left\{\begin{bmatrix} 1 \\ 0 \\ 0 \\ -1 \end{bmatrix}, \begin{bmatrix} 1 \\ -2 \\ 0 \\ 1 \end{bmatrix}, \begin{bmatrix} 1 \\ 1 \\ -3 \\ 1 \end{bmatrix}\right\}.$$

Now, since the equation $x+y+z+w=0$ can be written as

$$\begin{bmatrix} 1 \\ 1 \\ 1 \\ 1 \end{bmatrix} \cdot \begin{bmatrix} x \\ y \\ z \\ w \end{bmatrix} = 0,$$

the vector $\begin{bmatrix} 1 \\ 1 \\ 1 \\ 1 \end{bmatrix}$ is an eigenvector corresponding to the eigenvalue 5, as a consequence of Theorems 3.61 and 5.32.

We denote

$$\mathbf{v}_1 = \begin{bmatrix} 1 \\ 0 \\ 0 \\ -1 \end{bmatrix}, \quad \mathbf{v}_2 = \begin{bmatrix} 1 \\ -2 \\ 0 \\ 1 \end{bmatrix}, \quad \mathbf{v}_3 = \begin{bmatrix} 1 \\ 1 \\ -3 \\ 1 \end{bmatrix}, \quad \text{and} \quad \mathbf{v}_4 = \begin{bmatrix} 1 \\ 1 \\ 1 \\ 1 \end{bmatrix}$$

and calculate

$$\|\mathbf{v}_1\|^2 = 2, \quad \|\mathbf{v}_2\|^2 = 6, \quad \|\mathbf{v}_3\|^2 = 12, \quad \text{and} \quad \|\mathbf{v}_4\|^2 = 4.$$

The vectors

$$\mathbf{p}_1 = \begin{bmatrix} \frac{1}{\sqrt{2}} \\ 0 \\ 0 \\ -\frac{1}{\sqrt{2}} \end{bmatrix}, \quad \mathbf{p}_2 = \begin{bmatrix} \frac{1}{\sqrt{6}} \\ -\frac{2}{\sqrt{6}} \\ 0 \\ \frac{1}{\sqrt{6}} \end{bmatrix}, \quad \mathbf{p}_3 = \begin{bmatrix} \frac{1}{2\sqrt{3}} \\ \frac{1}{2\sqrt{3}} \\ -\frac{3}{2\sqrt{3}} \\ \frac{1}{2\sqrt{3}} \end{bmatrix}, \quad \text{and} \quad \mathbf{p}_4 = \begin{bmatrix} \frac{1}{2} \\ \frac{1}{2} \\ \frac{1}{2} \\ \frac{1}{2} \end{bmatrix}$$

form an orthonormal basis in $\mathbb{R}^4$ consisting of eigenvectors of the matrix A corresponding to the eigenvalues $\lambda = 1$, $\lambda = 1$, $\lambda = 1$, and $\lambda = 5$, respectively, and

$$A = \begin{bmatrix} \frac{1}{\sqrt{2}} & \frac{1}{\sqrt{6}} & \frac{1}{2\sqrt{3}} & \frac{1}{2} \\ 0 & -\frac{2}{\sqrt{6}} & \frac{1}{2\sqrt{3}} & \frac{1}{2} \\ 0 & 0 & -\frac{3}{2\sqrt{3}} & \frac{1}{2} \\ -\frac{1}{\sqrt{2}} & \frac{1}{\sqrt{6}} & \frac{1}{2\sqrt{3}} & \frac{1}{2} \end{bmatrix} \begin{bmatrix} 1 & 0 & 0 & 0 \\ 0 & 1 & 0 & 0 \\ 0 & 0 & 1 & 0 \\ 0 & 0 & 0 & 5 \end{bmatrix} \begin{bmatrix} \frac{1}{\sqrt{2}} & 0 & 0 & -\frac{1}{\sqrt{2}} \\ \frac{1}{\sqrt{6}} & -\frac{2}{\sqrt{6}} & 0 & \frac{1}{\sqrt{6}} \\ \frac{1}{2\sqrt{3}} & \frac{1}{2\sqrt{3}} & -\frac{3}{2\sqrt{3}} & \frac{1}{2\sqrt{3}} \\ \frac{1}{2} & \frac{1}{2} & \frac{1}{2} & \frac{1}{2} \end{bmatrix}$$

is an orthogonal diagonalization of A. □

The spectral decomposition of a symmetric matrix

In the last two sections we were interested in representing a matrix A in the form $A = PDP^{-1}$ where P is an orthogonal matrix and D is a diagonal matrix. The representation of a matrix presented in the next theorem is a consequence of the representation $A = PDP^{-1}$, but is expressed in a very different form. This form explains the geometric meaning of orthogonal diagonalization and is useful in applications.

Theorem 5.35. *If $\mathbf{v}_1, \ldots, \mathbf{v}_n$ is an orthogonal basis of eigenvectors corresponding to eigenvalues $\lambda_1, \ldots, \lambda_n$ of an $n \times n$ symmetric matrix A, then*

$$A = \lambda_1 \frac{1}{\|\mathbf{v}_1\|^2} \mathbf{v}_1 \mathbf{v}_1^T + \cdots + \lambda_n \frac{1}{\|\mathbf{v}_n\|^2} \mathbf{v}_n \mathbf{v}_n^T.$$

We present two proofs. The first one uses Theorem 5.32. The second one is based on Theorem 3.29.

First proof. Let

$$\mathbf{p}_1 = \frac{1}{\|\mathbf{v}_1\|} \mathbf{v}_1, \ldots, \mathbf{p}_n = \frac{1}{\|\mathbf{v}_n\|} \mathbf{v}_n.$$

First we note that

$$A = \begin{bmatrix} \mathbf{p}_1 & \dots & \mathbf{p}_n \end{bmatrix} \begin{bmatrix} \lambda_1 & 0 & \cdots & 0 \\ 0 & \lambda_2 & \cdots & 0 \\ \vdots & \vdots & \ddots & \vdots \\ 0 & 0 & \cdots & \lambda_n \end{bmatrix} \begin{bmatrix} \mathbf{p}_1^T \\ \vdots \\ \mathbf{p}_n^T \end{bmatrix}.$$

Now, if $\mathbf{x}$ is an arbitrary vector from $\mathbb{R}^n$, then we have

$$\begin{aligned} A\mathbf{x} &= \begin{bmatrix} \mathbf{p}_1 & \dots & \mathbf{p}_n \end{bmatrix} \begin{bmatrix} \lambda_1 & 0 & \cdots & 0 \\ 0 & \lambda_2 & \cdots & 0 \\ \vdots & \vdots & \ddots & \vdots \\ 0 & 0 & \cdots & \lambda_n \end{bmatrix} \begin{bmatrix} \mathbf{p}_1^T \\ \vdots \\ \mathbf{p}_n^T \end{bmatrix} \mathbf{x} \\ &= \begin{bmatrix} \mathbf{p}_1 & \dots & \mathbf{p}_n \end{bmatrix} \begin{bmatrix} \lambda_1 & 0 & \cdots & 0 \\ 0 & \lambda_2 & \cdots & 0 \\ \vdots & \vdots & \ddots & \vdots \\ 0 & 0 & \cdots & \lambda_n \end{bmatrix} \begin{bmatrix} \mathbf{p}_1^T\mathbf{x} \\ \vdots \\ \mathbf{p}_n^T\mathbf{x} \end{bmatrix} \\ &= \begin{bmatrix} \mathbf{p}_1 & \dots & \mathbf{p}_n \end{bmatrix} \begin{bmatrix} \lambda_1\mathbf{p}_1^T\mathbf{x} \\ \vdots \\ \lambda_n\mathbf{p}_n^T\mathbf{x} \end{bmatrix} \\ &= \lambda_1\mathbf{p}_1\left(\mathbf{p}_1^T\mathbf{x}\right) + \cdots + \lambda_n\mathbf{p}_n\left(\mathbf{p}_n^T\mathbf{x}\right) \\ &= \lambda_1\left(\mathbf{p}_1\mathbf{p}_1^T\right)\mathbf{x} + \cdots + \lambda_n\left(\mathbf{p}_n\mathbf{p}_n^T\right)\mathbf{x} \\ &= \left(\lambda_1\mathbf{p}_1\mathbf{p}_1^T + \cdots + \lambda_n\mathbf{p}_n\mathbf{p}_n^T\right)\mathbf{x}. \end{aligned}$$

Using Theorem 1.18, we obtain

$$A = \lambda_1\mathbf{p}_1\mathbf{p}_1^T + \cdots + \lambda_n\mathbf{p}_n\mathbf{p}_n^T,$$

which can be written as

$$A = \lambda_1\frac{1}{\|\mathbf{v}_1\|^2}\mathbf{v}_1\mathbf{v}_1^T + \cdots + \lambda_n\frac{1}{\|\mathbf{v}_n\|^2}\mathbf{v}_n\mathbf{v}_n^T.$$

□

Second proof. Since

$$\frac{1}{\|\mathbf{v}_1\|^2}\mathbf{v}_1\mathbf{v}_1^T + \cdots + \frac{1}{\|\mathbf{v}_n\|^2}\mathbf{v}_n\mathbf{v}_n^T$$

is the projection matrix on $\mathbb{R}^n$, we have

$$\frac{1}{\|\mathbf{v}_1\|^2}\mathbf{v}_1\mathbf{v}_1^T + \cdots + \frac{1}{\|\mathbf{v}_n\|^2}\mathbf{v}_n\mathbf{v}_n^T = I_n.$$

From this equality we get

$$A\left(\frac{1}{\|\mathbf{v}_1\|^2}\mathbf{v}_1\mathbf{v}_1^T + \cdots + \frac{1}{\|\mathbf{v}_n\|^2}\mathbf{v}_n\mathbf{v}_n^T\right) = AI_n = A,$$

which can be written as

$$\frac{1}{\|\mathbf{v}_1\|^2} A\mathbf{v}_1\mathbf{v}_1^T + \cdots + \frac{1}{\|\mathbf{v}_n\|^2} A\mathbf{v}_n\mathbf{v}_n^T = A.$$

Since

$$A\mathbf{v}_1 = \lambda_1\mathbf{v}_1, \ \ldots, \ A\mathbf{v}_n = \lambda_n\mathbf{v}_n,$$

we obtain the desired representation of the matrix A:

$$A = \lambda_1 \frac{1}{\|\mathbf{v}_1\|^2}\mathbf{v}_1\mathbf{v}_1^T + \cdots + \lambda_n \frac{1}{\|\mathbf{v}_n\|^2}\mathbf{v}_n\mathbf{v}_n^T.$$

□

Definition 5.36. If $\mathbf{v}_1, \ldots, \mathbf{v}_n$ is an orthogonal basis of eigenvectors corresponding to eigenvalues $\lambda_1, \ldots, \lambda_n$ of an $n \times n$ symmetric matrix A, then

$$A = \lambda_1 \frac{1}{\|\mathbf{v}_1\|^2}\mathbf{v}_1\mathbf{v}_1^T + \cdots + \lambda_n \frac{1}{\|\mathbf{v}_n\|^2}\mathbf{v}_n\mathbf{v}_n^T$$

is called the *spectral decomposition* of the matrix A.

Example 5.37. Find the spectral decomposition of the matrix

$$A = \begin{bmatrix} 2 & 1 & 1 & 1 \\ 1 & 2 & 1 & 1 \\ 1 & 1 & 2 & 1 \\ 1 & 1 & 1 & 2 \end{bmatrix}.$$

Solution. From the calculations in Example 5.34 the spectral decomposition of A is

$$A = \frac{1}{2}\mathbf{v}_1\mathbf{v}_1^T + \frac{1}{6}\mathbf{v}_2\mathbf{v}_2^T + \frac{1}{12}\mathbf{v}_3\mathbf{v}_3^T + \frac{5}{4}\mathbf{v}_4\mathbf{v}_4^T,$$

where

$$\mathbf{v}_1 = \begin{bmatrix} 1 \\ 0 \\ 0 \\ -1 \end{bmatrix}, \quad \mathbf{v}_2 = \begin{bmatrix} 1 \\ -2 \\ 0 \\ 1 \end{bmatrix}, \quad \mathbf{v}_3 = \begin{bmatrix} 1 \\ 1 \\ -3 \\ 1 \end{bmatrix}, \quad \text{and} \quad \mathbf{v}_4 = \begin{bmatrix} 1 \\ 1 \\ 1 \\ 1 \end{bmatrix}.$$

□

The following theorem complements Theorem 5.35. Note that we are not assuming that the vectors $\mathbf{v}_1, \ldots, \mathbf{v}_k$ form a basis of $\mathbb{R}^n$. The theorem is a consequence of Theorem 1.24.

Theorem 5.38. *If $\mathbf{v}_1, \dots, \mathbf{v}_k$ are nonzero orthogonal vectors in $\mathbb{R}^n$ and $\lambda_1, \dots, \lambda_k$ real numbers, then the matrix*

$$A = \lambda_1 \frac{1}{\|\mathbf{v}_1\|^2} \mathbf{v}_1 \mathbf{v}_1^T + \dots + \lambda_k \frac{1}{\|\mathbf{v}_k\|^2} \mathbf{v}_k \mathbf{v}_k^T$$

is symmetric, the numbers $\lambda_1, \dots, \lambda_k$ are eigenvalues of A, and the vectors $\mathbf{v}_1, \dots, \mathbf{v}_k$ are eigenvectors of A corresponding to the eigenvalues $\lambda_1, \dots, \lambda_k$.

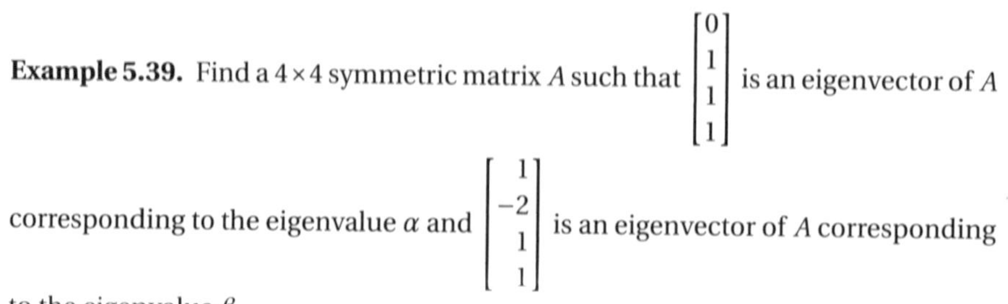

Example 5.39. Find a 4×4 symmetric matrix A such that $\begin{bmatrix} 0 \\ 1 \\ 1 \\ 1 \end{bmatrix}$ is an eigenvector of A corresponding to the eigenvalue α and $\begin{bmatrix} 1 \\ -2 \\ 1 \\ 1 \end{bmatrix}$ is an eigenvector of A corresponding to the eigenvalue β.

Solution.

$$\begin{aligned} A &= \frac{\alpha}{3} \begin{bmatrix} 0 \\ 1 \\ 1 \\ 1 \end{bmatrix} \begin{bmatrix} 0 & 1 & 1 & 1 \end{bmatrix} + \frac{\beta}{7} \begin{bmatrix} 1 \\ -2 \\ 1 \\ 1 \end{bmatrix} \begin{bmatrix} 1 & -2 & 1 & 1 \end{bmatrix} \\ &= \frac{\alpha}{3} \begin{bmatrix} 0 & 0 & 0 & 0 \\ 0 & 1 & 1 & 1 \\ 0 & 1 & 1 & 1 \\ 0 & 1 & 1 & 1 \end{bmatrix} + \frac{\beta}{7} \begin{bmatrix} 1 & -2 & 1 & 1 \\ -2 & 4 & -2 & -2 \\ 1 & -2 & 1 & 1 \\ 1 & -2 & 1 & 1 \end{bmatrix} \\ &= \begin{bmatrix} \frac{\beta}{7} & -\frac{2\beta}{7} & \frac{\beta}{7} & \frac{\beta}{7} \\ -\frac{2\beta}{7} & \frac{\alpha}{3} + \frac{4\beta}{7} & \frac{\alpha}{3} - \frac{2\beta}{7} & \frac{\alpha}{3} - \frac{2\beta}{7} \\ \frac{\beta}{7} & \frac{\alpha}{3} - \frac{2\beta}{7} & \frac{\alpha}{3} + \frac{\beta}{7} & \frac{\alpha}{3} + \frac{\beta}{7} \\ \frac{\beta}{7} & \frac{\alpha}{3} - \frac{2\beta}{7} & \frac{\alpha}{3} + \frac{\beta}{7} & \frac{\alpha}{3} + \frac{\beta}{7} \end{bmatrix}. \end{aligned}$$

□

Exercises 5.2

Determine a symmetric 2×2 matrix A that has the given eigenvalues α and β and the given vector $\mathbf{v}$ as an eigenvector corresponding to the eigenvalue α.

1. $\alpha = 3, \beta = 4, \mathbf{v} = \begin{bmatrix} 2 \\ 1 \end{bmatrix}$

2. $\alpha = 1, \beta = 2, \mathbf{v} = \begin{bmatrix} 1 \\ 1 \end{bmatrix}$

3. $\alpha = \lambda, \beta = \lambda, \mathbf{v} = \begin{bmatrix} x \\ y \end{bmatrix}$

4. $\alpha = 5, \beta = 0, \mathbf{v} = \begin{bmatrix} 1 \\ -2 \end{bmatrix}$

Determine a symmetric 3×3 matrix A that has the given eigenvalues α, β, and γ, the given vector $\mathbf{u}$ as an eigenvector corresponding to the eigenvalue α and the given vector $\mathbf{v}$ as an eigenvector corresponding to the eigenvalue β.

5. $\alpha = 1, \beta = 2, \gamma = 3, \mathbf{u} = \begin{bmatrix} 1 \\ 1 \\ 1 \end{bmatrix}, \mathbf{v} = \begin{bmatrix} 1 \\ -3 \\ 2 \end{bmatrix}$

6. $\alpha = 1, \beta = 3, \gamma = 0, \mathbf{u} = \begin{bmatrix} 1 \\ -1 \\ 0 \end{bmatrix}, \mathbf{v} = \begin{bmatrix} 2 \\ 2 \\ 1 \end{bmatrix}$

7. $\alpha = 1, \beta = 1, \gamma = 2, \mathbf{u} = \begin{bmatrix} 2 \\ 1 \\ 1 \end{bmatrix}, \mathbf{v} = \begin{bmatrix} 1 \\ -1 \\ -1 \end{bmatrix}$

8. $\alpha = 1, \beta = 1, \gamma = 0, \mathbf{u} = \begin{bmatrix} 1 \\ 1 \\ 1 \end{bmatrix}, \mathbf{v} = \begin{bmatrix} 1 \\ 1 \\ 0 \end{bmatrix}$

9. $\alpha = 0, \beta = 0, \gamma = 1, \mathbf{u} = \begin{bmatrix} 2 \\ 3 \\ 1 \end{bmatrix}, \mathbf{v} = \begin{bmatrix} 1 \\ 2 \\ 3 \end{bmatrix}$

10. $\alpha = 0, \beta = 0, \gamma = 3, \mathbf{u} = \begin{bmatrix} 1 \\ -1 \\ 1 \end{bmatrix}, \mathbf{v} = \begin{bmatrix} 1 \\ 1 \\ 1 \end{bmatrix}$

11. $\alpha = 2, \beta = \gamma = 1, \mathbf{u} = \begin{bmatrix} 1 \\ 3 \\ 1 \end{bmatrix}$

12. $\alpha = 1, \beta = \gamma = 0, \mathbf{u} = \begin{bmatrix} 2 \\ 1 \\ -1 \end{bmatrix}$

Determine a symmetric 4×4 matrix A that has the given eigenvalues α, β, γ, and δ, the given vector $\mathbf{u}$ as an eigenvector corresponding to the eigenvalue α, the given vector $\mathbf{v}$ as an eigenvector corresponding to the eigenvalue β and the given vector $\mathbf{w}$ as an eigenvector corresponding to the eigenvalue γ.

13. $\alpha = \beta = \gamma = 1, \delta = 0, \mathbf{u} = \begin{bmatrix} 2 \\ 1 \\ 1 \\ 1 \end{bmatrix}, \mathbf{v} = \begin{bmatrix} 1 \\ 2 \\ 1 \\ 1 \end{bmatrix}, \mathbf{w} = \begin{bmatrix} 1 \\ 1 \\ 2 \\ 1 \end{bmatrix}$

14. $\alpha = \beta = \gamma = 0, \delta = 2, \mathbf{u} = \begin{bmatrix} 1 \\ 1 \\ 1 \\ 1 \end{bmatrix}, \mathbf{v} = \begin{bmatrix} 1 \\ 1 \\ 0 \\ 0 \end{bmatrix}, \mathbf{w} = \begin{bmatrix} 1 \\ 1 \\ 0 \\ 0 \end{bmatrix}$

15. $\alpha = \beta = 1, \gamma = \delta = 0, \mathbf{u} = \begin{bmatrix} 1 \\ 1 \\ 1 \\ 1 \end{bmatrix}, \mathbf{v} = \begin{bmatrix} 1 \\ -1 \\ 1 \\ 1 \end{bmatrix}$

16. $\alpha = \beta = 0, \gamma = \delta = 2, \mathbf{u} = \begin{bmatrix} 1 \\ 0 \\ 0 \\ 1 \end{bmatrix}, \mathbf{v} = \begin{bmatrix} 1 \\ 1 \\ 1 \\ 0 \end{bmatrix}.$

Orthogonally diagonalize the given matrix A.

17. $A = \begin{bmatrix} 21 & 2 \\ 2 & 24 \end{bmatrix}$

18. $A = \begin{bmatrix} 9 & -5 \\ -5 & 9 \end{bmatrix}$

The given matrix A has the given eigenvalues α, β, and γ. Orthogonally diagonalize the matrix A.

19. $A = \begin{bmatrix} 10 & 5 & 2 \\ 5 & 5 & 3 \\ 2 & 3 & 2 \end{bmatrix}, \alpha = 14, \beta = 3, \gamma = 0$

20. $A = \begin{bmatrix} 3 & -1 & 1 \\ -1 & 3 & 1 \\ 1 & 1 & 1 \end{bmatrix}, \alpha = 4, \beta = 3, \gamma = 0$

21. $A = \begin{bmatrix} 7 & -5 & -3 \\ -5 & 7 & 3 \\ -3 & 3 & 15 \end{bmatrix}, \alpha = 2, \beta = 9, \gamma = 18$

22. $A = \begin{bmatrix} 14 & 3 & -10 \\ 3 & 31 & -3 \\ -10 & -3 & 14 \end{bmatrix}, \alpha = 4, \beta = 22, \gamma = 33$

23. $A = \begin{bmatrix} 22 & 4 & 20 \\ 4 & 28 & -10 \\ 20 & -10 & -20 \end{bmatrix}, \alpha = -30, \beta = 30, \gamma = 30$

24. $A = \begin{bmatrix} 3 & 0 & 1 \\ 0 & 2 & 0 \\ 1 & 0 & 3 \end{bmatrix}, \alpha = 2, \beta = 2, \gamma = 4$

25. $A = \begin{bmatrix} 14 & 2 & 4 \\ 2 & 17 & -2 \\ 4 & -2 & 14 \end{bmatrix}, \alpha = 9, \beta = 18, \gamma = 18$

26. $A = \begin{bmatrix} 29 & 13 & -13 \\ 13 & 29 & 13 \\ -13 & 13 & 29 \end{bmatrix}$, $\alpha = 42$, $\beta = 42$, $\gamma = 3$

27. $A = \begin{bmatrix} 7 & -5 & -3 \\ -5 & 7 & 3 \\ -3 & 3 & 15 \end{bmatrix}$, $\alpha = 2$, $\beta = 9$, $\gamma = 18$

28. $A = \begin{bmatrix} 14 & 3 & -10 \\ 3 & 31 & -3 \\ -10 & -3 & 14 \end{bmatrix}$, $\alpha = 4$, $\beta = 22$, $\gamma = 33$

29. $A = \begin{bmatrix} 22 & 4 & 20 \\ 4 & 28 & -10 \\ 20 & -10 & -20 \end{bmatrix}$, $\alpha = -30$, $\beta = 30$, $\gamma = 30$

30. $A = \begin{bmatrix} 29 & 13 & -13 \\ 13 & 29 & 13 \\ -13 & 13 & 29 \end{bmatrix}$, $\alpha = 42$, $\beta = 42$, $\gamma = 3$

Determine the spectral decomposition of the given matrix.

31. $\begin{bmatrix} 21 & 2 \\ 2 & 24 \end{bmatrix}$

32. $\begin{bmatrix} 9 & -5 \\ -5 & 9 \end{bmatrix}$

Orthogonally diagonalize the given matrix.

33. $\begin{bmatrix} 2 & 1 & 0 & 0 \\ 1 & 2 & 0 & 0 \\ 0 & 0 & 1 & 1 \\ 0 & 0 & 1 & 1 \end{bmatrix}$

34. $\begin{bmatrix} 3 & 1 & 0 & 0 \\ 1 & 3 & 0 & 0 \\ 0 & 0 & 1 & 4 \\ 0 & 0 & 4 & 1 \end{bmatrix}$

35. $\begin{bmatrix} 2 & 1 & 1 & 0 \\ 1 & 2 & 1 & 0 \\ 1 & 1 & 2 & 0 \\ 0 & 0 & 0 & 1 \end{bmatrix}$

36. $\begin{bmatrix} 3 & 1 & 0 & 1 \\ 1 & 3 & 0 & 1 \\ 0 & 0 & 1 & 0 \\ 1 & 1 & 0 & 3 \end{bmatrix}$

37. Find a 3×3 symmetric matrix A such that $\begin{bmatrix} 1 \\ 3 \\ 1 \end{bmatrix}$ is an eigenvector of A corresponding to the eigenvalue α and $\begin{bmatrix} 1 \\ 1 \\ -4 \end{bmatrix}$ is an eigenvector of A corresponding to the eigenvalue β.

38. Find a 3×3 symmetric matrix A such that $\begin{bmatrix} 2 \\ 1 \\ -1 \end{bmatrix}$ is an eigenvector of A corresponding to the eigenvalue α and $\begin{bmatrix} 1 \\ 0 \\ 2 \end{bmatrix}$ is an eigenvector of A corresponding to the eigenvalue β.

Chapter 6

Linear transformations

In this chapter we take another look at some familiar ideas from earlier chapters from a somewhat different point of view. While the new approach may seem less natural the first time we are exposed to it, it turns out to be quite useful when dealing with more advanced problems in linear algebra.

In Corollary 1.14 in Chapter 1 we observed that

$$A(\mathbf{x}+\mathbf{y}) = A\mathbf{x} + A\mathbf{y} \quad \text{and} \quad A(t\mathbf{x}) = tA\mathbf{x}$$

for any $m \times n$ matrix A, any $\mathbf{x}$ and $\mathbf{y}$ in $\mathbb{R}^n$, and any real number t. We also commented that, in some sense, these two properties characterize $m \times n$ matrices. The meaning of this comment will become clear in this chapter.

In Theorem 3.29 in Chapter 3 we show that an orthogonal projection on any subspace of $\mathbb{R}^n$ can be interpreted as matrix multiplication. It turns out that the same is true for rotations. Below we describe rotations in $\mathbb{R}^2$ and calculate the matrix corresponding to an arbitrary rotation in $\mathbb{R}^2$.

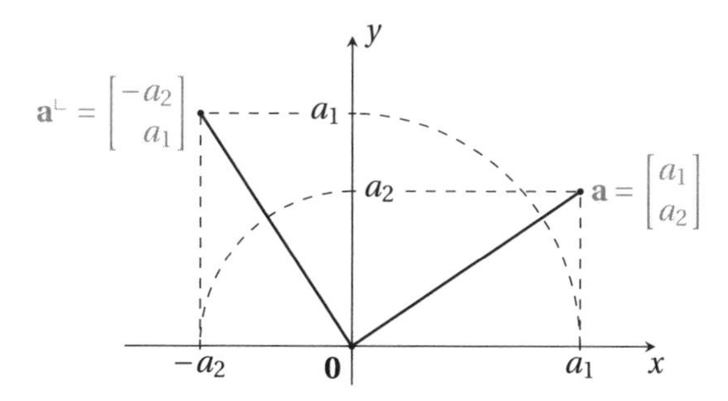

Figure 6.1: The perp operation.

First we introduce the "perp" operation in $\mathbb{R}^2$:

$$\mathbf{a}^{\llcorner} = \begin{bmatrix} a_1 \\ a_2 \end{bmatrix}^{\llcorner} = \begin{bmatrix} -a_2 \\ a_1 \end{bmatrix}.$$

The symbol $\mathbf{a}^{\llcorner}$ is read "**a** perp." Note that

$$\|\mathbf{a}^{\llcorner}\| = \|\mathbf{a}\|, \quad \mathbf{a}\cdot\mathbf{a}^{\llcorner} = 0, \quad \text{and} \quad (\mathbf{a}^{\llcorner})^{\llcorner} = -\mathbf{a}.$$

Moreover, if $\mathbf{a} \neq \begin{bmatrix} 0 \\ 0 \end{bmatrix}$ then $\mathbf{a}$ and $\mathbf{a}^{\llcorner}$ are linearly independent because $\det\begin{bmatrix}\mathbf{a} & \mathbf{a}^{\llcorner}\end{bmatrix} > 0$.

In what follows we assume that the reader is familiar with basic properties of the sine and cosine functions.

Suppose that $\mathbf{a}$ and $\mathbf{b}$ are nonzero vectors in $\mathbb{R}^2$. We will show that there are unique real numbers $t > 0$ and $0 \le \varphi < 2\pi$ such that

$$\mathbf{b} = t(\cos\varphi\,\mathbf{a} + \sin\varphi\,\mathbf{a}^{\llcorner}).$$

Since $\mathbf{a}$ and $\mathbf{a}^{\llcorner}$ are linearly independent, $\{\mathbf{a}, \mathbf{a}^{\llcorner}\}$ is a basis in $\mathbb{R}^2$ and thus there are unique real numbers p and q such that

$$\mathbf{b} = p\mathbf{a} + q\mathbf{a}^{\llcorner}.$$

Consequently,

$$\|\mathbf{b}\|^2 = \|p\mathbf{a} + q\mathbf{a}^{\llcorner}\|^2 = (p^2+q^2)\|\mathbf{a}\|^2$$

which gives us

$$p^2 + q^2 = \frac{\|\mathbf{b}\|^2}{\|\mathbf{a}\|^2},$$

since $\|\mathbf{a}\| > 0$. This means that the point (p, q) is a point on the circle of radius $\frac{\|\mathbf{b}\|}{\|\mathbf{a}\|}$ centered at the origin. Consequently, there is a unique angle $0 \le \varphi < 2\pi$ such that

$$p = \frac{\|\mathbf{b}\|}{\|\mathbf{a}\|}\cos\varphi \quad \text{and} \quad q = \frac{\|\mathbf{b}\|}{\|\mathbf{a}\|}\sin\varphi$$

and we have

$$\mathbf{b} = \frac{\|\mathbf{b}\|}{\|\mathbf{a}\|}\left(\cos\varphi\,\mathbf{a} + \sin\varphi\,\mathbf{a}^{\llcorner}\right).$$

The above formula can be interpreted as follows: When $\mathbf{a}$ is rotated counterclockwise about the origin by an angle φ and then multiplied by t, the result is $\mathbf{b}$.

Now we show that rotations in $\mathbb{R}^2$ can be described as matrix multiplication. Let $\mathbf{a} = \begin{bmatrix} a_1 \\ a_2 \end{bmatrix}$ and $\mathbf{b} = \begin{bmatrix} b_1 \\ b_2 \end{bmatrix}$ be vectors in $\mathbb{R}^2$ such that $\|\mathbf{a}\| = \|\mathbf{b}\| > 0$. Indeed, if $\mathbf{b}$ is obtained from $\mathbf{a}$ by a counterclockwise rotation about the origin by an angle φ, then

$$\begin{aligned}
\mathbf{b} &= \cos\varphi\,\mathbf{a} + \sin\varphi\,\mathbf{a}^{\llcorner} \\
&= \cos\varphi \begin{bmatrix} a_1 \\ a_2 \end{bmatrix} + \sin\varphi \begin{bmatrix} a_1 \\ a_2 \end{bmatrix}^{\llcorner} \\
&= \cos\varphi \begin{bmatrix} a_1 \\ a_2 \end{bmatrix} + \sin\varphi \begin{bmatrix} -a_2 \\ a_1 \end{bmatrix} \\
&= \begin{bmatrix} \cos\varphi & -\sin\varphi \\ \sin\varphi & \cos\varphi \end{bmatrix} \begin{bmatrix} a_1 \\ a_2 \end{bmatrix}.
\end{aligned}$$

6.1 The standard matrix of a linear transformation

Definition 6.1. A function $f : \mathbb{R}^n \to \mathbb{R}^m$ is called *linear* if it satisfies the following two conditions

(a) $f(\mathbf{x}+\mathbf{y}) = f(\mathbf{x}) + f(\mathbf{y})$ for every $\mathbf{x}$ and $\mathbf{y}$ in $\mathbb{R}^n$;

(b) $f(\alpha\mathbf{x}) = \alpha f(\mathbf{x})$ for every $\mathbf{x}$ in $\mathbb{R}^n$ and every real number α.

Linear functions from $\mathbb{R}^n$ to $\mathbb{R}^m$ are called *linear transformations*.

Note that, if $f : \mathbb{R}^n \to \mathbb{R}^m$ is a linear transformation, then

$$f(\alpha_1\mathbf{x}_1 + \cdots + \alpha_j\mathbf{x}_j) = \alpha_1 f(\mathbf{x}_1) + \cdots + \alpha_j f(\mathbf{x}_j)$$

for any vectors $\mathbf{x}_1, \dots, \mathbf{x}_j$ in $\mathbb{R}^n$ and any numbers $\alpha_1, \dots, \alpha_j$.

Theorem 6.2. *If A is an arbitrary $m \times n$ matrix, then the function $f : \mathbb{R}^n \to \mathbb{R}^m$ defined by $f(\mathbf{x}) = A\mathbf{x}$ is a linear transformation.*

Proof. This is Corollary 1.14 rephrased in terms of linear transformations. □

In the introduction to this chapter we pointed out that orthogonal projections and rotations about the origin in $\mathbb{R}^2$ are examples of linear transformations. Here are some other examples of linear transformations.

Example 6.3. Show that the identity function $\mathrm{id} : \mathbb{R}^n \to \mathbb{R}^n$, that is, the function $\mathrm{id}(\mathbf{x}) = \mathbf{x}$, is linear.

Solution. Since

$$\mathrm{id}(\mathbf{x}+\mathbf{y}) = \mathbf{x}+\mathbf{y} = \mathrm{id}(\mathbf{x}) + \mathrm{id}(\mathbf{y})$$

and

$$\mathrm{id}(\alpha\mathbf{x}) = \alpha\mathbf{x} = \alpha\,\mathrm{id}(\mathbf{x}),$$

id is a linear transformation. □

Example 6.4. Let $\mathbf{a}$ be a fixed vector in $\mathbb{R}^n$. Show that the function $f : \mathbb{R}^n \to \mathbb{R}$ defined by

$$f(\mathbf{x}) = \mathbf{x}\cdot\mathbf{a}$$

is a linear transformation.

Solution. Since

$$(\mathbf{x}+\mathbf{y})\cdot\mathbf{a}=\mathbf{x}\cdot\mathbf{a}+\mathbf{y}\cdot\mathbf{a}$$

and

$$(\alpha\mathbf{x})\cdot\mathbf{a}=\alpha(\mathbf{x}\cdot\mathbf{a}),$$

the function $f(\mathbf{x})=\mathbf{x}\cdot\mathbf{a}$ is a linear transformation. □

For our next example we need the following definition.

Definition 6.5. The *cross product* of vectors $\mathbf{x}=\begin{bmatrix}x\\y\\z\end{bmatrix}$ and $\mathbf{a}=\begin{bmatrix}a\\b\\c\end{bmatrix}$ in $\mathbb{R}^3$ is the vector

$$\begin{bmatrix}x\\y\\z\end{bmatrix}\times\begin{bmatrix}a\\b\\c\end{bmatrix}=\begin{bmatrix}yc-bz\\az-cx\\xb-ay\end{bmatrix}.$$

Example 6.6. Let $\mathbf{a}$ be a fixed vector in $\mathbb{R}^3$. Show that the function $f:\mathbb{R}^3\to\mathbb{R}^3$ defined by

$$f(\mathbf{x})=\mathbf{x}\times\mathbf{a}$$

is a linear transformation.

Solution. Since it is easy to verify that

$$(\mathbf{x}+\mathbf{y})\times\mathbf{a}=\mathbf{x}\times\mathbf{a}+\mathbf{y}\times\mathbf{a}$$

and

$$(\alpha\mathbf{x})\times\mathbf{a}=\alpha(\mathbf{x}\times\mathbf{a}),$$

the function $f(\mathbf{x})=\mathbf{x}\times\mathbf{a}$ is a linear transformation. □

Theorem 6.2 allows us to easily produce a lot of examples of linear transformations by taking an arbitrary matrix A of the proper size and defining $f(\mathbf{x})=A\mathbf{x}$. It may be somewhat unexpected that these are the only examples of linear transformations. In other words, every linear transformation can be defined by a matrix.

Theorem 6.7. *For every linear transformation $f : \mathbb{R}^n \to \mathbb{R}^m$ there is a unique $m \times n$ matrix A such that $f(\mathbf{x}) = A\mathbf{x}$ for all $\mathbf{x}$ in $\mathbb{R}^n$. The j-th column of the matrix A is $f(\mathbf{e}_j)$ where $\mathbf{e}_j$ is the j-th column of the $n \times n$ identity matrix, that is,*

$$A = \begin{bmatrix} f\left(\begin{bmatrix}1\\0\\\vdots\\0\end{bmatrix}\right) & f\left(\begin{bmatrix}0\\1\\\vdots\\0\end{bmatrix}\right) & \dots & f\left(\begin{bmatrix}0\\0\\\vdots\\1\end{bmatrix}\right)\end{bmatrix}.$$

In other words,

$$A = \begin{bmatrix} f(\mathbf{e}_1) & \dots & f(\mathbf{e}_n)\end{bmatrix},$$

where $\{\mathbf{e}_1, \dots, \mathbf{e}_n\}$ is the standard basis in $\mathbb{R}^n$.

We illustrate the idea of the proof for $m = 2$ and $n = 3$. Let $f : \mathbb{R}^3 \to \mathbb{R}^2$ be a linear transformation and let

$$f\left(\begin{bmatrix}1\\0\\0\end{bmatrix}\right) = \begin{bmatrix}a_{11}\\a_{21}\end{bmatrix},\quad f\left(\begin{bmatrix}0\\1\\0\end{bmatrix}\right) = \begin{bmatrix}a_{12}\\a_{22}\end{bmatrix},\quad f\left(\begin{bmatrix}0\\0\\1\end{bmatrix}\right) = \begin{bmatrix}a_{13}\\a_{23}\end{bmatrix}.$$

Since any vector $\mathbf{x} = \begin{bmatrix}x_1\\x_2\\x_3\end{bmatrix}$ in $\mathbb{R}^3$ can be written as

$$\mathbf{x} = \begin{bmatrix}x_1\\x_2\\x_3\end{bmatrix} = x_1\begin{bmatrix}1\\0\\0\end{bmatrix} + x_2\begin{bmatrix}0\\1\\0\end{bmatrix} + x_3\begin{bmatrix}0\\0\\1\end{bmatrix},$$

we have

$$\begin{aligned} f(\mathbf{x}) &= x_1 f\left(\begin{bmatrix}1\\0\\0\end{bmatrix}\right) + x_2 f\left(\begin{bmatrix}0\\1\\0\end{bmatrix}\right) + x_3 f\left(\begin{bmatrix}0\\0\\1\end{bmatrix}\right) \\ &= x_1\begin{bmatrix}a_{11}\\a_{21}\end{bmatrix} + x_2\begin{bmatrix}a_{12}\\a_{22}\end{bmatrix} + x_3\begin{bmatrix}a_{13}\\a_{23}\end{bmatrix} \\ &= \begin{bmatrix}a_{11}x_1 + a_{12}x_2 + a_{13}x_3\\a_{21}x_1 + a_{22}x_2 + a_{23}x_3\end{bmatrix} \\ &= \begin{bmatrix}a_{11} & a_{12} & a_{13}\\a_{21} & a_{22} & a_{23}\end{bmatrix}\begin{bmatrix}x_1\\x_2\\x_3\end{bmatrix} = A\mathbf{x} \end{aligned}$$

where A is the matrix $\begin{bmatrix}a_{11} & a_{12} & a_{13}\\a_{21} & a_{22} & a_{23}\end{bmatrix}$. □

Definition 6.8. Let $f:\mathbb{R}^n \to \mathbb{R}^m$ be a linear transformation. The unique $m\times n$ matrix A such that $f(\mathbf{x}) = A\mathbf{x}$ for all $\mathbf{x}$ in $\mathbb{R}^n$ is called the *standard matrix* of the linear transformation f.

Example 6.9. Find the standard matrix of the linear transformation $f:\mathbb{R}^3 \to \mathbb{R}^3$ defined by $f\left(\begin{bmatrix}x\\y\\z\end{bmatrix}\right) = \begin{bmatrix}x+2y\\y-3z\\4x-5z\end{bmatrix}$.

Solution. Since

$$f\left(\begin{bmatrix}1\\0\\0\end{bmatrix}\right) = \begin{bmatrix}1\\0\\4\end{bmatrix}, \quad f\left(\begin{bmatrix}0\\1\\0\end{bmatrix}\right) = \begin{bmatrix}2\\1\\0\end{bmatrix}, \quad \text{and} \quad f\left(\begin{bmatrix}0\\0\\1\end{bmatrix}\right) = \begin{bmatrix}0\\-3\\-5\end{bmatrix},$$

the standard matrix of f is

$$A = \begin{bmatrix}1&2&0\\0&1&-3\\4&0&-5\end{bmatrix}.$$

□

Example 6.10. Let $\begin{bmatrix}a\\b\\c\end{bmatrix}$ be a vector of $\mathbb{R}^3$. Find the standard matrix of the linear transformation $f:\mathbb{R}^3 \to \mathbb{R}^3$ defined by

$$f\left(\begin{bmatrix}x\\y\\z\end{bmatrix}\right) = \begin{bmatrix}x\\y\\z\end{bmatrix} \times \begin{bmatrix}a\\b\\c\end{bmatrix} = \begin{bmatrix}yc-bz\\az-cx\\xb-ay\end{bmatrix}.$$

Solution. Since

$$f\left(\begin{bmatrix}1\\0\\0\end{bmatrix}\right) = \begin{bmatrix}1\\0\\0\end{bmatrix} \times \begin{bmatrix}a\\b\\c\end{bmatrix} = \begin{bmatrix}0\\-c\\b\end{bmatrix}, \quad f\left(\begin{bmatrix}0\\1\\0\end{bmatrix}\right) = \begin{bmatrix}0\\1\\0\end{bmatrix} \times \begin{bmatrix}a\\b\\c\end{bmatrix} = \begin{bmatrix}c\\0\\-a\end{bmatrix},$$

$$\text{and} \quad f\left(\begin{bmatrix}0\\0\\1\end{bmatrix}\right) = \begin{bmatrix}0\\0\\1\end{bmatrix} \times \begin{bmatrix}a\\b\\c\end{bmatrix} = \begin{bmatrix}-b\\a\\0\end{bmatrix},$$

the standard matrix of f is

$$A = \begin{bmatrix} 0 & c & -b \\ -c & 0 & a \\ b & -a & 0 \end{bmatrix}.$$

□

Theorem 6.11. *If $f : \mathbb{R}^n \to \mathbb{R}^m$ and $g : \mathbb{R}^m \to \mathbb{R}^l$ are linear transformations and the standard matrices of f and g are A and B, respectively, then the composition $g \circ f$ is a linear transformation and the standard matrix of the composition $g \circ f$ is the matrix BA.*

Proof. Since

$$g \circ f(\mathbf{x}) = g(f(\mathbf{x})) = B(A\mathbf{x}) = (BA)\mathbf{x},$$

the standard matrix of the composition $g \circ f$ is the matrix BA. Moreover, since $g \circ f$ corresponds to multiplication by a matrix, it is a linear transformation. □

Example 6.12. Let $f : \mathbb{R}^2 \to \mathbb{R}^3$ be defined by $f\left(\begin{bmatrix} x \\ y \end{bmatrix}\right) = \begin{bmatrix} 2x \\ x+4y \\ 3x-y \end{bmatrix}$ and let $g : \mathbb{R}^3 \to \mathbb{R}$ be defined by $g\left(\begin{bmatrix} s \\ t \\ u \end{bmatrix}\right) = 2s + t + u$. Verify the result of Theorem 6.11 for f and g.

Solution. The standard matrix of f is $A = \begin{bmatrix} 2 & 0 \\ 1 & 4 \\ 3 & -1 \end{bmatrix}$ and the standard matrix of g is $B = \begin{bmatrix} 2 & 1 & 1 \end{bmatrix}$. Since

$$g\left(f\left(\begin{bmatrix} x \\ y \end{bmatrix}\right)\right) = 8x + 3y = \begin{bmatrix} 8 & 3 \end{bmatrix} \begin{bmatrix} x \\ y \end{bmatrix},$$

the standard matrix of the linear transformation $g \circ f$ is $\begin{bmatrix} 8 & 3 \end{bmatrix}$. Note that the same matrix is obtained as the product of the standard matrices of g and f:

$$BA = \begin{bmatrix} 2 & 1 & 1 \end{bmatrix} \begin{bmatrix} 2 & 0 \\ 1 & 4 \\ 3 & -1 \end{bmatrix} = \begin{bmatrix} 8 & 3 \end{bmatrix}.$$

□

Example 6.13. Let $f:\mathbb{R}^2 \to \mathbb{R}^2$ be defined by $f\left(\begin{bmatrix} x \\ y \end{bmatrix}\right) = \begin{bmatrix} 2x+y \\ 3y \end{bmatrix}$ and $g:\mathbb{R}^2 \to \mathbb{R}^2$ be defined by $g\left(\begin{bmatrix} s \\ t \end{bmatrix}\right) = \begin{bmatrix} s-t \\ s+t \end{bmatrix}$. Verify the result of Theorem 6.11 for f and g.

Solution. The standard matrix of f is $A = \begin{bmatrix} 2 & 1 \\ 0 & 3 \end{bmatrix}$ and the standard matrix of g is $B = \begin{bmatrix} 1 & -1 \\ 1 & 1 \end{bmatrix}$. Since

$$g\left(f\left(\begin{bmatrix} x \\ y \end{bmatrix}\right)\right) = g\left(\begin{bmatrix} 2x+y \\ 3y \end{bmatrix}\right) = \begin{bmatrix} 2x-2y \\ 2x+4y \end{bmatrix} = \begin{bmatrix} 2 & -2 \\ 2 & 4 \end{bmatrix}\begin{bmatrix} x \\ y \end{bmatrix}.$$

The standard matrix of $g \circ f$ is $\begin{bmatrix} 2 & -2 \\ 2 & 4 \end{bmatrix}$. The same matrix is obtained as the product of the standard matrices of g and f:

$$BA = \begin{bmatrix} 1 & -1 \\ 1 & 1 \end{bmatrix}\begin{bmatrix} 2 & 1 \\ 0 & 3 \end{bmatrix} = \begin{bmatrix} 2 & -2 \\ 2 & 4 \end{bmatrix}.$$

□

Theorem 6.14. *Let $f:\mathbb{R}^n \to \mathbb{R}^n$ be a linear transformation and let A be its standard matrix.*

(a) f is invertible if and only if the matrix A is invertible.

(b) If f is invertible, then f^{-1} is a linear transformation and its standard matrix is A^{-1}.

Proof. Assume that the matrix A is invertible and let $g:\mathbb{R}^m \to \mathbb{R}^n$ be the linear transformation that has A^{-1} as its standard matrix. Then

$$g \circ f(\mathbf{x}) = g(f(\mathbf{x})) = A^{-1}(A\mathbf{x}) = \mathbf{x}$$

and

$$f \circ g(\mathbf{x}) = f(g(\mathbf{x})) = A(A^{-1}\mathbf{x}) = \mathbf{x}.$$

This means that g is the inverse of f.

Now assume that the linear transformation f is invertible. If $A\mathbf{x} = \mathbf{0}$, then $f(\mathbf{x}) = \mathbf{0}$ and we must have $\mathbf{x} = \mathbf{0}$, because $f(\mathbf{0}) = \mathbf{0}$ and f is invertible. Since $A\mathbf{x} = \mathbf{0}$ implies $\mathbf{x} = \mathbf{0}$, the matrix A is invertible.

If f is invertible, then from the first part of the proof it is clear that $f^{-1}(\mathbf{x}) = A^{-1}\mathbf{x}$ for every $\mathbf{x}$ in $\mathbb{R}^n$. Moreover, since f^{-1} corresponds to matrix multiplication, it is a linear transformation. □

Example 6.15. Let $f : \mathbb{R}^2 \to \mathbb{R}^2$ be defined by $f\left(\begin{bmatrix} x \\ y \end{bmatrix}\right) = \begin{bmatrix} 2x+y \\ x-3y \end{bmatrix}$. Show that f is invertible and find the inverse of f.

Solution. The standard matrix of f is $A = \begin{bmatrix} 2 & 1 \\ 1 & 3 \end{bmatrix}$ and we have $A^{-1} = \frac{1}{5}\begin{bmatrix} 3 & -1 \\ -1 & 2 \end{bmatrix}$. Consequently, the linear transformation

$$g\left(\begin{bmatrix} x \\ y \end{bmatrix}\right) = \frac{1}{5}\begin{bmatrix} 3x-y \\ -x+2y \end{bmatrix}$$

is the inverse of f. □

Theorem 6.16. *Let $\{\mathbf{u}_1, \dots, \mathbf{u}_n\}$ be a basis of $\mathbb{R}^n$. For any vectors $\mathbf{c}_1, \dots, \mathbf{c}_n$ in $\mathbb{R}^m$ there is a unique linear transformation $f : \mathbb{R}^n \to \mathbb{R}^m$ such that*

$$f(\mathbf{u}_1) = \mathbf{c}_1, \dots, f(\mathbf{u}_n) = \mathbf{c}_n.$$

The standard matrix of that unique f is

$$A = \begin{bmatrix} \mathbf{c}_1 & \dots & \mathbf{c}_n \end{bmatrix}\begin{bmatrix} \mathbf{u}_1 & \dots & \mathbf{u}_n \end{bmatrix}^{-1}.$$

Proof. Since $\{\mathbf{u}_1, \dots, \mathbf{u}_n\}$ is a basis of $\mathbb{R}^n$, for every vector $\mathbf{x}$ in $\mathbb{R}^n$ there are unique numbers $x_1, \dots, x_n$ such that $\mathbf{x} = x_1\mathbf{u}_1 + \dots + x_n\mathbf{u}_n$. Since f is a linear transformation, we define

$$f(\mathbf{x}) = f(x_1\mathbf{u}_1 + \dots + x_n\mathbf{u}_n) = x_1\mathbf{c}_1 + \dots + x_n\mathbf{c}_n.$$

Clearly f is a linear transformation that is uniquely determined by the condition that $f(\mathbf{u}_1) = \mathbf{c}_1, \dots, f(\mathbf{u}_n) = \mathbf{c}_n$. If A is the standard matrix of f, then we have

$$A\begin{bmatrix} \mathbf{u}_1 & \dots & \mathbf{u}_n \end{bmatrix} = \begin{bmatrix} \mathbf{A}u_1 & \dots & \mathbf{A}u_n \end{bmatrix} = \begin{bmatrix} \mathbf{c}_1 & \dots & \mathbf{c}_n \end{bmatrix}.$$

Hence

$$A = \begin{bmatrix} \mathbf{c}_1 & \dots & \mathbf{c}_n \end{bmatrix}\begin{bmatrix} \mathbf{u}_1 & \dots & \mathbf{u}_n \end{bmatrix}^{-1},$$

because the matrix $\begin{bmatrix} \mathbf{u}_1 & \dots & \mathbf{u}_n \end{bmatrix}$ is invertible as a consequence of the fact that $\{\mathbf{u}_1, \dots, \mathbf{u}_n\}$ is a basis of $\mathbb{R}^n$. □

Example 6.17. Find a linear transformation $f : \mathbb{R}^2 \to \mathbb{R}^2$ such that $f\left(\begin{bmatrix} 1 \\ 2 \end{bmatrix}\right) = \begin{bmatrix} 1 \\ -1 \end{bmatrix}$ and $f\left(\begin{bmatrix} 3 \\ -1 \end{bmatrix}\right) = \begin{bmatrix} 2 \\ 3 \end{bmatrix}$.

Solution. The standard matrix A of f satisfies the matrix equation

$$A\begin{bmatrix}1 & 3\\ 2 & -1\end{bmatrix} = \begin{bmatrix}1 & 2\\ -1 & 3\end{bmatrix}.$$

Hence, using Theorem 6.16,

$$A = \begin{bmatrix}1 & 2\\ -1 & 3\end{bmatrix}\begin{bmatrix}1 & 3\\ 2 & -1\end{bmatrix}^{-1} = \begin{bmatrix}\frac{5}{7} & \frac{1}{7}\\ \frac{5}{7} & -\frac{6}{7}\end{bmatrix}$$

and thus

$$f\left(\begin{bmatrix}x_1\\ x_2\end{bmatrix}\right) = \begin{bmatrix}\frac{5}{7} & \frac{1}{7}\\ \frac{5}{7} & -\frac{6}{7}\end{bmatrix}\begin{bmatrix}x_1\\ x_2\end{bmatrix}.$$

□

Exercises 6.1

1. Show that the function $f : \mathbb{R}^4 \to \mathbb{R}$ defined by

$$f\left(\begin{bmatrix}x_1\\ x_2\\ x_3\\ x_4\end{bmatrix}\right) = 2x_1 - 3x_2 + x_3 + 2x_4$$

is a linear transformation.

2. Show that the function $f : \mathbb{R}^5 \to \mathbb{R}$ defined by

$$f\left(\begin{bmatrix}x_1\\ x_2\\ x_3\\ x_4\\ x_5\end{bmatrix}\right) = 3x_1 - x_2 + 7x_3 + 5x_4 + 8x_5$$

is a linear transformation.

3. Let $f : \mathbb{R}^2 \to \mathbb{R}^5$ be defined by $f\left(\begin{bmatrix}x\\ y\end{bmatrix}\right) = \begin{bmatrix}x\\ x+y\\ 2y\\ x-y\\ x+y\end{bmatrix}$. Show that f is a linear transformation.

4. Let $f : \mathbb{R}^3 \to \mathbb{R}^4$ be defined by $f\left(\begin{bmatrix}x\\ y\\ z\end{bmatrix}\right) = \begin{bmatrix}x+y+z\\ 3x+y-z\\ 2x+5y+4z\\ x+2y+7z\end{bmatrix}$. Show that f is a linear transformation.

5. Let $f:\mathbb{R}^2 \to \mathbb{R}^2$ be defined by $f\left(\begin{bmatrix} x \\ y \end{bmatrix}\right) = \begin{bmatrix} 2x+y \\ 3x-2y \end{bmatrix}$ and let $g:\mathbb{R}^2 \to \mathbb{R}$ be defined by $g\left(\begin{bmatrix} s \\ t \end{bmatrix}\right) = 2s+t$. Verify the result of Theorem 6.11 for f and g.

6. Let $f:\mathbb{R}^2 \to \mathbb{R}^3$ be defined by $f\left(\begin{bmatrix} x \\ y \end{bmatrix}\right) = \begin{bmatrix} x+y \\ x-3y \\ 2x+5y \end{bmatrix}$ and let $g:\mathbb{R}^3 \to \mathbb{R}$ be defined by $g\left(\begin{bmatrix} s \\ t \\ u \end{bmatrix}\right) = s-t+2u$. Verify the result of Theorem 6.11 for f and g.

7. Let $f:\mathbb{R}^2 \to \mathbb{R}^2$ be defined by $f\left(\begin{bmatrix} x \\ y \end{bmatrix}\right) = \begin{bmatrix} y \\ 2x-y \end{bmatrix}$ and let $g:\mathbb{R}^2 \to \mathbb{R}^2$ be defined by $g\left(\begin{bmatrix} s \\ t \end{bmatrix}\right) = \begin{bmatrix} 2s \\ s-t \end{bmatrix}$. Verify the result of Theorem 6.11 for f and g.

8. Let $f:\mathbb{R}^3 \to \mathbb{R}^3$ be defined by $f\left(\begin{bmatrix} x \\ y \\ z \end{bmatrix}\right) = \begin{bmatrix} y \\ z \\ x+y \end{bmatrix}$ and let $g:\mathbb{R}^3 \to \mathbb{R}^2$ be defined by $g\left(\begin{bmatrix} s \\ t \\ u \end{bmatrix}\right) = \begin{bmatrix} s+u \\ s+t \end{bmatrix}$. Verify the result of Theorem 6.11 for f and g.

9. Let $f:\mathbb{R}^2 \to \mathbb{R}^2$ be defined by $f\left(\begin{bmatrix} x \\ y \end{bmatrix}\right) = \begin{bmatrix} x+3y \\ 2x+5y \end{bmatrix}$. Show that f is invertible and find the inverse of f.

10. Let $f:\mathbb{R}^2 \to \mathbb{R}^2$ be defined by $f\left(\begin{bmatrix} x \\ y \end{bmatrix}\right) = \begin{bmatrix} y \\ 2x-5y \end{bmatrix}$. Show that f is invertible and find the inverse of f.

11. Let $f:\mathbb{R}^3 \to \mathbb{R}^3$ be defined by $f\left(\begin{bmatrix} x \\ y \\ z \end{bmatrix}\right) = \begin{bmatrix} x+z \\ y-z \\ x+y+z \end{bmatrix}$. Show that f is invertible and find the inverse of f.

12. Let $f:\mathbb{R}^3 \to \mathbb{R}^3$ be defined by $f\left(\begin{bmatrix} x \\ y \\ z \end{bmatrix}\right) = \begin{bmatrix} x+y+2z \\ y+z \\ x-y+3z \end{bmatrix}$. Show that f is invertible and find the inverse of f.

13. Find a linear transformation $f:\mathbb{R}^2 \to \mathbb{R}^2$ such that $f\left(\begin{bmatrix} 1 \\ 1 \end{bmatrix}\right) = \begin{bmatrix} 3 \\ 2 \end{bmatrix}$ and $f\left(\begin{bmatrix} 1 \\ 2 \end{bmatrix}\right) = \begin{bmatrix} 1 \\ 1 \end{bmatrix}$.

14. Find a linear transformation $f:\mathbb{R}^2 \to \mathbb{R}^2$ such that $f\left(\begin{bmatrix} 2 \\ -3 \end{bmatrix}\right) = \begin{bmatrix} 1 \\ 0 \end{bmatrix}$ and $f\left(\begin{bmatrix} 0 \\ 1 \end{bmatrix}\right) = \begin{bmatrix} 2 \\ 1 \end{bmatrix}$.

15. Find the standard matrix of the linear transformation $f:\mathbb{R}^3 \to \mathbb{R}^3$ such that $f\left(\begin{bmatrix} 1 \\ 0 \\ 0 \end{bmatrix}\right) = \begin{bmatrix} 1 \\ 2 \\ 1 \end{bmatrix}$, $f\left(\begin{bmatrix} 1 \\ 1 \\ 0 \end{bmatrix}\right) = \begin{bmatrix} 2 \\ 2 \\ 1 \end{bmatrix}$, and $f\left(\begin{bmatrix} 1 \\ 1 \\ 1 \end{bmatrix}\right) = \begin{bmatrix} 1 \\ 1 \\ 1 \end{bmatrix}$.

16. Find the standard matrix of the linear transformation $f:\mathbb{R}^3 \to \mathbb{R}^3$ such that $f\left(\begin{bmatrix} 2 \\ 1 \\ 1 \end{bmatrix}\right) = \begin{bmatrix} 1 \\ 0 \\ 1 \end{bmatrix}$, $f\left(\begin{bmatrix} 1 \\ 2 \\ 1 \end{bmatrix}\right) = \begin{bmatrix} 1 \\ 1 \\ 0 \end{bmatrix}$, and $f\left(\begin{bmatrix} 1 \\ 1 \\ 2 \end{bmatrix}\right) = \begin{bmatrix} 1 \\ 0 \\ 0 \end{bmatrix}$.

17. Find the standard matrix of the linear transformation $f:\mathbb{R}^3 \to \mathbb{R}^2$ such that $f\left(\begin{bmatrix} 2 \\ 1 \\ 1 \end{bmatrix}\right) = \begin{bmatrix} 1 \\ 1 \end{bmatrix}$, $f\left(\begin{bmatrix} 1 \\ 2 \\ 1 \end{bmatrix}\right) = \begin{bmatrix} 1 \\ 0 \end{bmatrix}$, and $f\left(\begin{bmatrix} 1 \\ 1 \\ 2 \end{bmatrix}\right) = \begin{bmatrix} 1 \\ 2 \end{bmatrix}$.

18. Find a linear transformation $f:\mathbb{R}^2 \to \mathbb{R}^3$ such that $f\left(\begin{bmatrix} 1 \\ 1 \end{bmatrix}\right) = \begin{bmatrix} 1 \\ 1 \\ 0 \end{bmatrix}$ and $f\left(\begin{bmatrix} 1 \\ 2 \end{bmatrix}\right) = \begin{bmatrix} 1 \\ 1 \\ 1 \end{bmatrix}$.

19. Show that a function $f:\mathbb{R}^n \to \mathbb{R}$ is a linear transformation if and only if there is a vector $\mathbf{a}$ in $\mathbb{R}^n$ such that $f(\mathbf{x}) = \mathbf{x}\cdot\mathbf{a}$ and that the vector $\mathbf{a}$ is uniquely determined by the function f.

20. Show that a function $f:\mathbb{R}^n \to \mathbb{R}^m$ is a linear transformation if and only if there are m vectors $\mathbf{a}_1,\dots,\mathbf{a}_m$ in $\mathbb{R}^n$ such that

$$f(\mathbf{x}) = \begin{bmatrix} \mathbf{x}\cdot\mathbf{a}_1 \\ \vdots \\ \mathbf{x}\cdot\mathbf{a}_m \end{bmatrix}$$

and that the vectors $\mathbf{a}_1,\dots,\mathbf{a}_m$ are uniquely determined by the function f.

6.2 Matrices associated with linear transformations

The standard matrix of a linear transformation $f : \mathbb{R}^n \to \mathbb{R}^m$ is closely related to the standard bases in $\mathbb{R}^n$ and $\mathbb{R}^m$. More precisely, if $\mathbf{x} = \begin{bmatrix} x_1 \\ \vdots \\ x_n \end{bmatrix}$ then the numbers $x_1, \dots, x_n$ are the coefficients of $\mathbf{x}$ in the standard basis of $\mathbb{R}^n$ and if $f(\mathbf{x}) = \begin{bmatrix} y_1 \\ \vdots \\ y_m \end{bmatrix}$ then the numbers $y_1, \dots, y_m$ are the coefficients of $f(\mathbf{x})$ in the standard basis of $\mathbb{R}^m$ and we have $\begin{bmatrix} y_1 \\ \vdots \\ y_m \end{bmatrix} = A \begin{bmatrix} x_1 \\ \vdots \\ x_n \end{bmatrix}$, where A is the standard matrix of f. In applications it is often more convenient to describe the input and the output of a linear transformation in terms of bases different from the standard basis. If $\{\mathbf{u}_1, \dots, \mathbf{u}_n\}$ is a basis in $\mathbb{R}^n$, $\{\mathbf{v}_1, \dots, \mathbf{v}_n\}$ a basis in $\mathbb{R}^m$, and

$$f(x_1\mathbf{u}_1 + \dots + x_n\mathbf{u}_n) = y_1\mathbf{v}_1 + \dots + y_m\mathbf{v}_m,$$

then there is no reason to expect that we still have $\begin{bmatrix} y_1 \\ \vdots \\ y_m \end{bmatrix} = A \begin{bmatrix} x_1 \\ \vdots \\ x_n \end{bmatrix}$. We need to find a new matrix that describes the action of f with respect to these bases. In this section we discuss how to find such a matrix. We use the observation that, if $\mathscr{B} = \{\mathbf{u}_1, \dots, \mathbf{u}_n\}$ is a basis in $\mathbb{R}^n$ and $U = \begin{bmatrix} \mathbf{u}_1 & \dots & \mathbf{u}_n \end{bmatrix}$, then every element of $\mathbb{R}^n$ can be written as $U\mathbf{x}$ for some $\mathbf{x}$ in $\mathbb{R}^n$.

Theorem 6.18. *Let $\mathscr{B} = \{\mathbf{u}_1, \dots, \mathbf{u}_n\}$ and $\mathscr{C} = \{\mathbf{v}_1, \dots, \mathbf{v}_m\}$ be bases in $\mathbb{R}^n$ and $\mathbb{R}^m$, respectively, and let $f : \mathbb{R}^n \to \mathbb{R}^m$ be the linear transformation such that*

$$f(\mathbf{u}_j) = b_{1j}\mathbf{v}_1 + \dots + b_{mj}\mathbf{v}_m = \begin{bmatrix} \mathbf{v}_1 & \dots & \mathbf{v}_m \end{bmatrix} \begin{bmatrix} b_{1j} \\ \vdots \\ b_{mj} \end{bmatrix}, \quad j = 1, \dots, n.$$

If $\mathbf{u} = x_1\mathbf{u}_1 + \dots + x_n\mathbf{u}_n$ is an arbitrary vector in $\mathbb{R}^n$, then we have

$$f(\mathbf{u}) = VB\mathbf{x}, \tag{6.1}$$

where $V = \begin{bmatrix} \mathbf{v}_1 & \cdots & \mathbf{v}_m \end{bmatrix}$, $B = \begin{bmatrix} b_{11} & \cdots & b_{1n} \\ \vdots & & \vdots \\ b_{m1} & \dots & b_{mn} \end{bmatrix}$, and $\mathbf{x} = \begin{bmatrix} x_1 \\ \vdots \\ x_n \end{bmatrix}$.

Proof for $n = 3$ and $m = 2$. The equalities

$$f(\mathbf{u}_1) = b_{11}\mathbf{v}_1 + b_{21}\mathbf{v}_2$$

$$f(\mathbf{u}_2) = b_{12}\mathbf{v}_1 + b_{22}\mathbf{v}_2$$

$$f(\mathbf{u}_3) = b_{13}\mathbf{v}_1 + b_{22}\mathbf{v}_2$$

are equivalent to

$$\begin{bmatrix} f(\mathbf{u}_1) & f(\mathbf{u}_2) & f(\mathbf{u}_3) \end{bmatrix} = \begin{bmatrix} \mathbf{v}_1 & \mathbf{v}_2 \end{bmatrix} \begin{bmatrix} b_{11} & b_{12} & b_{13} \\ b_{21} & b_{22} & b_{23} \end{bmatrix}.$$

Since f is a linear transformation, if $\mathbf{u} = x_1\mathbf{u}_1 + x_2\mathbf{u}_2 + x_3\mathbf{u}_3$ is an arbitrary vector in $\mathbb{R}^3$, then we have

$$\begin{aligned} f(\mathbf{u}) &= f(x_1\mathbf{u}_1 + x_2\mathbf{u}_2 + x_3\mathbf{u}_3) \\ &= x_1 f(\mathbf{u}_1) + x_2 f(\mathbf{u}_2) + x_3 f(\mathbf{u}_3) \\ &= \begin{bmatrix} f(\mathbf{u}_1) & f(\mathbf{u}_2) & f(\mathbf{u}_3) \end{bmatrix} \begin{bmatrix} x_1 \\ x_2 \\ x_3 \end{bmatrix} \\ &= \begin{bmatrix} \mathbf{v}_1 & \mathbf{v}_2 \end{bmatrix} \begin{bmatrix} b_{11} & b_{12} & b_{13} \\ b_{21} & b_{22} & b_{23} \end{bmatrix} \begin{bmatrix} x_1 \\ x_2 \\ x_3 \end{bmatrix}. \end{aligned}$$

□

Definition 6.19. Let $f : \mathbb{R}^n \to \mathbb{R}^m$ be a linear transformation and let $\mathscr{B} = \{\mathbf{u}_1, \dots, \mathbf{u}_n\}$ and $\mathscr{C} = \{\mathbf{v}_1, \dots, \mathbf{v}_m\}$ be bases in $\mathbb{R}^n$ and $\mathbb{R}^m$, respectively. The matrix B from Theorem 6.18 is called the *matrix of f relative to the bases $\mathscr{B}$ and $\mathscr{C}$*.
In case where $m = n$ and the basis $\mathscr{C}$ is the same as $\mathscr{B}$ the matrix B from Theorem 6.18 is called the *matrix of f relative to the basis $\mathscr{B}$* or simply the *$\mathscr{B}$-matrix of f*.

Example 6.20. Consider a basis $\mathscr{B} = \{\mathbf{u}_1, \mathbf{u}_2\}$ in $\mathbb{R}^2$ and a basis $\mathscr{C} = \{\mathbf{v}_1, \mathbf{v}_2, \mathbf{v}_3\}$ in $\mathbb{R}^3$. If $f : \mathbb{R}^2 \to \mathbb{R}^3$ is the linear transformation such that $f(\mathbf{u}_1) = 2\mathbf{v}_1 + \mathbf{v}_2 + \mathbf{v}_3$ and $f(\mathbf{u}_2) = \mathbf{v}_1 + 3\mathbf{v}_2 - \mathbf{v}_3$, find the matrix of the function f relative to the bases $\mathscr{B}$ and $\mathscr{C}$.

Solution. Using Theorem 6.18, the matrix of the function f relative to the bases $\mathscr{B}$ and $\mathscr{C}$ is

$$\begin{bmatrix} 2 & 1 \\ 1 & 3 \\ 1 & -1 \end{bmatrix},$$

since $f(\mathbf{u}_1) = [\mathbf{v}_1\mathbf{v}_2\mathbf{v}_3]\begin{bmatrix} 2 \\ 1 \\ 1 \end{bmatrix}$ and $f(\mathbf{u}_2) = [\mathbf{v}_1\mathbf{v}_2\mathbf{v}_3]\begin{bmatrix} 1 \\ 3 \\ -1 \end{bmatrix}$. □

Theorem 6.21. *Let $\mathscr{B} = \{\mathbf{u}_1, \ldots, \mathbf{u}_n\}$ and $\mathscr{C} = \{\mathbf{v}_1, \ldots, \mathbf{v}_m\}$ be bases for $\mathbb{R}^n$ and $\mathbb{R}^m$, respectively. If A is an $m \times n$ matrix, then the matrix B of the linear transformation $f(\mathbf{u}) = A\mathbf{u}$ relative to the bases $\mathscr{B}$ and $\mathscr{C}$ is*

$$B = V^{-1}AU,$$

where $U = \begin{bmatrix} \mathbf{u}_1 & \ldots & \mathbf{u}_n \end{bmatrix}$ and $V = \begin{bmatrix} \mathbf{v}_1 & \ldots & \mathbf{v}_m \end{bmatrix}$.

Proof. Let $f : \mathbb{R}^n \to \mathbb{R}^n$ be a linear transformation defined by $f(\mathbf{u}) = A\mathbf{u}$. For any $\mathbf{u} = U\mathbf{x}$, where $\mathbf{x}$ is an arbitrary vector in $\mathbb{R}^n$, we have

$$f(\mathbf{u}) = f(U\mathbf{x}) = AU\mathbf{x}.$$

On the other hand, by Theorem 6.18, we have

$$f(\mathbf{u}) = VB\mathbf{x}.$$

Consequently, by Theorem 1.18, we have

$$AU = VB$$

and thus

$$B = V^{-1}AU.$$

□

Example 6.22. Determine the matrix of the linear transformation

$$f\left(\begin{bmatrix} x_1 \\ x_2 \end{bmatrix}\right) = \begin{bmatrix} 1 & -4 \\ 0 & 1 \\ 3 & 1 \end{bmatrix}\begin{bmatrix} x_1 \\ x_2 \end{bmatrix}$$

relative to the bases $\left\{\begin{bmatrix} 1 \\ 2 \end{bmatrix}, \begin{bmatrix} 1 \\ 1 \end{bmatrix}\right\}$ and $\left\{\begin{bmatrix} 1 \\ 1 \\ 1 \end{bmatrix}, \begin{bmatrix} 1 \\ 1 \\ 0 \end{bmatrix}, \begin{bmatrix} 1 \\ 3 \\ 0 \end{bmatrix}\right\}$.

Solution. By Theorem 6.21, the matrix is

$$\begin{bmatrix}1&1&1\\1&1&3\\1&0&0\end{bmatrix}^{-1}\begin{bmatrix}1&-4\\0&1\\3&1\end{bmatrix}\begin{bmatrix}1&1\\2&1\end{bmatrix}.$$

Since

$$\begin{bmatrix}1&-4\\0&1\\3&1\end{bmatrix}\begin{bmatrix}1&1\\2&1\end{bmatrix}=\begin{bmatrix}-7&-3\\2&1\\5&4\end{bmatrix}$$

and

$$\begin{bmatrix}1&1&1\\1&1&3\\1&0&0\end{bmatrix}^{-1}=\begin{bmatrix}0&0&1\\\frac{3}{2}&-\frac{1}{2}&-1\\-\frac{1}{2}&\frac{1}{2}&0\end{bmatrix},$$

we have

$$\begin{bmatrix}1&1&1\\1&1&3\\1&0&0\end{bmatrix}^{-1}\left(\begin{bmatrix}1&-4\\0&1\\3&1\end{bmatrix}\begin{bmatrix}1&1\\2&1\end{bmatrix}\right)=\begin{bmatrix}0&0&1\\\frac{3}{2}&-\frac{1}{2}&-1\\-\frac{1}{2}&\frac{1}{2}&0\end{bmatrix}\begin{bmatrix}-7&-3\\2&1\\5&4\end{bmatrix}=\begin{bmatrix}5&4\\-\frac{33}{2}&-9\\\frac{9}{2}&2\end{bmatrix}.$$

□

From Theorem 6.21 we easily obtain the following useful result.

Corollary 6.23. *Let $\mathscr{B}=\{\mathbf{u}_1,\dots,\mathbf{u}_n\}$ and $\mathscr{C}=\{\mathbf{v}_1,\dots,\mathbf{v}_m\}$ be bases for $\mathbb{R}^n$ and $\mathbb{R}^m$, respectively. If A is an $m\times n$ matrix and B is the matrix of the linear transformation $f(\mathbf{u})=A\mathbf{u}$ relative to the bases $\mathscr{B}$ and $\mathscr{C}$, then*

$$A=\begin{bmatrix}\mathbf{v}_1&\dots&\mathbf{v}_m\end{bmatrix}B\begin{bmatrix}\mathbf{u}_1&\dots&\mathbf{u}_n\end{bmatrix}^{-1}.$$

Example 6.24. Consider the basis $\mathscr{B}=\{\mathbf{u}_1,\mathbf{u}_2\}$ in $\mathbb{R}^2$ where

$$\mathbf{u}_1=\begin{bmatrix}1\\1\end{bmatrix}\quad\text{and}\quad\mathbf{u}_2=\begin{bmatrix}-1\\3\end{bmatrix}$$

and the basis $\mathscr{C}=\{\mathbf{v}_1,\mathbf{v}_2,\mathbf{v}_3\}$ in $\mathbb{R}^3$ where

$$\mathbf{v}_1=\begin{bmatrix}1\\1\\0\end{bmatrix},\quad\mathbf{v}_2=\begin{bmatrix}1\\2\\0\end{bmatrix},\quad\text{and}\quad\mathbf{v}_3=\begin{bmatrix}1\\1\\1\end{bmatrix}.$$

If $f : \mathbb{R}^2 \to \mathbb{R}^3$ is the linear transformation such that

$$f(\mathbf{u}_1) = 2\mathbf{v}_1 + \mathbf{v}_2 + \mathbf{v}_3 \quad \text{and} \quad f(\mathbf{u}_2) = \mathbf{v}_1 + 3\mathbf{v}_2 - \mathbf{v}_3,$$

find the standard matrix for f.

Solution. The matrix of the function f relative to the bases $\mathscr{B}$ and $\mathscr{C}$ is $\begin{bmatrix} 2 & 1 \\ 1 & 3 \\ 1 & -1 \end{bmatrix}$. If A is the standard matrix for f, then

$$\begin{bmatrix} 2 & 1 \\ 1 & 3 \\ 1 & -1 \end{bmatrix} = \begin{bmatrix} \mathbf{v}_1 & \mathbf{v}_2 & \mathbf{v}_3 \end{bmatrix}^{-1} A \begin{bmatrix} \mathbf{u}_1 & \mathbf{u}_2 \end{bmatrix}$$

and thus

$$A = \begin{bmatrix} \mathbf{v}_1 & \mathbf{v}_2 & \mathbf{v}_3 \end{bmatrix} \begin{bmatrix} 2 & 1 \\ 1 & 3 \\ 1 & -1 \end{bmatrix} \begin{bmatrix} \mathbf{u}_1 & \mathbf{u}_2 \end{bmatrix}^{-1} = \begin{bmatrix} 1 & 1 & 1 \\ 1 & 2 & 1 \\ 0 & 0 & 1 \end{bmatrix} \begin{bmatrix} 2 & 1 \\ 1 & 3 \\ 1 & -1 \end{bmatrix} \begin{bmatrix} 1 & -1 \\ 1 & 3 \end{bmatrix}^{-1} = \begin{bmatrix} \frac{9}{4} & \frac{7}{4} \\ \frac{9}{4} & \frac{11}{4} \\ 1 & 0 \end{bmatrix}.$$

□

Example 6.25. We consider the basis $\mathscr{B} = \{\mathbf{a}, \mathbf{b}, \mathbf{a} \times \mathbf{b}\}$ of $\mathbb{R}^3$. Find the $\mathscr{B}$-matrix of the projection on the vector plane Span$\{\mathbf{a}, \mathbf{b}\}$.

Solution. Since

$$\text{proj}_{\text{Span}\{\mathbf{a},\mathbf{b}\}}(\mathbf{a}) = \mathbf{a}, \quad \text{proj}_{\text{Span}\{\mathbf{a},\mathbf{b}\}}(\mathbf{b}) = \mathbf{b}, \quad \text{and} \quad \text{proj}_{\text{Span}\{\mathbf{a},\mathbf{b}\}}(\mathbf{a} \times \mathbf{b}) = \mathbf{0},$$

the $\mathscr{B}$-matrix of the projection on the vector plane Span$\{\mathbf{a}, \mathbf{b}\}$ is

$$\begin{bmatrix} 1 & 0 & 0 \\ 0 & 1 & 0 \\ 0 & 0 & 0 \end{bmatrix}.$$

□

Example 6.26. Let $\mathscr{B} = \{\mathbf{a}, \mathbf{b}, \mathbf{a} \times \mathbf{b}\}$ be a basis in $\mathbb{R}^3$. The reflection $\text{refl}_{\text{Span}\{\mathbf{a},\mathbf{b}\}}(\mathbf{x})$ of a point $\mathbf{x}$ about the vector plane Span$\{\mathbf{a}, \mathbf{b}\}$ is defined by the equality

$$\frac{1}{2}\left(\mathbf{x} + \text{refl}_{\text{Span}\{\mathbf{a},\mathbf{b}\}}(\mathbf{x})\right) = \text{proj}_{\text{Span}\{\mathbf{a},\mathbf{b}\}}(\mathbf{x}).$$

Find the $\mathscr{B}$-matrix of the reflection about the vector plane $\operatorname{Span}\{\mathbf{a},\mathbf{b}\}$.

Solution. Since

$$\operatorname{refl}_{\operatorname{Span}\{\mathbf{a},\mathbf{b}\}}(\mathbf{a}) = \mathbf{a}, \quad \operatorname{refl}_{\operatorname{Span}\{\mathbf{a},\mathbf{b}\}}(\mathbf{b}) = \mathbf{b}, \quad \text{and} \quad \operatorname{refl}_{\operatorname{Span}\{\mathbf{a},\mathbf{b}\}}(\mathbf{a}\times\mathbf{b}) = -\mathbf{a}\times\mathbf{b},$$

the $\mathscr{B}$-matrix of the reflection about the vector plane $\operatorname{Span}\{\mathbf{a},\mathbf{b}\}$ is

$$\begin{bmatrix} 1 & 0 & 0 \\ 0 & 1 & 0 \\ 0 & 0 & -1 \end{bmatrix}.$$

□

Example 6.27. Consider the orthonormal basis $\mathscr{B} = \{\mathbf{a}, \mathbf{n}\times\mathbf{a}, \mathbf{n}\}$ of $\mathbb{R}^3$. If $f : \mathbb{R}^3 \to \mathbb{R}^3$ is a linear transformation such that

$$f(\mathbf{a}) = \cos\alpha\,\mathbf{a} + \sin\alpha\,(\mathbf{n}\times\mathbf{a}), \quad f(\mathbf{n}\times\mathbf{a}) = -\sin\alpha\,\mathbf{a} + \cos\alpha\,(\mathbf{n}\times\mathbf{a}), \quad \text{and} \quad f(\mathbf{n}) = \mathbf{n},$$

find the $\mathscr{B}$-matrix of the function f and verify that f is the rotation about the axis $\operatorname{Span}\{\mathbf{n}\}$ by angle α.

Solution. The $\mathscr{B}$-matrix of the function f is

$$\begin{bmatrix} \cos\alpha & -\sin\alpha & 0 \\ \sin\alpha & \cos\alpha & 0 \\ 0 & 0 & 1 \end{bmatrix}.$$

Since $\{\mathbf{a}, \mathbf{n}\times\mathbf{a}, \mathbf{n}\}$ is a basis of $\mathbb{R}^3$, every vector $\mathbf{x}$ in $\mathbb{R}^3$ can be written as

$$\mathbf{x} = x\mathbf{a} + y(\mathbf{n}\times\mathbf{a}) + z\mathbf{n},$$

for some numbers x, y, z. Then

$$f(\mathbf{x}) = f(x\mathbf{a} + y(\mathbf{n}\times\mathbf{a}) + z\mathbf{n}) = (x\cos\alpha - y\sin\alpha)\mathbf{a} + (x\sin\alpha + y\cos\alpha)(\mathbf{n}\times\mathbf{a}) + z\mathbf{n}$$

and

$$\begin{aligned} &\cos\alpha\mathbf{x} + (1-\cos\alpha)(\mathbf{x}\cdot\mathbf{n})\mathbf{n} + \sin\alpha\,(\mathbf{n}\times\mathbf{x}) \\ &\cos\alpha(x\mathbf{a} + y(\mathbf{n}\times\mathbf{a}) + z\mathbf{n}) + (1-\cos\alpha)z\mathbf{n} + x\sin\alpha\,(\mathbf{n}\times\mathbf{a}) - y\sin\alpha\,\mathbf{a} \\ &= (x\cos\alpha - y\sin\alpha)\mathbf{a} + (x\sin\alpha + y\cos\alpha)(\mathbf{n}\times\mathbf{a}) + z\mathbf{n}. \end{aligned}$$

□

Example 6.28. Let A be an $n \times n$ matrix and let $\mathscr{P} = \{\mathbf{p}_1, \dots, \mathbf{p}_n\}$ be a basis for $\mathbb{R}^n$ and let $P = \begin{bmatrix} \mathbf{p}_1 & \dots & \mathbf{p}_n \end{bmatrix}$. Show that the $\mathscr{P}$-matrix of the linear transformation $f(\mathbf{x}) = A\mathbf{x}$, where $A = PDP^{-1}\mathbf{x}$ for some $n \times n$ matrix D, is D.

Solution.

$$P^{-1}AP = P^{-1}PDP^{-1}P = D.$$

□

Exercises 6.2

1. Find the $\mathscr{B}$-matrix for the linear transformation $f\left(\begin{bmatrix} x_1 \\ x_2 \end{bmatrix}\right) = \begin{bmatrix} x_1 - x_2 \\ 2x_1 + 3x_2 \end{bmatrix}$, where $\mathscr{B} = \left\{\begin{bmatrix} 2 \\ 1 \end{bmatrix}, \begin{bmatrix} 1 \\ -1 \end{bmatrix}\right\}$.

2. Find the $\mathscr{B}$-matrix for the linear transformation $f\left(\begin{bmatrix} x_1 \\ x_2 \end{bmatrix}\right) = \begin{bmatrix} x_1 - x_2 \\ 2x_1 + 3x_2 \end{bmatrix}$, where $\mathscr{B} = \left\{\begin{bmatrix} 1 \\ 1 \end{bmatrix}, \begin{bmatrix} 2 \\ 3 \end{bmatrix}\right\}$.

3. Find the $\mathscr{B}$-matrix for the linear transformation $f\left(\begin{bmatrix} x_1 \\ x_2 \end{bmatrix}\right) = \begin{bmatrix} 2x_1 + 5x_2 \\ x_1 + 7x_2 \end{bmatrix}$, where $\mathscr{B} = \left\{\begin{bmatrix} 1 \\ 1 \end{bmatrix}, \begin{bmatrix} 1 \\ 3 \end{bmatrix}\right\}$.

4. Find the $\mathscr{B}$-matrix for the linear transformation $f\left(\begin{bmatrix} x_1 \\ x_2 \end{bmatrix}\right) = \begin{bmatrix} 3x_1 \\ x_1 + x_2 \end{bmatrix}$, where $\mathscr{B} = \left\{\begin{bmatrix} 0 \\ 1 \end{bmatrix}, \begin{bmatrix} 1 \\ 1 \end{bmatrix}\right\}$.

5. Find the $\mathscr{B}$-matrix for the linear transformation $f\left(\begin{bmatrix} x_1 \\ x_2 \\ x_3 \end{bmatrix}\right) = \begin{bmatrix} x_1 \\ x_1 + x_2 \\ x_1 + x_2 + x_3 \end{bmatrix}$, where $\mathscr{B} = \left\{\begin{bmatrix} 0 \\ 1 \\ 0 \end{bmatrix}, \begin{bmatrix} 1 \\ 1 \\ 1 \end{bmatrix}, \begin{bmatrix} 0 \\ 1 \\ 1 \end{bmatrix}\right\}$.

6. Find the $\mathscr{B}$-matrix for the linear transformation $f\left(\begin{bmatrix} x_1 \\ x_2 \\ x_3 \end{bmatrix}\right) = \begin{bmatrix} x_1 - x_2 \\ x_3 \\ x_2 + 5x_3 \end{bmatrix}$, where $\mathscr{B} = \left\{\begin{bmatrix} 2 \\ 1 \\ 0 \end{bmatrix}, \begin{bmatrix} 1 \\ 1 \\ 2 \end{bmatrix}, \begin{bmatrix} 1 \\ 0 \\ 0 \end{bmatrix}\right\}$.

7. Let $\mathbf{u}_1 = \begin{bmatrix} 2 \\ 1 \end{bmatrix}$ and $\mathbf{u}_2 = \begin{bmatrix} 1 \\ 2 \end{bmatrix}$. Then $\mathscr{B} = \{\mathbf{u}_1, \mathbf{u}_2\}$ is a basis in $\mathbb{R}^2$. If $f : \mathbb{R}^2 \to \mathbb{R}^2$ is a linear transformation such that $f(\mathbf{u}_1) = 5\mathbf{u}_1 + 7\mathbf{u}_2$ and $f(\mathbf{u}_2) = \mathbf{u}_1 + 4\mathbf{u}_2$, find the standard matrix of f.

8. Let $\mathbf{u}_1 = \begin{bmatrix} 2 \\ 1 \end{bmatrix}$ and $\mathbf{u}_2 = \begin{bmatrix} 1 \\ 1 \end{bmatrix}$. Then $\mathscr{B} = \{\mathbf{u}_1, \mathbf{u}_2\}$ is a basis in $\mathbb{R}^2$. If $f : \mathbb{R}^2 \to \mathbb{R}^2$ is a linear transformation such that $f(\mathbf{u}_1) = 2\mathbf{u}_1$ and $f(\mathbf{u}_2) = 5\mathbf{u}_2$, find the standard matrix of f.

9. Let $\mathbf{u}_1 = \begin{bmatrix} 2 \\ 1 \\ 1 \end{bmatrix}$, $\mathbf{u}_2 = \begin{bmatrix} 1 \\ 1 \\ 2 \end{bmatrix}$, and $\mathbf{u}_3 = \begin{bmatrix} 1 \\ -3 \\ 1 \end{bmatrix}$. Then $\mathscr{B} = \{\mathbf{u}_1, \mathbf{u}_2, \mathbf{u}_3\}$ is a basis in $\mathbb{R}^3$. If $f : \mathbb{R}^3 \to \mathbb{R}^3$ is a linear transformation such that $f(\mathbf{u}_1) = 3\mathbf{u}_1$, $f(\mathbf{u}_2) = 4\mathbf{u}_2$, and $f(\mathbf{u}_3) = \mathbf{u}_3$, find the standard matrix of f.

10. Let $\mathbf{u}_1 = \begin{bmatrix} 1 \\ 1 \\ 2 \end{bmatrix}$, $\mathbf{u}_2 = \begin{bmatrix} 2 \\ 2 \\ 1 \end{bmatrix}$, and $\mathbf{u}_3 = \begin{bmatrix} 1 \\ 1 \\ 0 \end{bmatrix}$. Then $\mathscr{B} = \{\mathbf{u}_1, \mathbf{u}_2, \mathbf{u}_3\}$ is a basis in $\mathbb{R}^3$. If $f : \mathbb{R}^3 \to \mathbb{R}^3$ is a linear transformation such that $f(\mathbf{u}_1) = \mathbf{u}_1$, $f(\mathbf{u}_2) = 5\mathbf{u}_2$, and $f(\mathbf{u}_3) = 4\mathbf{u}_3$, find the standard matrix of f.

11. Let $\mathbf{u}_1 = \begin{bmatrix} 1 \\ 1 \\ 1 \end{bmatrix}$, $\mathbf{u}_2 = \begin{bmatrix} 2 \\ 1 \\ -3 \end{bmatrix}$, and $\mathbf{u}_3 = \begin{bmatrix} 4 \\ -5 \\ 1 \end{bmatrix}$. Then $\mathscr{B} = \{\mathbf{u}_1, \mathbf{u}_2, \mathbf{u}_3\}$ is a basis in $\mathbb{R}^3$. If $f : \mathbb{R}^3 \to \mathbb{R}^3$ is a linear transformation such that $f(\mathbf{u}_1) = 2\mathbf{u}_1, f(\mathbf{u}_2) = \mathbf{u}_2$, and $f(\mathbf{u}_3) = 2\mathbf{u}_3$, find the standard matrix of f. Explain why this matrix is symmetric.

12. Let $\mathbf{u}_1 = \begin{bmatrix} 1 \\ 0 \\ 1 \end{bmatrix}$, $\mathbf{u}_2 = \begin{bmatrix} 1 \\ 1 \\ 0 \end{bmatrix}$, and $\mathbf{u}_3 = \begin{bmatrix} 1 \\ 2 \\ 3 \end{bmatrix}$. Then $\mathscr{B} = \{\mathbf{u}_1, \mathbf{u}_2, \mathbf{u}_3\}$ is a basis in $\mathbb{R}^3$. If $f : \mathbb{R}^3 \to \mathbb{R}^3$ is a linear transformation such that $f(\mathbf{u}_1) = \mathbf{u}_1 + \mathbf{u}_2 + \mathbf{u}_3, f(\mathbf{u}_2) = \mathbf{u}_2 + \mathbf{u}_3$, and $f(\mathbf{u}_3) = \mathbf{u}_3$, find the standard matrix of f.

13. Find the matrix for the linear transformation $f\left(\begin{bmatrix} x_1 \\ x_2 \end{bmatrix}\right) = \begin{bmatrix} x_1 + x_2 \\ x_1 - x_2 \\ 2x_1 \end{bmatrix}$ relative to the bases $\mathscr{B} = \left\{\begin{bmatrix} 1 \\ 3 \end{bmatrix}, \begin{bmatrix} 1 \\ -1 \end{bmatrix}\right\}$ and $\mathscr{C} = \left\{\begin{bmatrix} 2 \\ 1 \\ 0 \end{bmatrix}, \begin{bmatrix} 1 \\ 2 \\ 1 \end{bmatrix}, \begin{bmatrix} 0 \\ 0 \\ 1 \end{bmatrix}\right\}$.

14. Find the matrix for the linear transformation $f\left(\begin{bmatrix} x_1 \\ x_2 \\ x_3 \end{bmatrix}\right) = \begin{bmatrix} 5x_1 + x_2 + x_3 \\ x_1 - x_3 \end{bmatrix}$ relative to the bases $\mathscr{B} = \left\{\begin{bmatrix} 1 \\ 1 \\ 0 \end{bmatrix}, \begin{bmatrix} 1 \\ 0 \\ 1 \end{bmatrix}, \begin{bmatrix} 1 \\ 1 \\ 1 \end{bmatrix}\right\}$ and $\mathscr{C} = \left\{\begin{bmatrix} 2 \\ 1 \end{bmatrix}, \begin{bmatrix} 1 \\ 2 \end{bmatrix}\right\}$.

15. Consider the basis $\mathscr{B} = \{\mathbf{u}_1, \mathbf{u}_2\}$ in $\mathbb{R}^2$, where $\mathbf{u}_1 = \begin{bmatrix} 1 \\ 1 \end{bmatrix}$ and $\mathbf{u}_2 = \begin{bmatrix} 2 \\ 3 \end{bmatrix}$, and the basis $\mathscr{C} = \{\mathbf{v}_1, \mathbf{v}_2, \mathbf{v}_3\}$ in $\mathbb{R}^3$, where $\mathbf{v}_1 = \begin{bmatrix} 2 \\ 1 \\ 1 \end{bmatrix}$, $\mathbf{v}_2 = \begin{bmatrix} 1 \\ 2 \\ -2 \end{bmatrix}$, and $\mathbf{v}_3 = \begin{bmatrix} 0 \\ 1 \\ 1 \end{bmatrix}$. If $f : \mathbb{R}^2 \to \mathbb{R}^3$ is a linear transformation such that $f(\mathbf{u}_1) = \mathbf{v}_1 + 3\mathbf{v}_2 + 2\mathbf{v}_3$ and $f(\mathbf{u}_2) = 2\mathbf{v}_1 + \mathbf{v}_2 - \mathbf{v}_3$, find the standard matrix of f.

16. Consider the basis $\mathscr{B} = \{\mathbf{u}_1, \mathbf{u}_2\}$ in $\mathbb{R}^2$, where $\mathbf{u}_1 = \begin{bmatrix} 1 \\ 1 \end{bmatrix}$ and $\mathbf{u}_2 = \begin{bmatrix} -1 \\ 3 \end{bmatrix}$, and the basis $\mathscr{C} = \{\mathbf{v}_1, \mathbf{v}_2, \mathbf{v}_3\}$ in $\mathbb{R}^3$, where $\mathbf{v}_1 = \begin{bmatrix} 1 \\ 1 \\ 0 \end{bmatrix}, \begin{bmatrix} 1 \\ 2 \\ 0 \end{bmatrix}$, and $\mathbf{v}_3 = \begin{bmatrix} 1 \\ 1 \\ 1 \end{bmatrix}$. If $f : \mathbb{R}^2 \to \mathbb{R}^3$ is the linear transformation such that $f(\mathbf{u}_1) = 2\mathbf{v}_1 + \mathbf{v}_2 + \mathbf{v}_3$ and $f(\mathbf{u}_2) = \mathbf{v}_1 + 3\mathbf{v}_2 - \mathbf{v}_3$, find the standard matrix of f.

17. Consider the basis $\mathscr{B} = \{\mathbf{u}_1, \mathbf{u}_2, \mathbf{u}_3\}$ in $\mathbb{R}^3$, where $\mathbf{u}_1 = \begin{bmatrix} 1 \\ 1 \\ 1 \end{bmatrix}$, $\mathbf{u}_2 = \begin{bmatrix} 2 \\ 1 \\ -3 \end{bmatrix}$, and $\mathbf{u}_3 = \begin{bmatrix} 4 \\ -1 \\ 1 \end{bmatrix}$, and the basis $\mathscr{C} = \{\mathbf{v}_1, \mathbf{v}_2\}$ in $\mathbb{R}^2$, where $\mathbf{v}_1 = \begin{bmatrix} 2 \\ 1 \end{bmatrix}$ and $\mathbf{v}_2 = \begin{bmatrix} -1 \\ 2 \end{bmatrix}$. If $f : \mathbb{R}^3 \to \mathbb{R}^2$ is a linear transformation such that $f(\mathbf{u}_1) = 2\mathbf{v}_1$, $f(\mathbf{u}_2) = 7\mathbf{v}_2$, and $f(\mathbf{u}_3) = \mathbf{v}_1 + \mathbf{v}_2$, find the standard matrix of f.

18. Find the matrix for the linear transformation $f\left(\begin{bmatrix} x_1 \\ x_2 \end{bmatrix}\right) = \begin{bmatrix} x_1 + x_2 \\ x_1 - x_2 \\ x_1 \\ x_2 \end{bmatrix}$ relative to the bases $\mathscr{B} = \left\{\begin{bmatrix} 1 \\ 1 \end{bmatrix}, \begin{bmatrix} 1 \\ 0 \end{bmatrix}\right\}$ and $\mathscr{C} = \left\{\begin{bmatrix} 1 \\ 0 \\ 0 \\ 0 \end{bmatrix}, \begin{bmatrix} 0 \\ 1 \\ 0 \\ 0 \end{bmatrix}, \begin{bmatrix} 1 \\ 1 \\ 1 \\ 1 \end{bmatrix}, \begin{bmatrix} 0 \\ 0 \\ 0 \\ 1 \end{bmatrix}\right\}$.

Chapter 7

Jordan forms by examples

In Chapter 5 we studied matrices A that can be written in the form $A = PDP^{-1}$, where P is an invertible matrix and D is a diagonal matrix. Such a representation of a matrix can be very useful, but in applications we often encounter matrices that cannot be diagonalized in this way. In this chapter we present a method for representing matrices in a form similar to diagonalization that can be used for matrices that are not diagonalizable. We consider 2×2, 3×3, and 4×4 matrices. The idea presented here applies to larger matrices, but the general case will not be discussed in this book. Understanding the presented results and examples should provide a solid foundation for study of the general method.

Theorems and examples presented in this chapter give us an opportunity to use almost everything we learned so far, including subspaces, linear independence of vectors, basis for a subspace, dimension of a subspace, and the matrix of a linear transformation. While learning new ideas, we have a chance to review the material covered in previous chapters and deepen our understanding.

Because the main goal of this chapter is to practice linear algebra suitable for a first course in the subject we keep the theory to the minimum and present only particular cases of Jordan forms.

First we consider 2×2 matrices. Here the situation is relatively simple.

Theorem 7.1. *Let A be a 2×2 matrix and let α be a real number. If the characteristic polynomial of the matrix A is $(\lambda - \alpha)^2$ and*

$$\dim \mathbf{N}(A - \alpha I_2) = 1 \quad \textit{and} \quad \dim \mathbf{N}((A - \alpha I_2)^2) = 2,$$

then there is a basis $\mathscr{P} = \{\mathbf{p}_1, \mathbf{p}_2\}$ in $\mathbb{R}^2$ such that the $\mathscr{P}$-matrix of the linear transformation $f(\mathbf{x}) = A\mathbf{x}$ is $\begin{bmatrix} \alpha & 1 \\ 0 & \alpha \end{bmatrix}$ and consequently

$$A = \begin{bmatrix} \mathbf{p}_1 & \mathbf{p}_2 \end{bmatrix} \begin{bmatrix} \alpha & 1 \\ 0 & \alpha \end{bmatrix} \begin{bmatrix} \mathbf{p}_1 & \mathbf{p}_2 \end{bmatrix}^{-1}.$$

Proof. Let $\mathbf{u}$ be a vector in $\mathbf{N}((A - \alpha I_2)^2) = \mathbb{R}^2$ that is not in $\mathbf{N}(A - \alpha I_2)$, that is, the vector $\mathbf{u}$ satisfies $(A - \alpha I_2)^2 \mathbf{u} = \mathbf{0}$ and $(A - \alpha I_2)\mathbf{u} \neq \mathbf{0}$. Then the nonzero vector $(A - \alpha I_2)\mathbf{u}$ is in $\mathbf{N}(A - \alpha I_2)$ because $(A - \alpha I_2)(A - \alpha I_2)\mathbf{u} = (A - \alpha I_2)^2 \mathbf{u} = \mathbf{0}$, that is, the vector $(A - \alpha I_2)\mathbf{u}$ is an eigenvector corresponding to the eigenvalue α. We will show that the set $\mathscr{P} = \{(A - \alpha I_2)\mathbf{u}, \mathbf{u}\}$ has the desired properties.

To show that the vectors $(A - \alpha I_2)\mathbf{u}$ and $\mathbf{u}$ are linearly independent we assume that for some x_1 and x_2 we have

$$x_1(A - \alpha I_2)\mathbf{u} + x_2 \mathbf{u} = \mathbf{0}.$$

Then, since $(A - \alpha I_2)^2 \mathbf{u} = 0$, we get

$$x_1(A - \alpha I_2)^2 \mathbf{u} + x_2(A - \alpha I_2)\mathbf{u} = x_2(A - \alpha I_2)\mathbf{u} = \mathbf{0}$$

and thus $x_2 = 0$. Now $x_1 = 0$ follows because $(A - \alpha I_2)\mathbf{u} \neq \mathbf{0}$. Since the vectors $(A - \alpha I_2)\mathbf{u}$ and $\mathbf{u}$ are linearly independent, $\{(A - \alpha I_2)\mathbf{u}, \mathbf{u}\}$ is a basis in $\mathbb{R}^2$.

Now we calculate

$$A(A - \alpha I_2)\mathbf{u} = (A - \alpha I_2)^2 \mathbf{u} + \alpha(A - \alpha I_2)\mathbf{u} = \alpha(A - \alpha I_2)\mathbf{u}$$
$$A\mathbf{u} = (A - \alpha I_2)\mathbf{u} + \alpha \mathbf{u}.$$

If we let

$$\mathbf{p}_1 = (A - \alpha I_2)\mathbf{u} \quad \text{and} \quad \mathbf{p}_2 = \mathbf{u},$$

then we have

$$A\mathbf{p}_1 = \alpha \mathbf{p}_1$$
$$A\mathbf{p}_2 = \mathbf{p}_1 + \alpha \mathbf{p}_2.$$

Consequently, the $\{(A - \alpha I_2)\mathbf{u}, \mathbf{u}\}$-matrix of the linear transformation $f(\mathbf{x}) = A\mathbf{x}$ is $\begin{bmatrix} \alpha & 1 \\ 0 & \alpha \end{bmatrix}$ and, by Corollary 6.23, we have

$$A = \begin{bmatrix} \mathbf{p}_1 & \mathbf{p}_2 \end{bmatrix} \begin{bmatrix} \alpha & 1 \\ 0 & \alpha \end{bmatrix} \begin{bmatrix} \mathbf{p}_1 & \mathbf{p}_2 \end{bmatrix}^{-1}.$$

□

Definition 7.2. A 2×2 matrix is called a *standard Jordan canonical form* if it is a diagonal matrix or has the form

$$\begin{bmatrix} \alpha & 1 \\ 0 & \alpha \end{bmatrix}$$

for some real number α.

Example 7.3. Let $A = \begin{bmatrix} -3 & 9 \\ -1 & 3 \end{bmatrix}$. Find an invertible 2×2 matrix P and a standard Jordan canonical form J such that $A = PJP^{-1}$.

Solution. The characteristic polynomial of A is λ^2 and thus $\alpha = 0$. Since

$$\mathbf{N}(A - 0I_2) = \mathbf{N}\left(\begin{bmatrix} -3 & 9 \\ -1 & 3 \end{bmatrix}\right) = \text{Span}\left\{\begin{bmatrix} 3 \\ 1 \end{bmatrix}\right\}$$

and

$$\mathbf{N}((A - 0I_2)^2) = \mathbf{N}\left(\begin{bmatrix} 0 & 0 \\ 0 & 0 \end{bmatrix}\right) = \mathbb{R}^2,$$

we can use Theorem 7.1. We note that the vector $\begin{bmatrix} 1 \\ 0 \end{bmatrix}$ is not in $\text{Span}\left\{\begin{bmatrix} 3 \\ 1 \end{bmatrix}\right\}$ and

$$\begin{bmatrix} -3 & 9 \\ -1 & 3 \end{bmatrix}\begin{bmatrix} 1 \\ 0 \end{bmatrix} = \begin{bmatrix} -3 \\ -1 \end{bmatrix},$$

so we can take $P = \begin{bmatrix} -3 & 1 \\ -1 & 0 \end{bmatrix}$ and $J = \begin{bmatrix} 0 & 1 \\ 0 & 0 \end{bmatrix}$. □

Now we consider 3×3 matrices. The situation for such matrices is more complicated and we need to consider three different scenarios.

Theorem 7.4. *Let A be a 3×3 matrix and let α be a real number. If the characteristic polynomial of the matrix A is $-(\lambda-\alpha)^3$ and we have*

$$\dim \mathbf{N}(A-\alpha I_3)=1, \quad \dim \mathbf{N}((A-\alpha I_3)^2)=2, \quad \dim \mathbf{N}((A-\alpha I_3)^3)=3,$$

then there is a basis $\mathscr{P}=\{\mathbf{p}_1,\mathbf{p}_2,\mathbf{p}_3\}$ of $\mathbb{R}^3$ such that the $\mathscr{P}$-matrix of the linear transformation $f(\mathbf{x})=A\mathbf{x}$ is $\begin{bmatrix}\alpha & 1 & 0\\ 0 & \alpha & 1\\ 0 & 0 & \alpha\end{bmatrix}$ and consequently

$$A=\begin{bmatrix}\mathbf{p}_1 & \mathbf{p}_2 & \mathbf{p}_3\end{bmatrix}\begin{bmatrix}\alpha & 1 & 0\\ 0 & \alpha & 1\\ 0 & 0 & \alpha\end{bmatrix}\begin{bmatrix}\mathbf{p}_1 & \mathbf{p}_2 & \mathbf{p}_3\end{bmatrix}^{-1}.$$

Proof. Let $\mathbf{u}$ be a vector in $\mathbf{N}((A-\alpha I_3)^3)=\mathbb{R}^3$ which is not in $\mathbf{N}((A-\alpha I_3)^2)$. Then the nonzero vector $(A-\alpha I_3)\mathbf{u}$ is in $\mathbf{N}((A-\alpha I_3)^2)$ and the nonzero vector $(A-\alpha I_3)^2\mathbf{u}$ is in $\mathbf{N}(A-\alpha I_3)$. We will show that the set

$$\mathscr{P}=\{(A-\alpha I_3)^2\mathbf{u},(A-\alpha I_3)\mathbf{u},\mathbf{u}\}$$

is a basis with the desired properties.

To show that the vectors $(A-\alpha I_3)^2\mathbf{u}$, $(A-\alpha I_3)\mathbf{u}$, and $\mathbf{u}$ are linearly independent we assume that for some real numbers x_1,x_2,x_3 we have

$$x_1(A-\alpha I_3)^2\mathbf{u}+x_2(A-\alpha I_3)\mathbf{u}+x_3\mathbf{u}=\mathbf{0}.$$

Then

$$(A-\alpha I_3)^2\left(x_1(A-\alpha I_3)^2\mathbf{u}+x_2(A-\alpha I_3)\mathbf{u}+x_3\mathbf{u}\right)=x_3(A-\alpha I_3)^2\mathbf{u}=\mathbf{0}$$

and consequently $x_3=0$. Now, since

$$(A-\alpha I_3)\left(x_1(A-\alpha I_3)^2\mathbf{u}+x_2(A-\alpha I_3)\mathbf{u}\right)=x_2(A-\alpha I_3)^2\mathbf{u}=\mathbf{0},$$

we can conclude that $x_2=0$ and then that $x_1=0$. Since the vectors $(A-\alpha I_3)^2\mathbf{u}$, $(A-\alpha I_3)\mathbf{u}$, and $\mathbf{u}$ are linearly independent, the set $\{(A-\alpha I_3)^2\mathbf{u},(A-\alpha I_3)\mathbf{u},\mathbf{u}\}$ is a basis in $\mathbb{R}^3$.

We define

$$\mathbf{p}_1=(A-\alpha I_3)^2\mathbf{u}, \quad \mathbf{p}_2=(A-\alpha I_3)\mathbf{u}, \quad \text{and} \quad \mathbf{p}_3=\mathbf{u}.$$

Since

$$\begin{aligned}
A\mathbf{p}_1 &= A(A-\alpha I_3)^2\mathbf{u}=(A-\alpha I_3)^3\mathbf{u}+\alpha(A-\alpha I_3)^2\mathbf{u}=\alpha(A-\alpha I_3)^2=\alpha\mathbf{p}_1\\
A\mathbf{p}_2 &= A(A-\alpha I_3)\mathbf{u}=(A-\alpha I_3)^2\mathbf{u}+\alpha(A-\alpha I_3)\mathbf{u}=\mathbf{p}_1+\alpha\mathbf{p}_2\\
A\mathbf{p}_3 &= A\mathbf{u}=(A-\alpha I_3)\mathbf{u}+\alpha\mathbf{u}=\mathbf{p}_2+\alpha\mathbf{p}_3,
\end{aligned}$$

the $\{\mathbf{p}_1, \mathbf{p}_2, \mathbf{p}_3\}$-matrix of the linear transformation $f(\mathbf{x}) = A\mathbf{x}$ is $\begin{bmatrix} \alpha & 1 & 0 \\ 0 & \alpha & 1 \\ 0 & 0 & \alpha \end{bmatrix}$ and thus

$$A = \begin{bmatrix} \mathbf{p}_1 & \mathbf{p}_2 & \mathbf{p}_3 \end{bmatrix} \begin{bmatrix} \alpha & 1 & 0 \\ 0 & \alpha & 1 \\ 0 & 0 & \alpha \end{bmatrix} \begin{bmatrix} \mathbf{p}_1 & \mathbf{p}_2 & \mathbf{p}_3 \end{bmatrix}^{-1},$$

by Corollary 6.23. □

Theorem 7.5. *Let A be a 3×3 matrix and let α be a real number. If the characteristic polynomial of the matrix A is $-(\lambda - \alpha)^3$ and we have*

$$\dim \mathbf{N}(A - \alpha I_3) = 2 \quad \text{and} \quad \dim \mathbf{N}((A - \alpha I_3)^2) = 3,$$

then there is a basis $\mathscr{P} = \{\mathbf{p}_1, \mathbf{p}_2, \mathbf{p}_3\}$ of $\mathbb{R}^3$ such that the $\mathscr{P}$-matrix of the linear transformation $f(\mathbf{x}) = A\mathbf{x}$ is $\begin{bmatrix} \alpha & 1 & 0 \\ 0 & \alpha & 0 \\ 0 & 0 & \alpha \end{bmatrix}$ and consequently

$$A = \begin{bmatrix} \mathbf{p}_1 & \mathbf{p}_2 & \mathbf{p}_3 \end{bmatrix} \begin{bmatrix} \alpha & 1 & 0 \\ 0 & \alpha & 0 \\ 0 & 0 & \alpha \end{bmatrix} \begin{bmatrix} \mathbf{p}_1 & \mathbf{p}_2 & \mathbf{p}_3 \end{bmatrix}^{-1}.$$

Proof. Let $\mathbf{u}$ be a vector in $\mathbf{N}((A - \alpha I_3)^2)$ which is not in $\mathbf{N}(A - \alpha I_3)$. Then the nonzero vector $(A - \alpha I_3)\mathbf{u}$ is in $\mathbf{N}(A - \alpha I_3)$. Since $\dim \mathbf{N}(A - \alpha I_3) = 2$, we can choose a vector $\mathbf{v}$ in $\mathbf{N}(A - \alpha I_3)$ such that $\{(A - \alpha I_3)\mathbf{u}, \mathbf{v}\}$ is a basis of $\mathbf{N}(A - \alpha I_3)$. We will show that

$$\mathscr{P} = \{(A - \alpha I_3)\mathbf{u}, \mathbf{u}, \mathbf{v}\}$$

is a basis with the desired properties.

If

$$x_1(A - \alpha I_3)\mathbf{u} + x_2\mathbf{u} + x_3\mathbf{v} = \mathbf{0}$$

for some real numbers x_1, x_2, x_3, then

$$(A - \alpha I_3)(x_1(A - \alpha I_3)\mathbf{u} + x_2\mathbf{u} + x_3\mathbf{v}) = x_2(A - \alpha I_3)\mathbf{u} = \mathbf{0}$$

and consequently $x_2 = 0$. Now, since

$$x_1(A - \alpha I_3)\mathbf{u} + x_3\mathbf{v} = \mathbf{0},$$

we must have $x_1 = x_3 = 0$, because the vectors $(A - \alpha I_5)\mathbf{u}$ and $\mathbf{v}$ are linearly independent. This shows that the vectors $(A - \alpha I_3)\mathbf{u}$, $\mathbf{u}$ and $\mathbf{v}$ are linearly independent and consequently $\{(A - \alpha I_3)\mathbf{u}, \mathbf{u}, \mathbf{v}\}$ is a basis in $\mathbb{R}^3$.

For

$$\mathbf{p}_1 = (A - \alpha I_3)\mathbf{u}, \quad \mathbf{p}_2 = \mathbf{u}, \quad \text{and} \quad \mathbf{p}_3 = \mathbf{v}$$

we have

$$\begin{aligned} A\mathbf{p}_1 &= A(A-\alpha I_3)\mathbf{u} = (A-\alpha I_2)^2\mathbf{u} + \alpha(A-\alpha I_3)\mathbf{u} = \alpha(A-\alpha I_3)\mathbf{u} = \alpha\mathbf{p}_1 \\ A\mathbf{p}_2 &= A\mathbf{u} = (A-\alpha I_3)\mathbf{u} + \alpha\mathbf{u} = \mathbf{p}_1 + \alpha\mathbf{p}_2 \\ A\mathbf{p}_3 &= A\mathbf{v} = (A-\alpha I_3)\mathbf{v} + \alpha\mathbf{v} = \alpha\mathbf{v} = \alpha\mathbf{p}_3 \end{aligned}$$

which means that the $\{(A-\alpha I_3)\mathbf{u}, \mathbf{u}, \mathbf{v}\}$-matrix of the linear transformation $f(\mathbf{x}) = A\mathbf{x}$ is $\begin{bmatrix} \alpha & 1 & 0 \\ 0 & \alpha & 0 \\ 0 & 0 & \alpha \end{bmatrix}$. Consequently,

$$A = \begin{bmatrix} \mathbf{p}_1 & \mathbf{p}_2 & \mathbf{p}_3 \end{bmatrix} \begin{bmatrix} \alpha & 1 & 0 \\ 0 & \alpha & 0 \\ 0 & 0 & \alpha \end{bmatrix} \begin{bmatrix} \mathbf{p}_1 & \mathbf{p}_2 & \mathbf{p}_3 \end{bmatrix}^{-1},$$

by Corollary 6.23. □

Theorem 7.6. *Let A be a 3×3 matrix and let α and β be two distinct real numbers. If the characteristic polynomial of the matrix A is $-(\lambda-\alpha)^2(\lambda-\beta)$ and we have*

$$\dim \mathbf{N}(A-\alpha I_3) = 1, \quad \dim \mathbf{N}((A-\alpha I_3)^2) = 2, \quad \dim \mathbf{N}(A-\beta I_3) = 1,$$

then there is a basis $\mathscr{P} = \{\mathbf{p}_1, \mathbf{p}_2, \mathbf{p}_3\}$ of $\mathbb{R}^3$ such that the $\mathscr{P}$-matrix of the linear transformation $f(\mathbf{x}) = A\mathbf{x}$ is $\begin{bmatrix} \alpha & 1 & 0 \\ 0 & \alpha & 0 \\ 0 & 0 & \beta \end{bmatrix}$ and consequently

$$A = \begin{bmatrix} \mathbf{p}_1 & \mathbf{p}_2 & \mathbf{p}_3 \end{bmatrix} \begin{bmatrix} \alpha & 1 & 0 \\ 0 & \alpha & 0 \\ 0 & 0 & \beta \end{bmatrix} \begin{bmatrix} \mathbf{p}_1 & \mathbf{p}_2 & \mathbf{p}_3 \end{bmatrix}^{-1}.$$

Proof. Let $\mathbf{u}$ be a vector in $\mathbf{N}((A-\alpha I_3)^2)$ that is not in $\mathbf{N}(A-\alpha I_3)$ and let $\mathbf{v}$ be a nonzero vector in $\mathbf{N}(A-\beta I_3)$. Then the vectors $(A-\alpha I_3)\mathbf{u}$ and $\mathbf{u}$ are linearly independent, as in the proof of Theorem 7.1, and the vector $(A-\alpha I_3)\mathbf{u}$ is in $\mathbf{N}(A-\alpha I_3)$. We will show that

$$\mathscr{P} = \{(A-\alpha I_3)\mathbf{u}, \mathbf{u}, \mathbf{v}\}$$

is a basis with the desired properties.

If

$$x_1(A-\alpha I_3)\mathbf{u} + x_2\mathbf{u} + x_3\mathbf{v} = \mathbf{0},$$

for some real numbers x_1, x_2, x_3, then

$$(A-\alpha I_3)^2(x_1(A-\alpha I_3)\mathbf{u} + x_2\mathbf{u} + x_3\mathbf{v}) = x_3((A-\alpha I_3)^2)\mathbf{v} = \mathbf{0},$$

which gives us $x_3 = 0$, because

$$(A-\alpha I_3)^2\mathbf{v} = (A-\beta I_3)^2\mathbf{v} + 2(\beta-\alpha)(A-\beta I_3)\mathbf{v} + (\beta-\alpha)^2\mathbf{v} = (\beta-\alpha)^2\mathbf{v} \neq \mathbf{0}.$$

Now, since $x_1(A-\alpha I_3)\mathbf{u} + x_2\mathbf{u} = \mathbf{0}$ and the vectors $(A-\alpha I_3)\mathbf{u}$ and $\mathbf{u}$ are linearly independent, we get $x_1 = x_2 = 0$. This shows that the vectors $(A-\alpha I_3)\mathbf{u}$, $\mathbf{u}$ and $\mathbf{v}$ are linearly independent and consequently the set $\{(A-\alpha I_3)\mathbf{u}, \mathbf{u}, \mathbf{v}\}$ is a basis in $\mathbb{R}^3$.

For

$$\mathbf{p}_1 = (A-\alpha I_3)\mathbf{u}, \quad \mathbf{p}_2 = \mathbf{u}, \quad \text{and} \quad \mathbf{p}_3 = \mathbf{v}$$

we have

$$\begin{aligned} A\mathbf{p}_1 &= A(A-\alpha I_3)\mathbf{u} = (A-\alpha I_3)^2\mathbf{u} + \alpha(A-\alpha I_3)\mathbf{u} = \alpha(A-\alpha I_3)\mathbf{u} = \alpha\mathbf{p}_1 \\ A\mathbf{p}_2 &= A\mathbf{u} = (A-\alpha I_3)\mathbf{u} + \alpha\mathbf{u} = \mathbf{p}_1 + \alpha\mathbf{p}_2 \\ A\mathbf{p}_3 &= A\mathbf{v} = \beta\mathbf{v} = \beta\mathbf{p}_3 \end{aligned}$$

which means that the $\{(A-\alpha I_3)\mathbf{u}, \mathbf{u}, \mathbf{v}\}$-matrix of the linear transformation $f(\mathbf{x}) = A\mathbf{x}$ is $\begin{bmatrix} \alpha & 1 & 0 \\ 0 & \alpha & 0 \\ 0 & 0 & \beta \end{bmatrix}$. Consequently,

$$A = \begin{bmatrix} \mathbf{p}_1 & \mathbf{p}_2 & \mathbf{p}_3 \end{bmatrix} \begin{bmatrix} \alpha & 1 & 0 \\ 0 & \alpha & 0 \\ 0 & 0 & \beta \end{bmatrix} \begin{bmatrix} \mathbf{p}_1 & \mathbf{p}_2 & \mathbf{p}_3 \end{bmatrix}^{-1},$$

by Corollary 6.23. □

Definition 7.7. A 3×3 matrix is called a *standard Jordan canonical form* if it is a diagonal matrix or has one of the following three forms

$$\begin{bmatrix} \alpha & 1 & 0 \\ 0 & \alpha & 1 \\ 0 & 0 & \alpha \end{bmatrix}, \quad \begin{bmatrix} \alpha & 1 & 0 \\ 0 & \alpha & 0 \\ 0 & 0 & \alpha \end{bmatrix}, \quad \begin{bmatrix} \alpha & 1 & 0 \\ 0 & \alpha & 0 \\ 0 & 0 & \beta \end{bmatrix},$$

where α and β are distinct real numbers.

Note that, if

$$A = \begin{bmatrix} \mathbf{p}_1 & \mathbf{p}_2 & \mathbf{p}_3 \end{bmatrix} \begin{bmatrix} \alpha & 1 & 0 \\ 0 & \alpha & 0 \\ 0 & 0 & \alpha \end{bmatrix} \begin{bmatrix} \mathbf{p}_1 & \mathbf{p}_2 & \mathbf{p}_3 \end{bmatrix}^{-1},$$

then

$$A = \begin{bmatrix} \mathbf{p}_3 & \mathbf{p}_1 & \mathbf{p}_2 \end{bmatrix} \begin{bmatrix} \alpha & 0 & 0 \\ 0 & \alpha & 1 \\ 0 & 0 & \alpha \end{bmatrix} \begin{bmatrix} \mathbf{p}_3 & \mathbf{p}_1 & \mathbf{p}_2 \end{bmatrix}^{-1}.$$

Both matrices $\begin{bmatrix} \alpha & 1 & 0 \\ 0 & \alpha & 0 \\ 0 & 0 & \alpha \end{bmatrix}$ and $\begin{bmatrix} \alpha & 0 & 0 \\ 0 & \alpha & 1 \\ 0 & 0 & \alpha \end{bmatrix}$ are called Jordan canonical forms, but only the first one is called a standard Jordan canonical form. The same is true for matrices $\begin{bmatrix} \alpha & 1 & 0 \\ 0 & \alpha & 0 \\ 0 & 0 & \beta \end{bmatrix}$ and $\begin{bmatrix} \beta & 0 & 0 \\ 0 & \alpha & 1 \\ 0 & 0 & \alpha \end{bmatrix}$. It should be clear that standard Jordan canonical forms are sufficient for the purpose of representing a matrix A in the form $A = PJP^{-1}$, where P is an invertible matrix and J is a Jordan canonical form. In all examples in this book we use only standard Jordan canonical forms.

Example 7.8. The characteristic polynomial of the matrix

$$A = \begin{bmatrix} -2 & 4 & -1 \\ -1 & 1 & 1 \\ 0 & 1 & -2 \end{bmatrix}$$

is $-(\lambda+1)^3$. Find an invertible 3×3 matrix P and a 3×3 standard Jordan canonical form J such that $A = PJP^{-1}$.

Solution. First we find that

$$\mathbf{N}(A+I_3) = \mathbf{N}\left(\begin{bmatrix} -1 & 4 & -1 \\ -1 & 2 & 1 \\ 0 & 1 & -1 \end{bmatrix}\right) = \operatorname{Span}\left\{\begin{bmatrix} 3 \\ 1 \\ 1 \end{bmatrix}\right\},$$

$$\mathbf{N}((A+I_3)^2) = \mathbf{N}\left(\begin{bmatrix} -3 & 3 & 6 \\ -1 & 1 & 2 \\ -1 & 1 & 2 \end{bmatrix}\right) = \operatorname{Span}\left\{\begin{bmatrix} 1 \\ 1 \\ 0 \end{bmatrix}, \begin{bmatrix} 2 \\ 0 \\ 1 \end{bmatrix}\right\},$$

$$\mathbf{N}((A+I_3)^3) = \mathbb{R}^3,$$

which means that we can use Theorem 7.4.

Since the vector $\begin{bmatrix} 0 \\ 1 \\ 0 \end{bmatrix}$ is not in $\operatorname{Span}\left\{\begin{bmatrix} 1 \\ 1 \\ 0 \end{bmatrix}, \begin{bmatrix} 2 \\ 0 \\ 1 \end{bmatrix}\right\}$ and we have

$$\begin{bmatrix} -1 & 4 & -1 \\ -1 & 2 & 1 \\ 0 & 1 & -1 \end{bmatrix}\begin{bmatrix} 0 \\ 1 \\ 0 \end{bmatrix} = \begin{bmatrix} 4 \\ 2 \\ 1 \end{bmatrix}$$

and

$$\begin{bmatrix} -3 & 3 & 6 \\ -1 & 1 & 2 \\ -1 & 1 & 2 \end{bmatrix}\begin{bmatrix} 0 \\ 1 \\ 0 \end{bmatrix} = \begin{bmatrix} 3 \\ 1 \\ 1 \end{bmatrix},$$

we can take $P = \begin{bmatrix} 3 & 4 & 0 \\ 1 & 2 & 1 \\ 1 & 1 & 0 \end{bmatrix}$ and $J = \begin{bmatrix} -1 & 1 & 0 \\ 0 & -1 & 1 \\ 0 & 0 & -1 \end{bmatrix}$. □

Example 7.9. The characteristic polynomial of the matrix

$$A = \begin{bmatrix} 4 & -1 & 0 \\ 1 & 2 & 0 \\ 2 & -2 & 3 \end{bmatrix}$$

is $-(\lambda - 3)^3$. Find an invertible 3×3 matrix P and a 3×3 standard Jordan canonical form J such that $A = PJP^{-1}$.

Solution. Since

$$\mathbf{N}(A - 3I_3) = \mathbf{N}\left(\begin{bmatrix} 1 & -1 & 0 \\ 1 & -1 & 0 \\ 2 & -2 & 0 \end{bmatrix}\right) = \operatorname{Span}\left\{\begin{bmatrix} 1 \\ 1 \\ 0 \end{bmatrix}, \begin{bmatrix} 0 \\ 0 \\ 1 \end{bmatrix}\right\}$$

and

$$\mathbf{N}((A - 3I_3)^2) = \mathbb{R}^3,$$

we can use Theorem 7.5.

We note that the vector $\begin{bmatrix} 1 \\ 0 \\ 0 \end{bmatrix}$ is not in $\operatorname{Span}\left\{\begin{bmatrix} 1 \\ 1 \\ 0 \end{bmatrix}, \begin{bmatrix} 0 \\ 0 \\ 1 \end{bmatrix}\right\}$ and we have

$$\begin{bmatrix} 1 & -1 & 0 \\ 1 & -1 & 0 \\ 2 & -2 & 0 \end{bmatrix}\begin{bmatrix} 1 \\ 0 \\ 0 \end{bmatrix} = \begin{bmatrix} 1 \\ 1 \\ 2 \end{bmatrix}.$$

Since the vectors $\begin{bmatrix} 1 \\ 1 \\ 2 \end{bmatrix}$ and $\begin{bmatrix} 1 \\ 1 \\ 0 \end{bmatrix}$ are linearly independent, we can take $P = \begin{bmatrix} 1 & 1 & 1 \\ 1 & 0 & 1 \\ 2 & 0 & 0 \end{bmatrix}$ and $J = \begin{bmatrix} 3 & 1 & 0 \\ 0 & 3 & 0 \\ 0 & 0 & 3 \end{bmatrix}$. □

Example 7.10. The characteristic polynomial of the matrix

$$A = \begin{bmatrix} 0 & 5 & 1 \\ -1 & 4 & 1 \\ -1 & 2 & 3 \end{bmatrix}$$

is $-(\lambda-2)^2(\lambda-3)$. Find an invertible 3×3 matrix P and a 3×3 standard Jordan canonical form J such that $A = PJP^{-1}$.

Solution. Since

$$\mathbf{N}(A-3I_3) = \mathbf{N}\left(\begin{bmatrix} -3 & 5 & 1 \\ -1 & 1 & 1 \\ -1 & 2 & 0 \end{bmatrix}\right) = \operatorname{Span}\left\{\begin{bmatrix} 2 \\ 1 \\ 1 \end{bmatrix}\right\},$$

$$\mathbf{N}(A-2I_3) = \mathbf{N}\left(\begin{bmatrix} -2 & 5 & 1 \\ -1 & 2 & 1 \\ -1 & 2 & 1 \end{bmatrix}\right) = \operatorname{Span}\left\{\begin{bmatrix} 3 \\ 1 \\ 1 \end{bmatrix}\right\},$$

$$\mathbf{N}((A-2I_3)^2) = \mathbf{N}\left(\begin{bmatrix} -2 & 2 & 4 \\ -1 & 1 & 2 \\ -1 & 1 & 2 \end{bmatrix}\right) = \operatorname{Span}\left\{\begin{bmatrix} 1 \\ 1 \\ 0 \end{bmatrix}, \begin{bmatrix} 2 \\ 0 \\ 1 \end{bmatrix}\right\},$$

we can use Theorem 7.6.

The vector $\begin{bmatrix} 2 \\ 0 \\ 1 \end{bmatrix}$ is not in $\operatorname{Span}\left\{\begin{bmatrix} 3 \\ 1 \\ 1 \end{bmatrix}\right\}$ and we have

$$\begin{bmatrix} -2 & 5 & 1 \\ -1 & 2 & 1 \\ -1 & 2 & 1 \end{bmatrix}\begin{bmatrix} 2 \\ 0 \\ 1 \end{bmatrix} = \begin{bmatrix} -3 \\ -1 \\ -1 \end{bmatrix},$$

so we can take $P = \begin{bmatrix} -3 & 2 & 2 \\ -1 & 0 & 1 \\ -1 & 1 & 1 \end{bmatrix}$ and $J = \begin{bmatrix} 2 & 1 & 0 \\ 0 & 2 & 0 \\ 0 & 0 & 3 \end{bmatrix}$. □

Now we move to 4×4 matrices. For matrices of this size it is necessary to consider nine cases. The proofs of the next nine theorems are quite similar and are based on the same idea. We encourage the readers to try to prove the theorems on their own before checking the presented proof.

Theorem 7.11. *Let A be a 4×4 matrix and let α be a real number. If the characteristic polynomial of the matrix A is $(\lambda-\alpha)^4$ and we have*

$$\dim \mathbf{N}(A-\alpha I_4)=1, \quad \dim \mathbf{N}((A-\alpha I_4)^2)=2,$$

$$\dim \mathbf{N}((A-\alpha I_4)^3)=3, \quad \dim \mathbf{N}((A-\alpha I_4)^4)=4,$$

then there is a basis $\mathscr{P}=\{\mathbf{p}_1,\mathbf{p}_2,\mathbf{p}_3,\mathbf{p}_4\}$ in $\mathbb{R}^4$ such that the $\mathscr{P}$-matrix of the linear transformation $f(\mathbf{x})=A\mathbf{x}$ is $\begin{bmatrix}\alpha&1&0&0\\0&\alpha&1&0\\0&0&\alpha&1\\0&0&0&\alpha\end{bmatrix}$ and consequently

$$A=\begin{bmatrix}\mathbf{p}_1&\mathbf{p}_2&\mathbf{p}_3&\mathbf{p}_4\end{bmatrix}\begin{bmatrix}\alpha&1&0&0\\0&\alpha&1&0\\0&0&\alpha&1\\0&0&0&\alpha\end{bmatrix}\begin{bmatrix}\mathbf{p}_1&\mathbf{p}_2&\mathbf{p}_3&\mathbf{p}_4\end{bmatrix}^{-1}.$$

Proof. Let $\mathbf{u}$ be a vector in $\mathbf{N}((A-\alpha I_4)^4)=\mathbb{R}^4$ that is not in $\mathbf{N}((A-\alpha I_4)^3)$. Then the vector $(A-\alpha I_4)^3\mathbf{u}$ is a nonzero vector in $\mathbf{N}(A-\alpha I_4)$. We will show that the set

$$\left\{(A-\alpha I_4)^3\mathbf{u},(A-\alpha I_4)^2\mathbf{u},(A-\alpha I_4)\mathbf{u},\mathbf{u}\right\}$$

is a basis in $\mathbb{R}^4$ with the desired property.

If

$$x_1(A-\alpha I_4)^3\mathbf{u}+x_2(A-\alpha I_4)^2\mathbf{u}+x_3(A-\alpha I_4)\mathbf{u}+x_4\mathbf{u}=\mathbf{0} \tag{7.1}$$

for some real numbers x_1,x_2,x_3,x_4, then

$$x_4(A-\alpha I_4)^3\mathbf{u}=\mathbf{0}$$

and consequently $x_4=0$. Next, since

$$x_3(A-\alpha I_4)^3\mathbf{u}=\mathbf{0},$$

we have $x_3=0$. In the same way we get $x_2=x_1=0$. This shows that the vectors $(A-\alpha I_4)^3\mathbf{u}$, $(A-\alpha I_4)^2\mathbf{u}$, $(A-\alpha I_4)\mathbf{u}$, and $\mathbf{u}$ are linearly independent and, consequently, $\left\{(A-\alpha I_4)^3\mathbf{u},(A-\alpha I_4)^2\mathbf{u},(A-\alpha I_4)\mathbf{u},\mathbf{u}\right\}$ is a basis in $\mathbb{R}^4$.

Next we calculate

$$\begin{aligned}
&A(A-\alpha I_4)^3\mathbf{u}=(A-\alpha I_4)^4\mathbf{u}+\alpha(A-\alpha I_4)^3\mathbf{u}=\alpha(A-\alpha I_4)^3\mathbf{u},\\
&A(A-\alpha I_4)^2\mathbf{u}=(A-\alpha I_4)^3\mathbf{u}+\alpha(A-\alpha I_4)^2\mathbf{u},\\
&A(A-\alpha I_4)\mathbf{u}=(A-\alpha I_4)^2\mathbf{u}+\alpha(A-\alpha I_4)\mathbf{u}, \text{ and}\\
&A\mathbf{u}=(A-\alpha I_4)\mathbf{u}+\alpha\mathbf{u}.
\end{aligned}$$

If we let

$$\mathbf{p}_1=(A-\alpha I_4)^3\mathbf{u}, \quad \mathbf{p}_2=(A-\alpha I_4)^2\mathbf{u}, \quad \mathbf{p}_3=(A-\alpha I_4)\mathbf{u}, \quad \mathbf{p}_4=\mathbf{u},$$

then we have

$$\begin{aligned} A\mathbf{p}_1 &= \alpha\mathbf{p}_1, \\ A\mathbf{p}_2 &= \mathbf{p}_1 + \alpha\mathbf{p}_2, \\ A\mathbf{p}_3 &= \mathbf{p}_2 + \alpha\mathbf{p}_3, \\ A\mathbf{p}_4 &= \mathbf{p}_3 + \alpha\mathbf{p}_4. \end{aligned}$$

Consequently, the $\left\{(A-\alpha I_4)^3\mathbf{u}, (A-\alpha I_4)^2\mathbf{u}, (A-\alpha I_4)\mathbf{u}, \mathbf{u}\right\}$-matrix of the linear transformation $f(\mathbf{x}) = A\mathbf{x}$ is $\begin{bmatrix} \alpha & 1 & 0 & 0 \\ 0 & \alpha & 1 & 0 \\ 0 & 0 & \alpha & 1 \\ 0 & 0 & 0 & \alpha \end{bmatrix}$ and we have

$$A = \begin{bmatrix} \mathbf{p}_1 & \mathbf{p}_2 & \mathbf{p}_3 & \mathbf{p}_4 \end{bmatrix} \begin{bmatrix} \alpha & 1 & 0 & 0 \\ 0 & \alpha & 1 & 0 \\ 0 & 0 & \alpha & 1 \\ 0 & 0 & 0 & \alpha \end{bmatrix} \begin{bmatrix} \mathbf{p}_1 & \mathbf{p}_2 & \mathbf{p}_3 & \mathbf{p}_4 \end{bmatrix}^{-1},$$

by Corollary 6.23. □

Theorem 7.12. *Let A be a 4×4 matrix and let α be a real number. If the characteristic polynomial of the matrix A is $(\lambda - \alpha)^4$ and we have*

$$\dim \mathbf{N}(A - \alpha I_4) = 2, \quad \dim \mathbf{N}((A-\alpha I_4)^2) = 3, \quad \dim \mathbf{N}((A-\alpha I_4)^3) = 4,$$

then there is a basis $\mathcal{P} = \{\mathbf{p}_1, \mathbf{p}_2, \mathbf{p}_3, \mathbf{p}_4\}$ in $\mathbb{R}^4$ such that the $\mathcal{P}$-matrix of the linear transformation $f(\mathbf{x}) = A\mathbf{x}$ is $\begin{bmatrix} \alpha & 1 & 0 & 0 \\ 0 & \alpha & 1 & 0 \\ 0 & 0 & \alpha & 0 \\ 0 & 0 & 0 & \alpha \end{bmatrix}$ and consequently

$$A = \begin{bmatrix} \mathbf{p}_1 & \mathbf{p}_2 & \mathbf{p}_3 & \mathbf{p}_4 \end{bmatrix} \begin{bmatrix} \alpha & 1 & 0 & 0 \\ 0 & \alpha & 1 & 0 \\ 0 & 0 & \alpha & 0 \\ 0 & 0 & 0 & \alpha \end{bmatrix} \begin{bmatrix} \mathbf{p}_1 & \mathbf{p}_2 & \mathbf{p}_3 & \mathbf{p}_4 \end{bmatrix}^{-1}.$$

Proof. Let $\mathbf{u}$ be a vector in $\mathbf{N}((A-\alpha I_4)^3)$ that is not in $\mathbf{N}((A-\alpha I_4)^2)$. Then the nonzero vector $(A-\alpha I_4)^2\mathbf{u}$ is in $\mathbf{N}(A-\alpha I_4)$. Since $\dim \mathbf{N}(A-\alpha I_4) = 2$, we can choose a vector $\mathbf{v}$ such that $\{(A-\alpha I_4)^2\mathbf{u}, \mathbf{v}\}$ is a basis of $\mathbf{N}(A-\alpha I_4)$. We will show that

$$\mathcal{P} = \left\{(A-\alpha I_4)^2\mathbf{u}, (A-\alpha I_4)\mathbf{u}, \mathbf{u}, \mathbf{v}\right\}$$

is a basis in $\mathbb{R}^4$ with the desired property.

If

$$x_1(A-\alpha I_4)^2\mathbf{u} + x_2(A-\alpha I_4)\mathbf{u} + x_3\mathbf{u} + x_4\mathbf{v} = \mathbf{0} \tag{7.2}$$

for some real numbers x_1, x_2, x_3, x_4, then

$$x_3(A-\alpha I_4)^2\mathbf{u}=\mathbf{0}$$

and consequently $x_3 = 0$. Next, since

$$x_2(A-\alpha I_4)^2\mathbf{u}=\mathbf{0},$$

we get $x_2 = 0$. Now the equality (12.1) becomes

$$x_1(A-\alpha I_4)^2\mathbf{u}+x_4\mathbf{v}=0,$$

which gives us $x_1 = x_4 = 0$, because the vectors $(A-\alpha I_4)^2\mathbf{u}$ and $\mathbf{v}$ are linearly independent. This shows that the vectors $(A-\alpha I_4)^2\mathbf{u}$, $(A-\alpha I_4)\mathbf{u}$, $\mathbf{u}$, and $\mathbf{v}$ are linearly independent and consequently $\{(A-\alpha I_4)^2\mathbf{u}, (A-\alpha I_4)\mathbf{u}, \mathbf{u}, \mathbf{v}\}$ is a basis in $\mathbb{R}^4$.

Now we calculate

$$\begin{aligned}
&A(A-\alpha I_4)^2\mathbf{u}=(A-\alpha I_4)^3\mathbf{u}+\alpha(A-\alpha I_4)^2\mathbf{u}=\alpha(A-\alpha I_4)^2\mathbf{u},\\
&A(A-\alpha I_4)\mathbf{u}=(A-\alpha I_4)^2\mathbf{u}+\alpha(A-\alpha I_4)\mathbf{u},\\
&A\mathbf{u}=(A-\alpha I_4)\mathbf{u}+\alpha\mathbf{u}, \text{ and}\\
&A\mathbf{v}=\alpha\mathbf{v}.
\end{aligned}$$

If we let

$$\mathbf{p}_1=(A-\alpha I_4)^2\mathbf{u},\quad \mathbf{p}_2=(A-\alpha I_4)\mathbf{u},\quad \mathbf{p}_3=\mathbf{u},\quad \mathbf{p}_4=\mathbf{v},$$

then we have

$$\begin{aligned}
A\mathbf{p}_1&=\alpha\mathbf{p}_1,\\
A\mathbf{p}_2&=\mathbf{p}_1+\alpha\mathbf{p}_2,\\
A\mathbf{p}_3&=\mathbf{p}_2+\alpha\mathbf{p}_3,\\
A\mathbf{p}_4&=\alpha\mathbf{p}_4.
\end{aligned}$$

Consequently, the $\{(A-\alpha I_4)^2\mathbf{u}, (A-\alpha I_4)\mathbf{u}, \mathbf{u}, \mathbf{v}\}$-matrix of the linear transformation $f(\mathbf{x}) = A\mathbf{x}$ is $\begin{bmatrix}\alpha & 1 & 0 & 0\\ 0 & \alpha & 1 & 0\\ 0 & 0 & \alpha & 0\\ 0 & 0 & 0 & \alpha\end{bmatrix}$ and we have

$$A=\begin{bmatrix}\mathbf{p}_1 & \mathbf{p}_2 & \mathbf{p}_3 & \mathbf{p}_4\end{bmatrix}\begin{bmatrix}\alpha & 1 & 0 & 0\\ 0 & \alpha & 1 & 0\\ 0 & 0 & \alpha & 0\\ 0 & 0 & 0 & \alpha\end{bmatrix}\begin{bmatrix}\mathbf{p}_1 & \mathbf{p}_2 & \mathbf{p}_3 & \mathbf{p}_4\end{bmatrix}^{-1},$$

by Corollary 6.23. □

Theorem 7.13. *Let A be a 4×4 matrix and let α be a real number. If the characteristic polynomial of the matrix A is $(\lambda-\alpha)^4$ and we have*

$$\dim \mathbf{N}(A-\alpha I_4)=2 \quad \textit{and} \quad \dim \mathbf{N}((A-\alpha I_4)^2)=4,$$

then there is a basis $\mathscr{P}=\{\mathbf{p}_1,\mathbf{p}_2,\mathbf{p}_3,\mathbf{p}_4\}$ in $\mathbb{R}^4$ such that the $\mathscr{P}$-matrix of the linear transformation $f(\mathbf{x})=A\mathbf{x}$ is $\begin{bmatrix}\alpha&1&0&0\\0&\alpha&0&0\\0&0&\alpha&1\\0&0&0&\alpha\end{bmatrix}$ and consequently

$$A=\begin{bmatrix}\mathbf{p}_1&\mathbf{p}_2&\mathbf{p}_3&\mathbf{p}_4\end{bmatrix}\begin{bmatrix}\alpha&1&0&0\\0&\alpha&0&0\\0&0&\alpha&1\\0&0&0&\alpha\end{bmatrix}\begin{bmatrix}\mathbf{p}_1&\mathbf{p}_2&\mathbf{p}_3&\mathbf{p}_4\end{bmatrix}^{-1}.$$

Proof. Let $\{\mathbf{a},\mathbf{b}\}$ be a basis of eigenvectors corresponding to the eigenvalue α, that is a basis of $\mathbf{N}(A-\alpha I_4)$. Let $\mathbf{u}$ and $\mathbf{v}$ be linearly independent vectors such that $\{\mathbf{a},\mathbf{b},\mathbf{u},\mathbf{v}\}$ is a basis of $\mathbf{N}((A-\alpha I_4)^2)=\mathbb{R}^4$. We will show that

$$\mathscr{P}=\{(A-\alpha I_4)\mathbf{u},\mathbf{u},(A-\alpha I_4)\mathbf{v},\mathbf{v}\}$$

is a basis in $\mathbb{R}^4$ with the desired property.

If

$$x_1(A-\alpha I_4)\mathbf{u}+x_2\mathbf{u}+x_3(A-\alpha I_4)\mathbf{v}+x_4\mathbf{v}=\mathbf{0}$$

for some real numbers x_1,x_2,x_3,x_4, then

$$(A-\alpha I_4)(x_2\mathbf{u}+x_4\mathbf{v})=\mathbf{0}.$$

This means that the vector $x_2\mathbf{u}+x_4\mathbf{v}$ is in $\mathbf{N}(A-\alpha I_4)$ and consequently $x_2\mathbf{u}+x_4\mathbf{v}=\mathbf{0}$,because the vectors $\{\mathbf{a},\mathbf{b},\mathbf{u},\mathbf{v}\}$ are linearly independent. The equality $x_2\mathbf{u}+x_4\mathbf{v}=\mathbf{0}$ is equivalent to $x_2=x_4=0$, because the vectors $\mathbf{u}$ and $\mathbf{v}$ are linearly independent.

Next, since

$$(A-\alpha I_4)(x_1\mathbf{u}+x_3\mathbf{v})=\mathbf{0},$$

we get $x_1=x_3=0$, as before. This shows that the vectors $(A-\alpha I_4)\mathbf{u}$, $\mathbf{u}$, $(A-\alpha I_4)\mathbf{v}$, and $\mathbf{v}$ are linearly independent and consequently $\{(A-\alpha I_4)\mathbf{u},\mathbf{u},(A-\alpha I_4)\mathbf{v},\mathbf{v}\}$ is a basis in $\mathbb{R}^4$.

Now we calculate

$$\begin{aligned}
&A(A-\alpha I_4)\mathbf{u}=(A-\alpha I_4)^2\mathbf{u}+\alpha(A-\alpha I_4)\mathbf{u}=\alpha(A-\alpha I_4)\mathbf{u},\\
&A\mathbf{u}=(A-\alpha I_4)\mathbf{u}+\alpha\mathbf{u},\\
&A(A-\alpha I_4)\mathbf{v}=(A-\alpha I_4)^2\mathbf{v}+\alpha(A-\alpha I_4)\mathbf{v}=\alpha(A-\alpha I_4)\mathbf{v},\ \text{and}\\
&A\mathbf{v}=(A-\alpha I_4)\mathbf{v}+\alpha\mathbf{v}.
\end{aligned}$$

If we let

$$\mathbf{p}_1 = (A - \alpha I_4)\mathbf{u}, \quad \mathbf{p}_2 = \mathbf{u}, \quad \mathbf{p}_3 = (A - \alpha I_4)\mathbf{v}, \quad \mathbf{p}_4 = \mathbf{v},$$

then we have

$$\begin{aligned} A\mathbf{p}_1 &= \alpha\mathbf{p}_1, \\ A\mathbf{p}_2 &= \mathbf{p}_1 + \alpha\mathbf{p}_2, \\ A\mathbf{p}_3 &= \alpha\mathbf{p}_3, \\ A\mathbf{p}_4 &= \mathbf{p}_3 + \alpha\mathbf{p}_4. \end{aligned}$$

Consequently, the $\{(A-\alpha I_4)\mathbf{u}, \mathbf{u}, (A-\alpha I_4)\mathbf{v}, \mathbf{v}\}$-matrix of the linear transformation $f(\mathbf{x}) = A\mathbf{x}$ is $\begin{bmatrix} \alpha & 1 & 0 & 0 \\ 0 & \alpha & 0 & 0 \\ 0 & 0 & \alpha & 1 \\ 0 & 0 & 0 & \alpha \end{bmatrix}$ and we have

$$A = \begin{bmatrix} \mathbf{p}_1 & \mathbf{p}_2 & \mathbf{p}_3 & \mathbf{p}_4 \end{bmatrix} \begin{bmatrix} \alpha & 1 & 0 & 0 \\ 0 & \alpha & 0 & 0 \\ 0 & 0 & \alpha & 1 \\ 0 & 0 & 0 & \alpha \end{bmatrix} \begin{bmatrix} \mathbf{p}_1 & \mathbf{p}_2 & \mathbf{p}_3 & \mathbf{p}_4 \end{bmatrix}^{-1},$$

by Corollary 6.23. □

Theorem 7.14. *Let A be a 4×4 matrix and let α be a real number. If the characteristic polynomial of the matrix A is $(\lambda - \alpha)^4$ and we have*

$$\dim \mathbf{N}(A - \alpha I_4) = 3 \quad \textit{and} \quad \dim \mathbf{N}((A - \alpha I_4)^2) = 4,$$

then there is a basis $\mathcal{P} = \{\mathbf{p}_1, \mathbf{p}_2, \mathbf{p}_3, \mathbf{p}_4\}$ in $\mathbb{R}^4$ such that the $\mathcal{P}$-matrix of the linear transformation $f(\mathbf{x}) = A\mathbf{x}$ is $\begin{bmatrix} \alpha & 1 & 0 & 0 \\ 0 & \alpha & 0 & 0 \\ 0 & 0 & \alpha & 0 \\ 0 & 0 & 0 & \alpha \end{bmatrix}$ and consequently

$$A = \begin{bmatrix} \mathbf{p}_1 & \mathbf{p}_2 & \mathbf{p}_3 & \mathbf{p}_4 \end{bmatrix} \begin{bmatrix} \alpha & 1 & 0 & 0 \\ 0 & \alpha & 0 & 0 \\ 0 & 0 & \alpha & 0 \\ 0 & 0 & 0 & \alpha \end{bmatrix} \begin{bmatrix} \mathbf{p}_1 & \mathbf{p}_2 & \mathbf{p}_3 & \mathbf{p}_4 \end{bmatrix}^{-1}.$$

Proof. Let $\mathbf{u}$ be a vector in $\mathbf{N}((A - \alpha I_4)^2) = \mathbb{R}^4$ which is not in $\mathbf{N}((A - \alpha I_4))$. The nonzero vector $(A - \alpha I_4)\mathbf{u}$ is in $\mathbf{N}(A - \alpha I_4)$ and, since $\dim \mathbf{N}(A - \alpha I_4) = 3$, we can choose linearly independent vectors $\mathbf{v}$ and $\mathbf{w}$ in $\mathbf{N}(A - \alpha I_4)$ such that $\{(A - \alpha I_4)\mathbf{u}, \mathbf{v}, \mathbf{w}\}$ is a basis of the vector subspace $\mathbf{N}(A - \alpha I_4)$. We will show that

$$\mathcal{P} = \{(A - \alpha I_4)\mathbf{u}, \mathbf{u}, \mathbf{v}, \mathbf{w}\}$$

is a basis in $\mathbb{R}^4$ that has the desired property.

If

$$x_1(A-\alpha I_4)\mathbf{u}+x_2\mathbf{u}+x_3\mathbf{v}+x_4\mathbf{w}=\mathbf{0}$$

for some real numbers x_1, x_2, x_3, x_4, then

$$x_2(A-\alpha I_4)\mathbf{u}=\mathbf{0}$$

and consequently $x_2 = 0$. Next, since

$$x_1(A-\alpha I_4)\mathbf{u}+x_3\mathbf{v}+x_4\mathbf{w}=0,$$

we get $x_1 = x_3 = x_4 = 0$, because the vectors $(A-\alpha I_4)\mathbf{u}$, $\mathbf{v}$ and $\mathbf{w}$ are linearly independent. This shows that the vectors $(A-\alpha I_4)\mathbf{u}$, $\mathbf{u}$, $\mathbf{v}$, and $\mathbf{w}$ are linearly independent and consequently $\{(A-\alpha I_4)\mathbf{u},\mathbf{u},\mathbf{v},\mathbf{w}\}$ is a basis in $\mathbb{R}^4$.

Now we calculate

$$\begin{aligned}
&A(A-\alpha I_4)\mathbf{u}=(A-\alpha I_4)^2\mathbf{u}+\alpha(A-\alpha I_4)\mathbf{u}=\alpha(A-\alpha I_4)\mathbf{u},\\
&A\mathbf{u}=(A-\alpha I_4)\mathbf{u}+\alpha\mathbf{u},\\
&A\mathbf{v}=\alpha\mathbf{v}, \text{ and}\\
&A\mathbf{w}=\alpha\mathbf{w}.
\end{aligned}$$

If we let

$$\mathbf{p}_1=(A-\alpha I_4)\mathbf{u},\quad \mathbf{p}_2=\mathbf{u},\quad \mathbf{p}_3=\mathbf{v},\quad \mathbf{p}_4=\mathbf{w},$$

then we have

$$\begin{aligned}
A\mathbf{p}_1&=\alpha\mathbf{p}_1,\\
A\mathbf{p}_2&=\mathbf{p}_1+\alpha\mathbf{p}_2,\\
A\mathbf{p}_3&=\alpha\mathbf{p}_3,\\
A\mathbf{p}_4&=\alpha\mathbf{p}_4.
\end{aligned}$$

Consequently, the $\{(A-\alpha I_4)\mathbf{u},\mathbf{u},\mathbf{v},\mathbf{w}\}$-matrix of the linear transformation $f(\mathbf{x}) = A\mathbf{x}$ is $\begin{bmatrix}\alpha&1&0&0\\0&\alpha&0&0\\0&0&\alpha&0\\0&0&0&\alpha\end{bmatrix}$ and we have

$$A=\begin{bmatrix}\mathbf{p}_1&\mathbf{p}_2&\mathbf{p}_3&\mathbf{p}_4\end{bmatrix}\begin{bmatrix}\alpha&1&0&0\\0&\alpha&0&0\\0&0&\alpha&0\\0&0&0&\alpha\end{bmatrix}\begin{bmatrix}\mathbf{p}_1&\mathbf{p}_2&\mathbf{p}_3&\mathbf{p}_4\end{bmatrix}^{-1},$$

by Corollary 6.23. □

Theorem 7.15. *Let A be a 4×4 matrix and let α and β be two distinct real numbers. If the characteristic polynomial of the matrix A is $(\lambda-\alpha)^3(\lambda-\beta)$ and we have*

$$\dim \mathbf{N}(A-\alpha I_4)=1,\ \dim \mathbf{N}((A-\alpha I_4)^2)=2,$$

$$\dim \mathbf{N}((A-\alpha I_4)^3)=3,\ \dim \mathbf{N}(A-\beta I_4)=1,$$

then there is a basis $\mathscr{P}=\{\mathbf{p}_1,\mathbf{p}_2,\mathbf{p}_3,\mathbf{p}_4\}$ in $\mathbb{R}^4$ such that the $\mathscr{P}$-matrix of the linear transformation $f(\mathbf{x})=A\mathbf{x}$ is $\begin{bmatrix}\alpha&1&0&0\\0&\alpha&1&0\\0&0&\alpha&0\\0&0&0&\beta\end{bmatrix}$ and consequently

$$A=\begin{bmatrix}\mathbf{p}_1&\mathbf{p}_2&\mathbf{p}_3&\mathbf{p}_4\end{bmatrix}\begin{bmatrix}\alpha&1&0&0\\0&\alpha&1&0\\0&0&\alpha&0\\0&0&0&\beta\end{bmatrix}\begin{bmatrix}\mathbf{p}_1&\mathbf{p}_2&\mathbf{p}_3&\mathbf{p}_4\end{bmatrix}^{-1}.$$

Proof. Let $\mathbf{u}$ be a vector in $\mathbf{N}((A-\alpha I_4)^3)$ that is not in $\mathbf{N}((A-\alpha I_4)^2)$ and let $\mathbf{v}$ be a vector in $\mathbf{N}(A-\beta I_4)$. Then the nonzero vector $(A-\alpha I_4)^2\mathbf{u}$ is in $\mathbf{N}(A-\alpha I_4)$. We will show that

$$\mathscr{P}=\left\{(A-\alpha I_4)^2\mathbf{u},(A-\alpha I_4)\mathbf{u},\mathbf{u},\mathbf{v}\right\}$$

is a basis in $\mathbb{R}^4$ that has the desired property.

If

$$x_1(A-\alpha I_4)^2\mathbf{u}+x_2(A-\alpha I_4)\mathbf{u}+x_3\mathbf{u}+x_4\mathbf{v}=\mathbf{0}$$

for some real numbers x_1,x_2,x_3,x_4, then

$$x_4(A-\alpha I_4)^3\mathbf{v}=\mathbf{0}.$$

Since

$$\begin{aligned}(A-\alpha I_4)^3&=(A-\beta I_4+(\beta-\alpha)I_4)^3\\&=(A-\beta I_4)^3+3(\beta-\alpha)(A-\beta I_4)^2+3(\beta-\alpha)^2(A-\beta I_4)+(\beta-\alpha)^3 I_4,\end{aligned}$$

we can conclude that $x_4=0$. Now, from

$$x_3(A-\alpha I_4)^2\mathbf{u}=\mathbf{0}$$

we get $x_3=0$. Next, from

$$x_2(A-\alpha I_4)^2\mathbf{u}=\mathbf{0}$$

we get $x_2=0$. Finally, from

$$x_1(A-\alpha I_4)^2\mathbf{u}=0$$

we get $x_1=0$. This shows that the vectors $(A-\alpha I_4)^2\mathbf{u}$, $(A-\alpha I_4)\mathbf{u}$, $\mathbf{u}$, and $\mathbf{v}$ are linearly independent and consequently $\left\{(A-\alpha I_4)^2\mathbf{u},(A-\alpha I_4)\mathbf{u},\mathbf{u},\mathbf{v}\right\}$ is a basis in $\mathbb{R}^4$.

Now we calculate

$$\begin{aligned}
&A(A-\alpha I_4)^2\mathbf{u} = (A-\alpha I_4)^3\mathbf{u} + \alpha(A-\alpha I_4)^2\mathbf{u} = \alpha(A-\alpha I_4)^2\mathbf{u},\\
&A(A-\alpha I_4)\mathbf{u} = (A-\alpha I_4)^2\mathbf{u} + \alpha(A-\alpha I_4)\mathbf{u},\\
&A\mathbf{u} = (A-\alpha I_4)\mathbf{u} + \alpha\mathbf{u}, \text{ and}\\
&A\mathbf{v} = \beta\mathbf{v}.
\end{aligned}$$

If we let

$$\mathbf{p}_1 = (A-\alpha I_4)^2\mathbf{u}, \quad \mathbf{p}_2 = (A-\alpha I_4)\mathbf{u}, \quad \mathbf{p}_3 = \mathbf{u}, \quad \mathbf{p}_4 = \mathbf{v},$$

then we have

$$\begin{aligned}
A\mathbf{p}_1 &= \alpha\mathbf{p}_1,\\
A\mathbf{p}_2 &= \mathbf{p}_1 + \alpha\mathbf{p}_2,\\
A\mathbf{p}_3 &= \mathbf{p}_2 + \alpha\mathbf{p}_3,\\
A\mathbf{p}_4 &= \beta\mathbf{p}_4.
\end{aligned}$$

Consequently, the $\left\{(A-\alpha I_4)^2\mathbf{u}, (A-\alpha I_4)\mathbf{u}, \mathbf{u}, \mathbf{v}\right\}$-matrix of the linear transformation $f(\mathbf{x}) = A\mathbf{x}$ is $\begin{bmatrix} \alpha & 1 & 0 & 0\\ 0 & \alpha & 1 & 0\\ 0 & 0 & \alpha & 0\\ 0 & 0 & 0 & \beta \end{bmatrix}$ and we have

$$A = \begin{bmatrix} \mathbf{p}_1 & \mathbf{p}_2 & \mathbf{p}_3 & \mathbf{p}_4 \end{bmatrix} \begin{bmatrix} \alpha & 1 & 0 & 0\\ 0 & \alpha & 1 & 0\\ 0 & 0 & \alpha & 0\\ 0 & 0 & 0 & \beta \end{bmatrix} \begin{bmatrix} \mathbf{p}_1 & \mathbf{p}_2 & \mathbf{p}_3 & \mathbf{p}_4 \end{bmatrix}^{-1},$$

by Corollary 6.23. □

Theorem 7.16. *Let A be a 4×4 matrix and let α and β be two distinct real numbers. If the characteristic polynomial of the matrix A is $(\lambda-\alpha)^3(\lambda-\beta)$ and we have*

$$\dim \mathbf{N}(A-\alpha I_4) = 2, \ \dim \mathbf{N}((A-\alpha I_4)^2) = 3, \ \dim \mathbf{N}(A-\beta I_4) = 1,$$

then there is a basis $\mathscr{P} = \{\mathbf{p}_1, \mathbf{p}_2, \mathbf{p}_3, \mathbf{p}_4\}$ in $\mathbb{R}^4$ such that the $\mathscr{P}$-matrix of the linear transformation $f(\mathbf{x}) = A\mathbf{x}$ is $\begin{bmatrix} \alpha & 1 & 0 & 0\\ 0 & \alpha & 0 & 0\\ 0 & 0 & \alpha & 0\\ 0 & 0 & 0 & \beta \end{bmatrix}$ and consequently

$$A = \begin{bmatrix} \mathbf{p}_1 & \mathbf{p}_2 & \mathbf{p}_3 & \mathbf{p}_4 \end{bmatrix} \begin{bmatrix} \alpha & 1 & 0 & 0\\ 0 & \alpha & 0 & 0\\ 0 & 0 & \alpha & 0\\ 0 & 0 & 0 & \beta \end{bmatrix} \begin{bmatrix} \mathbf{p}_1 & \mathbf{p}_2 & \mathbf{p}_3 & \mathbf{p}_4 \end{bmatrix}^{-1}.$$

Proof. Let $\mathbf{u}$ be a vector in $\mathbf{N}((A-\alpha I_4)^2)$ which is not in $\mathbf{N}(A-\alpha I_4)$. Then the nonzero vector $(A-\alpha I_4)\mathbf{u}$ is in $\mathbf{N}(A-\alpha I_4)$ and, because $\dim \mathbf{N}(A-\alpha I_4) = 2$, we can choose a vector $\mathbf{v}$ such that $\{(A-\alpha I_4)\mathbf{u}, \mathbf{v}\}$ is a basis of $\mathbf{N}(A-\alpha I_4)$. Let $\mathbf{w}$ be a nonzero vector in $\mathbf{N}(A-\beta I_4)$. We will show that

$$\mathscr{P} = \{(A-\alpha I_4)\mathbf{u}, \mathbf{u}, \mathbf{v}, \mathbf{w}\}$$

is a basis in $\mathbb{R}^4$ with the desired property.

If

$$x_1(A-\alpha I_4)\mathbf{u} + x_2\mathbf{u} + x_3\mathbf{v} + x_4\mathbf{w} = \mathbf{0}$$

for some real numbers x_1, x_2, x_3, x_4, then

$$x_4(A-\alpha I_4)^2\mathbf{w} = \mathbf{0}$$

and consequently $x_4 = 0$, because

$$(A-\alpha I_4)^2\mathbf{w} = ((A-\beta I_4)^2 + 2(\beta-\alpha)(A-\beta I_4) + (\beta-\alpha)^2)\mathbf{w} = (\beta-\alpha)^2\mathbf{w}.$$

Now, since

$$x_1(A-\alpha I_4)\mathbf{u} + x_2\mathbf{u} + x_3\mathbf{v} = 0,$$

we get $x_2 = 0$ and then $x_1 = x_3 = 0$. This shows that the vectors $(A-\alpha I_3)\mathbf{u}$, $\mathbf{u}$, $\mathbf{v}$, and $\mathbf{w}$ are linearly independent and consequently $\{(A-\alpha I_4)\mathbf{u}, \mathbf{u}, \mathbf{v}, \mathbf{w}\}$ is a basis in $\mathbb{R}^4$.

Next we calculate

$$\begin{aligned}
&A(A-\alpha I_4)\mathbf{u} = (A-\alpha I_4)^2\mathbf{u} + \alpha(A-\alpha I_4)\mathbf{u} = \alpha(A-\alpha I_4)\mathbf{u},\\
&A\mathbf{u} = (A-\alpha I_4)\mathbf{u} + \alpha\mathbf{u},\\
&A\mathbf{v} = \alpha\mathbf{v}, \text{ and}\\
&A\mathbf{w} = \beta\mathbf{w}.
\end{aligned}$$

If we let

$$\mathbf{p}_1 = (A-\alpha I_4)^2\mathbf{u}, \quad \mathbf{p}_2 = (A-\alpha I_4)\mathbf{u}, \quad \mathbf{p}_3 = \mathbf{u}, \quad \mathbf{p}_4 = \mathbf{v},$$

then we have

$$\begin{aligned}
A\mathbf{p}_1 &= \alpha\mathbf{p}_1,\\
A\mathbf{p}_2 &= \mathbf{p}_1 + \alpha\mathbf{p}_2,\\
A\mathbf{p}_3 &= \alpha\mathbf{p}_3,\\
A\mathbf{p}_4 &= \beta\mathbf{p}_4.
\end{aligned}$$

Consequently, the $\{(A-\alpha I_4)\mathbf{u}, \mathbf{u}, \mathbf{v}, \mathbf{w}\}$-matrix of the linear transformation $f(\mathbf{x}) = A\mathbf{x}$ is $\begin{bmatrix} \alpha & 1 & 0 & 0\\ 0 & \alpha & 0 & 0\\ 0 & 0 & \alpha & 0\\ 0 & 0 & 0 & \beta \end{bmatrix}$ and we have

$$A = \begin{bmatrix}\mathbf{p}_1 & \mathbf{p}_2 & \mathbf{p}_3 & \mathbf{p}_4\end{bmatrix} \begin{bmatrix} \alpha & 1 & 0 & 0\\ 0 & \alpha & 0 & 0\\ 0 & 0 & \alpha & 0\\ 0 & 0 & 0 & \beta \end{bmatrix} \begin{bmatrix}\mathbf{p}_1 & \mathbf{p}_2 & \mathbf{p}_3 & \mathbf{p}_4\end{bmatrix}^{-1},$$

by Corollary 6.23. □

Theorem 7.17. *Let A be a 4×4 matrix and let α and β be two distinct real numbers. If the characteristic polynomial of the matrix A is $(\lambda-\alpha)^2(\lambda-\beta)^2$ and we have*

$$\dim \mathbf{N}(A-\alpha I_4)=1,\ \dim \mathbf{N}((A-\alpha I_4)^2)=2,$$

$$\dim \mathbf{N}(A-\beta I_4)=1,\ \dim \mathbf{N}((A-\beta I_4)^2)=2,$$

then there is a basis $\mathscr{P}=\{\mathbf{p}_1,\mathbf{p}_2,\mathbf{p}_3,\mathbf{p}_4\}$ in $\mathbb{R}^4$ such that the $\mathscr{P}$-matrix of the linear transformation $f(\mathbf{x})=A\mathbf{x}$ is $\begin{bmatrix}\alpha & 1 & 0 & 0\\ 0 & \alpha & 0 & 0\\ 0 & 0 & \beta & 1\\ 0 & 0 & 0 & \beta\end{bmatrix}$ and consequently

$$A=\begin{bmatrix}\mathbf{p}_1 & \mathbf{p}_2 & \mathbf{p}_3 & \mathbf{p}_4\end{bmatrix}\begin{bmatrix}\alpha & 1 & 0 & 0\\ 0 & \alpha & 0 & 0\\ 0 & 0 & \beta & 1\\ 0 & 0 & 0 & \beta\end{bmatrix}\begin{bmatrix}\mathbf{p}_1 & \mathbf{p}_2 & \mathbf{p}_3 & \mathbf{p}_4\end{bmatrix}^{-1}.$$

Proof. Let $\mathbf{u}$ be a vector in $\mathbf{N}((A-\alpha I_4)^2)$ that is not in $\mathbf{N}(A-\alpha I_4)$ and let $\mathbf{v}$ be a vector in $\mathbf{N}((A-\beta I_4)^2)$ that is not in $\mathbf{N}(A-\beta I_4)$. Then the nonzero vector $(A-\alpha I_4)\mathbf{u}$ is in $\mathbf{N}(A-\alpha I_4)$ and the nonzero vector $(A-\beta I_4)\mathbf{v}$ is in $\mathbf{N}(A-\beta I_4)$. We will show that

$$\mathscr{P}=\left\{(A-\alpha I_4)\mathbf{u},\mathbf{u},(A-\beta I_4)\mathbf{v},\mathbf{v}\right\}$$

is a basis in $\mathbb{R}^4$ that has the desired property.

If

$$x_1(A-\alpha I_4)\mathbf{u}+x_2\mathbf{u}+x_3(A-\beta I_4)\mathbf{v}+x_4\mathbf{v}=\mathbf{0}$$

for some real numbers x_1,x_2,x_3,x_4, then

$$x_1(A-\alpha I_4)^3\mathbf{u}+x_2(A-\alpha I_4)^2\mathbf{u}+x_3(A-\alpha I_4)^2(A-\beta I_4)\mathbf{v}+x_4(A-\alpha I_4)^2\mathbf{v}=\mathbf{0}$$

and thus

$$(A-\alpha I_4)^2(x_3(A-\beta I_4)\mathbf{v}+x_4\mathbf{v})=\mathbf{0}.$$

Since

$$(A-\alpha I_4)^2=(A-\beta I_4)^2+2(\beta-\alpha)(A-\beta I_4)+(\beta-\alpha)^2I_4,$$

the above equality becomes

$$(x_3(\beta-\alpha)^2+2x_4(\beta-\alpha))(A-\beta I_4)\mathbf{v}+x_4(\beta-\alpha)^2\mathbf{v}=\mathbf{0},$$

which gives us $x_3=x_4=0$, because the vectors $\mathbf{v}$ and $(A-\beta I_4)\mathbf{v}$ are linearly independent, which can be shown using the method presented in the proof of Theorem 7.1. Then we get $x_1=x_2=0$, again as in the proof of Theorem 7.1. This shows that the vectors $(A-\alpha I_4)\mathbf{u}$, $\mathbf{u}$, $(A-\beta I_4)\mathbf{v}$, and $\mathbf{v}$ are linearly independent and consequently $\{(A-\alpha I_4)\mathbf{u}, \mathbf{u}, (A-\beta I_4)\mathbf{v}, \mathbf{v}\}$ is a basis in $\mathbb{R}^4$.

Next we calculate

$$\begin{aligned}
&A(A-\alpha I_4)\mathbf{u} = (A-\alpha I_4)^2\mathbf{u} + \alpha(A-\alpha I_4)\mathbf{u} = \alpha(A-\alpha I_4)\mathbf{u},\\
&A\mathbf{u} = (A-\alpha I_4)\mathbf{u} + \alpha\mathbf{u},\\
&A(A-\beta I_4)\mathbf{v} = (A-\beta I_4)^2\mathbf{v} + \beta(A-\beta I_4)\mathbf{v} = \beta(A-\beta I_4)\mathbf{v}, \text{ and}\\
&A\mathbf{v} = (A-\beta I_4)\mathbf{v} + \beta\mathbf{v}.
\end{aligned}$$

If we let

$$\mathbf{p}_1 = (A-\alpha I_4)\mathbf{u}, \quad \mathbf{p}_2 = \mathbf{u}, \quad \mathbf{p}_3 = (A-\beta I_4)\mathbf{v}, \quad \mathbf{p}_4 = \mathbf{v},$$

then we have

$$\begin{aligned}
A\mathbf{p}_1 &= \alpha\mathbf{p}_1,\\
A\mathbf{p}_2 &= \mathbf{p}_1 + \alpha\mathbf{p}_2,\\
A\mathbf{p}_3 &= \beta\mathbf{p}_3,\\
A\mathbf{p}_4 &= \mathbf{p}_3 + \beta\mathbf{p}_4.
\end{aligned}$$

Consequently, the $\{(A-\alpha I_4)\mathbf{u}, \mathbf{u}, (A-\beta I_4)\mathbf{v}, \mathbf{v}\}$-matrix of the linear transformation $f(\mathbf{x}) = A\mathbf{x}$ is $\begin{bmatrix} \alpha & 1 & 0 & 0\\ 0 & \alpha & 0 & 0\\ 0 & 0 & \beta & 1\\ 0 & 0 & 0 & \beta \end{bmatrix}$ and we have

$$A = \begin{bmatrix} \mathbf{p}_1 & \mathbf{p}_2 & \mathbf{p}_3 & \mathbf{p}_4 \end{bmatrix} \begin{bmatrix} \alpha & 1 & 0 & 0\\ 0 & \alpha & 0 & 0\\ 0 & 0 & \beta & 1\\ 0 & 0 & 0 & \beta \end{bmatrix} \begin{bmatrix} \mathbf{p}_1 & \mathbf{p}_2 & \mathbf{p}_3 & \mathbf{p}_4 \end{bmatrix}^{-1},$$

by Corollary 6.23. □

Theorem 7.18. *Let A be a 4×4 matrix and let α and β be two distinct real numbers. If the characteristic polynomial of the matrix A is $(\lambda-\alpha)^2(\lambda-\beta)^2$ and we have*

$$\dim \mathbf{N}(A-\alpha I_4)=1,\ \dim \mathbf{N}((A-\alpha I_4)^2)=2,\ \dim \mathbf{N}(A-\beta I_4)=2,$$

then there is a basis $\mathscr{P}=\{\mathbf{p}_1,\mathbf{p}_2,\mathbf{p}_3,\mathbf{p}_4\}$ in $\mathbb{R}^4$ such that the $\mathscr{P}$-matrix of the linear transformation $f(\mathbf{x})=A\mathbf{x}$ is $\begin{bmatrix}\alpha&1&0&0\\0&\alpha&0&0\\0&0&\beta&0\\0&0&0&\beta\end{bmatrix}$ and consequently

$$A=\begin{bmatrix}\mathbf{p}_1&\mathbf{p}_2&\mathbf{p}_3&\mathbf{p}_4\end{bmatrix}\begin{bmatrix}\alpha&1&0&0\\0&\alpha&0&0\\0&0&\beta&0\\0&0&0&\beta\end{bmatrix}\begin{bmatrix}\mathbf{p}_1&\mathbf{p}_2&\mathbf{p}_3&\mathbf{p}_4\end{bmatrix}^{-1}.$$

Proof. Let $\mathbf{u}$ be a vector in $\mathbf{N}((A-\alpha I_4)^2)$ that is not in $\mathbf{N}(A-\alpha I_4)$ and let $\{\mathbf{v},\mathbf{w}\}$ be a basis in $\mathbf{N}(A-\beta I_4)$. Then the nonzero vector $(A-\alpha I_4)\mathbf{u}$ is in $\mathbf{N}(A-\alpha I_4)$. We will show that

$$\mathscr{P}=\{(A-\alpha I_4)\mathbf{u},\mathbf{u},\mathbf{v},\mathbf{w}\}$$

is a basis in $\mathbb{R}^4$ that has the desired property.

If

$$x_1(A-\alpha I_4)\mathbf{u}+x_2\mathbf{u}+x_3\mathbf{v}+x_4\mathbf{w}=\mathbf{0}$$

for some real numbers x_1,x_2,x_3,x_4, then

$$x_1(A-\alpha I_4)^3\mathbf{u}+x_2(A-\alpha I_4)^2\mathbf{u}+x_3(A-\alpha I_4)^2\mathbf{v}+x_4(A-\alpha I_4)^2\mathbf{w}=\mathbf{0}$$

and then

$$(A-\alpha I_4)^2(x_3\mathbf{v}+x_4\mathbf{w})=\mathbf{0}$$

which gives us $x_3=x_4=0$, because

$$(A-\alpha I_4)^2=(A-\beta I_4)^2+2(\beta-\alpha)(A-\beta I_4)+(\beta-\alpha)^2I_4.$$

Next we obtain $x_1=x_2=0$ as in the proof of Theorem 7.1. This shows that the vectors $(A-\alpha I_4)\mathbf{u}$, $\mathbf{u}$, $\mathbf{v}$, and $\mathbf{w}$ are linearly independent and consequently $\{(A-\alpha I_4)\mathbf{u},\mathbf{u},\mathbf{v},\mathbf{w}\}$ is a basis in $\mathbb{R}^4$.

Now we calculate

$$\begin{aligned}
&A(A-\alpha I_4)\mathbf{u}=(A-\alpha I_4)^2\mathbf{u}+\alpha(A-\alpha I_4)\mathbf{u})=\alpha(A-\alpha I_4)\mathbf{u},\\
&A\mathbf{u}=(A-\alpha I_4)\mathbf{u}+\alpha\mathbf{u},\\
&A\mathbf{v}=\beta\mathbf{v},\ \text{and}\\
&A\mathbf{w}=\beta\mathbf{w}.
\end{aligned}$$

If we let

$$\mathbf{p}_1 = (A - \alpha I_4)\mathbf{u}, \quad \mathbf{p}_2 = \mathbf{u}, \quad \mathbf{p}_3 = \mathbf{v}, \quad \mathbf{p}_4 = \mathbf{w},$$

then we have

$$\begin{aligned} A\mathbf{p}_1 &= \alpha\mathbf{p}_1, \\ A\mathbf{p}_2 &= \mathbf{p}_1 + \alpha\mathbf{p}_2, \\ A\mathbf{p}_3 &= \beta\mathbf{p}_3, \\ A\mathbf{p}_4 &= \beta\mathbf{p}_4. \end{aligned}$$

Consequently, the $\{(A - \alpha I_3)\mathbf{u}, \mathbf{u}, \mathbf{v}, \mathbf{w}\}$-matrix of the linear transformation $f(\mathbf{x}) = A\mathbf{x}$ is $\begin{bmatrix} \alpha & 1 & 0 & 0 \\ 0 & \alpha & 0 & 0 \\ 0 & 0 & \beta & 0 \\ 0 & 0 & 0 & \beta \end{bmatrix}$ and we have

$$A = \begin{bmatrix} \mathbf{p}_1 & \mathbf{p}_2 & \mathbf{p}_3 & \mathbf{p}_4 \end{bmatrix} \begin{bmatrix} \alpha & 1 & 0 & 0 \\ 0 & \alpha & 0 & 0 \\ 0 & 0 & \beta & 0 \\ 0 & 0 & 0 & \beta \end{bmatrix} \begin{bmatrix} \mathbf{p}_1 & \mathbf{p}_2 & \mathbf{p}_3 & \mathbf{p}_4 \end{bmatrix}^{-1},$$

by Corollary 6.23. □

Theorem 7.19. *Let A be a 4×4 matrix and let α, β, and γ be three distinct real numbers. If the characteristic polynomial of the matrix A is $(\lambda - \alpha)^2(\lambda - \beta)(\lambda - \gamma)$ and we have*

$$\dim \mathbf{N}(A - \alpha I_4) = 1, \ \dim \mathbf{N}((A - \alpha I_4)^2) = 2,$$

$$\dim \mathbf{N}(A - \beta I_4) = 1, \ \dim \mathbf{N}(A - \gamma I_4) = 1,$$

then there is a basis $\mathscr{P} = \{\mathbf{p}_1, \mathbf{p}_2, \mathbf{p}_3, \mathbf{p}_4\}$ in $\mathbb{R}^4$ such that the $\mathscr{P}$-matrix of the linear transformation $f(\mathbf{x}) = A\mathbf{x}$ is $\begin{bmatrix} \alpha & 1 & 0 & 0 \\ 0 & \alpha & 0 & 0 \\ 0 & 0 & \beta & 0 \\ 0 & 0 & 0 & \gamma \end{bmatrix}$ and consequently

$$A = \begin{bmatrix} \mathbf{p}_1 & \mathbf{p}_2 & \mathbf{p}_3 & \mathbf{p}_4 \end{bmatrix} \begin{bmatrix} \alpha & 1 & 0 & 0 \\ 0 & \alpha & 0 & 0 \\ 0 & 0 & \beta & 0 \\ 0 & 0 & 0 & \gamma \end{bmatrix} \begin{bmatrix} \mathbf{p}_1 & \mathbf{p}_2 & \mathbf{p}_3 & \mathbf{p}_4 \end{bmatrix}^{-1}.$$

Proof. Let $\mathbf{u}$ be a vector in $\mathbf{N}((A - \alpha I_4)^2)$ that is not in $\mathbf{N}(A - \alpha I_4)$ and let $\mathbf{v}$ and $\mathbf{w}$ be nonzero vectors in $\mathbf{N}(A - \beta I_4)$ and $\mathbf{N}(A - \gamma I_4)$, respectively. We will show that

$$\mathscr{P} = \{(A - \alpha I_4)\mathbf{u}, \mathbf{u}, \mathbf{v}, \mathbf{w}\}$$

is a basis in $\mathbb{R}^4$ that has the desired property.

If

$$x_1(A-\alpha I_4)\mathbf{u}+x_2\mathbf{u}+x_3\mathbf{v}+x_4\mathbf{w}=\mathbf{0}$$

for some real numbers x_1, x_2, x_3, x_4, then

$$(A-\alpha I_4)(x_3\mathbf{v}+x_4\mathbf{w})=\mathbf{0}.$$

Since

$$(A-\alpha I_4)\mathbf{v}=(A-\beta+(\beta-\alpha)I_4)\mathbf{v}=(\beta-\alpha)\mathbf{v}$$

and

$$(A-\alpha I_4)\mathbf{w}=(A-\gamma+(\gamma-\alpha)I_4)\mathbf{w}=(\gamma-\alpha)\mathbf{w},$$

we obtain $x_3 = x_4 = 0$, because two eigenvectors corresponding to different eigenvalues are linearly independent. Now we have

$$x_1(A-\alpha I_4)\mathbf{u}+x_2\mathbf{u}=0$$

which gives us $x_1 = x_2 = 0$ as in the proof of Theorem 7.1. This shows that the vectors $(A-\alpha I_4)\mathbf{u}$, $\mathbf{u}$, $\mathbf{v}$, and $\mathbf{w}$ are linearly independent and consequently $\{(A-\alpha I_4)\mathbf{u},\mathbf{u},\mathbf{v},\mathbf{w}\}$ is a basis in $\mathbb{R}^4$.

Now we calculate

$$\begin{aligned}
&A(A-\alpha I_4)\mathbf{u}=(A-\alpha I_4)^2\mathbf{u}+\alpha(A-\alpha I_4)\mathbf{u}=\alpha(A-\alpha I_4)\mathbf{u},\\
&A\mathbf{u}=(A-\alpha I_4)\mathbf{u}+\alpha\mathbf{u},\\
&A\mathbf{v}=\beta\mathbf{v}, \text{ and}\\
&A\mathbf{w}=\gamma\mathbf{w}.
\end{aligned}$$

If we let

$$\mathbf{p}_1=(A-\alpha I_4)\mathbf{u}, \quad \mathbf{p}_2=\mathbf{u}, \quad \mathbf{p}_3=\mathbf{v}, \quad \mathbf{p}_4=\mathbf{w},$$

then we have

$$\begin{aligned}
A\mathbf{p}_1&=\alpha\mathbf{p}_1,\\
A\mathbf{p}_2&=\mathbf{p}_1+\alpha\mathbf{p}_2,\\
A\mathbf{p}_3&=\beta\mathbf{p}_3,\\
A\mathbf{p}_4&=\gamma\mathbf{p}_4.
\end{aligned}$$

Consequently, the $\{(A-\alpha I_4)\mathbf{u},\mathbf{u},\mathbf{v},\mathbf{w}\}$-matrix of the linear transformation $f(\mathbf{x}) = A\mathbf{x}$ is $\begin{bmatrix}\alpha&1&0&0\\0&\alpha&0&0\\0&0&\beta&0\\0&0&0&\gamma\end{bmatrix}$ and we have

$$A=\begin{bmatrix}\mathbf{p}_1&\mathbf{p}_2&\mathbf{p}_3&\mathbf{p}_4\end{bmatrix}\begin{bmatrix}\alpha&1&0&0\\0&\alpha&0&0\\0&0&\beta&0\\0&0&0&\gamma\end{bmatrix}\begin{bmatrix}\mathbf{p}_1&\mathbf{p}_2&\mathbf{p}_3&\mathbf{p}_4\end{bmatrix}^{-1},$$

by Corollary 6.23. □

Definition 7.20. A 4×4 matrix is called a *standard Jordan canonical form* if it is a diagonal matrix or has of one of the following nine forms

$$\begin{bmatrix}\alpha & 1 & 0 & 0\\ 0 & \alpha & 1 & 0\\ 0 & 0 & \alpha & 1\\ 0 & 0 & 0 & \alpha\end{bmatrix}, \begin{bmatrix}\alpha & 1 & 0 & 0\\ 0 & \alpha & 1 & 0\\ 0 & 0 & \alpha & 0\\ 0 & 0 & 0 & \alpha\end{bmatrix}, \begin{bmatrix}\alpha & 1 & 0 & 0\\ 0 & \alpha & 0 & 0\\ 0 & 0 & \alpha & 1\\ 0 & 0 & 0 & \alpha\end{bmatrix}, \begin{bmatrix}\alpha & 1 & 0 & 0\\ 0 & \alpha & 0 & 0\\ 0 & 0 & \alpha & 0\\ 0 & 0 & 0 & \alpha\end{bmatrix},$$

$$\begin{bmatrix}\alpha & 1 & 0 & 0\\ 0 & \alpha & 1 & 0\\ 0 & 0 & \alpha & 0\\ 0 & 0 & 0 & \beta\end{bmatrix}, \begin{bmatrix}\alpha & 1 & 0 & 0\\ 0 & \alpha & 0 & 0\\ 0 & 0 & \alpha & 0\\ 0 & 0 & 0 & \beta\end{bmatrix}, \begin{bmatrix}\alpha & 1 & 0 & 0\\ 0 & \alpha & 0 & 0\\ 0 & 0 & \beta & 1\\ 0 & 0 & 0 & \beta\end{bmatrix}, \begin{bmatrix}\alpha & 1 & 0 & 0\\ 0 & \alpha & 0 & 0\\ 0 & 0 & \beta & 0\\ 0 & 0 & 0 & \beta\end{bmatrix},$$

$$\begin{bmatrix}\alpha & 1 & 0 & 0\\ 0 & \alpha & 0 & 0\\ 0 & 0 & \beta & 0\\ 0 & 0 & 0 & \gamma\end{bmatrix},$$

where α, β, and γ are distinct real numbers.

As in the case of 3×3 matrices, there are other 4×4 matrices that are called Jordan canonical forms, for example, $\begin{bmatrix}\alpha & 0 & 0 & 0\\ 0 & \alpha & 1 & 0\\ 0 & 0 & \alpha & 1\\ 0 & 0 & 0 & \alpha\end{bmatrix}$. Since, if for some 4×4 matrix A we have

$$A = \begin{bmatrix}\mathbf{p}_1 & \mathbf{p}_2 & \mathbf{p}_3 & \mathbf{p}_4\end{bmatrix}\begin{bmatrix}\alpha & 0 & 0 & 0\\ 0 & \alpha & 1 & 0\\ 0 & 0 & \alpha & 1\\ 0 & 0 & 0 & \alpha\end{bmatrix}\begin{bmatrix}\mathbf{p}_1 & \mathbf{p}_2 & \mathbf{p}_3 & \mathbf{p}_4\end{bmatrix}^{-1},$$

for some basis $\{\mathbf{p}_1, \mathbf{p}_2, \mathbf{p}_3, \mathbf{p}_4\}$ in $\mathbb{R}^4$, then we also have

$$A = \begin{bmatrix}\mathbf{p}_2 & \mathbf{p}_3 & \mathbf{p}_4 & \mathbf{p}_1\end{bmatrix}\begin{bmatrix}\alpha & 1 & 0 & 0\\ 0 & \alpha & 1 & 0\\ 0 & 0 & \alpha & 0\\ 0 & 0 & 0 & \alpha\end{bmatrix}\begin{bmatrix}\mathbf{p}_2 & \mathbf{p}_3 & \mathbf{p}_4 & \mathbf{p}_1\end{bmatrix}^{-1},$$

it is not necessary to include both on the list of standard Jordan canonical forms. From Theorems 7.11–7.19 it should be clear that every 4×4 matrix with real eigenvalues can be represented in the form $A = PJP^{-1}$, where P is an invertible matrix and J is a standard Jordan canonical form.

Example 7.21. The characteristic polynomial of the matrix

$$A = \begin{bmatrix} 3 & 0 & 1 & -2 \\ 2 & 2 & 1 & -3 \\ 2 & 0 & 3 & -3 \\ 1 & 0 & 1 & 0 \end{bmatrix}$$

is $(\lambda-2)^4$. Find an invertible 4×4 matrix P and a 4×4 standard Jordan canonical form J such that $A = PJP^{-1}$.

Solution. Since

$$\mathbf{N}(A-2I_4) = \mathbf{N}\left(\begin{bmatrix} 1 & 0 & 1 & -2 \\ 2 & 0 & 1 & -3 \\ 2 & 0 & 1 & -3 \\ 1 & 0 & 1 & -2 \end{bmatrix}\right) = \operatorname{Span}\left\{\begin{bmatrix} 1 \\ 0 \\ 1 \\ 1 \end{bmatrix}, \begin{bmatrix} 0 \\ 1 \\ 0 \\ 0 \end{bmatrix}\right\},$$

$$\mathbf{N}((A-2I_4)^2) = \mathbf{N}\left(\begin{bmatrix} 1 & 0 & 0 & -1 \\ 1 & 0 & 0 & -1 \\ 1 & 0 & 0 & -1 \\ 1 & 0 & 0 & -1 \end{bmatrix}\right) = \operatorname{Span}\left\{\begin{bmatrix} 0 \\ 1 \\ 0 \\ 0 \end{bmatrix}, \begin{bmatrix} 0 \\ 0 \\ 1 \\ 0 \end{bmatrix}, \begin{bmatrix} 1 \\ 0 \\ 0 \\ 1 \end{bmatrix}\right\}$$

and

$$\mathbf{N}((A-2I_4)^3) = \mathbf{N}\left(\begin{bmatrix} 0 & 0 & 0 & 0 \\ 0 & 0 & 0 & 0 \\ 0 & 0 & 0 & 0 \\ 0 & 0 & 0 & 0 \end{bmatrix}\right) = \mathbb{R}^4,$$

we can use Theorem 7.12.

The vector $\begin{bmatrix} 0 \\ 0 \\ 0 \\ 1 \end{bmatrix}$ is not in $\operatorname{Span}\left\{\begin{bmatrix} 0 \\ 1 \\ 0 \\ 0 \end{bmatrix}, \begin{bmatrix} 0 \\ 0 \\ 1 \\ 0 \end{bmatrix}, \begin{bmatrix} 1 \\ 0 \\ 0 \\ 1 \end{bmatrix}\right\}$ and we have

$$\begin{bmatrix} 1 & 0 & 1 & -2 \\ 2 & 0 & 1 & -3 \\ 2 & 0 & 1 & -3 \\ 1 & 0 & 1 & -2 \end{bmatrix}\begin{bmatrix} 0 \\ 0 \\ 0 \\ 1 \end{bmatrix} = \begin{bmatrix} -2 \\ -3 \\ -3 \\ -2 \end{bmatrix}$$

and

$$\begin{bmatrix} 1 & 0 & 0 & -1 \\ 1 & 0 & 0 & -1 \\ 1 & 0 & 0 & -1 \\ 1 & 0 & 0 & -1 \end{bmatrix}\begin{bmatrix} 0 \\ 0 \\ 0 \\ 1 \end{bmatrix} = \begin{bmatrix} -1 \\ -1 \\ -1 \\ -1 \end{bmatrix}.$$

Since the vectors $\begin{bmatrix} -1 \\ -1 \\ -1 \\ -1 \end{bmatrix}$ and $\begin{bmatrix} 0 \\ 1 \\ 0 \\ 0 \end{bmatrix}$ are linearly independent, we can take

$$P = \begin{bmatrix} -1 & -2 & 0 & 0 \\ -1 & -3 & 0 & 1 \\ -1 & -3 & 0 & 0 \\ -1 & -2 & 1 & 0 \end{bmatrix} \quad \text{and} \quad J = \begin{bmatrix} 2 & 1 & 0 & 0 \\ 0 & 2 & 1 & 0 \\ 0 & 0 & 2 & 0 \\ 0 & 0 & 0 & 2 \end{bmatrix}.$$

□

Example 7.22. The characteristic polynomial of the matrix

$$A = \begin{bmatrix} 5 & -1 & 0 & 2 \\ 0 & 3 & 0 & 4 \\ 0 & -1 & 5 & 2 \\ 0 & -1 & 0 & 7 \end{bmatrix}$$

is $(\lambda - 5)^4$. Find an invertible 4×4 matrix P and a 4×4 standard Jordan canonical form J such that $A = PJP^{-1}$.

Solution. Since

$$\mathbf{N}(A - 5I_4) = \mathbf{N}\left(\begin{bmatrix} 0 & -1 & 0 & 2 \\ 0 & -2 & 0 & 4 \\ 0 & -1 & 0 & 2 \\ 0 & -1 & 0 & 2 \end{bmatrix}\right) = \operatorname{Span}\left\{\begin{bmatrix} 1 \\ 0 \\ 0 \\ 0 \end{bmatrix}, \begin{bmatrix} 0 \\ 2 \\ 0 \\ 1 \end{bmatrix}, \begin{bmatrix} 0 \\ 0 \\ 1 \\ 0 \end{bmatrix}\right\}$$

and

$$\mathbf{N}((A - 5I_4)^2) = \mathbf{N}\left(\begin{bmatrix} 0 & 0 & 0 & 0 \\ 0 & 0 & 0 & 0 \\ 0 & 0 & 0 & 0 \\ 0 & 0 & 0 & 0 \end{bmatrix}\right) = \mathbb{R}^4,$$

we can use Theorem 7.14.

The vector $\begin{bmatrix} 0 \\ 1 \\ 0 \\ 0 \end{bmatrix}$ is not in $\operatorname{Span}\left\{\begin{bmatrix} 1 \\ 0 \\ 0 \\ 0 \end{bmatrix}, \begin{bmatrix} 0 \\ 2 \\ 0 \\ 1 \end{bmatrix}, \begin{bmatrix} 0 \\ 0 \\ 1 \\ 0 \end{bmatrix}\right\}$ and we have

$$\begin{bmatrix} 0 & -1 & 0 & 2 \\ 0 & -2 & 0 & 4 \\ 0 & -1 & 0 & 2 \\ 0 & -1 & 0 & 2 \end{bmatrix}\begin{bmatrix} 0 \\ 1 \\ 0 \\ 0 \end{bmatrix} = \begin{bmatrix} -1 \\ -2 \\ -1 \\ -1 \end{bmatrix}.$$

Since the vectors $\begin{bmatrix} -1 \\ -2 \\ -1 \\ -1 \end{bmatrix}$, $\begin{bmatrix} 1 \\ 0 \\ 0 \\ 0 \end{bmatrix}$ and $\begin{bmatrix} 0 \\ 0 \\ 1 \\ 0 \end{bmatrix}$ are linearly independent, we can take

$$P = \begin{bmatrix} -1 & 0 & 1 & 0 \\ -2 & 1 & 0 & 0 \\ -1 & 0 & 0 & 1 \\ -1 & 0 & 0 & 0 \end{bmatrix} \quad \text{and} \quad J = \begin{bmatrix} 5 & 1 & 0 & 0 \\ 0 & 5 & 0 & 0 \\ 0 & 0 & 5 & 0 \\ 0 & 0 & 0 & 5 \end{bmatrix}.$$

□

Example 7.23. The characteristic polynomial of the matrix

$$A = \begin{bmatrix} 5 & -6 & -10 & 11 \\ 10 & -11 & -20 & 21 \\ 5 & -6 & -10 & 11 \\ 10 & -11 & -20 & 21 \end{bmatrix}$$

is $\lambda^3(\lambda - 5)$. Find an invertible 4×4 matrix P and a 4×4 standard Jordan canonical form J such that $A = PJP^{-1}$.

Solution. Since

$$\mathbf{N}(A - 5I_4) = \mathbf{N}\left(\begin{bmatrix} 0 & -6 & -10 & 11 \\ 10 & -16 & -20 & 21 \\ 5 & -6 & -15 & 11 \\ 10 & -11 & -20 & 16 \end{bmatrix}\right) = \operatorname{Span}\left\{\begin{bmatrix} 1 \\ 2 \\ 1 \\ 2 \end{bmatrix}\right\},$$

$$\mathbf{N}(A) = \mathbf{N}\left(\begin{bmatrix} 5 & -6 & -10 & 11 \\ 10 & -11 & -20 & 21 \\ 5 & -6 & -10 & 11 \\ 10 & -11 & -20 & 21 \end{bmatrix}\right) = \operatorname{Span}\left\{\begin{bmatrix} -1 \\ 1 \\ 0 \\ 1 \end{bmatrix}, \begin{bmatrix} 2 \\ 0 \\ 1 \\ 0 \end{bmatrix}\right\}$$

and

$$\mathbf{N}(A^2) = \mathbf{N}\left(\begin{bmatrix} 25 & -25 & -50 & 50 \\ 50 & -50 & -100 & 100 \\ 25 & -25 & -50 & 50 \\ 50 & -50 & -100 & 100 \end{bmatrix}\right) = \operatorname{Span}\left\{\begin{bmatrix} 1 \\ 1 \\ 0 \\ 0 \end{bmatrix}, \begin{bmatrix} 2 \\ 0 \\ 1 \\ 0 \end{bmatrix}, \begin{bmatrix} -2 \\ 0 \\ 0 \\ 1 \end{bmatrix}\right\},$$

we can use Theorem 7.16.

The vector $\begin{bmatrix} 1 \\ 1 \\ 0 \\ 0 \end{bmatrix}$ is not in Span$\left\{\begin{bmatrix} -1 \\ 1 \\ 0 \\ 1 \end{bmatrix}, \begin{bmatrix} 2 \\ 0 \\ 1 \\ 0 \end{bmatrix}\right\}$ and we have

$$\begin{bmatrix} 5 & -6 & -10 & 11 \\ 10 & -11 & -20 & 21 \\ 5 & -6 & -10 & 11 \\ 10 & -11 & -20 & 21 \end{bmatrix}\begin{bmatrix} 1 \\ 1 \\ 0 \\ 0 \end{bmatrix} = \begin{bmatrix} -1 \\ -1 \\ -1 \\ -1 \end{bmatrix}.$$

Since the vectors $\begin{bmatrix} -1 \\ -1 \\ -1 \\ -1 \end{bmatrix}$ and $\begin{bmatrix} -1 \\ 1 \\ 0 \\ 1 \end{bmatrix}$ are linearly independent, we can take

$$P = \begin{bmatrix} -1 & 1 & -1 & 1 \\ -1 & 1 & 1 & 2 \\ -1 & 0 & 0 & 1 \\ -1 & 0 & 1 & 2 \end{bmatrix} \quad \text{and} \quad J = \begin{bmatrix} 0 & 1 & 0 & 0 \\ 0 & 0 & 0 & 0 \\ 0 & 0 & 0 & 0 \\ 0 & 0 & 0 & 5 \end{bmatrix}.$$

□

Example 7.24. The characteristic polynomial of the matrix

$$A = \begin{bmatrix} 3 & -4 & 4 & -1 \\ -3 & 7 & -7 & 4 \\ -4 & 4 & -5 & 5 \\ -4 & 0 & -4 & 7 \end{bmatrix}$$

is $(\lambda - 3)^4$. Find an invertible 4×4 matrix P and a 4×4 standard Jordan canonical form J such that $A = PJP^{-1}$.

Solution. Since

$$\mathbf{N}(A - 3I_4) = \mathbf{N}\left(\begin{bmatrix} 0 & -4 & 4 & -1 \\ -3 & 4 & -7 & 4 \\ -4 & 4 & -8 & 5 \\ -4 & 0 & -4 & 4 \end{bmatrix}\right) = \text{Span}\left\{\begin{bmatrix} -4 \\ 1 \\ 0 \\ -4 \end{bmatrix}, \begin{bmatrix} 3 \\ 0 \\ 1 \\ 4 \end{bmatrix}\right\}$$

and

$$\mathbf{N}((A - 3I_4)^2) = \mathbf{N}\left(\begin{bmatrix} 0 & 0 & 0 & 0 \\ 0 & 0 & 0 & 0 \\ 0 & 0 & 0 & 0 \\ 0 & 0 & 0 & 0 \end{bmatrix}\right) = \mathbb{R}^4,$$

we can use Theorem 7.13.

It is easy to see that the vectors $\begin{bmatrix} -4 \\ 1 \\ 0 \\ -4 \end{bmatrix}$, $\begin{bmatrix} 3 \\ 0 \\ 1 \\ 4 \end{bmatrix}$, $\begin{bmatrix} 1 \\ 0 \\ 0 \\ 0 \end{bmatrix}$, and $\begin{bmatrix} 0 \\ 0 \\ 0 \\ 1 \end{bmatrix}$ are linearly independent. Since

$$\begin{bmatrix} 0 & -4 & 4 & -1 \\ -3 & 4 & -7 & 4 \\ -4 & 4 & -8 & 5 \\ -4 & 0 & -4 & 4 \end{bmatrix} \begin{bmatrix} 1 \\ 0 \\ 0 \\ 0 \end{bmatrix} = \begin{bmatrix} 0 \\ -3 \\ -4 \\ -4 \end{bmatrix}$$

and

$$\begin{bmatrix} 0 & -4 & 4 & -1 \\ -3 & 4 & -7 & 4 \\ -4 & 4 & -8 & 5 \\ -4 & 0 & -4 & 4 \end{bmatrix} \begin{bmatrix} 0 \\ 0 \\ 0 \\ 1 \end{bmatrix} = \begin{bmatrix} -1 \\ 4 \\ 5 \\ 4 \end{bmatrix},$$

we can take

$$P = \begin{bmatrix} 0 & 1 & -1 & 0 \\ -3 & 0 & 4 & 0 \\ -4 & 0 & 5 & 0 \\ -4 & 0 & 4 & 1 \end{bmatrix} \quad \text{and} \quad J = \begin{bmatrix} 3 & 1 & 0 & 0 \\ 0 & 3 & 0 & 0 \\ 0 & 0 & 3 & 1 \\ 0 & 0 & 0 & 3 \end{bmatrix}.$$

□

Exercises 7.1

For the given matrix A find an invertible 2×2 matrix P and a 2×2 standard Jordan canonical form J such that $A = PJP^{-1}$.

1. $A = \begin{bmatrix} -2 & 9 \\ -4 & 10 \end{bmatrix}$

2. $A = \begin{bmatrix} -13 & 4 \\ -9 & -1 \end{bmatrix}$

For the given matrix A and its characteristic polynomial $p(\lambda)$ find an invertible 3×3 matrix P and a 3×3 standard Jordan canonical form J such that $A = PJP^{-1}$.

3. $A = \begin{bmatrix} -2 & 16 & -4 \\ -3 & 12 & -2 \\ -3 & 8 & 2 \end{bmatrix}$, $p(\lambda) = (\lambda - 4)^3$

4. $A = \begin{bmatrix} 2 & -1 & 0 \\ 4 & -1 & -1 \\ 4 & -1 & -1 \end{bmatrix}$, $p(\lambda) = -\lambda^3$

5. $A = \begin{bmatrix} 13 & -32 & 12 \\ 9 & -25 & 10 \\ 13 & -40 & 17 \end{bmatrix}$, $p(\lambda) = -(\lambda - 3)^2(\lambda + 1)$

6. $A = \begin{bmatrix} -2 & 3 & 0 \\ -11 & 10 & 1 \\ 27 & -19 & -3 \end{bmatrix}$, $p(\lambda) = -(\lambda-2)^2(\lambda-1)$

7. $A = \begin{bmatrix} -2 & 5 & -2 \\ -6 & 10 & -4 \\ -5 & 7 & -2 \end{bmatrix}$, $p(\lambda) = (\lambda-2)^3$

8. $A = \begin{bmatrix} -20 & 15 & 3 \\ -45 & 33 & 7 \\ 61 & -43 & -10 \end{bmatrix}$, $p(\lambda) = -(\lambda-1)^3$

9. $A = \begin{bmatrix} 4 & -3 & 2 \\ 5 & -4 & 2 \\ -5 & 3 & -3 \end{bmatrix}$, $p(\lambda) = -(\lambda+1)^3$

10. $A = \begin{bmatrix} -13 & 40 & -15 \\ -4 & 13 & -6 \\ -4 & 16 & -9 \end{bmatrix}$, $p(\lambda) = (\lambda+3)^3$

For the given matrix A and its characteristic polynomial $p(\lambda)$ find an invertible 4×4 matrix P and a 4×4 standard Jordan canonical form J such that $A = PJP^{-1}$.

11. $A = \begin{bmatrix} 3 & -2 & -3 & 3 \\ 4 & -7 & -9 & 13 \\ 2 & -3 & -3 & 5 \\ 4 & -8 & -9 & 14 \end{bmatrix}$, $p(\lambda) = (\lambda-1)^2(\lambda-2)(\lambda-3)$

12. $A = \begin{bmatrix} -4 & 6 & 10 & -4 \\ -3 & 5 & 5 & -2 \\ -3 & 3 & 7 & -2 \\ -2 & 2 & 3 & 1 \end{bmatrix}$, $p(\lambda) = (\lambda-2)^3(\lambda-3)$

13. $A = \begin{bmatrix} -5 & 0 & 4 & 3 \\ 0 & -4 & 2 & 4 \\ -3 & 1 & 1 & 1 \\ 2 & -2 & -1 & 0 \end{bmatrix}$, $p(\lambda) = (\lambda+2)^4$

14. $A = \begin{bmatrix} -1 & 0 & 4 & -1 \\ 9 & 2 & -8 & 3 \\ -3 & -1 & 6 & 0 \\ 13 & -4 & -12 & 9 \end{bmatrix}$, $p(\lambda) = (\lambda-4)^4$

15. Let A be a 5×5 matrix and let α be a real number such that

$$\dim \mathbf{N}(A-\alpha I_5) = 1,\ \dim \mathbf{N}(A-\alpha I_5)^2 = 2,\ \dim \mathbf{N}((A-\alpha I_5)^3) = 3,$$

$$\dim \mathbf{N}((A-\alpha I_5)^4) = 4, \text{ and } \dim \mathbf{N}((A-\alpha I_5)^5) = 5.$$

Show that there is a basis $\mathscr{P} = \{\mathbf{p}_1, \mathbf{p}_2, \mathbf{p}_3, \mathbf{p}_4, \mathbf{p}_5\}$ in $\mathbb{R}^5$ such that the $\mathscr{P}$-matrix of the linear transformation $f(\mathbf{x}) = A\mathbf{x}$ is $\begin{bmatrix} \alpha & 1 & 0 & 0 & 0 \\ 0 & \alpha & 1 & 0 & 0 \\ 0 & 0 & \alpha & 1 & 0 \\ 0 & 0 & 0 & \alpha & 1 \\ 0 & 0 & 0 & 0 & \alpha \end{bmatrix}$.

16. Let A be a 5×5 matrix and let α be a real number such that

$$\dim \mathbf{N}(A - \alpha I_5) = 2, \ \dim \mathbf{N}((A - \alpha I_5)^2) = 4, \ \dim \mathbf{N}((A - \alpha I_5)^3) = 5.$$

Show that there is a basis $\mathscr{P} = \{\mathbf{p}_1, \mathbf{p}_2, \mathbf{p}_3, \mathbf{p}_4, \mathbf{p}_5\}$ in $\mathbb{R}^5$ such that the $\mathscr{P}$-matrix of the linear transformation $f(\mathbf{x}) = A\mathbf{x}$ is $\begin{bmatrix} \alpha & 1 & 0 & 0 & 0 \\ 0 & \alpha & 1 & 0 & 0 \\ 0 & 0 & \alpha & 0 & 0 \\ 0 & 0 & 0 & \alpha & 1 \\ 0 & 0 & 0 & 0 & \alpha \end{bmatrix}$.

17. Let A be a 5×5 matrix and let α be a real number such that

$$\dim \mathbf{N}(A - \alpha I_5) = 3 \quad \text{and} \quad \dim \mathbf{N}((A - \alpha I_5)^2) = 5.$$

Show that there is a basis $\mathscr{P} = \{\mathbf{p}_1, \mathbf{p}_2, \mathbf{p}_3, \mathbf{p}_4, \mathbf{p}_5\}$ in $\mathbb{R}^5$ such that the $\mathscr{P}$-matrix of the linear transformation $f(\mathbf{x}) = A\mathbf{x}$ is $\begin{bmatrix} \alpha & 1 & 0 & 0 & 0 \\ 0 & \alpha & 0 & 0 & 0 \\ 0 & 0 & \alpha & 1 & 0 \\ 0 & 0 & 0 & \alpha & 0 \\ 0 & 0 & 0 & 0 & \alpha \end{bmatrix}$.

18. Let A be a 5×5 matrix and let α be a real number such that

$$\dim \mathbf{N}(A - \alpha I_5) = 2, \ \dim \mathbf{N}(A - \alpha I_5)^2 = 3,$$

$$\dim \mathbf{N}((A - \alpha I_5)^3) = 4, \ \dim \mathbf{N}((A - \alpha I_5)^4) = 5.$$

Show that there is a basis $\mathscr{P} = \{\mathbf{p}_1, \mathbf{p}_2, \mathbf{p}_3, \mathbf{p}_4, \mathbf{p}_5\}$ in $\mathbb{R}^5$ such that the $\mathscr{P}$-matrix of the linear transformation $f(\mathbf{x}) = A\mathbf{x}$ is $\begin{bmatrix} \alpha & 1 & 0 & 0 & 0 \\ 0 & \alpha & 1 & 0 & 0 \\ 0 & 0 & \alpha & 1 & 0 \\ 0 & 0 & 0 & \alpha & 0 \\ 0 & 0 & 0 & 0 & \alpha \end{bmatrix}$.

19. Show that there is a basis $\mathscr{P} = \{\mathbf{p}_1, \mathbf{p}_2, \mathbf{p}_3, \mathbf{p}_4, \mathbf{p}_5\}$ in $\mathbb{R}^5$ such that

$$\begin{bmatrix} 7 & 5 & 1 & -22 & 6 \\ 6 & 8 & 1 & -27 & 8 \\ 4 & 4 & 3 & -18 & 5 \\ 4 & 4 & 1 & -16 & 5 \\ 5 & 5 & 2 & -23 & 8 \end{bmatrix} = P \begin{bmatrix} 2 & 1 & 0 & 0 & 0 \\ 0 & 2 & 1 & 0 & 0 \\ 0 & 0 & 2 & 0 & 0 \\ 0 & 0 & 0 & 2 & 1 \\ 0 & 0 & 0 & 0 & 2 \end{bmatrix} P^{-1},$$

where $P = \begin{bmatrix} \mathbf{p}_1 & \mathbf{p}_2 & \mathbf{p}_3 & \mathbf{p}_4 & \mathbf{p}_5 \end{bmatrix}$.

20. Show that there is a basis $\mathscr{P} = \{\mathbf{p}_1, \mathbf{p}_2, \mathbf{p}_3, \mathbf{p}_4, \mathbf{p}_5\}$ in $\mathbb{R}^5$ such that

$$\begin{bmatrix} 6 & 4 & 0 & -17 & 5 \\ 5 & 7 & 0 & -22 & 7 \\ 3 & 3 & 2 & -13 & 4 \\ 3 & 3 & 0 & -11 & 4 \\ 3 & 3 & 0 & -13 & 6 \end{bmatrix} = P \begin{bmatrix} 2 & 1 & 0 & 0 & 0 \\ 0 & 2 & 0 & 0 & 0 \\ 0 & 0 & 2 & 1 & 0 \\ 0 & 0 & 0 & 2 & 0 \\ 0 & 0 & 0 & 0 & 2 \end{bmatrix} P^{-1},$$

where $P = \begin{bmatrix} \mathbf{p}_1 & \mathbf{p}_2 & \mathbf{p}_3 & \mathbf{p}_4 & \mathbf{p}_5 \end{bmatrix}$.

Chapter 8

Singular value decomposition

In Chapter 7 we discussed a decomposition of square matrices that is similar to diagonalization, but can be applied to matrices that are not diagonalizable. In this chapter we consider a decomposition of matrices that are not square matrices that is also similar to diagonalization.

The singular value decomposition has many important applications in engineering. To illustrate how it can be used, consider the following example. If a vector $\mathbf{u}$ is in $\mathbb{R}^{300}$ and a vector $\mathbf{v}$ is in $\mathbb{R}^{100}$, then the product $\mathbf{u}\mathbf{v}^T$ is a matrix with 30,000 entries. On the other hand, the total number of entries in vectors $\mathbf{u}$ and $\mathbf{v}$ is 400. If we could write a matrix as a sum of products of vectors, we would need significantly fewer numbers to describe the matrix. The singular value decomposition gives us exactly this kind of representation of a matrix.

In our presentation of the singular decomposition of matrices we use most of the tools developed in Chapters 1–6, even more so than Chapter 7. It gives us an excellent opportunity to review the material from those chapters and show the ideas introduced there in action.

8.1 The outer product expansion of a matrix

We first consider an idea related to the singular value decomposition, namely the outer product expansion of a matrix.

Definition 8.1. Let $\mathbf{u}$ be a vector in $\mathbb{R}^m$ and let $\mathbf{v}$ be a vector in $\mathbb{R}^n$. The product

$$\mathbf{u}\mathbf{v}^T$$

is called the *outer product* of the vectors $\mathbf{u}$ and $\mathbf{v}$.

Example 8.2. For the vectors $\mathbf{u} = \begin{bmatrix} 2 \\ 1 \\ 5 \end{bmatrix}$ and $\mathbf{v} = \begin{bmatrix} 4 \\ 3 \end{bmatrix}$ we have

$$\mathbf{u}\mathbf{v}^T = \begin{bmatrix} 8 & 6 \\ 4 & 3 \\ 20 & 15 \end{bmatrix} \quad \text{and} \quad \mathbf{v}\mathbf{u}^T = \begin{bmatrix} 8 & 4 & 20 \\ 6 & 3 & 15 \end{bmatrix} = (\mathbf{u}\mathbf{v}^T)^T.$$

Suppose that we know that a given matrix A can be written as

$$A = \sigma_1 \mathbf{u}_1 \mathbf{v}_1^T + \sigma_2 \mathbf{u}_2 \mathbf{v}_2^T. \tag{8.1}$$

How can we obtain such a representation of the matrix A? More precisely, how can we find numbers σ_1 and σ_2 and vectors $\mathbf{u}_1$, $\mathbf{v}_1$, $\mathbf{u}_2$, and $\mathbf{v}_2$ such that the equation (8.1) holds? It's not even clear where to begin. Let's try to learn more about the equation (8.1) by carefully analyzing it.

First we note that, without loss of generality, we can assume that

$$\|\mathbf{v}_1\| = \|\mathbf{v}_2\| = \|\mathbf{u}_1\| = \|\mathbf{u}_2\| = 1,$$

since we can always adjust σ_1 and σ_2 if necessary. Similarly, we can assume that $\sigma_1 > 0$ and $\sigma_2 > 0$, because we can change the sign of one or both of the vectors $\mathbf{u}_1$ and $\mathbf{u}_2$, if necessary.

Since the expansion (8.1) is similar to the spectral decomposition of a symmetric matrix, we will suppose that

$$\mathbf{v}_1 \cdot \mathbf{v}_2 = 0 \quad \text{and} \quad \mathbf{u}_1 \cdot \mathbf{u}_2 = 0.$$

Now, with all these assumptions, the equation (8.1) gives us

$$A\mathbf{v}_1 = \sigma_1 \mathbf{u}_1 \quad \text{and} \quad A\mathbf{v}_2 = \sigma_2 \mathbf{u}_2.$$

Since, by the equation (8.1), the transpose of A is

$$A^T = \sigma_1 \mathbf{v}_1 \mathbf{u}_1^T + \sigma_2 \mathbf{v}_2 \mathbf{u}_2^T,$$

we also have

$$A^T \mathbf{u}_1 = \sigma_1 \mathbf{v}_1 \quad \text{and} \quad A^T \mathbf{u}_2 = \sigma_2 \mathbf{v}_2.$$

From the above equalities we get

$$A^T A\mathbf{v}_1 = \sigma_1^2 \mathbf{v}_1, \quad A^T A\mathbf{v}_2 = \sigma_2^2 \mathbf{v}_2.$$

Our investigation of the equation (8.1) can be summarised as follows: If the equation (8.1) holds, then the vectors $\mathbf{v}_1$ and $\mathbf{v}_2$ must be eigenvectors of the symmetric matrix $A^T A$ corresponding to the eigenvalues σ_1^2 and σ_2^2, respectively. This observation is not limited to a representation with two terms. We formulate the general result in the next theorem.

Theorem 8.3. *Let A be an $m \times n$ nonzero matrix. Suppose*

$$A = \sigma_1 \mathbf{u}_1 \mathbf{v}_1^T + \dots + \sigma_r \mathbf{u}_r \mathbf{v}_r^T,$$

where

(a) *r is a positive integer such that $r \le m$ and $r \le n$,*

(b) $\sigma_1 \ge \dots \ge \sigma_r > 0$,

(c) *$\mathbf{u}_1, \dots \mathbf{u}_r$ are orthonormal vectors in $\mathbb{R}^m$, and*

(d) *$\mathbf{v}_1, \dots, \mathbf{v}_r$ are orthonormal vectors in $\mathbb{R}^n$.*

Then

$$A^T A \mathbf{v}_1 = \sigma_1^2 \mathbf{v}_1, \dots, A^T A \mathbf{v}_r = \sigma_r^2 \mathbf{v}_r.$$

In other words, the vectors $\mathbf{v}_1, \dots, \mathbf{v}_r$ are eigenvectors of the symmetric matrix $A^T A$ corresponding to the eigenvalues $\sigma_1^2, \dots, \sigma_r^2$, respectively.

It is our goal to devise a general method for representing a matrix as a sum of terms of the form $\sigma_j \mathbf{u}_j \mathbf{v}_j^T$, but first we address the question of existence of such a representation. It turns out that it is possible for every matrix, which is the main result of this chapter.

Theorem 8.4. *For an arbitrary nonzero $m \times n$ matrix A there exist*

(a) *a positive integer r such that $r \le m$ and $r \le n$,*

(b) *real numbers $\sigma_1 \ge \dots \ge \sigma_r > 0$,*

(c) *orthonormal vectors $\mathbf{u}_1, \dots, \mathbf{u}_r$ in $\mathbb{R}^m$, and*

(d) *orthonormal vectors $\mathbf{v}_1, \dots, \mathbf{v}_r$ in $\mathbb{R}^n$,*

such that

$$A = \sigma_1 \mathbf{u}_1 \mathbf{v}_1^T + \dots + \sigma_r \mathbf{u}_r \mathbf{v}_r^T.$$

Proof of a particular case. We illustrate the method of proof by considering a case when $n = 5$. All the essential ideas are present in this case, so it should be clear how prove the general case.

Let A be an $m \times 5$ matrix and let $\{\mathbf{v}_1, \mathbf{v}_2, \mathbf{v}_3, \mathbf{v}_4, \mathbf{v}_5\}$ be a basis of orthonormal eigenvectors of the symmetric 5×5 matrix $A^T A$. If λ_j is the eigenvalue corresponding to the eigenvector $\mathbf{v}_j$, then

$$0 \le \|A\mathbf{v}_j\|^2 = (A\mathbf{v}_j)\cdot(A\mathbf{v}_j) = (A^T A\mathbf{v}_j)\cdot\mathbf{v}_j = \lambda_j \mathbf{v}_j\cdot\mathbf{v}_j = \lambda_j\|\mathbf{v}_j\|^2 = \lambda_j.$$

This means that

$$\lambda_j = \|A\mathbf{v}_j\|^2 \ge 0$$

for $j = 1,\dots,5$. For $i \ne j$ we have

$$(A\mathbf{v}_i)\cdot(A\mathbf{v}_j) = (A^T A\mathbf{v}_i)\cdot\mathbf{v}_j = \lambda_i\mathbf{v}_i\cdot\mathbf{v}_j = 0.$$

If we denote $\sigma_j = \sqrt{\lambda_j}$, then we have

$$\|A\mathbf{v}_j\| = \sigma_j$$

for $j = 1,\dots,5$. Suppose that

$$\sigma_1 \ge \sigma_2 \ge \sigma_3 > 0 \quad \text{and} \quad \sigma_4 = \sigma_5 = 0.$$

If we take

$$\mathbf{u}_1 = \frac{1}{\|A\mathbf{v}_1\|}A\mathbf{v}_1 = \frac{1}{\sigma_1}A\mathbf{v}_1, \quad \mathbf{u}_2 = \frac{1}{\|A\mathbf{v}_2\|}A\mathbf{v}_2 = \frac{1}{\sigma_2}A\mathbf{v}_2, \quad \mathbf{u}_3 = \frac{1}{\|A\mathbf{v}_3\|}A\mathbf{v}_3 = \frac{1}{\sigma_3}A\mathbf{v}_3,$$

then we have

$$A\mathbf{v}_1 = \sigma_1\mathbf{u}_1, \quad A\mathbf{v}_2 = \sigma_2\mathbf{u}_2, \quad A\mathbf{v}_3 = \sigma_3\mathbf{u}_3$$

and

$$A\mathbf{v}_4 = \mathbf{0}, \quad A\mathbf{v}_5 = \mathbf{0}.$$

Note that

$$\|\mathbf{u}_1\| = \|\mathbf{u}_2\| = \|\mathbf{u}_3\| = 1$$

and

$$\mathbf{u}_1\cdot\mathbf{u}_2 = \mathbf{u}_1\cdot\mathbf{u}_3 = \mathbf{u}_2\cdot\mathbf{u}_3 = 0,$$

since $(A\mathbf{v}_i)\cdot(A\mathbf{v}_j) = 0$ for $i \ne j$. Consequently,

$$A\mathbf{v}_1\mathbf{v}_1^T = \sigma_1\mathbf{u}_1\mathbf{v}_1^T, \quad A\mathbf{v}_2\mathbf{v}_2^T = \sigma_2\mathbf{u}_2\mathbf{v}_2^T, \quad A\mathbf{v}_3\mathbf{v}_3^T = \sigma_3\mathbf{u}_3\mathbf{v}_3^T \tag{8.2}$$

and

$$A\mathbf{v}_4\mathbf{v}_4^T = A\mathbf{v}_5\mathbf{v}_5^T = \mathbf{0}. \tag{8.3}$$

By adding the equalities (8.2) and (8.3) we get

$$A\left(\mathbf{v}_1\mathbf{v}_1^T + \mathbf{v}_2\mathbf{v}_2^T + \mathbf{v}_3\mathbf{v}_3^T + \mathbf{v}_4\mathbf{v}_4^T + \mathbf{v}_5\mathbf{v}_5^T\right) = \sigma_1\mathbf{u}_1\mathbf{v}_1^T + \sigma_2\mathbf{u}_2\mathbf{v}_2^T + \sigma_3\mathbf{u}_3\mathbf{v}_3^T. \tag{8.4}$$

Since $\{\mathbf{v}_1, \mathbf{v}_2, \mathbf{v}_3, \mathbf{v}_4, \mathbf{v}_5\}$ is an orthonormal basis in $\mathbb{R}^5$, the matrix

$$\mathbf{v}_1\mathbf{v}_1^T + \mathbf{v}_2\mathbf{v}_2^T + \mathbf{v}_3\mathbf{v}_3^T + \mathbf{v}_4\mathbf{v}_4^T + \mathbf{v}_5\mathbf{v}_5^T$$

is the matrix of the orthogonal projection on

$$\operatorname{Span}\{\mathbf{v}_1, \mathbf{v}_2, \mathbf{v}_3, \mathbf{v}_4, \mathbf{v}_5\} = \mathbb{R}^5.$$

But this means that

$$\mathbf{v}_1\mathbf{v}_1^T + \mathbf{v}_2\mathbf{v}_2^T + \mathbf{v}_3\mathbf{v}_3^T + \mathbf{v}_4\mathbf{v}_4^T + \mathbf{v}_5\mathbf{v}_5^T = \begin{bmatrix} 1 & 0 & 0 & 0 & 0 \\ 0 & 1 & 0 & 0 & 0 \\ 0 & 0 & 1 & 0 & 0 \\ 0 & 0 & 0 & 1 & 0 \\ 0 & 0 & 0 & 0 & 1 \end{bmatrix}.$$

Now the equation (8.4) can be written as

$$A = A \begin{bmatrix} 1 & 0 & 0 & 0 & 0 \\ 0 & 1 & 0 & 0 & 0 \\ 0 & 0 & 1 & 0 & 0 \\ 0 & 0 & 0 & 1 & 0 \\ 0 & 0 & 0 & 0 & 1 \end{bmatrix} = \sigma_1\mathbf{u}_1\mathbf{v}_1^T + \sigma_2\mathbf{u}_2\mathbf{v}_2^T + \sigma_3\mathbf{u}_3\mathbf{v}_3^T,$$

which gives us the desired representation of A. Note that in our case $r = 3$. □

Definition 8.5. The numbers

$$\sigma_1 \geq \cdots \geq \sigma_n \geq 0$$

such that $\sigma_j = \sqrt{\lambda_j}$, where $\lambda_1, \ldots, \lambda_n$ are the eigenvalues of the matrix $A^T A$, are called the *singular values* of the matrix A. The representation of a matrix A in the form

$$A = \sigma_1\mathbf{u}_1\mathbf{v}_1^T + \cdots + \sigma_r\mathbf{u}_r\mathbf{v}_r^T$$

is called the *outer product expansion* or the *outer product decomposition* of the matrix A.

Note that, while the theorem tells us that every matrix A has an outer product expansion, the proof of the theorem gives us an effective way of finding such a representation. It takes us through the steps necessary to obtain the outer product expansion. In practice, we often slightly alter the order of these steps. In the proof we assume from the beginning that the eigenvectors $\mathbf{v}_1, \ldots, \mathbf{v}_j$ are unit vectors. When we find eigenvectors of a specific matrix, they are usually not unit vectors. Clearly, we can normalize them right away, but this is often not practical. Since the fact that the vectors are unit vectors is not essential until the end of the proof, we can postpone normalizing the vectors until it's necessary. This approach leads to the following method for constructing the outer product expansion of a matrix A.

Step 1 Calculate $A^T A$.

Step 2 Find the eigenvalues $\lambda_1 \geq \cdots \geq \lambda_n$ of $A^T A$.

Step 3 Find orthogonal eigenvectors $\mathbf{V}_1, \dots, \mathbf{V}_n$ corresponding to the eigenvalues $\lambda_1, \dots, \lambda_n$.

Step 4 Write

$$I_n = \frac{1}{\|\mathbf{V}_1\|^2}\mathbf{V}_1\mathbf{V}_1^T + \cdots + \frac{1}{\|\mathbf{V}_n\|^2}\mathbf{V}_n\mathbf{V}_n^T.$$

Step 5 Multiply both sides of the above equality by the matrix A to get

$$A = \frac{1}{\|\mathbf{V}_1\|^2}(A\mathbf{V}_1)\mathbf{V}_1^T + \cdots + \frac{1}{\|\mathbf{V}_n\|^2}(A\mathbf{V}_n)\mathbf{V}_n^T$$

Step 6 Define $\sigma_j = \sqrt{\lambda_j}$ and

$$\mathbf{v}_j = \frac{1}{\|\mathbf{V}_1\|}\mathbf{V}_j \quad \text{and} \quad \mathbf{u}_j = \frac{1}{\sigma_j\|\mathbf{V}_j\|}A\mathbf{V}_j$$

for those j's for which $\sigma_j > 0$.

Step 7 Write the outer product expansion of A:

$$A = \sigma_1\mathbf{u}_1\mathbf{v}_1^T + \cdots + \sigma_r\mathbf{u}_r\mathbf{v}_r^T.$$

Note that, since $\|AV_j\| = \sigma_j\|V_j\|$, we have $\|\mathbf{u}_j\| = 1$ for those j's for which $\sigma_j > 0$.

Example 8.6. Find the outer product expansion of the matrix $A = \begin{bmatrix} 3 & -1 \\ 3 & 4 \\ 1 & 3 \\ -3 & 1 \\ 3 & -1 \end{bmatrix}$.

Solution.

Step 1 $A^T A = \begin{bmatrix} 37 & 6 \\ 6 & 28 \end{bmatrix}$.

Step 2 The eigenvalues of the symmetric matrix $A^T A = \begin{bmatrix} 37 & 6 \\ 6 & 28 \end{bmatrix}$ are 40 and 25.

Step 3 $\begin{bmatrix} 2 \\ 1 \end{bmatrix}$ is an eigenvector corresponding to the eigenvalue 40 and $\begin{bmatrix} 1 \\ -2 \end{bmatrix}$ is an eigenvector corresponding to the eigenvalue 25.

Step 4 $\begin{bmatrix} 1 & 0 \\ 0 & 1 \end{bmatrix} = \frac{1}{5}\begin{bmatrix} 2 \\ 1 \end{bmatrix}\begin{bmatrix} 2 & 1 \end{bmatrix} + \frac{1}{5}\begin{bmatrix} 1 \\ -2 \end{bmatrix}\begin{bmatrix} 1 & -2 \end{bmatrix}.$

Step 5

$$A = \frac{1}{5}\begin{bmatrix} 3 & -1 \\ 3 & 4 \\ 1 & 3 \\ -3 & 1 \\ 3 & -1 \end{bmatrix}\begin{bmatrix} 2 \\ 1 \end{bmatrix}\begin{bmatrix} 2 & 1 \end{bmatrix} + \frac{1}{5}\begin{bmatrix} 3 & -1 \\ 3 & 4 \\ 1 & 3 \\ -3 & 1 \\ 3 & -1 \end{bmatrix}\begin{bmatrix} 1 \\ -2 \end{bmatrix}\begin{bmatrix} 1 & -2 \end{bmatrix}$$

$$= \frac{1}{5}\begin{bmatrix} 5 \\ 10 \\ 5 \\ -5 \\ 5 \end{bmatrix}\begin{bmatrix} 2 & 1 \end{bmatrix} + \frac{1}{5}\begin{bmatrix} 5 \\ -5 \\ -5 \\ -5 \\ 5 \end{bmatrix}\begin{bmatrix} 1 & -2 \end{bmatrix}.$$

Step 6 $\sigma_1 = \sqrt{40}$ and $\sigma_2 = 5$ and

$$\mathbf{u}_1 = \frac{1}{\sqrt{40}\sqrt{5}}\begin{bmatrix} 5 \\ 10 \\ 5 \\ -5 \\ 5 \end{bmatrix} = \frac{1}{\sqrt{8}}\begin{bmatrix} 1 \\ 2 \\ 1 \\ -1 \\ 1 \end{bmatrix}, \quad \mathbf{u}_2 = \frac{1}{5\sqrt{5}}\begin{bmatrix} 5 \\ -5 \\ -5 \\ -5 \\ 5 \end{bmatrix} = \frac{1}{\sqrt{5}}\begin{bmatrix} 1 \\ -1 \\ -1 \\ -1 \\ 1 \end{bmatrix},$$

$$\mathbf{v}_1 = \frac{1}{\sqrt{5}}\begin{bmatrix} 2 \\ 1 \end{bmatrix}, \quad \mathbf{v}_2 = \frac{1}{\sqrt{5}}\begin{bmatrix} 1 \\ -2 \end{bmatrix}.$$

Step 7 The outer product expansion of A is

$$A = \sqrt{40}\left(\frac{1}{\sqrt{8}}\begin{bmatrix} 1 \\ 2 \\ 1 \\ -1 \\ 1 \end{bmatrix}\right)\left(\frac{1}{\sqrt{5}}\begin{bmatrix} 2 & 1 \end{bmatrix}\right) + 5\left(\frac{1}{\sqrt{5}}\begin{bmatrix} 1 \\ -1 \\ -1 \\ -1 \\ 1 \end{bmatrix}\right)\left(\frac{1}{\sqrt{5}}\begin{bmatrix} 1 & -2 \end{bmatrix}\right).$$

The singular values of the matrix A are $\sigma_1 = \sqrt{40}$ and $\sigma_2 = 5$. □

It is easy to obtain the outer product expansion of the transpose of a matrix A if we already have the outer product expansion of the matrix A.

Theorem 8.7. *Let A be an $m \times n$ matrix with the outer product expansion*

$$A = \sigma_1 \mathbf{u}_1 \mathbf{v}_1^T + \cdots + \sigma_r \mathbf{u}_r \mathbf{v}_r^T.$$

Then the vector $\mathbf{u}_k$ is an eigenvector of the matrix AA^T corresponding to the eigenvalue σ_k^2 and we have

$$A^T \mathbf{u}_k = \sigma_k \mathbf{v}_k,$$

for $1 \le k \le r$. The outer product expansion of the matrix A^T is

$$A^T = \sigma_1 \mathbf{v}_1 \mathbf{u}_1^T + \cdots + \sigma_r \mathbf{v}_r \mathbf{u}_r^T.$$

Proof. By applying the transpose operation to both sides of the equation

$$A = \sigma_1 \mathbf{u}_1 \mathbf{v}_1^T + \cdots + \sigma_r \mathbf{u}_r \mathbf{v}_r^T$$

we obtain

$$A^T = \sigma_1 \mathbf{v}_1 \mathbf{u}_1^T + \cdots + \sigma_r \mathbf{v}_r \mathbf{u}_r^T.$$

Moreover,

$$A^T \mathbf{u}_k = \sigma_k \mathbf{v}_k \mathbf{u}_k^T \mathbf{u}_k = \sigma_k \mathbf{v}_k$$

and

$$AA^T \mathbf{u}_k = A(\sigma_k \mathbf{v}_k) = \sigma_k A \mathbf{v}_k = \sigma_k^2 \mathbf{u}_k.$$

□

If A is an $m \times n$ matrix and $m < n$, then the matrix $A^T A$ is bigger than the matrix AA^T. If we need to find the outer product expansion of A, we can use the above theorem to reduce the amount of calculations by finding first the outer product expansion of A^T and then obtaining from it the outer product expansion of A. The next example illustrates such a situation.

Example 8.8. Find the outer product expansion of the matrix

$$A = \begin{bmatrix} 4 & 2 & 0 \\ 1 & 2 & \sqrt{3} \end{bmatrix}.$$

Solution. The outer product expansion of the matrix

$$B = A^T = \begin{bmatrix} 4 & 1 \\ 2 & 2 \\ 0 & \sqrt{3} \end{bmatrix}$$

is

$$B = 2\sqrt{6}\mathbf{u}_1 \mathbf{v}_1^T + 2\mathbf{u}_2 \mathbf{v}_2^T,$$

where

$$\mathbf{v}_1 = \frac{1}{\sqrt{5}}\begin{bmatrix} 2 \\ 1 \end{bmatrix}, \quad \mathbf{v}_2 = \frac{1}{\sqrt{5}}\begin{bmatrix} 1 \\ -2 \end{bmatrix}$$

and

$$\mathbf{u}_1 = \frac{1}{2\sqrt{10}}\begin{bmatrix} 3\sqrt{3} \\ 2\sqrt{3} \\ 1 \end{bmatrix}, \quad \mathbf{u}_2 = \frac{1}{\sqrt{5}}\begin{bmatrix} 1 \\ -1 \\ -\sqrt{3} \end{bmatrix}.$$

Consequently, the outer product expansion of the matrix A is

$$A = B^T = 2\sqrt{6}\mathbf{v}_1\mathbf{u}_1^T + 2\mathbf{v}_2\mathbf{u}_2^T,$$

by Theorem 8.7. □

The outer product expansion and the rank of a matrix

Theorem 8.9. *Let A be an $m \times n$ matrix . If the outer product expansion of the matrix A is*

$$A = \sigma_1\mathbf{u}_1\mathbf{v}_1^T + \cdots + \sigma_r\mathbf{u}_r\mathbf{v}_r^T,$$

then

(a) $\{\mathbf{u}_1, \ldots \mathbf{u}_r\}$ *is an orthonormal basis of* $\mathbf{C}(A)$*;*

(b) $\{\mathbf{v}_1, \ldots \mathbf{v}_r\}$ *is an orthonormal basis of* $\mathbf{C}(A^T)$*.*

Proof of a particular case. We give the proof for $m = 4$, $n = 5$, and $r = 3$.

Let A be a 4×5 matrix. Suppose that $\mathbf{v}_1, \mathbf{v}_2, \mathbf{v}_3, \mathbf{v}_4, \mathbf{v}_5$ is an orthonormal basis of eigenvectors of the matrix A^TA and that the outer product expansion of the matrix A is

$$A = \sigma_1\mathbf{u}_1\mathbf{v}_1^T + \sigma_2\mathbf{u}_2\mathbf{v}_2^T + \sigma_3\mathbf{u}_3\mathbf{v}_3^T.$$

Every vector in $\mathbf{C}(A)$ is of the form $A\begin{bmatrix} x_1 \\ x_2 \\ x_3 \\ x_4 \\ x_5 \end{bmatrix}$. From the outer product decomposition of A we obtain

$$A\begin{bmatrix}x_1\\x_2\\x_3\\x_4\\x_5\end{bmatrix} = \sigma_1\mathbf{u}_1\mathbf{v}_1^T\begin{bmatrix}x_1\\x_2\\x_3\\x_4\\x_5\end{bmatrix} + \sigma_2\mathbf{u}_2\mathbf{v}_2^T\begin{bmatrix}x_1\\x_2\\x_3\\x_4\\x_5\end{bmatrix} + \sigma_3\mathbf{u}_3\mathbf{v}_3^T\begin{bmatrix}x_1\\x_2\\x_3\\x_4\\x_5\end{bmatrix}$$

$$= \left(\sigma_1\mathbf{v}_1^T\begin{bmatrix}x_1\\x_2\\x_3\\x_4\\x_5\end{bmatrix}\right)\mathbf{u}_1 + \left(\sigma_2\mathbf{v}_2^T\begin{bmatrix}x_1\\x_2\\x_3\\x_4\\x_5\end{bmatrix}\right)\mathbf{u}_2 + \left(\sigma_3\mathbf{v}_3^T\begin{bmatrix}x_1\\x_2\\x_3\\x_4\\x_5\end{bmatrix}\right)\mathbf{u}_3.$$

Consequently, the vectors $\mathbf{u}_1$, $\mathbf{u}_2$, $\mathbf{u}_3$ span the subspace $\mathbf{C}(A)$. Since the vectors $\mathbf{u}_1$, $\mathbf{u}_2$, $\mathbf{u}_3$ are orthonormal, they are linearly independent and consequently the set $\{\mathbf{u}_1, \mathbf{u}_2, \mathbf{u}_3\}$ is an orthonormal basis of the subspace $\mathbf{C}(A)$.

Using the outer product expansion of the matrix A^T

$$A^T = \sigma_1\mathbf{v}_1\mathbf{u}_1^T + \sigma_2\mathbf{v}_2\mathbf{u}_2^T + \sigma_3\mathbf{v}_3\mathbf{u}_3^T$$

we can show, as above, that $\{\mathbf{v}_1, \mathbf{v}_2, \mathbf{v}_3\}$ is an orthonormal basis of $\mathbf{C}(A^T)$. □

The pseudoinverse of a matrix

In Chapter 3 we discuss least squares solutions of equations the form $A\mathbf{x} = \mathbf{b}$ that do not have exact solutions. The outer product expansion of the matrix A gives us an elegant way of describing least squares solutions of the equation $A\mathbf{x} = \mathbf{b}$.

Theorem 8.10. *Let A be an $m \times n$ nonzero matrix with the outer product expansion*

$$A = \sigma_1\mathbf{u}_1\mathbf{v}_1^T + \cdots + \sigma_r\mathbf{u}_r\mathbf{v}_r^T$$

and let $\mathbf{b}$ be a vector in $\mathbb{R}^m$. If $\{\mathbf{v}_1, \dots, \mathbf{v}_n\}$ is an orthonormal basis of eigenvectors of the matrix A^TA, then every least square solution of the equation

$$A\mathbf{x} = \mathbf{b}$$

is of the form

$$\mathbf{x} = \frac{1}{\sigma_1}(\mathbf{u}_1^T\mathbf{b})\mathbf{v}_1 + \cdots + \frac{1}{\sigma_r}(\mathbf{u}_r^T\mathbf{b})\mathbf{v}_r + x_{r+1}\mathbf{v}_{r+1} + \cdots + x_n\mathbf{v}_n$$

where $x_{r+1}, \dots, x_n$ are arbitrary real numbers. There is a unique least square solution of minimal length, which is

$$\mathbf{x} = \frac{1}{\sigma_1}(\mathbf{u}_1^T\mathbf{b})\mathbf{v}_1 + \cdots + \frac{1}{\sigma_r}(\mathbf{u}_r^T\mathbf{b})\mathbf{v}_r = \left(\frac{1}{\sigma_1}\mathbf{v}_1\mathbf{u}_1^T + \cdots + \frac{1}{\sigma_r}\mathbf{v}_r\mathbf{u}_r^T\right)\mathbf{b}.$$

Proof of a particular case. We give a proof for $r = 3$ and $n = 5$.

Assume that the outer product expansion of A is

$$A = \sigma_1 \mathbf{u}_1 \mathbf{v}_1^T + \sigma_2 \mathbf{u}_2 \mathbf{v}_2^T + \sigma_3 \mathbf{u}_3 \mathbf{v}_3^T$$

and that $\{\mathbf{v}_1, \mathbf{v}_2, \mathbf{v}_3, \mathbf{v}_4, \mathbf{v}_5\}$ is an orthonormal basis of eigenvectors of the matrix $A^T A$. A vector $\mathbf{x}$ in $\mathbb{R}^5$ is a least square solution of the equation $A\mathbf{x} = \mathbf{b}$ if it satisfies the equation

$$A\mathbf{x} = \operatorname{proj}_{\mathbf{C}(A)}(\mathbf{b}).$$

Since $\{\mathbf{v}_1, \mathbf{v}_2, \mathbf{v}_3, \mathbf{v}_4, \mathbf{v}_5\}$ is a basis of $\mathbb{R}^5$, we can write

$$\mathbf{x} = x_1 \mathbf{v}_1 + x_2 \mathbf{v}_2 + x_3 \mathbf{v}_3 + x_4 \mathbf{v}_4 + x_5 \mathbf{v}_5.$$

Then the equation $A\mathbf{x} = \operatorname{proj}_{\mathbf{C}(A)}(\mathbf{b})$ becomes

$$A(x_1 \mathbf{v}_1 + x_2 \mathbf{v}_2 + x_3 \mathbf{v}_3 + x_4 \mathbf{v}_4 + x_5 \mathbf{v}_5) = \operatorname{proj}_{\mathbf{C}(A)}(\mathbf{b}).$$

Since

$$A(x_1 \mathbf{v}_1) = x_1 \sigma_1 \mathbf{u}_1, \quad A(x_2 \mathbf{v}_2) = x_2 \sigma_2 \mathbf{u}_2, \quad A(x_3 \mathbf{v}_3) = x_3 \sigma_3 \mathbf{u}_3,$$

and

$$A(x_4 \mathbf{v}_4) = A(x_5 \mathbf{v}_5) = \mathbf{0},$$

we have

$$x_1 \sigma_1 \mathbf{u}_1 + x_2 \sigma_2 \mathbf{u}_2 + x_3 \sigma_3 \mathbf{u}_3 = \operatorname{proj}_{\mathbf{C}(A)}(\mathbf{b}).$$

Now, since $\{\mathbf{u}_1, \mathbf{u}_2, \mathbf{u}_3\}$ is an orthonormal basis of $\mathbf{C}(A)$ and

$$x_1 \sigma_1 = \mathbf{b} \cdot \mathbf{u}_1 = \mathbf{u}_1^T \mathbf{b}, \quad x_2 \sigma_2 = \mathbf{b} \cdot \mathbf{u}_2 = \mathbf{u}_2^T \mathbf{b}, \quad x_3 \sigma_3 = \mathbf{b} \cdot \mathbf{u}_3 = \mathbf{u}_3^T \mathbf{b},$$

we conclude that

$$\mathbf{x} = \frac{1}{\sigma_1}(\mathbf{u}_1^T \mathbf{b})\mathbf{v}_1 + \frac{1}{\sigma_2}(\mathbf{u}_2^T \mathbf{b})\mathbf{v}_2 + \frac{1}{\sigma_3}(\mathbf{u}_3^T \mathbf{b})\mathbf{v}_3 + x_4 \mathbf{v}_4 + x_5 \mathbf{v}_5.$$

From the Pythagorean Theorem we get

$$\left\| \frac{1}{\sigma_1}(\mathbf{u}_1^T \mathbf{b})\mathbf{v}_1 + \frac{1}{\sigma_2}(\mathbf{u}_2^T \mathbf{b})\mathbf{v}_2 + \frac{1}{\sigma_3}(\mathbf{u}_3^T \mathbf{b})\mathbf{v}_3 + x_4 \mathbf{v}_4 + x_5 \mathbf{v}_5 \right\|^2$$

$$= \frac{1}{\sigma_1^2}|\mathbf{u}_1^T \mathbf{b}|^2 + \frac{1}{\sigma_2^2}|\mathbf{u}_2^T \mathbf{b}|^2 + \frac{1}{\sigma_3^2}|\mathbf{u}_3^T \mathbf{b}|^2 + x_4^2 + x_5^2.$$

It is clear that the norm is minimized if $x_4 = x_5 = 0$. Consequently,

$$\mathbf{x} = \frac{1}{\sigma_1}(\mathbf{u}_1^T \mathbf{b})\mathbf{v}_1 + \frac{1}{\sigma_2}(\mathbf{u}_2^T \mathbf{b})\mathbf{v}_2 + \frac{1}{\sigma_3}(\mathbf{u}_3^T \mathbf{b})\mathbf{v}_3$$

is the unique least square solution of minimal norm. Moreover, since

$$(\mathbf{u}^T \mathbf{b})\mathbf{v} = (\mathbf{u} \cdot \mathbf{b})\mathbf{v} = \mathbf{v}(\mathbf{u} \cdot \mathbf{b}) = \mathbf{v}(\mathbf{u}^T \mathbf{b}) = (\mathbf{v}\mathbf{u}^T)\mathbf{b},$$

for any vectors $\mathbf{u}$ and $\mathbf{b}$ in $\mathbb{R}^m$ and $\mathbf{v}$ in $\mathbb{R}^n$, we have

$$\frac{1}{\sigma_1}(\mathbf{u}_1^T \mathbf{b})\mathbf{v}_1 + \frac{1}{\sigma_2}(\mathbf{u}_2^T \mathbf{b})\mathbf{v}_2 + \frac{1}{\sigma_3}(\mathbf{u}_3^T \mathbf{b})\mathbf{v}_3 = \left(\frac{1}{\sigma_1}\mathbf{v}_1 \mathbf{u}_1^T + \frac{1}{\sigma_2}\mathbf{v}_2 \mathbf{u}_2^T + \frac{1}{\sigma_3}\mathbf{v}_3 \mathbf{u}_3^T\right)\mathbf{b}.$$

□

If A is an invertible square matrix, then the unique solution of the equation $A\mathbf{x} = \mathbf{b}$ is $\mathbf{x} = A^{-1}\mathbf{b}$. If A is an arbitrary matrix, then the unique least square solution of minimal length of the equation $A\mathbf{x} = \mathbf{b}$ is $\mathbf{x} = \left(\frac{1}{\sigma_1}\mathbf{v}_1\mathbf{u}_1^T + \cdots + \frac{1}{\sigma_r}\mathbf{v}_r\mathbf{u}_r^T\right)\mathbf{b}$. So the matrix $\frac{1}{\sigma_1}\mathbf{v}_1\mathbf{u}_1^T + \cdots + \frac{1}{\sigma_r}\mathbf{v}_r\mathbf{u}_r^T$ plays a somewhat similar role as the inverse of A. We could say that it is a generalization of the inverse of a matrix.

Definition 8.11. Let A be a $m \times n$ nonzero matrix with the outer product expansion

$$A = \sigma_1\mathbf{u}_1\mathbf{v}_1^T + \cdots + \sigma_r\mathbf{u}_r\mathbf{v}_r^T.$$

The $n \times m$ matrix

$$A^+ = \frac{1}{\sigma_1}\mathbf{v}_1\mathbf{u}_1^T + \cdots + \frac{1}{\sigma_r}\mathbf{v}_r\mathbf{u}_r^T$$

is called the *pseudoinverse* of the matrix A or the *Moore-Penrose inverse* of the matrix A.

Example 8.12. Find the pseudoinverse of the matrix $A = \begin{bmatrix} 1 & 2 \\ 1 & 1 \\ -1 & 1 \\ 2 & 1 \end{bmatrix}$.

Solution. The outer product expansion of the matrix A is

$$A = \sqrt{11}\mathbf{u}_1\mathbf{v}_1^T + \sqrt{3}\mathbf{u}_2\mathbf{v}_2^T,$$

where

$$\mathbf{v}_1 = \frac{1}{\sqrt{2}}\begin{bmatrix} 1 \\ 1 \end{bmatrix}, \quad \mathbf{v}_2 = \frac{1}{\sqrt{2}}\begin{bmatrix} 1 \\ -1 \end{bmatrix}, \quad \mathbf{u}_1 = \frac{1}{\sqrt{22}}\begin{bmatrix} 3 \\ 2 \\ 0 \\ 3 \end{bmatrix}, \quad \mathbf{u}_2 = \frac{1}{\sqrt{6}}\begin{bmatrix} -1 \\ 0 \\ -2 \\ 1 \end{bmatrix}.$$

Thus the pseudoinverse of A is

$$A^+ = \frac{1}{\sqrt{11}}\mathbf{v}_1\mathbf{u}_1^T + \frac{1}{\sqrt{3}}\mathbf{v}_2\mathbf{u}_2^T = \frac{1}{33}\begin{bmatrix} -1 & 3 & -11 & 10 \\ 10 & 3 & 11 & -1 \end{bmatrix}.$$

□

Example 8.13. Use the pseudoinverse of the matrix $A = \begin{bmatrix} 3 & 1 \\ 9 & 3 \\ -3 & -1 \\ 12 & 4 \end{bmatrix}$ to find the least square solution of minimal norm of the equation

$$A \begin{bmatrix} x \\ y \end{bmatrix} = \begin{bmatrix} 270 \\ 270 \\ 270 \\ 270 \end{bmatrix}.$$

Solution. Since the outer expansion of the matrix A is

$$3\sqrt{30}\mathbf{u}_1\mathbf{v}_1^T,$$

where

$$\mathbf{v}_1 = \frac{1}{\sqrt{10}} \begin{bmatrix} 3 \\ 1 \end{bmatrix} \quad \text{and} \quad \mathbf{u}_1 = \frac{1}{3\sqrt{3}} \begin{bmatrix} 1 \\ 3 \\ -1 \\ 4 \end{bmatrix},$$

the pseudoinverse A^+ of the matrix A is

$$\frac{1}{3\sqrt{30}}\mathbf{v}_1\mathbf{u}_1^T = \begin{bmatrix} \frac{1}{90} & \frac{1}{30} & -\frac{1}{90} & \frac{2}{45} \\ \frac{1}{270} & \frac{1}{90} & -\frac{1}{270} & \frac{2}{135} \end{bmatrix}.$$

Thus the least square solution of minimal norm of the equation $A \begin{bmatrix} x \\ y \end{bmatrix} = \begin{bmatrix} 270 \\ 270 \\ 270 \\ 270 \end{bmatrix}$ is

$$\begin{bmatrix} x \\ y \end{bmatrix} = A^+ \begin{bmatrix} 270 \\ 270 \\ 270 \\ 270 \end{bmatrix} = \begin{bmatrix} \frac{1}{90} & \frac{1}{30} & -\frac{1}{90} & \frac{2}{45} \\ \frac{1}{270} & \frac{1}{90} & -\frac{1}{270} & \frac{2}{135} \end{bmatrix} \begin{bmatrix} 270 \\ 270 \\ 270 \\ 270 \end{bmatrix} = \begin{bmatrix} 21 \\ 7 \end{bmatrix}.$$

□

Exercises 8.1

Calculate the outer products $\mathbf{u}\mathbf{v}^T$ and $\mathbf{v}\mathbf{u}^T$ for the given vectors $\mathbf{u}$ and $\mathbf{v}$.

1. $\mathbf{u} = \begin{bmatrix} 5 \\ 2 \end{bmatrix}$ and $\begin{bmatrix} -4 \\ 9 \end{bmatrix}$

2. $\mathbf{u} = \begin{bmatrix} 2 \\ 1 \\ 4 \end{bmatrix}$ and $\mathbf{v} = \begin{bmatrix} 3 \\ -2 \\ 8 \end{bmatrix}$

3. $\mathbf{u} = \begin{bmatrix} 1 \\ 2 \\ 5 \end{bmatrix}$ and $\mathbf{v} = \begin{bmatrix} 2 \\ 1 \\ 3 \\ 7 \end{bmatrix}$

4. $\mathbf{u} = \begin{bmatrix} 3 \\ 2 \end{bmatrix}$ and $\mathbf{v} = \begin{bmatrix} 5 \\ 1 \\ 1 \\ 7 \\ 4 \end{bmatrix}$

Find the outer product expansion of the given matrix A.

5. $A = \begin{bmatrix} -2 & 2 \\ 1 & 1 \end{bmatrix}$

6. $A = \begin{bmatrix} 4 & 0 \\ 3 & 5 \end{bmatrix}$

7. $A = \begin{bmatrix} 3 & 1 & 1 \\ 3 & 1 & 1 \\ 3 & 1 & 1 \end{bmatrix}$

8. $A = \begin{bmatrix} 1 & 2 \\ 1 & 2 \\ -1 & -2 \\ 1 & 2 \end{bmatrix}$

9. $A = \begin{bmatrix} 2 & 1 & \sqrt{5} \\ 1 & 3 & 0 \end{bmatrix}$

10. $A = \begin{bmatrix} 3 & 1 \\ 0 & -5 \\ -3 & -1 \end{bmatrix}$

11. $A = \begin{bmatrix} 3 & 1 \\ 9 & -2 \\ 5 & -5 \\ -3 & 4 \end{bmatrix}$

12. $A = \begin{bmatrix} 1 & 2 \\ 2 & 1 \\ 1 & -1 \\ 1 & 1 \end{bmatrix}$

13. $A = \begin{bmatrix} 3 & 1 \\ 1 & 1 \\ 1 & 1 \\ \sqrt{2} & -\sqrt{2} \end{bmatrix}$

14. $A = \begin{bmatrix} 1 & 2 \\ 2 & 1 \\ 1 & -1 \\ 1 & 1 \end{bmatrix}$

15. $A = \begin{bmatrix} -1 & -1 \\ 1 & -1 \\ 1 & 1 \\ -1 & 1 \\ 4 & 1 \end{bmatrix}$

16. $A = \begin{bmatrix} 1 & 2 \\ 1 & 0 \\ 0 & 1 \\ 2 & 0 \\ 0 & -2 \end{bmatrix}$

17. $A = \begin{bmatrix} 3 & 1 & 2 & 1 & 1 \\ 1 & 2 & -1 & 2 & -3 \end{bmatrix}$

18. $A = \begin{bmatrix} 3 & 1 & 1 & -1 \\ 1 & 2 & -3 & -2 \end{bmatrix}$

19. Prove Theorem 8.10 for $r = 2$ and $n = 3$.

20. Prove Theorem 8.10 for $r = 1$ and $n = 4$.

Find the pseudoinverse of the given matrix A.

21. $A = \begin{bmatrix} 3 & -3 \\ 1 & -1 \end{bmatrix}$

22. $A = \begin{bmatrix} 8 & 10 \\ 4 & 5 \end{bmatrix}$

23. $A = \begin{bmatrix} 1 & 2 \\ 1 & 3 \\ -3 & 1 \end{bmatrix}$

24. $A = \begin{bmatrix} 2 & 1 & \sqrt{5} \\ 1 & 3 & 0 \end{bmatrix}$

25. $A = \begin{bmatrix} 3 & 1 \\ 1 & 1 \\ 1 & -1 \\ 0 & 4 \end{bmatrix}$

26. $A = \begin{bmatrix} 3 & 5 \\ 4 & 4 \\ 4 & 0 \\ 3 & 3 \end{bmatrix}$

27. $A = \begin{bmatrix} 1 & 2 & 0 \\ 2 & 4 & 0 \\ 1 & 0 & 1 \end{bmatrix}$

28. $A = \begin{bmatrix} 1 & 1 & 1 \\ 1 & 1 & 1 \\ 2 & 2 & 2 \\ 1 & 1 & 1 \end{bmatrix}$

29. Show that for an arbitrary nonzero matrix A we have $AA^+ = \operatorname{proj}_{\mathbf{C}(A)}$.

30. Show that $A^+ = A^{-1}$ for every invertible matrix A.

31. Show that $A^T AA^+ = A^T$.

32. Show that $A^+ A = \operatorname{proj}_{\mathbf{C}(A^T)}$.

8.2 Singular value decomposition

The singular value decomposition of a matrix is closely related to the outer product expansion. First we prove a theorem that guarantees the existence of the singular value decomposition for any matrix.

Theorem 8.14. *Let A be an $m\times n$ matrix and let $\{\mathbf{v}_1,\dots,\mathbf{v}_n\}$ be an orthonormal basis of eigenvectors of the matrix A^TA. If the outer product expansion of the matrix A is*

$$A=\sigma_1\mathbf{u}_1\mathbf{v}_1^T+\cdots+\sigma_r\mathbf{u}_r\mathbf{v}_r^T,$$

then there is an orthonormal basis $\{\mathbf{u}_1,\dots,\mathbf{u}_m\}$ of $\mathbb{R}^m$ such that

$$A=\begin{bmatrix}\mathbf{u}_1 & \dots & \mathbf{u}_m\end{bmatrix}\Sigma\begin{bmatrix}\mathbf{v}_1^T\\ \vdots\\ \mathbf{v}_n^T\end{bmatrix}$$

where

$$\Sigma=\begin{bmatrix}\sigma_1 & 0 & \dots & 0 & 0 & \dots & 0\\ 0 & \sigma_2 & \dots & 0 & 0 & \dots & 0\\ \vdots & & \ddots & & \vdots & & \vdots\\ 0 & 0 & \dots & \sigma_r & 0 & \dots & 0\\ 0 & 0 & \dots & 0 & 0 & \dots & 0\\ \vdots & \vdots & & \vdots & \vdots & & \vdots\\ 0 & 0 & \dots & 0 & 0 & \dots & 0\end{bmatrix}$$

is the $m\times n$ matrix with the singular values $\sigma_1,\dots,\sigma_r$ of A along the main diagonal in the first r columns and 0 everywhere else.

Proof of a particular case. We illustrate the method of the proof for a 4×5 matrix A with the outer product expansion

$$A=\sigma_1\mathbf{u}_1\mathbf{v}_1^T+\sigma_2\mathbf{u}_2\mathbf{v}_2^T.$$

Let $\{\mathbf{v}_1,\mathbf{v}_2,\mathbf{v}_3,\mathbf{v}_4,\mathbf{v}_5\}$ be an orthonormal basis in $\mathbb{R}^5$ consisting of eigenvectors of the symmetric matrix A^TA and let $\{\mathbf{u}_3,\mathbf{u}_4\}$ be an orthonormal basis of the subspace $(\mathbf{C}(A))^\perp=\mathbf{N}(A^T)$. From

$$A=\begin{bmatrix}\mathbf{u}_1 & \mathbf{u}_2\end{bmatrix}\begin{bmatrix}\sigma_1 & 0\\ 0 & \sigma_2\end{bmatrix}\begin{bmatrix}\mathbf{v}_1^T\\ \mathbf{v}_2^T\end{bmatrix}$$

and

$$\begin{bmatrix}\mathbf{u}_1 & \mathbf{u}_2\end{bmatrix}\begin{bmatrix}\sigma_1 & 0\\ 0 & \sigma_2\end{bmatrix}=\begin{bmatrix}\mathbf{u}_1 & \mathbf{u}_2 & \mathbf{u}_3 & \mathbf{u}_4\end{bmatrix}\begin{bmatrix}\sigma_1 & 0\\ 0 & \sigma_2\\ 0 & 0\\ 0 & 0\end{bmatrix},$$

we obtain

$$A=\begin{bmatrix}\mathbf{u}_1 & \mathbf{u}_2\end{bmatrix}\begin{bmatrix}\sigma_1 & 0\\ 0 & \sigma_2\end{bmatrix}\begin{bmatrix}\mathbf{v}_1^T\\ \mathbf{v}_2^T\end{bmatrix}=\begin{bmatrix}\mathbf{u}_1 & \mathbf{u}_2 & \mathbf{u}_3 & \mathbf{u}_4\end{bmatrix}\begin{bmatrix}\sigma_1 & 0\\ 0 & \sigma_2\\ 0 & 0\\ 0 & 0\end{bmatrix}\begin{bmatrix}\mathbf{v}_1^T\\ \mathbf{v}_2^T\end{bmatrix}.$$

Since we also have

$$\begin{bmatrix}\sigma_1 & 0\\ 0 & \sigma_2\\ 0 & 0\\ 0 & 0\end{bmatrix}\begin{bmatrix}\mathbf{v}_1^T\\ \mathbf{v}_2^T\end{bmatrix} = \begin{bmatrix}\sigma_1 & 0 & 0 & 0 & 0\\ 0 & \sigma_2 & 0 & 0 & 0\\ 0 & 0 & 0 & 0 & 0\\ 0 & 0 & 0 & 0 & 0\end{bmatrix}\begin{bmatrix}\mathbf{v}_1^T\\ \mathbf{v}_2^T\\ \mathbf{v}_3^T\\ \mathbf{v}_4^T\\ \mathbf{v}_5^T\end{bmatrix},$$

so we can conclude that

$$A = \begin{bmatrix}\mathbf{u}_1 & \mathbf{u}_2 & \mathbf{u}_3 & \mathbf{u}_4\end{bmatrix}\begin{bmatrix}\sigma_1 & 0 & 0 & 0 & 0\\ 0 & \sigma_2 & 0 & 0 & 0\\ 0 & 0 & 0 & 0 & 0\\ 0 & 0 & 0 & 0 & 0\end{bmatrix}\begin{bmatrix}\mathbf{v}_1^T\\ \mathbf{v}_2^T\\ \mathbf{v}_3^T\\ \mathbf{v}_4^T\\ \mathbf{v}_5^T\end{bmatrix}.$$

□

Definition 8.15. By the *singular value decomposition* of an $m \times n$ matrix A we mean the representation of A in the form

$$A = \begin{bmatrix}\mathbf{u}_1 & \dots & \mathbf{u}_m\end{bmatrix}\Sigma\begin{bmatrix}\mathbf{v}_1^T\\ \vdots\\ \mathbf{v}_n^T\end{bmatrix},$$

where $\{\mathbf{u}_1,\dots,\mathbf{u}_m\}$ is an orthonormal basis in $\mathbb{R}^m$, $\{\mathbf{v}_1,\dots,\mathbf{v}_n\}$ is an orthonormal basis in $\mathbb{R}^n$, and

$$\Sigma = \begin{bmatrix}\sigma_1 & 0 & \dots & 0 & 0 & \dots & 0\\ 0 & \sigma_2 & \dots & 0 & 0 & \dots & 0\\ \vdots & & \ddots & & \vdots & & \vdots\\ 0 & 0 & \dots & \sigma_r & 0 & \dots & 0\\ 0 & 0 & \dots & 0 & 0 & \dots & 0\\ \vdots & \vdots & & \vdots & \vdots & & \vdots\\ 0 & 0 & \dots & 0 & 0 & \dots & 0\end{bmatrix}$$

is the $m \times n$ matrix with the singular values $\sigma_1,\dots,\sigma_r$ of A along the main diagonal in the first r columns and 0 everywhere else.

Example 8.16. Suppose that A is a 5×4 matrix with the outer product decomposition

$$A = \sigma_1 \mathbf{u}_1 \mathbf{v}_1^T + \sigma_2 \mathbf{u}_2 \mathbf{v}_2^T + \sigma_3 \mathbf{u}_3 \mathbf{v}_3^T.$$

What is the singular value decomposition of the matrix A?

Solution. The singular value decomposition of A is

$$A = \begin{bmatrix} \mathbf{u}_1 & \mathbf{u}_2 & \mathbf{u}_3 & \mathbf{u}_4 & \mathbf{u}_5 \end{bmatrix} \begin{bmatrix} \sigma_1 & 0 & 0 & 0 \\ 0 & \sigma_2 & 0 & 0 \\ 0 & 0 & \sigma_3 & 0 \\ 0 & 0 & 0 & 0 \\ 0 & 0 & 0 & 0 \end{bmatrix} \begin{bmatrix} \mathbf{v}_1^T \\ \mathbf{v}_2^T \\ \mathbf{v}_3^T \\ \mathbf{v}_4^T \end{bmatrix},$$

where $\{\mathbf{u}_4, \mathbf{u}_5\}$ is an orthonormal basis of the subspace $(\mathbf{C}(A))^\perp$ and $\{\mathbf{v}_4\}$ is a basis of the subspace $\mathbf{N}(A) = (\mathbf{C}(A^T))^\perp$. □

Example 8.17. Find the singular value decomposition of the matrix

$$A = \begin{bmatrix} 0 & -2 & 1 \\ -2 & 0 & -1 \\ 2 & 0 & 1 \\ 0 & -2 & 1 \end{bmatrix}.$$

Solution. The eigenvalues of the matrix

$$A^T A = \begin{bmatrix} 8 & 0 & 4 \\ 0 & 8 & -4 \\ 4 & -4 & 4 \end{bmatrix}$$

are 12, 8, and 0. The vectors $\begin{bmatrix} 1 \\ -1 \\ 1 \end{bmatrix}$, $\begin{bmatrix} 1 \\ 1 \\ 0 \end{bmatrix}$, and $\begin{bmatrix} -1 \\ 1 \\ 2 \end{bmatrix}$ are eigenvectors corresponding the eigenvalues 12, 8, and 0, respectively. The vectors

$$A \begin{bmatrix} 1 \\ -1 \\ 1 \end{bmatrix} = \begin{bmatrix} 3 \\ -3 \\ 3 \\ 3 \end{bmatrix}, \quad A \begin{bmatrix} 1 \\ 1 \\ 0 \end{bmatrix} = \begin{bmatrix} -2 \\ -2 \\ 2 \\ -2 \end{bmatrix}, \quad \text{and} \quad A \begin{bmatrix} -1 \\ 1 \\ 2 \end{bmatrix} = \begin{bmatrix} 0 \\ 0 \\ 0 \\ 0 \end{bmatrix}$$

are orthogonal. Since

$$A\begin{bmatrix} \frac{1}{\sqrt{3}} \\ -\frac{1}{\sqrt{3}} \\ \frac{1}{\sqrt{3}} \end{bmatrix} = \sqrt{12}\begin{bmatrix} \frac{1}{2} \\ -\frac{1}{2} \\ \frac{1}{2} \\ \frac{1}{2} \end{bmatrix} \quad \text{and} \quad A\begin{bmatrix} \frac{1}{\sqrt{2}} \\ \frac{1}{\sqrt{2}} \\ 0 \end{bmatrix} = \sqrt{8}\begin{bmatrix} -\frac{1}{2} \\ -\frac{1}{2} \\ -\frac{1}{2} \\ \frac{1}{2} \end{bmatrix},$$

we have

$$A\begin{bmatrix} \frac{1}{\sqrt{3}} & \frac{1}{\sqrt{2}} \\ -\frac{1}{\sqrt{3}} & \frac{1}{\sqrt{2}} \\ \frac{1}{\sqrt{3}} & 0 \end{bmatrix} = \begin{bmatrix} \frac{1}{2} & -\frac{1}{2} \\ -\frac{1}{2} & -\frac{1}{2} \\ \frac{1}{2} & -\frac{1}{2} \\ \frac{1}{2} & \frac{1}{2} \end{bmatrix}\begin{bmatrix} \sqrt{12} & 0 \\ 0 & \sqrt{8} \end{bmatrix}.$$

To determine an orthonormal basis of the subspace $(\mathbf{C}(A))^{\perp}$ we first solve the system

$$\begin{cases} x - y + z + w = 0 \\ -x - y - z + w = 0 \end{cases}.$$

Since

$$\begin{bmatrix} 1 & -1 & 1 & 1 \\ -1 & -1 & -1 & 1 \end{bmatrix} \sim \begin{bmatrix} 1 & 0 & 1 & 0 \\ 0 & 1 & 0 & -1 \end{bmatrix},$$

the general solution of the system is

$$\begin{bmatrix} -z \\ w \\ z \\ w \end{bmatrix} = z\begin{bmatrix} -1 \\ 0 \\ 1 \\ 0 \end{bmatrix} + w\begin{bmatrix} 0 \\ 1 \\ 0 \\ 1 \end{bmatrix}.$$

The vectors $\begin{bmatrix} -1 \\ 0 \\ 1 \\ 0 \end{bmatrix}$ and $\begin{bmatrix} 0 \\ 1 \\ 0 \\ 1 \end{bmatrix}$ form a basis of the subspace $(\mathbf{C}(A))^{\perp}$ and they are orthogonal. After normalizing these two vectors we obtain the singular value decomposition of A:

$$A = \begin{bmatrix} \frac{1}{2} & -\frac{1}{2} & -\frac{1}{\sqrt{2}} & 0 \\ -\frac{1}{2} & -\frac{1}{2} & 0 & \frac{1}{\sqrt{2}} \\ \frac{1}{2} & -\frac{1}{2} & \frac{1}{\sqrt{2}} & 0 \\ \frac{1}{2} & \frac{1}{2} & 0 & \frac{1}{\sqrt{2}} \end{bmatrix}\begin{bmatrix} \sqrt{15} & 0 & 0 \\ 0 & \sqrt{12} & 0 \\ 0 & 0 & 0 \\ 0 & 0 & 0 \end{bmatrix}\begin{bmatrix} \frac{1}{\sqrt{3}} & -\frac{1}{\sqrt{3}} & \frac{1}{\sqrt{3}} \\ \frac{1}{\sqrt{2}} & \frac{1}{\sqrt{2}} & 0 \\ -\frac{1}{\sqrt{6}} & \frac{1}{\sqrt{6}} & \frac{2}{\sqrt{6}} \end{bmatrix}.$$

□

Example 8.18. Find the singular value decomposition of the matrix

$$A = \begin{bmatrix} 1 & 2 \\ 0 & 2 \\ 3 & 1 \\ 0 & 1 \end{bmatrix}.$$

Solution. The eigenvalues of the matrix

$$A^T A = \begin{bmatrix} 10 & 5 \\ 5 & 10 \end{bmatrix}$$

are 15 and 5. The vectors $\begin{bmatrix} 1 \\ 1 \end{bmatrix}$ and $\begin{bmatrix} 1 \\ -1 \end{bmatrix}$ are eigenvectors corresponding to the eigenvalues 15 and 5, respectively. The vectors

$$A\begin{bmatrix} 1 \\ 1 \end{bmatrix} = \begin{bmatrix} 3 \\ 2 \\ 4 \\ 1 \end{bmatrix} \quad \text{and} \quad A\begin{bmatrix} 1 \\ -1 \end{bmatrix} = \begin{bmatrix} -1 \\ -2 \\ 2 \\ -1 \end{bmatrix}$$

are orthogonal. Since

$$A\begin{bmatrix} \frac{1}{\sqrt{2}} \\ \frac{1}{\sqrt{2}} \end{bmatrix} = \sqrt{15}\begin{bmatrix} \frac{3}{\sqrt{30}} \\ \frac{2}{\sqrt{30}} \\ \frac{4}{\sqrt{30}} \\ \frac{1}{\sqrt{30}} \end{bmatrix} \quad \text{and} \quad A\begin{bmatrix} \frac{1}{\sqrt{2}} \\ -\frac{1}{\sqrt{2}} \end{bmatrix} = \sqrt{5}\begin{bmatrix} -\frac{1}{\sqrt{10}} \\ -\frac{2}{\sqrt{10}} \\ \frac{2}{\sqrt{10}} \\ -\frac{1}{\sqrt{10}} \end{bmatrix},$$

we have

$$A\begin{bmatrix} \frac{1}{\sqrt{2}} & \frac{1}{\sqrt{2}} \\ -\frac{1}{\sqrt{2}} & \frac{1}{\sqrt{2}} \end{bmatrix} = \begin{bmatrix} \frac{3}{\sqrt{30}} & -\frac{1}{\sqrt{10}} \\ \frac{2}{\sqrt{30}} & -\frac{2}{\sqrt{10}} \\ \frac{4}{\sqrt{30}} & \frac{2}{\sqrt{10}} \\ \frac{1}{\sqrt{30}} & -\frac{1}{\sqrt{10}} \end{bmatrix} \begin{bmatrix} \sqrt{15} & 0 \\ 0 & \sqrt{5} \end{bmatrix}.$$

To determine an orthonormal basis of the subspace $(\mathbf{C}(A))^\perp$ we first solve the system

$$\begin{cases} 3x + 2y + 4z + w = 0 \\ -x - 2y + 2z - w = 0 \end{cases}.$$

Since

$$\begin{bmatrix} 3 & 2 & 4 & 1 \\ -1 & -2 & 2 & -1 \end{bmatrix} \sim \begin{bmatrix} 1 & 0 & 3 & 0 \\ 0 & 1 & -\frac{5}{2} & \frac{1}{2} \end{bmatrix},$$

the general solution of the system is

$$\begin{bmatrix} -3z \\ \frac{5}{2}z - \frac{1}{2}w \\ z \\ w \end{bmatrix} = \frac{z}{2}\begin{bmatrix} -6 \\ 5 \\ 2 \\ 0 \end{bmatrix} + \frac{w}{2}\begin{bmatrix} 0 \\ -1 \\ 0 \\ 2 \end{bmatrix}.$$

The vectors $\begin{bmatrix} -6 \\ 5 \\ 2 \\ 0 \end{bmatrix}$ and $\begin{bmatrix} 0 \\ -1 \\ 0 \\ 2 \end{bmatrix}$ form a basis of the subspace $(\mathbf{C}(A))^{\perp}$, but they are not orthogonal. To deal with that we use the Gram-Schmidt process and get

$$(\mathbf{C}(A))^{\perp} = \operatorname{Span}\left\{\begin{bmatrix} -6 \\ 5 \\ 2 \\ 0 \end{bmatrix}, \begin{bmatrix} 0 \\ -1 \\ 0 \\ 2 \end{bmatrix}\right\} = \operatorname{Span}\left\{\begin{bmatrix} -6 \\ 5 \\ 2 \\ 0 \end{bmatrix}, \begin{bmatrix} 3 \\ -2 \\ -1 \\ -1 \end{bmatrix}\right\}.$$

After normalizing the vectors we finally obtain the singular value decomposition of A:

$$A = \begin{bmatrix} \frac{3}{\sqrt{30}} & -\frac{1}{\sqrt{10}} & 0 & \frac{3}{\sqrt{15}} \\ \frac{2}{\sqrt{30}} & -\frac{2}{\sqrt{10}} & -\frac{1}{\sqrt{5}} & -\frac{2}{\sqrt{15}} \\ \frac{4}{\sqrt{30}} & \frac{2}{\sqrt{10}} & 0 & -\frac{1}{\sqrt{15}} \\ \frac{1}{\sqrt{30}} & -\frac{1}{\sqrt{10}} & \frac{2}{\sqrt{5}} & -\frac{1}{\sqrt{15}} \end{bmatrix} \begin{bmatrix} \sqrt{15} & 0 \\ 0 & \sqrt{5} \\ 0 & 0 \\ 0 & 0 \end{bmatrix} \begin{bmatrix} \frac{1}{\sqrt{2}} & -\frac{1}{\sqrt{2}} \\ \frac{1}{\sqrt{2}} & \frac{1}{\sqrt{2}} \end{bmatrix}.$$

□

In the last example, substantial effort was spent on finding the vectors $\begin{bmatrix} 0 \\ -\frac{1}{\sqrt{5}} \\ 0 \\ \frac{2}{\sqrt{5}} \end{bmatrix}$ and $\begin{bmatrix} \frac{3}{\sqrt{15}} \\ -\frac{2}{\sqrt{15}} \\ -\frac{1}{\sqrt{15}} \\ -\frac{1}{\sqrt{15}} \end{bmatrix}$ in order to complete the matrix $\begin{bmatrix} \frac{3}{\sqrt{30}} & -\frac{1}{\sqrt{10}} & 0 & \frac{3}{\sqrt{15}} \\ \frac{2}{\sqrt{30}} & -\frac{2}{\sqrt{10}} & -\frac{1}{\sqrt{5}} & -\frac{2}{\sqrt{15}} \\ \frac{4}{\sqrt{30}} & \frac{2}{\sqrt{10}} & 0 & -\frac{1}{\sqrt{15}} \\ \frac{1}{\sqrt{30}} & -\frac{1}{\sqrt{10}} & \frac{2}{\sqrt{5}} & -\frac{1}{\sqrt{15}} \end{bmatrix}$. We could argue that it was a wasted effort since in the product

$$\begin{bmatrix} \frac{3}{\sqrt{30}} & -\frac{1}{\sqrt{10}} & 0 & \frac{3}{\sqrt{15}} \\ \frac{2}{\sqrt{30}} & -\frac{2}{\sqrt{10}} & -\frac{1}{\sqrt{5}} & -\frac{2}{\sqrt{15}} \\ \frac{4}{\sqrt{30}} & \frac{2}{\sqrt{10}} & 0 & -\frac{1}{\sqrt{15}} \\ \frac{1}{\sqrt{30}} & -\frac{1}{\sqrt{10}} & \frac{2}{\sqrt{5}} & -\frac{1}{\sqrt{15}} \end{bmatrix} \begin{bmatrix} \sqrt{15} & 0 \\ 0 & \sqrt{5} \\ 0 & 0 \\ 0 & 0 \end{bmatrix}$$

the last two columns are multiplied by 0, so it really does not matter what vectors we put there. Moreover, note that

$$A = \begin{bmatrix} \frac{3}{\sqrt{30}} & -\frac{1}{\sqrt{10}} \\ \frac{2}{\sqrt{30}} & -\frac{2}{\sqrt{10}} \\ \frac{4}{\sqrt{30}} & \frac{2}{\sqrt{10}} \\ \frac{1}{\sqrt{30}} & -\frac{1}{\sqrt{10}} \end{bmatrix} \begin{bmatrix} \sqrt{15} & 0 \\ 0 & \sqrt{5} \end{bmatrix} \begin{bmatrix} \frac{1}{\sqrt{2}} & -\frac{1}{\sqrt{2}} \\ \frac{1}{\sqrt{2}} & \frac{1}{\sqrt{2}} \end{bmatrix}.$$

This observation leads to the following definition.

Definition 8.19. By the *compact singular value decomposition* of an $m \times n$ matrix A we mean the representation of A in the form

$$A = \begin{bmatrix} \mathbf{u}_1 & \dots & \mathbf{u}_r \end{bmatrix} D \begin{bmatrix} \mathbf{v}_1^T \\ \vdots \\ \mathbf{v}_r^T \end{bmatrix},$$

where $\{\mathbf{u}_1, \dots, \mathbf{u}_r\}$ is an orthonormal basis of the subspace $\mathbf{C}(A)$, $\{\mathbf{v}_1, \dots, \mathbf{v}_r\}$ is an orthonormal basis of the subspace $\mathbf{C}(A^T)$, and

$$D = \begin{bmatrix} \sigma_1 & 0 & \dots & 0 \\ 0 & \sigma_2 & \dots & 0 \\ \vdots & & \ddots & \vdots \\ 0 & 0 & \dots & \sigma_r \end{bmatrix}$$

is the $r \times r$ diagonal matrix with the singular values $\sigma_1, \dots, \sigma_r$ of A on the main diagonal.

Example 8.20. Suppose that A is a 6×5 matrix with the outer product decomposition

$$A = \sigma_1 \mathbf{u}_1 \mathbf{v}_1^T + \sigma_2 \mathbf{u}_2 \mathbf{v}_2^T + \sigma_3 \mathbf{u}_3 \mathbf{v}_3^T.$$

What is the compact singular value decomposition of the matrix A?

Solution. The compact singular value decomposition of A is

$$A = \begin{bmatrix} \mathbf{u}_1 & \mathbf{u}_2 & \mathbf{u}_3 \end{bmatrix} \begin{bmatrix} \sigma_1 & 0 & 0 \\ 0 & \sigma_2 & 0 \\ 0 & 0 & \sigma_3 \end{bmatrix} \begin{bmatrix} \mathbf{v}_1^T \\ \mathbf{v}_2^T \\ \mathbf{v}_3^T \end{bmatrix}.$$

□

As the above example illustrates, the outer product decomposition and the compact singular value decomposition of a matrix A are just two different ways of expressing the same decomposition of A.

Example 8.21. Find the compact singular value decomposition of the matrix

$$A = \begin{bmatrix} 0 & -2 & 1 \\ -2 & 0 & -1 \\ 2 & 0 & 1 \\ 0 & -2 & 1 \end{bmatrix}.$$

Solution. In Example 8.17 we show that

$$A \begin{bmatrix} \frac{1}{\sqrt{3}} & \frac{1}{\sqrt{2}} \\ -\frac{1}{\sqrt{3}} & \frac{1}{\sqrt{2}} \\ \frac{1}{\sqrt{3}} & 0 \end{bmatrix} = \begin{bmatrix} \frac{1}{2} & -\frac{1}{2} \\ -\frac{1}{2} & -\frac{1}{2} \\ \frac{1}{2} & -\frac{1}{2} \\ \frac{1}{2} & \frac{1}{2} \end{bmatrix} \begin{bmatrix} \sqrt{12} & 0 \\ 0 & \sqrt{8} \end{bmatrix}.$$

Consequently, the compact singular value decomposition of A is

$$A = \begin{bmatrix} \frac{1}{2} & -\frac{1}{2} \\ -\frac{1}{2} & -\frac{1}{2} \\ \frac{1}{2} & -\frac{1}{2} \\ \frac{1}{2} & \frac{1}{2} \end{bmatrix} \begin{bmatrix} \sqrt{12} & 0 \\ 0 & \sqrt{8} \end{bmatrix} \begin{bmatrix} \frac{1}{\sqrt{3}} & -\frac{1}{\sqrt{3}} & \frac{1}{\sqrt{3}} \\ \frac{1}{\sqrt{2}} & \frac{1}{\sqrt{2}} & 0 \end{bmatrix}.$$

□

Example 8.22. Find the compact singular value decomposition of the matrix

$$A = \begin{bmatrix} 3 & -1 \\ 3 & 4 \\ 1 & 3 \\ -3 & 1 \\ 3 & -1 \end{bmatrix}.$$

Solution. In Example 8.6 we show that the outer product expansion of A is

$$A = \sqrt{40}\left(\frac{1}{\sqrt{8}}\begin{bmatrix}1\\2\\1\\-1\\1\end{bmatrix}\right)\left(\frac{1}{\sqrt{5}}\begin{bmatrix}2 & 1\end{bmatrix}\right) + 5\left(\frac{1}{\sqrt{5}}\begin{bmatrix}1\\-1\\-1\\-1\\1\end{bmatrix}\right)\left(\frac{1}{\sqrt{5}}\begin{bmatrix}1 & -2\end{bmatrix}\right).$$

Consequently, the compact singular value decomposition of A is

$$A = \begin{bmatrix}\mathbf{u}_1 & \mathbf{u}_2\end{bmatrix}\begin{bmatrix}\sqrt{40} & 0\\ 0 & 5\end{bmatrix}\begin{bmatrix}\mathbf{v}_1^T\\ \mathbf{v}_2^T\end{bmatrix},$$

where

$$\mathbf{u}_1 = \frac{1}{\sqrt{8}}\begin{bmatrix}1\\2\\1\\-1\\1\end{bmatrix},\quad \mathbf{u}_2 = \frac{1}{\sqrt{5}}\begin{bmatrix}1\\-1\\-1\\-1\\1\end{bmatrix},\quad \mathbf{v}_1 = \frac{1}{\sqrt{5}}\begin{bmatrix}2\\1\end{bmatrix},\quad \text{and}\quad \mathbf{v}_2 = \frac{1}{\sqrt{5}}\begin{bmatrix}1\\-2\end{bmatrix}.$$

□

The singular value decomposition and bases of fundamental spaces of matrices

Recall that by the fundamental subspaces of an $m \times n$ matrix A we mean the following four subspaces: $\mathbf{C}(A)$ = the column space of A, $\mathbf{N}(A)$ = the null space of A, $\mathbf{C}(A^T)$ = the column space of A^T, and $\mathbf{N}(A^T)$ = the null space of A^T. It turns out that there is an important connection between the vectors in the singular value decomposition of the matrix A and bases of fundamental spaces of A.

Theorem 8.23. *If A is an $m \times n$ matrix that has a singular value decomposition as in Definition 8.15, then*

(a) $\{\mathbf{u}_1, \dots, \mathbf{u}_r\}$ *is an orthonormal basis of* $\mathbf{C}(A)$,

(b) $\{\mathbf{v}_1, \dots, \mathbf{v}_r\}$ *is an orthonormal basis of* $\mathbf{C}(A^T)$,

(c) $\{\mathbf{v}_{r+1}, \dots, \mathbf{v}_n\}$ *is an orthonormal basis of* $\mathbf{N}(A)$, *and*

(d) $\{\mathbf{u}_{r+1}, \dots, \mathbf{u}_m\}$ *is an orthonormal basis of* $\mathbf{N}(A^T)$.

Proof. Properties (a) and (b) are consequences of Theorem 8.9.

The set $\{\mathbf{v}_{r+1},\dots,\mathbf{v}_n\}$ is an orthonormal basis of the eigenspace of the matrix A^TA corresponding to the eigenvalue 0, that is $\mathbf{N}(A^TA)$. Since

$$\mathbf{x}^TA^TA\mathbf{x} = \|A\mathbf{x}\|^2,$$

we have $\mathbf{N}(A^TA) = \mathbf{N}(A)$.

Finally, since $\{\mathbf{u}_{r+1},\dots,\mathbf{u}_m\}$ is an orthonormal basis of $(\mathbf{C}(A))^\perp$ and $\mathbf{N}(A^T) = (\mathbf{C}(A))^\perp$, $\{\mathbf{u}_{r+1},\dots,\mathbf{u}_m\}$ is an orthonormal basis of $\mathbf{N}(A^T)$. □

Example 8.24. Suppose that

$$A = \begin{bmatrix}\mathbf{u}_1 & \mathbf{u}_2 & \mathbf{u}_3 & \mathbf{u}_4 & \mathbf{u}_5\end{bmatrix}\begin{bmatrix}\sigma_1 & 0 & 0 & 0\\ 0 & \sigma_2 & 0 & 0\\ 0 & 0 & \sigma_3 & 0\\ 0 & 0 & 0 & \sigma_4\\ 0 & 0 & 0 & 0\end{bmatrix}\begin{bmatrix}\mathbf{v}_1^T\\ \mathbf{v}_2^T\\ \mathbf{v}_3^T\\ \mathbf{v}_4^T\end{bmatrix}$$

and that $\sigma_1 \geq \sigma_2 > 0$ and $\sigma_3 = \sigma_4 = 0$. Find bases of the subspaces $\mathbf{C}(A)$, $\mathbf{N}(A^T)$, $\mathbf{C}(A^T)$, $\mathbf{N}(A)$, and the rank of A.

Solution.

$$\begin{aligned}\mathbf{C}(A) &= \operatorname{Span}\{\mathbf{u}_1,\mathbf{u}_2\}\\ \mathbf{N}(A^T) &= \operatorname{Span}\{\mathbf{u}_3,\mathbf{u}_4,\mathbf{u}_5\}\\ \mathbf{C}(A^T) &= \operatorname{Span}\{\mathbf{v}_1,\mathbf{v}_2\}\\ \mathbf{N}(A) &= \operatorname{Span}\{\mathbf{v}_3,\mathbf{v}_4\}\\ \operatorname{Rank} A &= 2.\end{aligned}$$

□

Exercises 8.2

1. Suppose that A is a 4×4 matrix with the outer product expansion

$$A = \sigma_1\mathbf{u}_1\mathbf{v}_1^T + \sigma_2\mathbf{u}_2\mathbf{v}_2^T + \sigma_3\mathbf{u}_3\mathbf{v}_3^T.$$

Find the singular value decomposition of A.

2. Suppose that A is a 3×7 matrix with the outer product expansion

$$A = \sigma_1\mathbf{u}_1\mathbf{v}_1^T + \sigma_2\mathbf{u}_2\mathbf{v}_2^T + \sigma_3\mathbf{u}_3\mathbf{v}_3^T.$$

Find the singular value decomposition of A.

3. Suppose that A is a 5×2 matrix with the outer product expansion

$$A = \sigma_1 \mathbf{u}_1 \mathbf{v}_1^T + \sigma_2 \mathbf{u}_2 \mathbf{v}_2^T.$$

Find the singular value decomposition of A.

4. Suppose that A is a 2×5 matrix with the outer product decomposition

$$A = \sigma_1 \mathbf{u}_1 \mathbf{v}_1^T + \sigma_2 \mathbf{u}_2 \mathbf{v}_2^T.$$

Find the singular value decomposition of A.

Find the singular value decomposition of the given matrix A.

5. $A = \begin{bmatrix} 3 & 5 \\ -4 & 0 \end{bmatrix}$

6. $A = \begin{bmatrix} 1 & 0 \\ 1 & \sqrt{2} \end{bmatrix}$

7. $A = \begin{bmatrix} 3 & -3 \\ 1 & -1 \end{bmatrix}$

8. $A = \begin{bmatrix} 5 & -2 \\ -5 & 2 \end{bmatrix}$

9. $A = \begin{bmatrix} 1 & 1 & -3 \\ 2 & 3 & 1 \end{bmatrix}$

10. $A = \begin{bmatrix} 1 & 4 & 0 \\ 3 & 1 & \sqrt{7} \end{bmatrix}$

11. $A = \begin{bmatrix} -3 & 2 & 1 \\ -1 & -2 & 3 \\ 1 & 1 & 1 \end{bmatrix}$

12. $A = \begin{bmatrix} 1 & \sqrt{2} & -\sqrt{2} \\ 2 & \sqrt{2} & \sqrt{2} \end{bmatrix}$

13. $A = \begin{bmatrix} 2 & -1 & 0 \\ -1 & -1 & -3 \\ 2 & -1 & 0 \end{bmatrix}$

14. $A = \begin{bmatrix} 1 & 1 & 1 \\ 1 & 1 & 1 \\ 1 & 1 & 1 \\ 1 & 1 & 1 \end{bmatrix}$

15. $A = \begin{bmatrix} 3 & 1 \\ 1 & 1 \\ 1 & -1 \\ 0 & 4 \end{bmatrix}$

16. $A = \begin{bmatrix} 3 & 5 \\ 4 & 4 \\ 4 & 0 \\ 3 & 3 \end{bmatrix}$

17. $A = \begin{bmatrix} 3 & 1 & 1 & 0 \\ 1 & 1 & -1 & 4 \end{bmatrix}$

18. $A = \begin{bmatrix} 1 & 2 & 1 & 1 \\ 2 & 1 & -1 & 1 \end{bmatrix}$

In Exercises 19 and 20 find numbers d_1, d_2, d_3, d_4 and k_1, k_2, k_3, k_4 such that

(a) $\mathbf{C}(A)$ is a subspace of $\mathbb{R}^{k_1}$ of dimension d_1,

(b) $\mathbf{C}(A^T)$ is a subspace of $\mathbb{R}^{k_2}$ of dimension d_2,

(c) $\mathbf{N}(A)$ is a subspace of $\mathbb{R}^{k_3}$ of dimension d_3, and

(d) $\mathbf{N}(A^T)$ is a subspace of $\mathbb{R}^{k_4}$ of dimension d_4.

19. A is a 8×5 matrix with the outer product representation

$$A = 11\mathbf{u}_1\mathbf{v}_1^T + 5\mathbf{u}_2\mathbf{v}_2^T + 4\mathbf{u}_3\mathbf{v}_3^T.$$

20. A is a 3×13 matrix with the outer product representation

$$A = 117\mathbf{u}_1\mathbf{v}_1^T + 9\mathbf{u}_2\mathbf{v}_2^T.$$

21. If

$$A = \begin{bmatrix} \mathbf{u}_1 & \mathbf{u}_2 & \mathbf{u}_3 & \mathbf{u}_4 & \mathbf{u}_5 \end{bmatrix} \begin{bmatrix} \sigma_1 & 0 & 0 & 0 \\ 0 & \sigma_2 & 0 & 0 \\ 0 & 0 & \sigma_3 & 0 \\ 0 & 0 & 0 & \sigma_4 \\ 0 & 0 & 0 & 0 \end{bmatrix} \begin{bmatrix} \mathbf{v}_1^T \\ \mathbf{v}_2^T \\ \mathbf{v}_3^T \\ \mathbf{v}_4^T \end{bmatrix},$$

where $\sigma_1 > 0$ and $\sigma_2 = \sigma_3 = \sigma_4 = 0$, find bases in the fundamental subspaces $\mathbf{C}(A)$, $\mathbf{C}(A^T)$, $\mathbf{N}(A)$, and $\mathbf{N}(A^T)$.

22. If

$$A = \begin{bmatrix} \mathbf{u}_1 & \mathbf{u}_2 & \mathbf{u}_3 \end{bmatrix} \begin{bmatrix} \sigma_1 & 0 & 0 & 0 \\ 0 & \sigma_2 & 0 & 0 \\ 0 & 0 & \sigma_3 & 0 \end{bmatrix} \begin{bmatrix} \mathbf{v}_1^T \\ \mathbf{v}_2^T \\ \mathbf{v}_3^T \\ \mathbf{v}_4^T \end{bmatrix},$$

where $\sigma_1 > 0$, $\sigma_2 > 0$, and $\sigma_3 = 0$, find bases in the fundamental subspaces $\mathbf{C}(A)$, $\mathbf{C}(A^T)$, $\mathbf{N}(A)$, and $\mathbf{N}(A^T)$.

Chapter 9

Quadratic forms and positive definite matrices

In this chapter we introduce some ideas that are important from the point of view of mathematics as well as applications. The chapter also provides an opportunity to practice applying many of the tools we learned in the first eight chapters of this book.

Quadratic forms

By a quadratic form we mean a polynomial in several variables with all terms of degree two. For example,

$$x^2 - 7y^2 + 4xy \quad \text{and} \quad 3x^2 - 5y^2 + 7z^2 + 6xy - 10xz - 2yz$$

are quadratic forms in two and three variables, respectively. Note that

$$x^2 - 7y^2 + 4xy = \begin{bmatrix} x & y \end{bmatrix} \begin{bmatrix} 1 & 2 \\ 2 & -7 \end{bmatrix} \begin{bmatrix} x \\ y \end{bmatrix}$$

and

$$3x^2 - 5y^2 + 7z^2 + 6xy - 10xz - 2yz = \begin{bmatrix} x & y & z \end{bmatrix} \begin{bmatrix} 3 & 3 & -5 \\ 3 & -5 & -1 \\ -5 & -1 & 7 \end{bmatrix} \begin{bmatrix} x \\ y \\ z \end{bmatrix}$$

and that in both cases the matrix is symmetric. This connection allows us to use tools of linear algebra to study quadratic forms. Since every quadratic form can be described by a symmetric matrix as shown above, it is natural to define quadratic forms in terms of symmetric matrices.

Definition 9.1. By a *quadratic form* we mean a function $f : \mathbb{R}^n \to \mathbb{R}$ defined by $f(\mathbf{x}) = \mathbf{x}^T A\mathbf{x}$, where A is a $n \times n$ symmetric matrix.

Example 9.2. Find the quadratic form associated with the matrix

$$A = \begin{bmatrix} 7 & 5 \\ 5 & 2 \end{bmatrix}.$$

Solution.

$$f\left(\begin{bmatrix} x \\ y \end{bmatrix}\right) = \begin{bmatrix} x & y \end{bmatrix} \begin{bmatrix} 7 & 5 \\ 5 & 2 \end{bmatrix} \begin{bmatrix} x \\ y \end{bmatrix} = \begin{bmatrix} 7x+5y & 5x+2y \end{bmatrix} \begin{bmatrix} x \\ y \end{bmatrix} = 7x^2 + 2y^2 + 10xy.$$

□

Example 9.3. Find the quadratic form associated with the matrix

$$A = \begin{bmatrix} 2 & 4 & 7 \\ 4 & 5 & 11 \\ 7 & 11 & 3 \end{bmatrix}.$$

Solution.

$$\begin{aligned} f\left(\begin{bmatrix} x \\ y \\ z \end{bmatrix}\right) &= \begin{bmatrix} x & y & z \end{bmatrix} \begin{bmatrix} 2 & 4 & 7 \\ 4 & 5 & 11 \\ 7 & 11 & 3 \end{bmatrix} \begin{bmatrix} x \\ y \\ z \end{bmatrix} \\ &= \begin{bmatrix} 2x+4y+7z & 4x+5y+11z & 7x+11y+3z \end{bmatrix} \begin{bmatrix} x \\ y \\ z \end{bmatrix} \\ &= 2x^2 + 5y^2 + 3z^2 + 8xy + 14xz + 22yz. \end{aligned}$$

□

Example 9.4. Find a symmetric 3×3 matrix A such that

$$5x^2 + y^2 - 4z^2 + 2xy + 8xz - 18yz = \begin{bmatrix} x & y & z \end{bmatrix} A \begin{bmatrix} x \\ y \\ z \end{bmatrix}.$$

Solution.

$$A = \begin{bmatrix} 5 & 1 & 4 \\ 1 & 1 & -9 \\ 4 & -9 & -4 \end{bmatrix}.$$

□

The following classification of symmetric matrices is important in studying properties of quadratic forms. In other words, the described properties of symmetric matrices provide useful information about the associated polynomials.

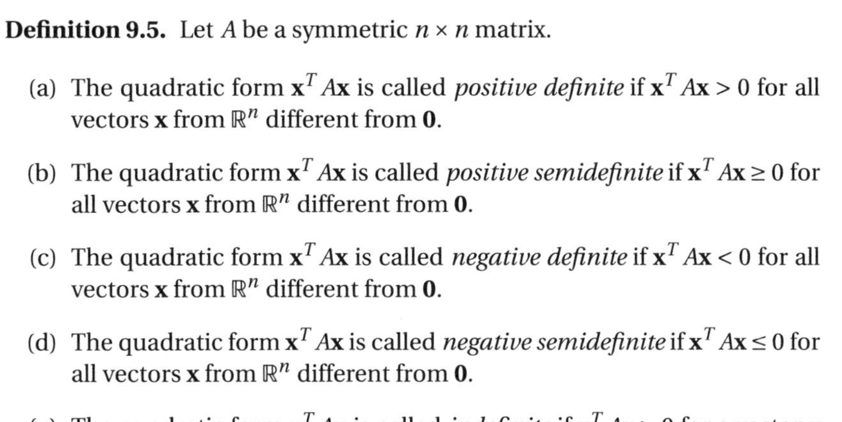

Definition 9.5. Let A be a symmetric $n \times n$ matrix.

(a) The quadratic form $\mathbf{x}^T A\mathbf{x}$ is called *positive definite* if $\mathbf{x}^T A\mathbf{x} > 0$ for all vectors $\mathbf{x}$ from $\mathbb{R}^n$ different from $\mathbf{0}$.

(b) The quadratic form $\mathbf{x}^T A\mathbf{x}$ is called *positive semidefinite* if $\mathbf{x}^T A\mathbf{x} \geq 0$ for all vectors $\mathbf{x}$ from $\mathbb{R}^n$ different from $\mathbf{0}$.

(c) The quadratic form $\mathbf{x}^T A\mathbf{x}$ is called *negative definite* if $\mathbf{x}^T A\mathbf{x} < 0$ for all vectors $\mathbf{x}$ from $\mathbb{R}^n$ different from $\mathbf{0}$.

(d) The quadratic form $\mathbf{x}^T A\mathbf{x}$ is called *negative semidefinite* if $\mathbf{x}^T A\mathbf{x} \leq 0$ for all vectors $\mathbf{x}$ from $\mathbb{R}^n$ different from $\mathbf{0}$.

(e) The quadratic form $\mathbf{x}^T A\mathbf{x}$ is called *indefinite* if $\mathbf{x}^T A\mathbf{x} > 0$ for a vector $\mathbf{x}$ from $\mathbb{R}^n$ different from $\mathbf{0}$ and $\mathbf{y}^T A\mathbf{y} < 0$ for a vector $\mathbf{y}$ from $\mathbb{R}^n$ different from $\mathbf{0}$.

Example 9.6. If $A = \begin{bmatrix} 1 & -1 & 0 \\ -1 & 2 & -1 \\ 0 & -1 & 5 \end{bmatrix}$, then for any $\mathbf{x} = \begin{bmatrix} x_1 \\ x_2 \\ x_3 \end{bmatrix}$ we have

$$\begin{aligned} \mathbf{x}^T A\mathbf{x} &= \begin{bmatrix} x_1 & x_2 & x_3 \end{bmatrix} \begin{bmatrix} 1 & -1 & 0 \\ -1 & 2 & -1 \\ 0 & -1 & 5 \end{bmatrix} \begin{bmatrix} x_1 \\ x_2 \\ x_3 \end{bmatrix} \\ &= \begin{bmatrix} x_1 & x_2 & x_3 \end{bmatrix} \begin{bmatrix} x_1 - x_2 \\ -x_1 + 2x_2 - x_3 \\ -x_2 + 5x_3 \end{bmatrix} \\ &= x_1^2 - 2x_1x_2 + 2x_2^2 - 2x_2x_3 + 5x_3^2 \\ &= (x_1 - x_2)^2 + (x_2 - x_3)^2 + 4x_3^2. \end{aligned}$$

Since

$$(x_1 - x_2)^2 + (x_2 - x_3)^2 + 4x_3^2 > 0$$

whenever at least one of the numbers x_1, x_2, and x_3 is different from 0, the quadratic form $\mathbf{x}^T A\mathbf{x}$ is positive definite.

It turns out that the properties of symmetric matrices described in Definition 9.5 are closely related to the eigenvalues. To classify quadratic forms in terms of the eigenvalues we will use the following result.

Theorem 9.7. *If A is an $n \times n$ symmetric matrix with eigenvalues $\lambda_1, \dots, \lambda_n$ and P is an orthogonal $n \times n$ matrix such that*

$$A = P \begin{bmatrix} \lambda_1 & & 0 \\ & \ddots & \\ 0 & & \lambda_n \end{bmatrix} P^T,$$

then

$$\mathbf{x}^T A \mathbf{x} = y_1^2 \lambda_1 + \cdots + y_n^2 \lambda_n,$$

where

$$\begin{bmatrix} y_1 \\ \vdots \\ y_n \end{bmatrix} = P^T \mathbf{x}.$$

Proof for $n = 3$. We have to show that, if A is a 3×3 symmetric matrix with eigenvalues $\lambda_1, \lambda_2, \lambda_3$ and P is an orthogonal 3×3 matrix such that

$$A = P \begin{bmatrix} \lambda_1 & 0 & 0 \\ 0 & \lambda_2 & 0 \\ 0 & 0 & \lambda_3 \end{bmatrix} P^T,$$

then

$$\begin{bmatrix} x_1 & x_2 & x_3 \end{bmatrix} A \begin{bmatrix} x_1 \\ x_2 \\ x_3 \end{bmatrix} = y_1^2 \lambda_1 + y_2^2 \lambda_2 + y_3^2 \lambda_3,$$

where

$$\begin{bmatrix} y_1 \\ y_2 \\ y_3 \end{bmatrix} = P^T \begin{bmatrix} x_1 \\ x_2 \\ x_3 \end{bmatrix}.$$

First we note that

$$\begin{bmatrix} x_1 & x_2 & x_3 \end{bmatrix} A \begin{bmatrix} x_1 \\ x_2 \\ x_3 \end{bmatrix} = \begin{bmatrix} x_1 & x_2 & x_3 \end{bmatrix} P \begin{bmatrix} \lambda_1 & 0 & 0 \\ 0 & \lambda_2 & 0 \\ 0 & 0 & \lambda_3 \end{bmatrix} P^T \begin{bmatrix} x_1 \\ x_2 \\ x_3 \end{bmatrix}.$$

If

$$\begin{bmatrix} y_1 \\ y_2 \\ y_3 \end{bmatrix} = P^T \begin{bmatrix} x_1 \\ x_2 \\ x_3 \end{bmatrix},$$

then

$$\begin{bmatrix} y_1 & y_2 & y_3 \end{bmatrix} = \begin{bmatrix} x_1 & x_2 & x_3 \end{bmatrix} P$$

and thus

$$\begin{bmatrix} x_1 & x_2 & x_3 \end{bmatrix} P \begin{bmatrix} \lambda_1 & 0 & 0 \\ 0 & \lambda_2 & 0 \\ 0 & 0 & \lambda_3 \end{bmatrix} P^T \begin{bmatrix} x_1 \\ x_2 \\ x_3 \end{bmatrix} = \begin{bmatrix} y_1 & y_2 & y_3 \end{bmatrix} \begin{bmatrix} \lambda_1 & 0 & 0 \\ 0 & \lambda_2 & 0 \\ 0 & 0 & \lambda_3 \end{bmatrix} \begin{bmatrix} y_1 \\ y_2 \\ y_3 \end{bmatrix}$$
$$= y_1^2\lambda_1 + y_2^2\lambda_2 + y_3^2\lambda_3.$$

□

Using Theorem 9.7 we can prove the following important result connecting the properties of a symmetric matrix A described in Definition 9.5 with the position of the eigenvalues of A on the real axis.

Theorem 9.8. *Let A be a symmetric $n \times n$ matrix with eigenvalues $\lambda_1, \dots, \lambda_n$.*

(a) *The quadratic form $\mathbf{x}^T A\mathbf{x}$ is positive definite if and only if all its eigenvalues are positive, that is $\lambda_1 > 0, \dots, \lambda_n > 0$.*

(b) *The quadratic form $\mathbf{x}^T A\mathbf{x}$ is positive semidefinite if and only if all its eigenvalues are nonnegative, that is $\lambda_1 \ge 0, \dots, \lambda_n \ge 0$.*

(c) *The quadratic form $\mathbf{x}^T A\mathbf{x}$ is negative definite if and only if all its eigenvalues are negative, that is $\lambda_1 < 0, \dots, \lambda_n < 0$.*

(d) *The quadratic form $\mathbf{x}^T A\mathbf{x}$ is negative semidefinite if and only if all its eigenvalues are nonpositive, that is $\lambda_1 \le 0, \dots, \lambda_n \le 0$.*

(e) *The quadratic form $\mathbf{x}^T A\mathbf{x}$ is indefinite if and only if at least one eigenvalue of A is positive and at least one eigenvalue of A is negative.*

Proof for $n = 3$. We only prove that a quadratic form $\mathbf{x}^T A\mathbf{x}$ is positive definite if and only if the eigenvalues of the matrix A are positive. The other proofs are similar.

We have to prove that, if A is a 3×3 symmetric matrix with eigenvalues $\lambda_1, \lambda_2, \lambda_3$, then we have $\lambda_1 > 0, \lambda_2 > 0, \lambda_3 > 0$ if and only if

$$\begin{bmatrix} x_1 & x_2 & x_3 \end{bmatrix} A \begin{bmatrix} x_1 \\ x_2 \\ x_3 \end{bmatrix} > 0$$

for all $\begin{bmatrix} x_1 \\ x_2 \\ x_3 \end{bmatrix} \neq \begin{bmatrix} 0 \\ 0 \\ 0 \end{bmatrix}$.

According to Theorem 9.7, there is an orthogonal 3×3 matrix P such that if

$$\begin{bmatrix} y_1 \\ y_2 \\ y_3 \end{bmatrix} = P^T \begin{bmatrix} x_1 \\ x_2 \\ x_3 \end{bmatrix},$$

then

$$\begin{bmatrix} x_1 & x_2 & x_3 \end{bmatrix} A \begin{bmatrix} x_1 \\ x_2 \\ x_3 \end{bmatrix} = y_1^2\lambda_1 + y_2^2\lambda_2 + y_3^2\lambda_3.$$

If $\lambda_1 > 0, \lambda_2 > 0, \lambda_3 > 0$ and $\begin{bmatrix} x_1 \\ x_2 \\ x_3 \end{bmatrix} \neq \begin{bmatrix} 0 \\ 0 \\ 0 \end{bmatrix}$, then

$$\begin{bmatrix} y_1 \\ y_2 \\ y_3 \end{bmatrix} = P^T \begin{bmatrix} x_1 \\ x_2 \\ x_3 \end{bmatrix} \neq \begin{bmatrix} 0 \\ 0 \\ 0 \end{bmatrix}$$

and consequently

$$\begin{bmatrix} x_1 & x_2 & x_3 \end{bmatrix} A \begin{bmatrix} x_1 \\ x_2 \\ x_3 \end{bmatrix} = y_1^2\lambda_1 + y_2^2\lambda_2 + y_3^2\lambda_3 > 0.$$

Now assume that

$$\begin{bmatrix} x_1 & x_2 & x_3 \end{bmatrix} A \begin{bmatrix} x_1 \\ x_2 \\ x_3 \end{bmatrix} > 0$$

for all $\begin{bmatrix} x_1 \\ x_2 \\ x_3 \end{bmatrix} \neq \begin{bmatrix} 0 \\ 0 \\ 0 \end{bmatrix}$. If we take $\begin{bmatrix} x_1 \\ x_2 \\ x_3 \end{bmatrix} = P \begin{bmatrix} 1 \\ 0 \\ 0 \end{bmatrix}$, then we have $\begin{bmatrix} x_1 \\ x_2 \\ x_3 \end{bmatrix} \neq \begin{bmatrix} 0 \\ 0 \\ 0 \end{bmatrix}$ and consequently

$$\begin{aligned} \begin{bmatrix} x_1 & x_2 & x_3 \end{bmatrix} A \begin{bmatrix} x_1 \\ x_2 \\ x_3 \end{bmatrix} &= \begin{bmatrix} 1 & 0 & 0 \end{bmatrix} P^T P \begin{bmatrix} \lambda_1 & 0 & 0 \\ 0 & \lambda_2 & 0 \\ 0 & 0 & \lambda_3 \end{bmatrix} P^T P \begin{bmatrix} 1 \\ 0 \\ 0 \end{bmatrix} \\ &= \begin{bmatrix} 1 & 0 & 0 \end{bmatrix} \begin{bmatrix} \lambda_1 & 0 & 0 \\ 0 & \lambda_2 & 0 \\ 0 & 0 & \lambda_3 \end{bmatrix} \begin{bmatrix} 1 \\ 0 \\ 0 \end{bmatrix} = \lambda_1 > 0. \end{aligned}$$

Using a similar argument we can show that $\lambda_2 > 0$ and $\lambda_3 > 0$. □

The general form of a quadratic form on $\mathbb{R}^2$ is

$$\begin{bmatrix} x & y \end{bmatrix} A \begin{bmatrix} x \\ y \end{bmatrix} = ax^2 + by^2 + cxy,$$

where a, b, c are arbitrary real numbers. The terms ax^2 and by^2 are called *quadratic* and cxy is called the *cross-product* term. Similarly, the general form of a quadratic form on $\mathbb{R}^3$ is

$$\begin{bmatrix} x & y & z \end{bmatrix} A \begin{bmatrix} x \\ y \\ z \end{bmatrix} = ax^2 + by^2 + cz^2 + dxy + exz + fyz,$$

where a, b, c, d, e, f are arbitrary real numbers. In this case we have up to three quadratic terms, namely ax^2, by^2, and cz^2, and up to three cross-product terms, namely dxy, exz, fyz.

Note that, if A is a diagonal matrix, then the quadratic form $\mathbf{x}^T A\mathbf{x}$ has no cross-product terms. Since an orthogonal $n \times n$ matrix P corresponds to a change of basis in $\mathbb{R}^n$, the representation

$$A = P \begin{bmatrix} \lambda_1 & & 0 \\ & \ddots & \\ 0 & & \lambda_n \end{bmatrix} P^T$$

in Theorem 9.7 can be used to find new variables in $\mathbb{R}^n$ for which the quadratic form has no cross-product terms.

Example 9.9. Classify the quadratic form $2x^2 + 17y^2 + 8xy$ and find a change of variables $\begin{bmatrix} x \\ y \end{bmatrix} = P \begin{bmatrix} x' \\ y' \end{bmatrix}$ such that the quadratic form expressed in these new variables has no cross-product term.

Solution. We have

$$2x^2 + 4y^2 + 8xy = \begin{bmatrix} x & y \end{bmatrix} \begin{bmatrix} 2 & 4 \\ 4 & 17 \end{bmatrix} \begin{bmatrix} x \\ y \end{bmatrix}$$

and

$$\begin{bmatrix} 2 & 4 \\ 4 & 17 \end{bmatrix} = P \begin{bmatrix} 1 & 0 \\ 0 & 18 \end{bmatrix} P^T,$$

where

$$P = \begin{bmatrix} \frac{4}{\sqrt{17}} & \frac{1}{\sqrt{17}} \\ -\frac{1}{\sqrt{17}} & \frac{4}{\sqrt{17}} \end{bmatrix}.$$

Since both eigenvalues are positive, the quadratic form is positive definite. In terms of the new variables x' and y' the quadratic form becomes

$$\begin{bmatrix} x' & y' \end{bmatrix} \begin{bmatrix} 1 & 0 \\ 0 & 18 \end{bmatrix} \begin{bmatrix} x' \\ y' \end{bmatrix} = (x')^2 + 18(y')^2.$$

□

Example 9.10. Classify the quadratic form $x^2+y^2+7z^2+8xy-4xz-4yz$ and find a change of variables $\begin{bmatrix} x \\ y \\ z \end{bmatrix} = P \begin{bmatrix} x' \\ y' \\ z' \end{bmatrix}$ such that the quadratic form expressed in these new variables has no cross-product term.

Solution. We have

$$x^2+y^2+7z^2+8xy-4xz-4yz = \begin{bmatrix} x & y & z \end{bmatrix} \begin{bmatrix} 1 & 4 & -2 \\ 4 & 1 & -2 \\ -2 & -2 & 7 \end{bmatrix} \begin{bmatrix} x \\ y \\ z \end{bmatrix}$$

and

$$\begin{bmatrix} 1 & 4 & -2 \\ 4 & 1 & -2 \\ -2 & -2 & 7 \end{bmatrix} = P \begin{bmatrix} 9 & 0 & 0 \\ 0 & 3 & 0 \\ 0 & 0 & -3 \end{bmatrix} P^T,$$

where

$$P = \begin{bmatrix} \frac{1}{\sqrt{6}} & \frac{1}{\sqrt{3}} & \frac{1}{\sqrt{2}} \\ \frac{1}{\sqrt{6}} & \frac{1}{\sqrt{3}} & -\frac{1}{\sqrt{2}} \\ -\frac{2}{\sqrt{6}} & \frac{1}{\sqrt{3}} & 0 \end{bmatrix}.$$

Since two eigenvalues are positive and one is negative, the quadratic form is undefinite. In terms of the new variables x', y', and z', the quadratic form becomes

$$9(x')^2+3(y')^2-3(z')^2.$$

□

The Cholesky decomposition

Definition 9.11. An $n \times n$ symmetric matrix A is called *positive definite* if

$$\mathbf{x}^T A\mathbf{x} > 0$$

for all vectors $\mathbf{x}$ from $\mathbb{R}^n$ different from $\mathbf{0}$.

Note that the quadratic form $\mathbf{x}^T A\mathbf{x}$ is positive definite if and only if the matrix A is positive definite. Theorem 9.13 gives us an interesting characterization of positive definite matrices. It can be used to easily generate examples of positive definite matrices.

Definition 9.12. An $n \times n$ matrix is called *upper diagonal* if all entries below the main diagonal are 0. An upper diagonal matrix has the form

$$\begin{bmatrix} a_{11} & a_{12} & \cdots & a_{1n} \\ 0 & a_{22} & \cdots & a_{2n} \\ \vdots & \vdots & \ddots & \vdots \\ 0 & 0 & \cdots & a_{nn} \end{bmatrix}.$$

Theorem 9.13. *Let A be a symmetric $n \times n$ matrix. The following conditions are equivalent:*

(a) *A is positive definite;*

(b) *There is an upper triangular $n \times n$ matrix M with positive elements on the main diagonal such that*

$$A = M^T M.$$

Proof for $n = 2$. First we show that (a) implies (b). If the matrix $A = \begin{bmatrix} a_{11} & a_{21} \\ a_{21} & a_{22} \end{bmatrix}$ is positive definite, then

$$a_{11} = \begin{bmatrix} 1 & 0 \end{bmatrix} \begin{bmatrix} a_{11} & a_{21} \\ a_{21} & a_{22} \end{bmatrix} \begin{bmatrix} 1 \\ 0 \end{bmatrix} > 0$$

and

$$\begin{bmatrix} a_{11} & a_{21} \\ a_{21} & a_{22} \end{bmatrix} - \begin{bmatrix} \sqrt{a_{11}} \\ \frac{a_{21}}{\sqrt{a_{11}}} \end{bmatrix} \begin{bmatrix} \sqrt{a_{11}} & \frac{a_{21}}{\sqrt{a_{11}}} \end{bmatrix} = \begin{bmatrix} a_{11} & a_{21} \\ a_{21} & a_{22} \end{bmatrix} - \begin{bmatrix} a_{11} & a_{21} \\ a_{21} & \frac{a_{21}^2}{a_{11}} \end{bmatrix} = \begin{bmatrix} 0 & 0 \\ 0 & a_{22} - \frac{a_{21}^2}{a_{11}} \end{bmatrix}.$$

Consequently,

$$\begin{bmatrix} x_1 & x_2 \end{bmatrix} \begin{bmatrix} a_{11} & a_{21} \\ a_{21} & a_{22} \end{bmatrix} \begin{bmatrix} x_1 \\ x_2 \end{bmatrix} - \begin{bmatrix} x_1 & x_2 \end{bmatrix} \begin{bmatrix} \sqrt{a_{11}} \\ \frac{a_{21}}{\sqrt{a_{11}}} \end{bmatrix} \begin{bmatrix} \sqrt{a_{11}} & \frac{a_{21}}{\sqrt{a_{11}}} \end{bmatrix} \begin{bmatrix} x_1 \\ x_2 \end{bmatrix}$$

$$= \begin{bmatrix} x_1 & x_2 \end{bmatrix} \begin{bmatrix} 0 & 0 \\ 0 & a_{22} - \frac{a_{21}^2}{a_{11}} \end{bmatrix} \begin{bmatrix} x_1 \\ x_2 \end{bmatrix},$$

which can be written as

$$\begin{bmatrix} x_1 & x_2 \end{bmatrix} \begin{bmatrix} a_{11} & a_{21} \\ a_{21} & a_{22} \end{bmatrix} \begin{bmatrix} x_1 \\ x_2 \end{bmatrix} - \left(\sqrt{a_{11}} x_1 + \frac{a_{21}}{\sqrt{a_{11}}} x_2 \right)^2 = \left(a_{22} - \frac{a_{21}^2}{a_{11}} \right) x_2^2.$$

Thus the matrix $\begin{bmatrix} a_{11} & a_{21} \\ a_{21} & a_{22} \end{bmatrix}$ is positive definite if and only if $a_{11} > 0$ and $a_{22} - \frac{a_{21}^2}{a_{11}} > 0$. Now it is easy to verify that

$$\begin{bmatrix} a_{11} & a_{21} \\ a_{21} & a_{22} \end{bmatrix} = \begin{bmatrix} \sqrt{a_{11}} & 0 \\ \frac{a_{21}}{\sqrt{a_{11}}} & \sqrt{a_{22} - \frac{a_{21}^2}{a_{11}}} \end{bmatrix} \begin{bmatrix} \sqrt{a_{11}} & \frac{a_{21}}{\sqrt{a_{11}}} \\ 0 & \sqrt{a_{22} - \frac{a_{21}^2}{a_{11}}} \end{bmatrix}.$$

To prove that (b) implies (a) it suffices to observe that

$$\mathbf{x}^T A\mathbf{x} = \mathbf{x}^T M^T M\mathbf{x} = \|M\mathbf{x}\|^2$$

and that $\|M\mathbf{x}\| > 0$ if $\mathbf{x} \neq \mathbf{0}$, which is a consequence of the fact that the columns of the matrix M are linearly independent. This completes the proof for $n = 2$. □

Proof for $n = 3$. Let $A = \begin{bmatrix} a_{11} & a_{21} & a_{31} \\ a_{21} & a_{22} & a_{32} \\ a_{31} & a_{32} & a_{33} \end{bmatrix}$ be a positive definite matrix. We prove that $a_{11} > 0$ as in the case when $n = 2$. Then we note that

$$\begin{aligned} &\begin{bmatrix} a_{11} & a_{21} & a_{31} \\ a_{21} & a_{22} & a_{32} \\ a_{31} & a_{32} & a_{33} \end{bmatrix} - \begin{bmatrix} \sqrt{a_{11}} \\ \frac{a_{21}}{\sqrt{a_{11}}} \\ \frac{a_{31}}{\sqrt{a_{11}}} \end{bmatrix} \begin{bmatrix} \sqrt{a_{11}} & \frac{a_{21}}{\sqrt{a_{11}}} & \frac{a_{31}}{\sqrt{a_{11}}} \end{bmatrix} \\ &= \begin{bmatrix} a_{11} & a_{21} & a_{31} \\ a_{21} & a_{22} & a_{32} \\ a_{31} & a_{32} & a_{33} \end{bmatrix} - \begin{bmatrix} a_{11} & a_{21} & a_{31} \\ a_{21} & \frac{a_{21}^2}{a_{11}} & \frac{a_{21}a_{31}}{a_{11}} \\ a_{31} & \frac{a_{21}a_{31}}{a_{11}} & \frac{a_{31}^2}{a_{11}} \end{bmatrix} \\ &= \begin{bmatrix} 0 & 0 & 0 \\ 0 & a_{22} - \frac{a_{21}^2}{a_{11}} & a_{32} - \frac{a_{21}a_{31}}{a_{11}} \\ 0 & a_{32} - \frac{a_{21}a_{31}}{a_{11}} & a_{33} - \frac{a_{31}^2}{a_{11}} \end{bmatrix}. \end{aligned}$$

Consequently,

$$\begin{aligned} &\begin{bmatrix} x_1 & x_2 & x_3 \end{bmatrix} \begin{bmatrix} a_{11} & a_{21} & a_{31} \\ a_{21} & a_{22} & a_{32} \\ a_{31} & a_{32} & a_{33} \end{bmatrix} \begin{bmatrix} x_1 \\ x_2 \\ x_3 \end{bmatrix} - \begin{bmatrix} x_1 & x_2 & x_3 \end{bmatrix} \begin{bmatrix} \sqrt{a_{11}} \\ \frac{a_{21}}{\sqrt{a_{11}}} \\ \frac{a_{31}}{\sqrt{a_{11}}} \end{bmatrix} \begin{bmatrix} \sqrt{a_{11}} & \frac{a_{21}}{\sqrt{a_{11}}} & \frac{a_{31}}{\sqrt{a_{11}}} \end{bmatrix} \begin{bmatrix} x_1 \\ x_2 \\ x_3 \end{bmatrix} \\ &= \begin{bmatrix} x_1 & x_2 & x_3 \end{bmatrix} \begin{bmatrix} 0 & 0 & 0 \\ 0 & a_{22} - \frac{a_{21}^2}{a_{11}} & a_{32} - \frac{a_{21}a_{31}}{a_{11}} \\ 0 & a_{32} - \frac{a_{21}a_{31}}{a_{11}} & a_{33} - \frac{a_{31}^2}{a_{11}} \end{bmatrix} \begin{bmatrix} x_1 \\ x_2 \\ x_3 \end{bmatrix}, \end{aligned}$$

which can be written as

$$\begin{bmatrix} x_1 & x_2 & x_3 \end{bmatrix} \begin{bmatrix} a_{11} & a_{21} & a_{31} \\ a_{21} & a_{22} & a_{32} \\ a_{31} & a_{32} & a_{33} \end{bmatrix} \begin{bmatrix} x_1 \\ x_2 \\ x_3 \end{bmatrix} - \left(\sqrt{a_{11}}x_1 + \frac{a_{21}}{\sqrt{a_{11}}}x_2 + \frac{a_{31}}{\sqrt{a_{11}}}x_3\right)^2$$

$$= \begin{bmatrix} x_2 & x_3 \end{bmatrix} \begin{bmatrix} a_{22} - \frac{a_{21}^2}{a_{11}} & a_{32} - \frac{a_{21}a_{31}}{a_{11}} \\ a_{32} - \frac{a_{21}a_{31}}{a_{11}} & a_{33} - \frac{a_{31}^2}{a_{11}} \end{bmatrix} \begin{bmatrix} x_2 \\ x_3 \end{bmatrix}.$$

This equality shows that the matrix $\begin{bmatrix} a_{11} & a_{21} & a_{31} \\ a_{21} & a_{22} & a_{32} \\ a_{31} & a_{32} & a_{33} \end{bmatrix}$ is positive definite if and only if $a_{11} > 0$ and the matrix $\begin{bmatrix} a_{22} - \frac{a_{21}^2}{a_{11}} & a_{32} - \frac{a_{21}a_{31}}{a_{11}} \\ a_{32} - \frac{a_{21}a_{31}}{a_{11}} & a_{33} - \frac{a_{31}^2}{a_{11}} \end{bmatrix}$ is positive definite. Now, if we denote

$$\begin{bmatrix} b_{11} & b_{21} \\ b_{21} & b_{22} \end{bmatrix} = \begin{bmatrix} a_{22} - \frac{a_{21}^2}{a_{11}} & a_{32} - \frac{a_{21}a_{31}}{a_{11}} \\ a_{32} - \frac{a_{21}a_{31}}{a_{11}} & a_{33} - \frac{a_{31}^2}{a_{11}} \end{bmatrix},$$

then obtain

$$\begin{bmatrix} a_{11} & a_{21} & a_{31} \\ a_{21} & a_{22} & a_{32} \\ a_{31} & a_{32} & a_{33} \end{bmatrix} = \begin{bmatrix} \sqrt{a_{11}} & 0 & 0 \\ \frac{a_{21}}{\sqrt{a_{11}}} & \sqrt{b_{11}} & 0 \\ \frac{a_{31}}{\sqrt{a_{11}}} & \frac{b_{21}}{\sqrt{b_{11}}} & \sqrt{c} \end{bmatrix} \begin{bmatrix} \sqrt{a_{11}} & \frac{a_{21}}{\sqrt{a_{11}}} & \frac{a_{31}}{\sqrt{a_{11}}} \\ 0 & \sqrt{b_{11}} & \frac{b_{21}}{\sqrt{b_{11}}} \\ 0 & 0 & \sqrt{c} \end{bmatrix},$$

where $c = b_{22} - \frac{b_{21}^2}{b_{11}}$. Indeed, we have

$$\begin{bmatrix} \sqrt{a_{11}} & 0 & 0 \\ \frac{a_{21}}{\sqrt{a_{11}}} & \sqrt{b_{11}} & 0 \\ \frac{a_{31}}{\sqrt{a_{11}}} & \frac{b_{21}}{\sqrt{b_{11}}} & \sqrt{c} \end{bmatrix} \begin{bmatrix} \sqrt{a_{11}} & \frac{a_{21}}{\sqrt{a_{11}}} & \frac{a_{31}}{\sqrt{a_{11}}} \\ 0 & \sqrt{b_{11}} & \frac{b_{21}}{\sqrt{b_{11}}} \\ 0 & 0 & \sqrt{c} \end{bmatrix}$$

$$= \left(\begin{bmatrix} \sqrt{a_{11}} & 0 & 0 \\ \frac{a_{21}}{\sqrt{a_{11}}} & 0 & 0 \\ \frac{a_{31}}{\sqrt{a_{11}}} & 0 & 0 \end{bmatrix} + \begin{bmatrix} 0 & 0 & 0 \\ 0 & \sqrt{b_{11}} & 0 \\ 0 & \frac{b_{21}}{\sqrt{b_{11}}} & \sqrt{c} \end{bmatrix} \right) \left(\begin{bmatrix} \sqrt{a_{11}} & \frac{a_{21}}{\sqrt{a_{11}}} & \frac{a_{31}}{\sqrt{a_{11}}} \\ 0 & 0 & 0 \\ 0 & 0 & 0 \end{bmatrix} + \begin{bmatrix} 0 & 0 & 0 \\ 0 & \sqrt{b_{11}} & \frac{b_{21}}{\sqrt{b_{11}}} \\ 0 & 0 & \sqrt{c} \end{bmatrix} \right)$$

$$= \begin{bmatrix} \sqrt{a_{11}} & 0 & 0 \\ \frac{a_{21}}{\sqrt{a_{11}}} & 0 & 0 \\ \frac{a_{31}}{\sqrt{a_{11}}} & 0 & 0 \end{bmatrix} \begin{bmatrix} \sqrt{a_{11}} & \frac{a_{21}}{\sqrt{a_{11}}} & \frac{a_{31}}{\sqrt{a_{11}}} \\ 0 & 0 & 0 \\ 0 & 0 & 0 \end{bmatrix} + \begin{bmatrix} 0 & 0 & 0 \\ 0 & \sqrt{b_{11}} & 0 \\ 0 & \frac{b_{21}}{\sqrt{b_{11}}} & \sqrt{c} \end{bmatrix} \begin{bmatrix} 0 & 0 & 0 \\ 0 & \sqrt{b_{11}} & \frac{b_{21}}{\sqrt{b_{11}}} \\ 0 & 0 & \sqrt{c} \end{bmatrix}$$

$$= \begin{bmatrix} a_{11} & a_{21} & a_{31} \\ a_{21} & \frac{a_{21}^2}{a_{11}} & \frac{a_{21}a_{31}}{a_{11}} \\ a_{31} & \frac{a_{21}a_{31}}{a_{11}} & \frac{a_{31}^2}{a_{11}} \end{bmatrix} + \begin{bmatrix} 0 & 0 & 0 \\ 0 & a_{22} - \frac{a_{21}^2}{a_{11}} & a_{32} - \frac{a_{21}a_{31}}{a_{11}} \\ 0 & a_{32} - \frac{a_{21}a_{31}}{a_{11}} & a_{33} - \frac{a_{31}^2}{a_{11}} \end{bmatrix}$$

$$= \begin{bmatrix} a_{11} & a_{21} & a_{31} \\ a_{21} & a_{22} & a_{32} \\ a_{31} & a_{32} & a_{33} \end{bmatrix}.$$

This proves that (a) implies (b).

We prove that (b) implies (a) as in the proof for $n = 2$. □

Definition 9.14. The decomposition of a symmetric $n \times n$ matrix A in the form

$$A = M^T M,$$

where M is an upper triangular $n \times n$ matrix with positive elements on the main diagonal, is called the *Cholesky decomposition* of A.

Theorem 9.13 says that a symmetric $n \times n$ matrix is positive definite if and only if it has a Cholesky decomposition.

Example 9.15. Find the Cholesky decomposition of the matrix $A = \begin{bmatrix} 3 & 1 \\ 1 & 5 \end{bmatrix}$.

Proof. We follow the method of the proof of Theorem 9.13 for $n = 2$. If we take $\mathbf{v} = \begin{bmatrix} \sqrt{3} \\ \frac{1}{\sqrt{3}} \end{bmatrix}$, then

$$\mathbf{v}\mathbf{v}^T = \begin{bmatrix} 3 & 1 \\ 1 & \frac{1}{3} \end{bmatrix} \quad \text{and} \quad A - \mathbf{v}\mathbf{v}^T = \begin{bmatrix} 0 & 0 \\ 0 & \frac{14}{3} \end{bmatrix}.$$

Consequently, the Cholesky decomposition of A is

$$A = \begin{bmatrix} \sqrt{3} & 0 \\ \frac{\sqrt{3}}{3} & \frac{\sqrt{42}}{3} \end{bmatrix} \begin{bmatrix} \sqrt{3} & \frac{\sqrt{3}}{3} \\ 0 & \frac{\sqrt{42}}{3} \end{bmatrix}.$$

□

Example 9.16. Show that the matrix $A = \begin{bmatrix} 2 & 3 \\ 3 & 4 \end{bmatrix}$ is not positive definite.

Proof. We follow the method of the proof of Theorem 9.13 for $n = 2$. If we take $\mathbf{v} = \begin{bmatrix} \sqrt{2} \\ \frac{3}{\sqrt{2}} \end{bmatrix}$, then

$$\mathbf{v}\mathbf{v}^T = \begin{bmatrix} 2 & 3 \\ 3 & \frac{9}{2} \end{bmatrix} \quad \text{and} \quad A - \mathbf{v}\mathbf{v}^T = \begin{bmatrix} 0 & 0 \\ 0 & -\frac{1}{2} \end{bmatrix}.$$

Consequently, the matrix A is not positive definite because $-\frac{1}{2} < 0$. □

Example 9.17. Find the Cholesky decomposition of the matrix

$$A = \begin{bmatrix} 2 & 3 & 1 \\ 3 & 7 & 2 \\ 1 & 2 & 4 \end{bmatrix}.$$

Proof. First we consider the vector $\mathbf{v} = \begin{bmatrix} \sqrt{2} \\ \frac{3}{\sqrt{2}} \\ \frac{1}{\sqrt{2}} \end{bmatrix}$ and calculate

$$\mathbf{v}\mathbf{v}^T = \begin{bmatrix} 2 & 3 & 1 \\ 3 & \frac{9}{2} & \frac{3}{2} \\ 1 & \frac{3}{2} & \frac{1}{2} \end{bmatrix} \quad \text{and} \quad A - \mathbf{v}\mathbf{v}^T = \begin{bmatrix} 0 & 0 & 0 \\ 0 & \frac{5}{2} & \frac{1}{2} \\ 0 & \frac{1}{2} & \frac{7}{2} \end{bmatrix}.$$

Then we consider the vector $\mathbf{w} = \begin{bmatrix} \frac{\sqrt{5}}{\sqrt{2}} \\ \frac{1}{2} \cdot \frac{\sqrt{2}}{\sqrt{5}} \end{bmatrix} = \begin{bmatrix} \frac{\sqrt{5}}{\sqrt{2}} \\ \frac{1}{\sqrt{10}} \end{bmatrix}$ and calculate

$$\begin{bmatrix} \frac{5}{2} & \frac{1}{2} \\ \frac{1}{2} & \frac{7}{2} \end{bmatrix} - \mathbf{w}\mathbf{w}^T = \begin{bmatrix} 0 & 0 \\ 0 & \frac{17}{5} \end{bmatrix}.$$

Consequently, the Cholesky decomposition of A is

$$A = \begin{bmatrix} \sqrt{2} & 0 & 0 \\ \frac{3}{\sqrt{2}} & \frac{\sqrt{5}}{\sqrt{2}} & 0 \\ \frac{1}{\sqrt{2}} & \frac{1}{\sqrt{10}} & \sqrt{\frac{17}{5}} \end{bmatrix} \begin{bmatrix} \sqrt{2} & \frac{3}{\sqrt{2}} & \frac{1}{\sqrt{2}} \\ 0 & \frac{\sqrt{5}}{\sqrt{2}} & \frac{1}{\sqrt{10}} \\ 0 & 0 & \sqrt{\frac{17}{5}} \end{bmatrix}.$$

□

Exercises 9.1

Determine if the given matrix A is positive definite.

1. $A = \begin{bmatrix} 2 & 3 \\ 3 & 4 \end{bmatrix}$

2. $A = \begin{bmatrix} -14 & 2 \\ 2 & -11 \end{bmatrix}$

3. $A = \begin{bmatrix} 14 & 2 \\ 2 & 11 \end{bmatrix}$

4. $A = \begin{bmatrix} -4 & 2 \\ 2 & -1 \end{bmatrix}$

5. $A = \begin{bmatrix} 2 & 1 & 1 \\ 1 & 1 & 3 \\ 1 & 3 & 1 \end{bmatrix}$

6. $A=\begin{bmatrix} -5 & 0 & -1 \\ 0 & -2 & 0 \\ -1 & 0 & -5 \end{bmatrix}$

7. $A=\begin{bmatrix} 5 & 1 & -4 \\ 1 & 5 & 4 \\ -4 & 4 & 8 \end{bmatrix}$

8. $A=\begin{bmatrix} 2 & 1 & 2 \\ 1 & 2 & -2 \\ 2 & -2 & -1 \end{bmatrix}$

9. $A=\begin{bmatrix} -4 & 1 & 1 \\ 1 & -7 & 2 \\ 1 & 2 & -7 \end{bmatrix}$

10. $A=\begin{bmatrix} -2 & -1 & -1 \\ -1 & -5 & 4 \\ -1 & 4 & -5 \end{bmatrix}$

Find the Cholesky decomposition of the given matrix A.

11. $A=\begin{bmatrix} 5 & 1 \\ 1 & 1 \end{bmatrix}$

12. $A=\begin{bmatrix} 1 & 1 \\ 1 & 4 \end{bmatrix}$

13. $A=\begin{bmatrix} 7 & -5 \\ -5 & 4 \end{bmatrix}$

14. $A=\begin{bmatrix} 2 & 3 \\ 3 & 5 \end{bmatrix}$

15. $A=\begin{bmatrix} 5 & 2 & 1 \\ 2 & 1 & 1 \\ 1 & 1 & 4 \end{bmatrix}$

16. $\begin{bmatrix} 1 & -2 & 1 \\ -2 & 5 & 1 \\ 1 & 1 & 11 \end{bmatrix}$

17. $A=\begin{bmatrix} 1 & -2 & 1 \\ -2 & 7 & 1 \\ 1 & 1 & 5 \end{bmatrix}$

18. $\begin{bmatrix} 1 & -2 & 1 \\ -2 & 7 & 1 \\ 1 & 1 & 5 \end{bmatrix}$

19. Show that a matrix $\begin{bmatrix} a_{11} & a_{21} \\ a_{21} & a_{22} \end{bmatrix}$ is positive definite if and only if

$$a_{11} > 0 \quad \text{and} \quad \det \begin{bmatrix} a_{11} & a_{21} \\ a_{21} & a_{22} \end{bmatrix} > 0.$$

20. Show that a matrix $A=\begin{bmatrix} 7 & -11 \\ -3 & 5 \end{bmatrix}$ is positive definite.

21. Show that a matrix $\begin{bmatrix} a_{11} & a_{21} & a_{31} \\ a_{21} & a_{22} & a_{32} \\ a_{31} & a_{32} & a_{33} \end{bmatrix}$ is positive definite if and only if

$$a_{11} > 0, \quad \det \begin{bmatrix} a_{11} & a_{21} \\ a_{21} & a_{22} \end{bmatrix} > 0, \quad \text{and} \quad \det \begin{bmatrix} a_{11} & a_{21} & a_{31} \\ a_{21} & a_{22} & a_{32} \\ a_{31} & a_{32} & a_{33} \end{bmatrix} > 0.$$

22. Show that a matrix $\begin{bmatrix} 1 & -2 & 1 \\ -2 & 7 & 1 \\ 1 & 1 & 5 \end{bmatrix}$ is positive definite.

Find a change of variables $\begin{bmatrix} x \\ y \end{bmatrix} = P \begin{bmatrix} x' \\ y' \end{bmatrix}$ such that the given quadratic form expressed in these new variables has no cross-product term.

23. $6x^2 + 3y^2 + 4xy$

24. $-7x^2 + 17y^2 - 18xy$

25. $-11x^2 - 19y^2 + 6xy$

26. $x^2 + 4y^2 - 4xy$

Find a change of variables $\begin{bmatrix} x \\ y \\ z \end{bmatrix} = P \begin{bmatrix} x' \\ y' \\ z' \end{bmatrix}$ such that the given quadratic form expressed in these new variables has no cross-product term.

27. $7x^2 + 7y^2 + 10z^2 + 2xy + 4xz + 4yz$

28. $\begin{bmatrix} x & y & z \end{bmatrix} \begin{bmatrix} 4 & 2 & 2 \\ 2 & 7 & -5 \\ 2 & -5 & 7 \end{bmatrix} \begin{bmatrix} x \\ y \\ z \end{bmatrix}$

29. $\begin{bmatrix} x & y & z \end{bmatrix} \begin{bmatrix} 2 & 3 & -3 \\ 3 & 10 & 1 \\ -3 & 1 & 10 \end{bmatrix} \begin{bmatrix} x \\ y \\ z \end{bmatrix}$

30. $\begin{bmatrix} x & y & z \end{bmatrix} \begin{bmatrix} 4 & -1 & -1 \\ -1 & 7 & -2 \\ -1 & -2 & 7 \end{bmatrix} \begin{bmatrix} x \\ y \\ z \end{bmatrix}$

31. Use the Cholesky decomposition to show that any 3×3 positive definite matrix A can be written as

$$A = \begin{bmatrix} 1 & 0 & 0 \\ l_{21} & 1 & 0 \\ l_{31} & l_{32} & 1 \end{bmatrix} \begin{bmatrix} d_1 & 0 & 0 \\ 0 & d_2 & 0 \\ 0 & 0 & d_3 \end{bmatrix} \begin{bmatrix} 1 & l_{21} & l_{31} \\ 0 & 1 & l_{32} \\ 0 & 0 & 1 \end{bmatrix},$$

where $d_1 > 0$, $d_2 > 0$, and $d_3 > 0$.

32. Let A be a symmetric $n \times n$ matrix. Show that the following conditions are equivalent:

    (a) A is positive definite;

    (b) There is an $n \times n$ matrix B with linearly independent columns such that $A = B^T B$.

33. If

$$A = \begin{bmatrix} 1 & 0 & 0 \\ l_{21} & 1 & 0 \\ l_{31} & l_{32} & 1 \end{bmatrix} \begin{bmatrix} d_1 & 0 & 0 \\ 0 & d_2 & 0 \\ 0 & 0 & d_3 \end{bmatrix} \begin{bmatrix} 1 & l_{21} & l_{31} \\ 0 & 1 & l_{32} \\ 0 & 0 & 1 \end{bmatrix},$$

where $d_1 > 0$, $d_2 > 0$, and $d_3 > 0$, show that A is positive definite and find the Cholesky decomposition of A.

Chapter 10

Vector spaces

In this chapter we show that instead of the space $\mathbb{R}^n$ we can consider other sets like the set of all matrices of a given size or the set of all real-valued functions defined on a common domain. It turns out that most results formulated and proven about the space $\mathbb{R}^n$ remain valid in such sets.

10.1 General Vector Spaces

We denote by $\mathscr{M}_{m\times n}$ the set of all matrices with m rows and n columns. For any matrices A and B in $\mathscr{M}_{m\times n}$ and any real number c the sum $A+B$ and the product cA are defined in Chapter 1.

We also denote by $\mathscr{F}_D$ the set of all real-valued functions defined on a common domain D. If f and g are functions from $\mathscr{F}_D$ and c is a real number, then we define

$$(f+g)(t) = f(t) + g(t) \quad \text{and} \quad (cf)(t) = cf(t)$$

for every t in the domain D.

First we ask the following question: What do $\mathbb{R}^n$, $\mathscr{M}_{m\times n}$, and $\mathscr{F}_D$ have in common? In order to answer this question we denote by $\mathscr{L}$ one of the sets $\mathbb{R}^n$, $\mathscr{M}_{m\times n}$, or $\mathscr{F}_D$. Here is the answer to our question:

- for every two elements $\mathbf{v}$ and $\mathbf{w}$ in $\mathscr{L}$ we can define in $\mathscr{L}$ the element $\mathbf{v}+\mathbf{w}$ and
- for every element $\mathbf{v}$ in $\mathscr{L}$ and every real number c we can define in $\mathscr{L}$ the element $c\mathbf{v}$

in such a way that the following properties hold:

1. for every two elements $\mathbf{v}$ and $\mathbf{w}$ in $\mathscr{L}$ we have

$$\mathbf{v}+\mathbf{w} = \mathbf{w}+\mathbf{v}$$

2. for every three elements $\mathbf{u}, \mathbf{v}$ and $\mathbf{w}$ in $\mathscr{L}$ we have

$$\mathbf{u}+(\mathbf{v}+\mathbf{w}) = (\mathbf{u}+\mathbf{w})+\mathbf{v}$$

3. there is an element $\mathbf{0}$ in $\mathscr{L}$ such that for every element $\mathbf{v}$ in $\mathscr{L}$ we have

$$\mathbf{0} + \mathbf{v} = \mathbf{v}$$

4. for every element $\mathbf{v}$ in $\mathscr{L}$ there is an element $-\mathbf{v}$ such that

$$\mathbf{v} + (-\mathbf{v}) = (-\mathbf{v}) + \mathbf{v} = \mathbf{0}$$

5. for every element $\mathbf{v}$ in $\mathscr{L}$

$$1\mathbf{v} = \mathbf{v}$$

6. for every element $\mathbf{v}$ in $\mathscr{L}$ and every real numbers c_1 and c_2 we have

$$(c_1 c_2)\mathbf{v} = c_1(c_2\mathbf{v})$$

7. for every element $\mathbf{v}$ in $\mathscr{L}$ and every real numbers c_1 and c_2 we have

$$(c_1 + c_2)\mathbf{v} = c_1\mathbf{v} + c_2\mathbf{v}$$

8. for every two elements $\mathbf{v}$ and $\mathbf{w}$ in $\mathscr{L}$ and every real number c we have

$$c(\mathbf{v} + \mathbf{w}) = c\mathbf{v} + c\mathbf{w}$$

A set $\mathscr{L}$ where operations of addition of elements and multiplication by real numbers are defined is called a *vector space* if the above eight properties hold.

We note that it is essential that for every two elements $\mathbf{v}$ and $\mathbf{w}$ in $\mathscr{L}$ the element $\mathbf{v} + \mathbf{w}$ is an element of $\mathscr{L}$ and for every element $\mathbf{v}$ in $\mathscr{L}$ and every real number c the element $c\mathbf{v}$ is an element of $\mathscr{L}$.

Sometimes elements of $\mathscr{L}$ will be referred to as vectors. In the remainder of this chapter we will assume that $\mathscr{L}$ is a vector space.

We begin with the definitions of linear combination, linear dependence and independence, and linear span of elements of $\mathscr{L}$. Note that these definitions are the same as those for the vectors of $\mathbb{R}^n$.

Definition 10.1. Let $\mathbf{v}_1, \dots, \mathbf{v}_j$ be vectors in a vector space $\mathscr{L}$. A vector of the form

$$c_1\mathbf{v}_1 + \cdots + c_j\mathbf{v}_j,$$

where $c_1, \dots, c_j$ are arbitrary real numbers, is called a *linear combination* of the vectors $\mathbf{v}_1, \dots, \mathbf{v}_j$.

Definition 10.2. Vectors $\mathbf{v}_1, \dots, \mathbf{v}_j$ in a vector space $\mathscr{L}$ are *linearly dependent* if and only if the equation

$$x_1\mathbf{v}_1 + \cdots + x_j\mathbf{v}_j = \mathbf{0}$$

has a nontrivial solution, that is, a solution such that at least one of the numbers $x_1, \dots, x_j$ is different from 0.

Definition 10.3. Vectors $\mathbf{v}_1, \dots, \mathbf{v}_j$ in a vector space $\mathscr{L}$ are *linearly independent* if and only if the only solution of the equation

$$x_1\mathbf{v}_1 + \cdots + x_j\mathbf{v}_j = \mathbf{0}$$

is the trivial solution, that is $x_1 = \cdots = x_j = 0$.

Example 10.4. Show that the functions

$$p_0(t) = 1,\ p_1(t) = t,\ p_2(t) = t^2, \dots,\ p_n(t) = t^n$$

are linearly independent vectors in $\mathscr{F}_{\mathbb{R}}$.

Solution. We show that the functions p_0, p_1, p_2 are linearly independent vectors in $\mathscr{F}_{\mathbb{R}}$. The general case can be shown in the same way.

If we assume that

$$x_1 p_0 + x_2 p_1 + x_3 p_2 = \mathbf{0},$$

then we have

$$x_1 + x_2 t + x_3 t^2 = 0$$

for every t in $\mathbb{R}$. By differentiating the above equality with respect to t we get

$$x_2 + 2x_3 t = 0$$

and by differentiating it again we get

$$2x_3 = 0.$$

This means that $x_2 p_0 + 2x_3 p_1 = \mathbf{0}$ and $x_3 p_0 = \mathbf{0}$, which gives us $x_3 = x_2 = x_1 = 0$, proving linear independence of the functions p_0, p_1, p_2. $\square$

Definition 10.5. A nonempty subset $\mathscr{V}$ of a vector space $\mathscr{L}$ which satisfies the following two conditions

(a) If $\mathbf{v}$ is a vector in $\mathscr{V}$ and c a real number, then the vector $c\mathbf{v}$ is in $\mathscr{V}$,

(b) If $\mathbf{v}$ and $\mathbf{w}$ are vectors in $\mathscr{V}$, then the vector $\mathbf{v}+\mathbf{w}$ is in $\mathscr{V}$,

is called a *subspace* of $\mathscr{L}$.

Example 10.6. The set $\mathscr{V} = \{\mathbf{0}\}$ is a subspace.

Example 10.7. Show that the set $\mathscr{D}_{\mathbb{R}}$ of all real-valued functions defined on $\mathbb{R}$ that are differentiable at every point is a subspace of $\mathscr{F}_{\mathbb{R}}$.

Solution. This is an immediate consequence of the fact that the sum $f+g$ of differentiable functions f and g is a differentiable function and that the product cf of a real number c and a differentiable function f is a differentiable function. □

In the definition of a vector space $\mathscr{L}$ we assume that there is an element $\mathbf{0}$ in $\mathscr{L}$ such that for every element $\mathbf{v}$ in $\mathscr{L}$ we have $\mathbf{0}+\mathbf{v}=\mathbf{v}$. It can be easily shown that such an element $\mathbf{0}$ must be unique. Indeed, if we have $\mathbf{0}_1+\mathbf{v}=\mathbf{v}$ and $\mathbf{0}_2+\mathbf{v}=\mathbf{v}$ for every element $\mathbf{v}$ in $\mathscr{L}$, then

$$\mathbf{0}_1 = \mathbf{0}_2+\mathbf{0}_1 = \mathbf{0}_1+\mathbf{0}_2 = \mathbf{0}_2.$$

Similarly, we assume that for every element $\mathbf{v}$ in $\mathscr{L}$ there is an element $-\mathbf{v}$ such that $\mathbf{v}+(-\mathbf{v}) = (-\mathbf{v})+\mathbf{v} = \mathbf{0}$. To show that the element $-\mathbf{v}$ is unique, assume that $\mathbf{v}+\mathbf{x}=\mathbf{x}+\mathbf{v}=\mathbf{0}$ and $\mathbf{v}+\mathbf{y}=\mathbf{y}+\mathbf{v}=\mathbf{0}$ for some vectors $\mathbf{x}$ and $\mathbf{y}$ in $\mathscr{L}$. Then

$$\mathbf{x} = \mathbf{x}+\mathbf{0} = \mathbf{x}+(\mathbf{v}+\mathbf{y}) = (\mathbf{x}+\mathbf{v})+\mathbf{y} = \mathbf{0}+\mathbf{y} = \mathbf{y}.$$

These two properties are used in the proof of the next theorem.

Theorem 10.8. *If $\mathscr{V}$ is subspace of a vector space $\mathscr{L}$, then $\mathbf{0}$ is in $\mathscr{V}$ and the vector $-\mathbf{v}$ is in $\mathscr{V}$ for every vector $\mathbf{v}$ in $\mathscr{V}$.*

Proof. Let $\mathbf{v}$ be a vector in $\mathscr{V}$. Since

$$\mathbf{v}+0\cdot\mathbf{v} = 1\cdot\mathbf{v}+0\cdot\mathbf{v} = (1+0)\mathbf{v} = 1\cdot\mathbf{v} = \mathbf{v},$$

we have

$$0\cdot\mathbf{v} = \mathbf{0}$$

and thus $\mathbf{0}$ is in $\mathscr{V}$ by (a) in Definition 10.5. Similarly, since

$$\mathbf{v}+(-1)\cdot\mathbf{v}=1\cdot\mathbf{v}+(-1)\cdot\mathbf{v}=(1+(-1))\mathbf{v}=0\cdot\mathbf{v}=\mathbf{0},$$

we have

$$(-1)\cdot\mathbf{v}=-\mathbf{v}.$$

□

Theorem 10.9. *Every subspace $\mathscr{V}$ of the vector space $\mathscr{L}$ is a vector space.*

Proof. For every two elements $\mathbf{v}$ and $\mathbf{w}$ in $\mathscr{V}$ the element $\mathbf{v}+\mathbf{w}$ is in $\mathscr{V}$ and for every element $\mathbf{v}$ in $\mathscr{V}$ and every real number c the element $c\mathbf{v}$ is in $\mathscr{L}$. Moreover, since the eight conditions in the definition of a vector space are true for all elements of $\mathscr{L}$, they are true for the elements of $\mathscr{V}$. Consequently, $\mathscr{V}$ is a vector space. □

Definition 10.10. Let $\mathbf{v}_1,\dots,\mathbf{v}_k$ be vectors in $\mathscr{L}$. The set of all linear combinations of the vectors $\mathbf{v}_1,\dots,\mathbf{v}_k$, that is, the set of all possible sums of the form

$$x_1\mathbf{v}_1+\cdots+x_k\mathbf{v}_k,$$

where $x_1,\dots,x_k$ are arbitrary real numbers, is denoted by

$$\operatorname{Span}\{\mathbf{v}_1,\dots,\mathbf{v}_k\}$$

and called the linear span (or simply span) of vectors $\mathbf{v}_1,\dots,\mathbf{v}_k$. If $\mathscr{V}=\operatorname{Span}\{\mathbf{v}_1,\dots,\mathbf{v}_k\}$, we say that $\{\mathbf{v}_1,\dots,\mathbf{v}_k\}$ is a spanning set for $\mathscr{V}$.

Theorem 10.11. *The set $\mathscr{V}=\operatorname{Span}\{\mathbf{v}_1,\dots,\mathbf{v}_n\}$ is a subspace of $\mathscr{L}$ for any vectors $\mathbf{v}_1,\dots,\mathbf{v}_n$ in $\mathscr{L}$.*

Proof. We show, as in the proof for Theorem 2.20, that for every two elements $\mathbf{v}$ and $\mathbf{w}$ in $\mathscr{V}$ the element $\mathbf{v}+\mathbf{w}$ is an element of $\mathscr{V}$ and for every element $\mathbf{v}$ in $\mathscr{V}$ and every real number c the element $c\mathbf{v}$ is an element of $\mathscr{V}$. □

Example 10.12. The set $\mathbb{P}_n=\operatorname{Span}\{p_0,p_1,p_2,\dots,p_n\}$ is the set of all polynomials of degree at most n. It is easy to verify that $\mathbb{P}_n$ is a vector subspace of $\mathscr{D}_{\mathbb{R}}$ and consequently also a subspace of $\mathscr{F}_{\mathbb{R}}$.

Example 10.13. The set of all solutions of the differential equation

$$y' - 3y = 0$$

is $\operatorname{Span}\{\exp_3\}$, where the function $\exp_3 : \mathbb{R} \to \mathbb{R}$ is defined by $\exp_3(t) = e^{3t}$. This is a subspace of $\mathscr{D}_{\mathbb{R}}$.

Example 10.14. The set of all solutions of the differential equation

$$y'' + y = 0$$

is $\operatorname{Span}\{C, S\}$ where the functions $C : \mathbb{R} \to \mathbb{R}$ and $S : \mathbb{R} \to \mathbb{R}$ are defined by $C(t) = \cos t$ and $S(t) = \sin t$. This is also a subspace of $\mathscr{D}_{\mathbb{R}}$.

We note that sometimes, when it is clear from the context what we mean, we simply write e^{3t}, $\cos t$, and $\sin t$ instead of $\exp_3$, C, and S.

The next result is essential for the proof of Theorem 10.26.

Theorem 10.15. *Let* $\mathbf{v}_1, \dots, \mathbf{v}_k$ *be linearly independent vectors in a vector space* $\mathscr{L}$ *and let* $\mathbf{x}$ *be vector in* $\operatorname{Span}\{\mathbf{v}_1, \dots, \mathbf{v}_k\}$. *If*

$$\mathbf{x} = c_1\mathbf{v}_1 + \cdots + c_k\mathbf{v}_k = d_1\mathbf{v}_1 + \cdots + d_k\mathbf{v}_k, \tag{10.1}$$

then we must have

$$c_1 = d_1, c_2 = d_2, \dots, c_k = d_k.$$

Proof. The equation (10.1) yields

$$(c_1 - d_1)\mathbf{v}_1 + \cdots + (c_k - d_k)\mathbf{v}_k = \mathbf{0},$$

and thus

$$c_1 - d_1 = \cdots = c_k - d_k = 0,$$

by linear independence of the vectors $\mathbf{v}_1, \dots, \mathbf{v}_k$. □

Definition 10.16. Let $\mathscr{V}$ be a vector subspace of $\mathscr{L}$. A collection of vectors $\{\mathbf{v}_1, \dots, \mathbf{v}_k\}$ in $\mathscr{V}$ is a *basis* in $\mathscr{V}$ if the following two conditions are satisfied:

(a) the vectors $\mathbf{v}_1, \dots, \mathbf{v}_k$ are linearly independent;

(b) $\mathscr{V} = \operatorname{Span}\{\mathbf{v}_1, \dots, \mathbf{v}_k\}$.

Definition 10.17. Let $\mathcal{V} = \operatorname{Span}\{\mathbf{v}_1, \dots, \mathbf{v}_k\}$ where the vectors $\mathbf{v}_1, \dots, \mathbf{v}_k$ are linearly independent (so $\{\mathbf{v}_1, \dots, \mathbf{v}_k\}$ is a basis in $\mathcal{V}$). If $\mathbf{x}$ a vector in $\mathcal{V}$, then the unique numbers $c_1, \dots, c_k$ such that $\mathbf{x} = c_1\mathbf{v}_1 + \cdots + c_k\mathbf{v}_k$ are called the coordinates of $\mathbf{x}$ in the basis $\{\mathbf{v}_1, \dots, \mathbf{v}_k\}$.

Now we will discuss the dimension of a vector space which is of fundamental importance in linear algebra. Before we can give the definition of the dimension of a vector space we need to prove several results about bases in vector spaces.

First we show how Theorem 2.44 and its proof can be adapted to the general case. Up to now all matrices considered here had entries that were specific real numbers, like

$$\begin{bmatrix} 1 & \pi & -4 \\ 0.23 & \frac{1}{3} & \sqrt{7} \end{bmatrix},$$

or unspecified real numbers, like

$$\begin{bmatrix} a & b \\ c & d \\ e & f \end{bmatrix},$$

or both, like

$$\begin{bmatrix} a & 1 & 2 \\ 0 & b & 3 \\ 0 & 0 & c \end{bmatrix}.$$

In the proof of the theorem below we consider matrices with entries in a vector space $\mathcal{L}$. Actually, we used matrices whose entries were vectors quite frequently in previous chapters. It was often convenient to write a matrix

$$A = \begin{bmatrix} a_{11} & a_{12} & \dots & a_{1n} \\ a_{21} & a_{22} & \dots & a_{2n} \\ \vdots & \vdots & & \vdots \\ a_{m1} & a_{m2} & \dots & a_{mn} \end{bmatrix}$$

in the form

$$\begin{bmatrix} \mathbf{v}_1 & \mathbf{v}_2 & \dots & \mathbf{v}_n \end{bmatrix}$$

where $\mathbf{v}_1, \mathbf{v}_2, \dots, \mathbf{v}_n$ were the column vectors of the matrix A, that is,

$$\mathbf{v}_1 = \begin{bmatrix} a_{11} \\ a_{21} \\ \vdots \\ a_{m1} \end{bmatrix}, \mathbf{v}_2 = \begin{bmatrix} a_{12} \\ a_{22} \\ \vdots \\ a_{m2} \end{bmatrix}, \dots, \mathbf{v}_m = \begin{bmatrix} a_{1n} \\ a_{2n} \\ \vdots \\ a_{mn} \end{bmatrix}.$$

Note that the matrix $\begin{bmatrix} \mathbf{v}_1 & \mathbf{v}_2 & \dots & \mathbf{v}_n \end{bmatrix}$ can be interpreted as a $1 \times n$ matrix whose entries are vectors. We never used such an interpretation, because it was not necessary, but in the proof of the next theorem this is exactly what is needed. We consider matrices of the form $\begin{bmatrix} \mathbf{v}_1 & \mathbf{v}_2 & \dots & \mathbf{v}_n \end{bmatrix}$, where $\mathbf{v}_1, \mathbf{v}_2, \dots, \mathbf{v}_n$ are elements of a vector space $\mathcal{L}$.

Actually, we can consider an arbitrary matrix

$$A = \begin{bmatrix} \mathbf{v}_{11} & \mathbf{v}_{12} & \dots & \mathbf{v}_{1n} \\ \mathbf{v}_{21} & \mathbf{v}_{22} & \dots & \mathbf{v}_{2n} \\ \vdots & \vdots & & \vdots \\ \mathbf{v}_{m1} & \mathbf{v}_{m2} & \dots & \mathbf{v}_{mn} \end{bmatrix}$$

where all entries $\mathbf{v}_{ij}$ are elements of a vector space $\mathscr{L}$. The operations of addition and multiplication by real numbers are defined as for matrices with real entries (see Definitions 1.4 and 1.7) and they have the same algebraic properties.

Theorem 10.18. *If a subspace* $\operatorname{Span}\{\mathbf{v}_1,\dots,\mathbf{v}_k\}$ *of a vector space* $\mathscr{L}$ *contains* k *linearly independent vectors* $\mathbf{w}_1,\dots,\mathbf{w}_k$*, then the vectors* $\mathbf{v}_1,\dots,\mathbf{v}_k$ *are linearly independent and*

$$\operatorname{Span}\{\mathbf{v}_1,\dots,\mathbf{v}_k\} = \operatorname{Span}\{\mathbf{w}_1,\dots,\mathbf{w}_k\}.$$

Proof. For a matrix $\begin{bmatrix}\mathbf{u}_1 & \dots & \mathbf{u}_k\end{bmatrix}$ with entries in $\mathscr{L}$ and a vector $\mathbf{x} = \begin{bmatrix} x_1 \\ \vdots \\ x_k \end{bmatrix}$ in $\mathbb{R}^n$ we define

$$\begin{bmatrix}\mathbf{u}_1 & \dots & \mathbf{u}_k\end{bmatrix}\mathbf{x} = x_1\mathbf{u}_1 + \dots + x_n\mathbf{u}_n.$$

Similarly, for a matrix $\begin{bmatrix}\mathbf{u}_1 & \dots & \mathbf{u}_k\end{bmatrix}$ with entries in $\mathscr{L}$ and a $k\times k$ matrix $A = \begin{bmatrix}\mathbf{a}_1 & \dots & \mathbf{a}_k\end{bmatrix}$ with entries in $\mathbb{R}$ we define

$$\begin{bmatrix}\mathbf{u}_1 & \dots & \mathbf{u}_k\end{bmatrix} A = \begin{bmatrix}\begin{bmatrix}\mathbf{u}_1 & \dots & \mathbf{u}_k\end{bmatrix}\mathbf{a}_1 & \dots & \begin{bmatrix}\mathbf{u}_1 & \dots & \mathbf{u}_k\end{bmatrix}\mathbf{a}_k\end{bmatrix}.$$

We note that $\begin{bmatrix}\begin{bmatrix}\mathbf{u}_1 & \dots & \mathbf{u}_k\end{bmatrix}\mathbf{a}_1 & \dots & \begin{bmatrix}\mathbf{u}_1 & \dots & \mathbf{u}_k\end{bmatrix}\mathbf{a}_k\end{bmatrix}$ is a matrix with entries in $\mathscr{L}$. We can verify, as in Chapter 1, that for every row matrix $\begin{bmatrix}\mathbf{u}_1 & \dots & \mathbf{u}_k\end{bmatrix}$ with entries in $\mathscr{L}$, every vector $\mathbf{x} = \begin{bmatrix} x_1 \\ \vdots \\ x_k \end{bmatrix}$ in $\mathbb{R}^n$, and every $k \times k$ matrices A and B with entries in $\mathbb{R}$ we have

$$\left(\begin{bmatrix}\mathbf{u}_1 & \dots & \mathbf{u}_k\end{bmatrix} A\right)\mathbf{x} = \begin{bmatrix}\mathbf{u}_1 & \dots & \mathbf{u}_k\end{bmatrix}(A\mathbf{x})$$

and

$$\left(\begin{bmatrix}\mathbf{u}_1 & \dots & \mathbf{u}_k\end{bmatrix} A\right)B = \begin{bmatrix}\mathbf{u}_1 & \dots & \mathbf{u}_k\end{bmatrix}(AB).$$

Now we consider the matrices $\begin{bmatrix}\mathbf{v}_1 & \dots & \mathbf{v}_k\end{bmatrix}$ and $\begin{bmatrix}\mathbf{w}_1 & \dots & \mathbf{w}_k\end{bmatrix}$ with entries in $\mathscr{L}$ and proceed as in the proof of Theorem 2.44. □

Theorem 10.19. *Let* $\{\mathbf{v}_1,\dots,\mathbf{v}_k\}$ *and* $\{\mathbf{w}_1,\dots,\mathbf{w}_m\}$ *be bases in a vector subspace* $\mathscr{V}$ *of a vector space* $\mathscr{L}$*. Then* $k = m$*.*

Proof. Without loss of generality we can assume that $m > k$. Since the vectors $\mathbf{w}_1, \dots, \mathbf{w}_k$ are linearly independent we have

$$\text{Span}\{\mathbf{w}_1, \dots, \mathbf{w}_k\} = \text{Span}\{\mathbf{v}_1, \dots, \mathbf{v}_k\} = \mathscr{V},$$

by Theorem 10.18. But this contradicts the fact that the vectors $\mathbf{w}_1, \dots, \mathbf{w}_m$ are linearly independent, since

$$\text{Span}\{\mathbf{w}_1, \dots, \mathbf{w}_m\} = \mathscr{V} = \text{Span}\{\mathbf{w}_1, \dots, \mathbf{w}_k\}.$$

□

Theorem 10.20. *If $\mathscr{W} = \text{Span}\{\mathbf{w}_1, \dots, \mathbf{w}_n\}$, where $\mathbf{w}_1, \dots, \mathbf{w}_n$ are linearly independent vectors in a vector space $\mathscr{L}$, and $\mathscr{V}$ is a subspace of $\mathscr{W}$, then there are linearly independent vectors $\mathbf{v}_1, \dots, \mathbf{v}_k$ in $\mathscr{W}$ for some $k \le n$ such that*

$$\mathscr{V} = \text{Span}\{\mathbf{v}_1, \dots, \mathbf{v}_k\}.$$

Proof of a particular case. Suppose $n = 4$.

If $\mathscr{V}$ contains a vector $\mathbf{v}_1 \neq \mathbf{0}$, then $\mathscr{V}$ must contain $\text{Span}\{\mathbf{v}_1\}$. If $\mathscr{V} = \text{Span}\{\mathbf{v}_1\}$, then we are done.

If $\mathscr{V} \neq \text{Span}\{\mathbf{v}_1\}$, then $\mathscr{V}$ contains a vector $\mathbf{v}_2$ which is not in $\text{Span}\{\mathbf{v}_1\}$. The subspace $\mathscr{V}$ must contain $\text{Span}\{\mathbf{v}_1, \mathbf{v}_2\}$ and the vectors $\mathbf{v}_1$ and $\mathbf{v}_2$ are linearly independent. If $\mathscr{V} = \text{Span}\{\mathbf{v}_1, \mathbf{v}_2\}$, then we are done.

If $\mathscr{V} \neq \text{Span}\{\mathbf{v}_1, \mathbf{v}_2\}$, then $\mathscr{V}$ contains a vector $\mathbf{v}_3$ which is not in $\text{Span}\{\mathbf{v}_1, \mathbf{v}_2\}$. The subspace $\mathscr{V}$ must contain $\text{Span}\{\mathbf{v}_1, \mathbf{v}_2, \mathbf{v}_3\}$ and the vectors $\mathbf{v}_1$, $\mathbf{v}_2$, and $\mathbf{v}_3$ are linearly independent. If $\mathscr{V} = \text{Span}\{\mathbf{v}_1, \mathbf{v}_2, \mathbf{v}_3\}$, then we are done.

If $\mathscr{V} \neq \text{Span}\{\mathbf{v}_1, \mathbf{v}_2, \mathbf{v}_3\}$, then the subspace $\mathscr{V}$ must contain a vector $\mathbf{v}_4$ which is not in $\text{Span}\{\mathbf{v}_1, \mathbf{v}_2, \mathbf{v}_3\}$ and $\mathscr{V}$ must contain $\text{Span}\{\mathbf{v}_1, \mathbf{v}_2, \mathbf{v}_3, \mathbf{v}_4\}$ and the vectors $\mathbf{v}_1$, $\mathbf{v}_2$, $\mathbf{v}_3$, and $\mathbf{v}_4$ are linearly independent. But then we have

$$\mathscr{V} = \text{Span}\{\mathbf{v}_1, \mathbf{v}_2, \mathbf{v}_3, \mathbf{v}_4\} = \text{Span}\{\mathbf{w}_1, \mathbf{w}_2, \mathbf{w}_3, \mathbf{w}_4\}$$

by Theorem 10.18. □

Theorem 10.21. *Let $\mathscr{V}$ be a vector subspace of a vector space $\mathscr{L}$. If $\mathscr{V}$ has a basis with k elements, then for any collection of vectors $\{\mathbf{v}_1, \dots, \mathbf{v}_k\}$ in $\mathscr{V}$, the following three conditions are equivalent:*

(a) *The set $\{\mathbf{v}_1, \dots, \mathbf{v}_k\}$ is basis of $\mathscr{V}$;*

(b) *The vectors $\mathbf{v}_1, \dots, \mathbf{v}_k$ are linearly independent;*

(c) $\mathscr{V} = \text{Span}\{\mathbf{v}_1, \dots, \mathbf{v}_k\}$.

Proof. Suppose that $\{\mathbf{w}_1,\dots,\mathbf{w}_k\}$ is a basis of $\mathscr{V}$.

It is clear that (a) implies (b).

Now, if the vectors $\mathbf{v}_1,\dots,\mathbf{v}_k$ are linearly independent, then

$$\text{Span}\{\mathbf{v}_1,\dots,\mathbf{v}_k\} = \text{Span}\{\mathbf{w}_1,\dots,\mathbf{w}_k\} = \mathscr{V},$$

by Theorem 10.18. This proves that (b) implies (c).

Finally, if $\mathscr{V} = \text{Span}\{\mathbf{v}_1,\dots,\mathbf{v}_k\}$, then the linearly independent vectors $\mathbf{w}_1,\dots,\mathbf{w}_k$ are in $\text{Span}\{\mathbf{v}_1,\dots,\mathbf{v}_k\}$ and, again by Theorem 10.18, the vectors $\mathbf{v}_1,\dots,\mathbf{v}_k$ are linearly independent. This means that the set $\{\mathbf{v}_1,\dots,\mathbf{v}_k\}$ is a basis of $\mathscr{V}$, proving that (c) implies (a). □

Corollary 10.22. *Let $\mathbf{v}_1,\dots,\mathbf{v}_k$ be linearly independent vectors in a vector space $\mathscr{L}$. Then any $k+1$ vectors in* $\text{Span}\{\mathbf{v}_1,\dots,\mathbf{v}_k\}$ *are linearly dependent.*

Proof. Let $\mathbf{w}_1,\dots,\mathbf{w}_k,\mathbf{w}_{k+1}$ be vectors in $\text{Span}\{\mathbf{v}_1,\dots,\mathbf{v}_k\}$.

If the vectors $\mathbf{w}_1,\dots,\mathbf{w}_k$ are linearly dependent, we are done.

If the vectors $\mathbf{w}_1,\dots,\mathbf{w}_k$ are linearly independent, then the set $\{\mathbf{w}_1,\dots,\mathbf{w}_k\}$ is a basis in $\text{Span}\{\mathbf{v}_1,\dots,\mathbf{v}_k\}$, by Theorem 10.21. Consequently, the vector $\mathbf{w}_{k+1}$ is a linear combination of vectors $\mathbf{w}_1,\dots,\mathbf{w}_k$. But then the vectors $\mathbf{w}_1,\dots,\mathbf{w}_k,\mathbf{w}_{k+1}$ are linearly dependent. □

Note that the above corollary applies to any number of vectors greater than k. More precisely, if $\mathbf{v}_1,\dots,\mathbf{v}_k$ are linearly independent vectors in a vector space $\mathscr{L}$, then any m vectors in $\text{Span}\{\mathbf{v}_1,\dots,\mathbf{v}_k\}$ are linearly dependent as long as $m > k$.

The results proven above justify the following definition.

Definition 10.23. Let $\mathscr{V}$ be a subspace of a vector space $\mathscr{L}$. If $\mathscr{V}$ has a basis with k elements, where k is an integer ≥ 1, then k is called the *dimension* of $\mathscr{V}$ and we write $\dim \mathscr{V} = k$.

We also define $\dim\{\mathbf{0}\} = 0$.

Example 10.24. The dimension of the subspace $\text{Span}\{p_0, p_1, p_2,\dots,p_n\}$ is $n+1$, that is

$$\dim \text{Span}\{p_0, p_1, p_2,\dots,p_n\} = n+1,$$

because the functions $p_0, p_1, p_2,\dots,p_n$ are linearly independent.

Definition 10.25. Let $\mathscr{L}$ and $\mathscr{M}$ be vector spaces. The function $f : \mathscr{V} \to \mathscr{M}$ is called a *linear transformation* if it has the following properties

(a) $f(c\mathbf{x}) = cf(\mathbf{x})$ for any vector $\mathbf{x}$ in $\mathscr{V}$ and any real number c;

(b) $f(\mathbf{x}+\mathbf{y}) = f(\mathbf{x}) + f(\mathbf{y})$ for any vectors $\mathbf{x}$ and $\mathbf{y}$ in $\mathscr{V}$.

The meaning of the following theorem is that, from the point of view of linear algebra, any subspace $\mathscr{V}$ of a vector space $\mathscr{L}$, can be treated as $\mathbb{R}^n$, where n is the dimension of $\mathscr{V}$.

Theorem 10.26. *Let $\mathbf{v}_1, \dots, \mathbf{v}_n$ be linearly independent vectors in $\mathscr{L}$ and let $\mathscr{V} = \operatorname{Span}\{\mathbf{v}_1, \dots, \mathbf{v}_n\}$. The function $f : \mathscr{V} \to \mathbb{R}^n$ defined by*

$$f(x_1\mathbf{v}_2 + \cdots + x_n\mathbf{v}_n) = \begin{bmatrix} x_1 \\ \vdots \\ x_n \end{bmatrix}$$

is linear, one-to-one and onto.

Proof. Note that the function f is well-defined and is one-to-one, by Theorem 10.15. Since for any $\begin{bmatrix} x_1 \\ \vdots \\ x_n \end{bmatrix}$ in $\mathbb{R}^n$ there is a vector $\mathbf{x}$ in $\mathscr{V}$ such that $f(\mathbf{x}) = \begin{bmatrix} x_1 \\ \vdots \\ x_n \end{bmatrix}$, namely $\mathbf{x} = x_1\mathbf{v}_2 + \cdots + x_n\mathbf{v}_n$, the function f is onto.

If $\mathbf{x}$ is a vector in $\mathscr{V}$, then

$$\mathbf{x} = x_1\mathbf{v}_1 + \cdots + x_k\mathbf{v}_k,$$

for some real numbers $x_1, \dots, x_k$. For any real number c we have

$$c\mathbf{x} = c(x_1\mathbf{v}_1 + \cdots + x_k\mathbf{v}_k) = (cx_1)\mathbf{v}_1 + \cdots + (cx_k)\mathbf{v}_k,$$

which means that

$$f(c\mathbf{x}) = \begin{bmatrix} cx_1 \\ \vdots \\ cx_n \end{bmatrix} = c\begin{bmatrix} x_1 \\ \vdots \\ x_n \end{bmatrix} = cf(\mathbf{x}).$$

Similarly, if $\mathbf{x}$ and $\mathbf{y}$ are vectors in $\mathscr{V}$, then

$$\mathbf{x} = x_1\mathbf{v}_1 + \cdots + x_k\mathbf{v}_k \quad \text{and} \quad \mathbf{y} = y_1\mathbf{v}_1 + \cdots + y_k\mathbf{v}_k,$$

for some real numbers $x_1, \dots, x_k$ and $y_1, \dots, y_k$. Since

$$\mathbf{x}+\mathbf{y} = x_1\mathbf{v}_1 + \cdots + x_k\mathbf{v}_k + y_1\mathbf{v}_1 + \cdots + y_k\mathbf{v}_k = (x_1+y_1)\mathbf{v}_1 + \cdots + (x_k+y_k)\mathbf{v}_k,$$

we get

$$f(\mathbf{x}+\mathbf{y}) = \begin{bmatrix} x_1 + y_1 \\ \vdots \\ x_n + y_n \end{bmatrix} = \begin{bmatrix} x_1 \\ \vdots \\ x_n \end{bmatrix} + \begin{bmatrix} y_1 \\ \vdots \\ y_n \end{bmatrix} = f(\mathbf{x}) + f(\mathbf{y}).$$

□

Note that for the function f in the above theorem is a linear transformation and we have $f(\mathbf{0}) = \begin{bmatrix} 0 \\ \vdots \\ 0 \end{bmatrix}$.

The following corollary allows us to reduce the linear independence of vectors from $\mathscr{L}$ to linear independence of vectors from $\mathbb{R}^n$.

Corollary 10.27. *Let $\mathbf{v}_1, \dots, \mathbf{v}_n$ be linearly independent vectors in a vector space $\mathscr{L}$, $\mathcal{V} = \operatorname{Span}\{\mathbf{v}_1, \dots, \mathbf{v}_n\}$, and let f be the function defined in Theorem 10.26. For any collection of vectors $\{\mathbf{w}_1, \dots, \mathbf{w}_k\}$ in $\mathcal{V}$ the following two conditions are equivalent:*

(a) *The vectors $\mathbf{w}_1, \dots, \mathbf{w}_k$ are linearly independent;*

(b) *The vectors $f(\mathbf{w}_1), \dots, f(\mathbf{w}_k)$ are linearly independent.*

Proof. This is a consequence of the fact that the following two conditions are equivalent:

(α) $x_1\mathbf{w}_1 + \cdots + x_k\mathbf{w}_k = \mathbf{0}$;

(β) $x_1 f(\mathbf{w}_1) + \cdots + x_k f(\mathbf{w}_k) = \begin{bmatrix} 0 \\ \vdots \\ 0 \end{bmatrix}$.

□

Example 10.28. Show that the matrices

$$\begin{bmatrix} 1 & 2 \\ 0 & 4 \\ 2 & 0 \end{bmatrix}, \quad \begin{bmatrix} 2 & 3 \\ 0 & 1 \\ 5 & 0 \end{bmatrix}, \quad \text{and} \quad \begin{bmatrix} 3 & 7 \\ 0 & 19 \\ 5 & 0 \end{bmatrix}$$

are linearly dependent.

Solution. The matrices $\begin{bmatrix} 1 & 2 \\ 0 & 4 \\ 2 & 0 \end{bmatrix}$, $\begin{bmatrix} 2 & 3 \\ 0 & 1 \\ 5 & 0 \end{bmatrix}$, and $\begin{bmatrix} 3 & 7 \\ 0 & 19 \\ 5 & 0 \end{bmatrix}$ are in

$$\operatorname{Span}\left\{ \begin{bmatrix} 1 & 0 \\ 0 & 0 \\ 0 & 0 \end{bmatrix}, \begin{bmatrix} 0 & 1 \\ 0 & 0 \\ 0 & 0 \end{bmatrix}, \begin{bmatrix} 0 & 0 \\ 0 & 1 \\ 0 & 0 \end{bmatrix}, \begin{bmatrix} 0 & 0 \\ 0 & 0 \\ 1 & 0 \end{bmatrix} \right\}.$$

If we let

$$\mathbf{v}_1 = \begin{bmatrix} 1 & 0 \\ 0 & 0 \\ 0 & 0 \end{bmatrix}, \mathbf{v}_2 = \begin{bmatrix} 0 & 1 \\ 0 & 0 \\ 0 & 0 \end{bmatrix}, \mathbf{v}_3 = \begin{bmatrix} 0 & 0 \\ 0 & 1 \\ 0 & 0 \end{bmatrix}, \text{ and } \mathbf{v}_4 = \begin{bmatrix} 0 & 0 \\ 0 & 0 \\ 1 & 0 \end{bmatrix},$$

then, using the function f in Theorem 10.26, we have

$$f\left(\begin{bmatrix} 1 & 2 \\ 0 & 4 \\ 2 & 0 \end{bmatrix}\right) = \begin{bmatrix} 1 \\ 2 \\ 4 \\ 2 \end{bmatrix}, f\left(\begin{bmatrix} 2 & 3 \\ 0 & 1 \\ 5 & 0 \end{bmatrix}\right) = \begin{bmatrix} 2 \\ 3 \\ 1 \\ 5 \end{bmatrix}, \text{ and } f\left(\begin{bmatrix} 3 & 7 \\ 0 & 19 \\ 5 & 0 \end{bmatrix}\right) = \begin{bmatrix} 3 \\ 7 \\ 19 \\ 5 \end{bmatrix}.$$

Now our result is a consequence of the fact that

$$\begin{bmatrix} 1 & 2 & 3 \\ 2 & 3 & 7 \\ 4 & 1 & 19 \\ 2 & 5 & 5 \end{bmatrix} \sim \begin{bmatrix} 1 & 0 & 5 \\ 0 & 1 & -1 \\ 0 & 0 & 0 \\ 0 & 0 & 0 \end{bmatrix},$$

by Corollary 10.27. □

Example 10.29. Show that the polynomials p, q, r defined by

$$p(t) = 2 + 5t + t^2, \; q(t) = 1 + 3t + 4t^2, \text{ and } r(t) = 4 + 2t - 7t^2$$

are linearly independent.

Solution. The polynomials p, q, and r are in $\operatorname{Span}\{p_0, p_1, p_2\}$.

If we let

$$\mathbf{v}_1 = p_0, \mathbf{v}_2 = p_1, \text{ and } \mathbf{v}_3 = p_2,$$

then, using the function f in Theorem 10.26, we have

$$f(p) = \begin{bmatrix} 2 \\ 5 \\ 1 \end{bmatrix}, f(q) = \begin{bmatrix} 1 \\ 3 \\ 4 \end{bmatrix}, \text{ and } f(r) = \begin{bmatrix} 4 \\ 2 \\ -7 \end{bmatrix}.$$

Now the result is a consequence of the fact that

$$\begin{bmatrix} 2 & 1 & 4 \\ 5 & 3 & 2 \\ 1 & 4 & -7 \end{bmatrix} \sim \begin{bmatrix} 1 & 0 & 0 \\ 0 & 1 & 0 \\ 0 & 0 & 1 \end{bmatrix}.$$

□

Exercises 10.1

1. Show that the matrices $A = \begin{bmatrix} 1 & 2 \\ 3 & 1 \end{bmatrix}$, $B = \begin{bmatrix} 3 & 1 \\ 4 & 3 \end{bmatrix}$, and $C = \begin{bmatrix} 4 & 3 \\ 7 & 4 \end{bmatrix}$ are linearly dependent and write C as a linear combination of A and B.

2. Show that the matrices $A = \begin{bmatrix} 1 & 3 \\ 1 & 2 \end{bmatrix}$, $B = \begin{bmatrix} 2 & 1 \\ 1 & 5 \end{bmatrix}$, and $C = \begin{bmatrix} 3 & -1 \\ 1 & 8 \end{bmatrix}$ are linearly dependent and write C as a linear combination of A and B.

3. Show that the matrices

$$A = \begin{bmatrix} 1 & 3 & 4 \\ 2 & 1 & 3 \\ 3 & 4 & 7 \\ 1 & 3 & 4 \end{bmatrix}, \quad B = \begin{bmatrix} 1 & 0 & 1 \\ 0 & 1 & 1 \\ 0 & 0 & 0 \\ 0 & 0 & 0 \end{bmatrix}, \quad \text{and} \quad C = \begin{bmatrix} 1 & 6 & 7 \\ 4 & 1 & 5 \\ 6 & 8 & 14 \\ 2 & 6 & 8 \end{bmatrix}$$

   are linearly dependent and write C as a linear combination of A and B.

4. Show that the matrices

$$A = \begin{bmatrix} 1 & 2 & 1 \\ 2 & 4 & 1 \\ 3 & 1 & 2 \end{bmatrix}, \quad B = \begin{bmatrix} 2 & 0 & 2 \\ 1 & 1 & 1 \\ -1 & 1 & 1 \end{bmatrix}, \quad \text{and} \quad C = \begin{bmatrix} -1 & 10 & -1 \\ 7 & 17 & 2 \\ 18 & 2 & 7 \end{bmatrix}$$

   are linearly dependent and write C as a linear combination of A and B.

5. Let $\mathbf{u}$, $\mathbf{v}$, and $\mathbf{w}$ be linearly independent vectors in a vector space $\mathscr{L}$. Find a number s such that the vectors $2\mathbf{u} + 3\mathbf{v} + \mathbf{w}$, $\mathbf{u} + \mathbf{v} + 4\mathbf{w}$, and $7\mathbf{u} + 9\mathbf{v} + s\mathbf{w}$ are linearly dependent.

6. Let $\mathbf{u}$, $\mathbf{v}$, and $\mathbf{w}$ be linearly independent vectors in a vector space $\mathscr{L}$. Find a number s such that the vectors $\mathbf{u} + 5\mathbf{v} + 7\mathbf{w}$, $3\mathbf{u} - \mathbf{v} + 2\mathbf{w}$, and $3\mathbf{u} + c\mathbf{v} + 5\mathbf{w}$ are linearly dependent.

7. Let $\mathbf{a}$, $\mathbf{b}$, $\mathbf{c}$, and $\mathbf{d}$ be linearly independent vectors in a vector space $\mathscr{L}$. Find numbers s and t such that the vectors $\mathbf{a} + 2\mathbf{b} + \mathbf{c} + 2\mathbf{d}$, $3\mathbf{a} + 5\mathbf{b} + 2\mathbf{c} + \mathbf{d}$, and $\mathbf{a} + s\mathbf{b} + t\mathbf{c} + \mathbf{d}$ are linearly dependent.

8. Let $\mathbf{a}$, $\mathbf{b}$, $\mathbf{c}$, and $\mathbf{d}$ be linearly independent vectors in a vector space $\mathscr{L}$. Find numbers s and t such that the vectors $2\mathbf{a}+\mathbf{b}-2\mathbf{c}+\mathbf{d}$, $\mathbf{a}+3\mathbf{b}+\mathbf{c}-\mathbf{d}$, and $\mathbf{a}+s\mathbf{b}+2\mathbf{c}+t\mathbf{d}$ are linearly dependent.

9. Show that the set of all matrices of the form

$$\begin{bmatrix} a & b \\ b & a \end{bmatrix}$$

is a subspace of $\mathscr{M}_{2\times 2}$ and determine a basis of this subspace.

10. Show that the set of all matrices of the form

$$\begin{bmatrix} a & b & c \\ b & b & a \end{bmatrix}$$

is a subspace of $\mathscr{M}_{2\times 3}$ and determine a basis of this subspace.

11. Let $\mathbf{u}$, $\mathbf{v}$, and $\mathbf{w}$ be linearly independent vectors in a vector space $\mathscr{L}$. Show that the set $\{2\mathbf{u}+\mathbf{v}+\mathbf{w}, \mathbf{u}+2\mathbf{v}+\mathbf{w}, \mathbf{u}+\mathbf{v}+2\mathbf{w}\}$ is a basis in $\operatorname{Span}\{\mathbf{u},\mathbf{v},\mathbf{w}\}$.

12. Let $\mathbf{u}$, $\mathbf{v}$, and $\mathbf{w}$ be linearly independent vectors in a vector space $\mathscr{L}$. Show that the set $\{4\mathbf{u}+\mathbf{v}+2\mathbf{w}, 2\mathbf{u}+\mathbf{v}+\mathbf{w}, \mathbf{u}+\mathbf{v}+\mathbf{w}\}$ is a basis in $\operatorname{Span}\{\mathbf{u},\mathbf{v},\mathbf{w}\}$.

13. Let $\mathbf{a}$, $\mathbf{b}$, $\mathbf{c}$, and $\mathbf{d}$ be linearly independent vectors in a vector space $\mathscr{L}$. Find a number s such that the vectors $\mathbf{a}-\mathbf{b}+\mathbf{c}+\mathbf{d}$, $\mathbf{a}+\mathbf{b}+4\mathbf{c}+3\mathbf{d}$, $5\mathbf{a}+3\mathbf{b}+\mathbf{c}+3\mathbf{d}$, and $\mathbf{a}+\mathbf{b}+\mathbf{c}+s\mathbf{d}$ are linearly dependent.

14. Let $\mathbf{a}$, $\mathbf{b}$, $\mathbf{c}$, and $\mathbf{d}$ be linearly independent vectors in a vector space $\mathscr{L}$. Find a number s such that the vectors $2\mathbf{a}+\mathbf{b}+3\mathbf{c}-\mathbf{d}$, $\mathbf{a}+\mathbf{b}+\mathbf{c}+\mathbf{d}$, $2\mathbf{a}+s\mathbf{b}+2\mathbf{c}+\mathbf{d}$, and $2\mathbf{a}+\mathbf{b}+\mathbf{c}+2\mathbf{d}$ are linearly dependent.

15. Let $\mathbf{a}$, $\mathbf{b}$, $\mathbf{c}$, $\mathbf{d}$, and $\mathbf{p}$ be vectors in a vector space $\mathscr{L}$. If the vector $\mathbf{p}$ is nonzero and is in $\operatorname{Span}\{\mathbf{a},\mathbf{b},\mathbf{c},\mathbf{d}\}$, show that the vector $\mathbf{a}$ is in $\operatorname{Span}\{\mathbf{b},\mathbf{c},\mathbf{d},\mathbf{p}\}$ or the vector $\mathbf{b}$ is in $\operatorname{Span}\{\mathbf{a},\mathbf{c},\mathbf{d},\mathbf{p}\}$ or the vector $\mathbf{c}$ is in $\operatorname{Span}\{\mathbf{a},\mathbf{b},\mathbf{d},\mathbf{p}\}$ or the vector $\mathbf{d}$ is in $\operatorname{Span}\{\mathbf{a},\mathbf{b},\mathbf{c},\mathbf{p}\}$.

16. Let $\mathbf{a}$, $\mathbf{p}$, $\mathbf{q}$, $\mathbf{r}$, and $\mathbf{s}$ be vectors in a vector space $\mathscr{L}$. If the vectors $\mathbf{p}$, $\mathbf{q}$, $\mathbf{r}$, and $\mathbf{s}$ are linearly independent and the vector $\mathbf{s}$ is in $\operatorname{Span}\{\mathbf{a},\mathbf{p},\mathbf{q},\mathbf{r}\}$, show that the vector $\mathbf{a}$ is in $\operatorname{Span}\{\mathbf{p},\mathbf{q},\mathbf{r},\mathbf{s}\}$.

17. Let $\mathbf{a}$, $\mathbf{b}$, $\mathbf{p}$, $\mathbf{q}$, and $\mathbf{r}$ be vectors in a vector space $\mathscr{L}$. If the vectors $\mathbf{p}$, $\mathbf{q}$, and $\mathbf{r}$ are linearly independent and that the vector $\mathbf{r}$ is in $\operatorname{Span}\{\mathbf{a},\mathbf{b},\mathbf{p},\mathbf{q}\}$, show that the vector $\mathbf{a}$ is in $\operatorname{Span}\{\mathbf{b},\mathbf{p},\mathbf{q},\mathbf{r}\}$ or the vector $\mathbf{b}$ is in $\operatorname{Span}\{\mathbf{a},\mathbf{p},\mathbf{q},\mathbf{r}\}$.

18. Let $\mathbf{a}$, $\mathbf{b}$, $\mathbf{c}$, $\mathbf{p}$, and $\mathbf{q}$ be vectors in a vector space $\mathscr{L}$. If the vectors $\mathbf{p}$ and $\mathbf{q}$ are linearly independent and that the vector $\mathbf{q}$ is in $\operatorname{Span}\{\mathbf{a},\mathbf{b},\mathbf{c},\mathbf{p}\}$, show that the vector $\mathbf{a}$ is in $\operatorname{Span}\{\mathbf{b},\mathbf{c},\mathbf{p},\mathbf{q}\}$ or the vector $\mathbf{b}$ is in $\operatorname{Span}\{\mathbf{a},\mathbf{c},\mathbf{p},\mathbf{q}\}$ or the vector $\mathbf{c}$ is in $\operatorname{Span}\{\mathbf{a},\mathbf{b},\mathbf{p},\mathbf{q}\}$.

19. Show that the space of solutions of the differential equation

$$y'' - 7y' + 12y = 0$$

is Span$\{f, g\}$, where the functions $f : \mathbb{R} \to \mathbb{R}$ and $g : \mathbb{R} \to \mathbb{R}$ are defined by $f(t) = e^{3t}$ and $g(t) = e^{4t}$, and that $\{f, g\}$ is a basis of this subspace of $\mathscr{D}_{\mathbb{R}}$.

20. Show that the space of solutions of the differential equation

$$y'' - 9y' + 14y = 0$$

is Span$\{f, g\}$, where the functions $f : \mathbb{R} \to \mathbb{R}$ and $g : \mathbb{R} \to \mathbb{R}$ are defined by $f(t) = e^{2t}$ and $g(t) = e^{7t}$, and that $\{f, g\}$ is a basis of this subspace of $\mathscr{D}_{\mathbb{R}}$.

21. Show that the space of solutions of the differential equation

$$y'' - 4y' + 4y = 0$$

is Span$\{f, g\}$, where the functions $f : \mathbb{R} \to \mathbb{R}$ and $g : \mathbb{R} \to \mathbb{R}$ are defined by $f(t) = e^{2t}$ and $g(t) = te^{2t}$, and that $\{f, g\}$ is a basis of this subspace of $\mathscr{D}_{\mathbb{R}}$.

22. Show that the space of solutions of the differential equation

$$y'' + 10y' + 25y = 0$$

is Span$\{f, g\}$, where the functions $f : \mathbb{R} \to \mathbb{R}$ and $g : \mathbb{R} \to \mathbb{R}$ are defined by $f(t) = e^{5t}$ and $g(t) = te^{5t}$, and that $\{f, g\}$ is a basis of this subspace of $\mathscr{D}_{\mathbb{R}}$.

23. Show that the space of solutions of the differential equation

$$y'' - 2y' + 2y = 0$$

is Span$\{f, g\}$, where the functions $f : \mathbb{R} \to \mathbb{R}$ and $g : \mathbb{R} \to \mathbb{R}$ are defined by $f(t) = e^t \cos t$ and $g(t) = e^t \sin t$, and that $\{f, g\}$ is a basis of this subspace of $\mathscr{D}_{\mathbb{R}}$.

24. Show that the space of solutions of the differential equation

$$y'' - 4y' + 13y = 0$$

is Span$\{f, g\}$, where the functions $f : \mathbb{R} \to \mathbb{R}$ and $g : \mathbb{R} \to \mathbb{R}$ are defined by $f(t) = e^{2t} \cos 3t$ and $g(t) = e^{2t} \sin 3t$, and that $\{f, g\}$ is a basis of this subspace of $\mathscr{D}_{\mathbb{R}}$.

25. Let $\mathscr{L}$ and $\mathscr{M}$ be vector spaces and let $f : \mathscr{L} \to \mathscr{M}$ and $g : \mathscr{L} \to \mathscr{M}$ be linear transformations. Show that the function $f + g : \mathscr{L} \to \mathscr{M}$ defined by $(f + g)(\mathbf{v}) = f(\mathbf{v}) + g(\mathbf{v})$ is a linear transformation.

26. Let $\mathscr{L}$ and $\mathscr{M}$ be vector spaces, t a real number, and let $f : \mathscr{L} \to \mathscr{M}$ be a linear transformation. Show that the function $cf : \mathscr{L} \to \mathscr{M}$ defined by $(tf)(\mathbf{v}) = tf(\mathbf{v})$ is a linear transformation.

27. Let $\mathscr{K}$, $\mathscr{L}$, and $\mathscr{M}$ be vector spaces and let $f : \mathscr{K} \to \mathscr{L}$ and $g : \mathscr{L} \to \mathscr{M}$ be linear transformations. Show that the function $h : \mathscr{K} \to \mathscr{M}$ defined by $h(\mathbf{v}) = g(f(\mathbf{v}))$ is a linear transformation.

28. Let $\mathscr{L}$ and $\mathscr{M}$ be vector spaces. Show that the set of all linear transformation $f : \mathscr{L} \to \mathscr{M}$ is a vector space with the operations defined in Exercises 25 and 26.

29. Let $\mathscr{L}$ and $\mathscr{M}$ be vector spaces and let $f : \mathscr{L} \to \mathscr{M}$ be a linear transformation. Show that the set of all vectors $\mathbf{v}$ in $\mathscr{L}$ such that $f(\mathbf{v}) = \mathbf{0}$ is a subspace of $\mathscr{L}$. We call this subspace the *kernel* of f and denote it by $\ker f$. Note that, if $\mathscr{L} = \mathbb{R}^n$ and $\mathscr{M} = \mathbb{R}^m$ and the function f is defined by $f(\mathbf{v}) = A\mathbf{v}$ where A is an $m \times n$ matrix, then $\ker f = \mathbf{N}(A)$.

30. Let $\mathscr{L}$ and $\mathscr{M}$ be vector spaces and let $f : \mathscr{L} \to \mathscr{M}$ be a linear transformation. Show that the set of all vectors $\mathbf{w}$ in $\mathscr{M}$ such that $f(\mathbf{v}) = \mathbf{w}$ for some vector in $\mathscr{L}$ is a subspace of $\mathscr{M}$. We call this subspace the *range* of f and denote it by $f(\mathscr{L})$. Note that, if $\mathscr{L} = \mathbb{R}^n$ and $\mathscr{M} = \mathbb{R}^m$ and the function f is defined by $f(\mathbf{v}) = A\mathbf{v}$ where A is a $m \times n$ matrix, then $f(\mathscr{L}) = \mathbf{C}(A)$.

31. Show that the function $f : \mathscr{D}_{\mathbb{R}} \to \mathscr{D}_{\mathbb{R}}$ defined by $f(y) = y'''$ is a linear transformation and determine $\ker f$.

32. Show that the function $f : \mathscr{D}_{\mathbb{R}} \to \mathscr{D}_{\mathbb{R}}$ defined by $f(y) = y'' - 2\alpha y' + \alpha^2 y$, where α is a real number, is a linear transformation and determine $\ker f$.

33. Let $\mathscr{L}$ and $\mathscr{M}$ be vector spaces and let $f : \mathscr{L} \to \mathscr{M}$ be a linear transformation. If the dimension of the subspace $f(\mathscr{L})$ is 3 and the dimension of the subspace $\ker f$ is 2, show that the dimension of $\mathscr{L}$ is 5.

34. Let $\mathscr{L}$ and $\mathscr{M}$ be vector spaces and let $f : \mathscr{L} \to \mathscr{M}$ be a linear transformation. If the dimension of the subspace $f(\mathscr{L})$ is 2 and the dimension of the subspace $\ker f$ is 3, show that the dimension of $\mathscr{L}$ is 5.

10.2 Inner Product Spaces

The dot product plays an important role in the linear algebra of $\mathbb{R}^n$ and is an irreplaceable tool for many problems in $\mathbb{R}^n$. It turns out that the idea can be extended to vector spaces. In the context of general vector spaces the dot product is usually called the inner product. As we will see, almost everything we learned about the dot product remains true for inner products.

Definition 10.30. By an *inner product* on a vector space $\mathscr{L}$ we mean a function which associates with every pair of vectors $\mathbf{u}$ and $\mathbf{v}$ from $\mathscr{L}$ a real number $\langle \mathbf{u}, \mathbf{v} \rangle$ such that for all vectors $\mathbf{u}$, $\mathbf{v}$, and $\mathbf{w}$ in $\mathscr{L}$ and all real numbers c we have

1. $\langle \mathbf{u}, \mathbf{v} \rangle = \langle \mathbf{v}, \mathbf{u} \rangle$,
2. $\langle \mathbf{u}, \mathbf{v} + \mathbf{w} \rangle = \langle \mathbf{u}, \mathbf{v} \rangle + \langle \mathbf{u}, \mathbf{w} \rangle$,
3. $\langle c\mathbf{u}, \mathbf{v} \rangle = c\langle \mathbf{u}, \mathbf{v} \rangle$,
4. $\langle \mathbf{u}, \mathbf{u} \rangle \geq 0$,
5. $\langle \mathbf{u}, \mathbf{u} \rangle = 0$ if and only if $\mathbf{u} = \mathbf{0}$.

   A vector space $\mathscr{L}$ with an inner product is called an *inner product space.*

The dot product in $\mathbb{R}^n$ was defined as a specific function on pairs of vectors, namely,

$$\begin{bmatrix} u_1 \\ u_2 \\ \vdots \\ u_n \end{bmatrix} \cdot \begin{bmatrix} v_1 \\ v_2 \\ \vdots \\ v_n \end{bmatrix} = u_1 v_1 + u_2 v_2 + \cdots + u_n v_n.$$

The dot product is an example of an inner product in $\mathbb{R}^n$, but it is not the only possible inner product in $\mathbb{R}^n$. It is easy to verify that

$$\left\langle \begin{bmatrix} u_1 \\ u_2 \\ \vdots \\ u_n \end{bmatrix}, \begin{bmatrix} v_1 \\ v_2 \\ \vdots \\ v_n \end{bmatrix} \right\rangle = \alpha_1 u_1 v_1 + \alpha_2 u_2 v_2 + \cdots + \alpha_n u_n v_n$$

is an inner product in $\mathbb{R}^n$ for any positive real numbers $\alpha_1, \alpha_2, \ldots, \alpha_n$. Such inner product is referred to as a *weighted dot product* or a *weighted inner product.*

Example 10.31. The vector space $\mathbb{R}^3$ is an inner product space with the inner product defined by

$$\left\langle \begin{bmatrix} u_1 \\ u_2 \\ u_3 \end{bmatrix}, \begin{bmatrix} v_1 \\ v_2 \\ v_3 \end{bmatrix} \right\rangle = 5u_1 v_1 + \frac{1}{2} u_2 v_2 + \sqrt{7} u_3 v_3.$$

Example 10.32. The vector space $\mathcal{M}_{2\times 2}$ is a inner product space with the inner product defined by

$$\left\langle \begin{bmatrix} u_1 & u_2 \\ u_3 & u_4 \end{bmatrix}, \begin{bmatrix} v_1 & v_2 \\ v_3 & v_4 \end{bmatrix} \right\rangle = u_1 v_1 + u_2 v_2 + u_3 v_3 + u_4 v_4.$$

Example 10.33. The vector space $\mathscr{C}_{[a,b]}$ is a inner product space with the inner product defined by

$$\langle f, g\rangle = \int_a^b f(t)g(t)dt.$$

In $\mathbb{R}^n$ we define the dot product and the norm independently and then we discover that there is a nice connection between them. In inner product spaces we define the norm using that connection.

Definition 10.34. By the *norm* of a vector $\mathbf{u}$ in an inner product space $\mathscr{L}$ we mean the number

$$\|\mathbf{u}\| = \sqrt{\langle \mathbf{u}, \mathbf{u}\rangle}.$$

Example 10.35. The norm of the function $f(t) = t$ in the inner product space $\mathscr{C}_{[0,1]}$ is

$$\|f\| = \sqrt{\int_0^1 t^2 dt} = \frac{1}{\sqrt{3}}.$$

Definition 10.36. By the *distance* between vectors $\mathbf{u}$ and $\mathbf{v}$ in an inner product space $\mathscr{L}$ we mean the number

$$d(\mathbf{u}, \mathbf{v}) = \|\mathbf{u} - \mathbf{v}\|.$$

The distance defined this way has a very clear geometric meaning in $\mathbb{R}^2$ or $\mathbb{R}^3$, but it does not really have a geometric interpretation in general inner product spaces.

Example 10.37. The distance between the functions $f(t) = t$ and $g(t) = t^2$ in the inner product space $\mathscr{C}_{[0,1]}$ is

$$d(f,g) = \|f-g\| = \sqrt{\int_0^1 (t-t^2)^2 dt} = \sqrt{\int_0^1 (t^2-2t^3+t^4) dt} = \frac{1}{\sqrt{30}}.$$

Definition 10.38. Two vectors $\mathbf{u}$ and $\mathbf{v}$ in an inner product space $\mathscr{L}$ are called *orthogonal* if

$$\|\mathbf{u}+\mathbf{v}\| = \|\mathbf{u}-\mathbf{v}\|.$$

Vectors in $\mathbb{R}^2$ or $\mathbb{R}^3$ are orthogonal, if they are perpendicular. While orthogonality in inner product spaces does not have such a geometric interpretation, it has the same algebraic properties and is very useful in applications.

Theorem 10.39. *Vectors* $\mathbf{u}$ *and* $\mathbf{v}$ *in an inner product space* $\mathscr{L}$ *are orthogonal if and only if* $\langle \mathbf{u}, \mathbf{v} \rangle = 0$.

Proof. Since

$$\begin{aligned}
\|\mathbf{u}+\mathbf{v}\|^2 &= \langle \mathbf{u}+\mathbf{v}, \mathbf{u}+\mathbf{v} \rangle \\
&= \langle \mathbf{u}+\mathbf{v}, \mathbf{u} \rangle + \langle \mathbf{u}+\mathbf{v}, \mathbf{v} \rangle \\
&= \langle \mathbf{u}, \mathbf{u}+\mathbf{v} \rangle + \langle \mathbf{v}, \mathbf{u}+\mathbf{v} \rangle \\
&= \langle \mathbf{u}, \mathbf{u} \rangle + \langle \mathbf{u}, \mathbf{v} \rangle + \langle \mathbf{v}, \mathbf{u} \rangle + \langle \mathbf{v}, \mathbf{v} \rangle \\
&= \|\mathbf{u}\|^2 + 2\langle \mathbf{u}, \mathbf{v} \rangle + \|\mathbf{v}\|^2
\end{aligned}$$

and, similarly,

$$\|\mathbf{u}-\mathbf{v}\|^2 = \|\mathbf{u}\|^2 - 2\langle \mathbf{u}, \mathbf{v} \rangle + \|\mathbf{v}\|^2,$$

we have $\|\mathbf{u}+\mathbf{v}\| = \|\mathbf{u}-\mathbf{v}\|$ if and only if $\langle \mathbf{u}, \mathbf{v} \rangle = 0$. □

Example 10.40. Show that, if m and n are different nonnegative integers, the functions $f(t) = \sin mt$ and $g(t) = \cos nt$ in the space of continuous functions on $[0, 2\pi]$ are orthogonal with the inner product

$$\langle f, g \rangle = \int_0^{2\pi} f(t)g(t)\, dt.$$

Solution. Using the trigonometric identity

$$\sin\alpha\cos\beta = \frac{1}{2}(\sin(\alpha+\beta) + \sin(\alpha-\beta))$$

we get

$$\begin{aligned}\langle f, g\rangle &= \int_0^{2\pi} \sin mt \cos nt\, dt \\ &= \frac{1}{2}\int_0^{2\pi} (\sin(m+n)t + \sin(m-n)t)\, dt \\ &= -\frac{1}{2}\left[\frac{\cos(m+n)t}{m+n} + \frac{\cos(m-n)t}{m-n}\right]_0^{2\pi} = 0.\end{aligned}$$

□

Best approximations and projections in inner product spaces

We have seen that the dot product played a prominent role when we talked about best approximations and projections in $\mathbb{R}^n$. The situation is very similar in general inner product spaces. With the exception of Theorem 10.45, the proofs of the theorems that follow are the same as in the case of $\mathbb{R}^n$.

Definition 10.41. Let $\mathbf{u}_1, \mathbf{u}_2, \ldots, \mathbf{u}_k$ and $\mathbf{b}$ be vectors in an inner product space $\mathscr{L}$.

(a) A vector $\mathbf{p}$ in $\operatorname{Span}\{\mathbf{u}_1, \ldots, \mathbf{u}_k\}$ is called the *best approximation* to the vector $\mathbf{b}$ by vectors from $\operatorname{Span}\{\mathbf{u}_1, \ldots, \mathbf{u}_k\}$ if

$$\|\mathbf{b} - \mathbf{p}\| < \|\mathbf{b} - \mathbf{q}\|$$

for every $\mathbf{q}$ in $\operatorname{Span}\{\mathbf{u}_1, \ldots, \mathbf{u}_k\}$ such that $\mathbf{q} \neq \mathbf{p}$.

(b) A vector $\mathbf{p}$ in $\operatorname{Span}\{\mathbf{u}_1, \ldots, \mathbf{u}_k\}$ is called an *orthogonal projection* of the vector $\mathbf{b}$ on $\operatorname{Span}\{\mathbf{u}_1, \ldots, \mathbf{u}_k\}$ if

$$\langle \mathbf{b} - \mathbf{p}, \mathbf{v}\rangle = 0$$

for every vector $\mathbf{v}$ in $\operatorname{Span}\{\mathbf{u}_1, \ldots, \mathbf{u}_k\}$.

Theorem 10.42. *Let* $\mathbf{u}_1, \mathbf{u}_2, \ldots, \mathbf{u}_k$ *and* $\mathbf{b}$ *be vectors in an inner product space* $\mathscr{L}$. *The following conditions are equivalent:*

(a) $\mathbf{p}$ *in* $\operatorname{Span}\{\mathbf{u}_1, \ldots, \mathbf{u}_k\}$ *is an orthogonal projection of the vector* $\mathbf{b}$ *on* $\operatorname{Span}\{\mathbf{u}_1, \ldots, \mathbf{u}_k\}$;

(b) $\mathbf{p}$ *is the best approximation to the vector* $\mathbf{b}$ *by vectors from* $\operatorname{Span}\{\mathbf{u}_1, \ldots, \mathbf{u}_k\}$.

Theorem 10.43. *Let* $\mathbf{u}_1, \mathbf{u}_2, \ldots, \mathbf{u}_k$ *be vectors in an inner product space* $\mathscr{L}$. *The following conditions are equivalent:*

(a) *The vector*

$$x_1\mathbf{u}_1 + x_2\mathbf{u}_2 + \cdots + x_k\mathbf{u}_k$$

is the projection of a vector $\mathbf{b}$ *on* $\operatorname{Span}\{\mathbf{u}_1, \ldots, \mathbf{u}_k\}$;

(b) *The numbers* $x_1, x_2, \ldots, x_k$ *satisfy the following system of equations*

$$\begin{cases} x_1\langle\mathbf{u}_1,\mathbf{u}_1\rangle & + & x_2\langle\mathbf{u}_2,\mathbf{u}_1\rangle & + & \cdots & + & x_k\langle\mathbf{u}_k,\mathbf{u}_1\rangle & = & \langle\mathbf{b},\mathbf{u}_1\rangle \\ x_1\langle\mathbf{u}_1,\mathbf{u}_2\rangle & + & x_2\langle\mathbf{u}_2,\mathbf{u}_2\rangle & + & \cdots & + & x_k\langle\mathbf{u}_k,\mathbf{u}_2\rangle & = & \langle\mathbf{b},\mathbf{u}_2\rangle \\ & \vdots & & \vdots & & \vdots & & \vdots & \\ x_1\langle\mathbf{u}_1,\mathbf{u}_k\rangle & + & x_2\langle\mathbf{u}_2,\mathbf{u}_k\rangle & + & \cdots & + & x_k\langle\mathbf{u}_k,\mathbf{u}_k\rangle & = & \langle\mathbf{b},\mathbf{u}_k\rangle \end{cases}.$$

Corollary 10.44. *Let* $\mathbf{u}_1, \mathbf{u}_2, \ldots, \mathbf{u}_k$ *and* $\mathbf{b}$ *be vectors in an inner product space* $\mathscr{L}$. *The following conditions are equivalent:*

(a) *The numbers* $x_1, x_2, \ldots, x_k$ *minimize the norm*

$$\|\mathbf{b} - x_1\mathbf{u}_1 - x_2\mathbf{u}_2 - \cdots - x_k\mathbf{u}_k\|,$$

that is, the vector $x_1\mathbf{u}_1 + x_2\mathbf{u}_2 + \cdots + x_k\mathbf{u}_k$ *is the best approximation to the vector* $\mathbf{b}$ *by vectors from* $\operatorname{Span}\{\mathbf{u}_1, \ldots, \mathbf{u}_k\}$;

(b) *The numbers* $x_1, x_2, \ldots, x_k$ *satisfy the following system of equations*

$$\begin{cases} x_1\langle\mathbf{u}_1,\mathbf{u}_1\rangle & + & x_2\langle\mathbf{u}_2,\mathbf{u}_1\rangle & + & \cdots & + & x_k\langle\mathbf{u}_k,\mathbf{u}_1\rangle & = & \langle\mathbf{b},\mathbf{u}_1\rangle \\ x_1\langle\mathbf{u}_1,\mathbf{u}_2\rangle & + & x_2\langle\mathbf{u}_2,\mathbf{u}_2\rangle & + & \cdots & + & x_k\langle\mathbf{u}_k,\mathbf{u}_2\rangle & = & \langle\mathbf{b},\mathbf{u}_2\rangle \\ & \vdots & & \vdots & & \vdots & & \vdots & \\ x_1\langle\mathbf{u}_1,\mathbf{u}_k\rangle & + & x_2\langle\mathbf{u}_2,\mathbf{u}_k\rangle & + & \cdots & + & x_k\langle\mathbf{u}_k,\mathbf{u}_k\rangle & = & \langle\mathbf{b},\mathbf{u}_k\rangle \end{cases}.$$

Theorem 10.45. *Let* $\mathbf{u}_1, \mathbf{u}_2, \ldots, \mathbf{u}_k$ *and* $\mathbf{b}$ *be vectors in an inner product space* $\mathscr{L}$. *The orthogonal projection of* $\mathbf{b}$ *on the subspace* $\operatorname{Span}\{\mathbf{u}_1, \ldots, \mathbf{u}_k\}$ *always exists.*

Proof for $k = 3$. We can suppose that the vectors $\mathbf{u}_1, \mathbf{u}_2, \mathbf{u}_3$ are linearly independent. It is enough to prove that the system in Theorem 10.43 has a unique solution, or equivalently, that the matrix

$$\begin{bmatrix} \langle \mathbf{u}_1, \mathbf{u}_1 \rangle & \langle \mathbf{u}_2, \mathbf{u}_1 \rangle & \langle \mathbf{u}_3, \mathbf{u}_1 \rangle \\ \langle \mathbf{u}_1, \mathbf{u}_2 \rangle & \langle \mathbf{u}_2, \mathbf{u}_2 \rangle & \langle \mathbf{u}_3, \mathbf{u}_2 \rangle \\ \langle \mathbf{u}_1, \mathbf{u}_3 \rangle & \langle \mathbf{u}_2, \mathbf{u}_3 \rangle & \langle \mathbf{u}_3, \mathbf{u}_3 \rangle \end{bmatrix}$$

is invertible. If

$$\begin{bmatrix} \langle \mathbf{u}_1, \mathbf{u}_1 \rangle & \langle \mathbf{u}_2, \mathbf{u}_1 \rangle & \langle \mathbf{u}_3, \mathbf{u}_1 \rangle \\ \langle \mathbf{u}_1, \mathbf{u}_2 \rangle & \langle \mathbf{u}_2, \mathbf{u}_2 \rangle & \langle \mathbf{u}_3, \mathbf{u}_2 \rangle \\ \langle \mathbf{u}_1, \mathbf{u}_3 \rangle & \langle \mathbf{u}_2, \mathbf{u}_3 \rangle & \langle \mathbf{u}_3, \mathbf{u}_3 \rangle \end{bmatrix} \begin{bmatrix} x \\ y \\ z \end{bmatrix} = \begin{bmatrix} 0 \\ 0 \\ 0 \end{bmatrix},$$

then we have

$$\begin{aligned} \langle x\mathbf{u}_1 + y\mathbf{u}_2 + z\mathbf{u}_3, \mathbf{u}_1 \rangle &= 0, \\ \langle x\mathbf{u}_1 + y\mathbf{u}_2 + z\mathbf{u}_3, \mathbf{u}_2 \rangle &= 0, \\ \langle x\mathbf{u}_1 + y\mathbf{u}_2 + z\mathbf{u}_3, \mathbf{u}_3 \rangle &= 0. \end{aligned}$$

Hence

$$\langle x\mathbf{u}_1 + y\mathbf{u}_2 + z\mathbf{u}_3, x\mathbf{u}_1 + y\mathbf{u}_2 + z\mathbf{u}_3 \rangle = \|x\mathbf{u}_1 + y\mathbf{u}_2 + z\mathbf{u}_3\|^2 = 0$$

and thus

$$x\mathbf{u}_1 + y\mathbf{u}_2 + z\mathbf{u}_3 = \mathbf{0}.$$

Since the vectors $\mathbf{u}_1, \mathbf{u}_2, \mathbf{u}_3$ are linearly independent, we get $x = y = z = 0$. This completes the proof by Theorem 2.17. □

The orthogonal projection of a vector $\mathbf{b}$ in an inner product space $\mathscr{L}$ on a subspace $\mathscr{V}$ is denoted by $\operatorname{proj}_{\mathscr{V}} \mathbf{b}$.

Example 10.46. Consider the vector space of continuous functions defined on the interval $[0, 1]$ with the inner product $\langle f, g \rangle = \int_0^1 f(t)g(t)\,dt$. Find the best approximation of the function e^t by functions from $\operatorname{Span}\{1, t\}$.

Solution. By Theorem 10.43, we need to solve the system

$$\begin{cases} x_1 \int_0^1 1\,dt & + & x_2 \int_0^1 t\,dt & = & \int_0^1 e^t\,dt \\ x_1 \int_0^1 t\,dt & + & x_2 \int_0^1 t^2\,dt & = & \int_0^1 te^t\,dt \end{cases}$$

which, after calculating the integrals, becomes

$$\begin{cases} x_1 & + & \frac{1}{2}x_2 & = & e - 1 \\ \frac{1}{2}x_1 & + & \frac{1}{3}x_2 & = & 1 \end{cases}.$$

The unique solution of the system is $x_1 = 4e - 10$ and $x_2 = 18 - 6e$. Consequently, the best approximation is $4e - 10 + (18 - 6e)t$. This means that

$$\int_0^1 (e^t - (4e - 10 + (18 - 6e)t))^2 dt \le \int_0^1 (e^t - (a + bt))^2 dt$$

for arbitrary real numbers a and b. □

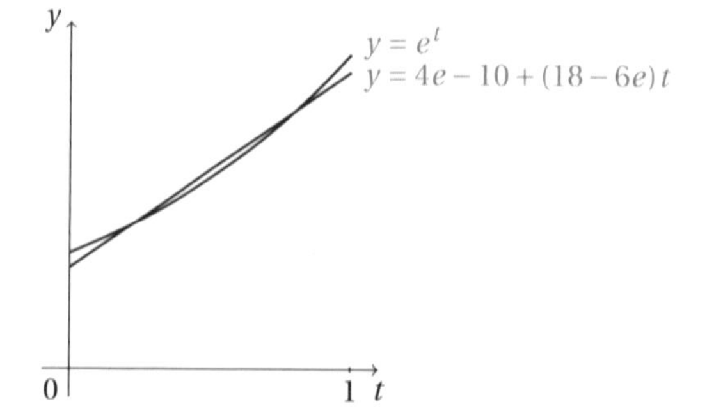

Figure 10.1: The best approximation of the function $f(t) = e^t$ by functions from Span$\{1, t\}$.

Definition 10.47. Let $\mathbf{u}_1, \dots, \mathbf{u}_k$ be vectors in an inner product space $\mathscr{L}$. We say that the set $\{\mathbf{u}_1, \dots, \mathbf{u}_k\}$ is an *orthogonal set* if $\langle \mathbf{u}_i, \mathbf{u}_j \rangle = 0$ for all $i, j = 1, \dots, k$ such that $i \neq j$.

Example 10.48. It is easy to verify, as in Example 10.40, that the set of functions

$$1, \sin t, \dots, \sin nt, \cos t, \dots, \cos nt$$

is an orthogonal set in the inner product space $\mathscr{C}_{[0,2\pi]}$ with the inner product

$$\langle f, g \rangle = \int_0^{2\pi} f(t)g(t)\, dt.$$

Theorem 10.49. *If* $\{\mathbf{u}_1, \dots, \mathbf{u}_k\}$ *is an orthogonal set of nonzero vectors in an inner product space* $\mathscr{L}$*, then the vectors* $\mathbf{u}_1, \dots, \mathbf{u}_k$ *are linearly independent.*

Theorem 10.50. *Let $\{\mathbf{u}_1, \dots, \mathbf{u}_k\}$ be an orthogonal set of nonzero vectors in an inner product space $\mathscr{L}$ and let $\mathscr{U} = \operatorname{Span}\{\mathbf{u}_1, \dots, \mathbf{u}_k\}$. Then*

$$\operatorname{proj}_{\mathscr{U}} \mathbf{b} = \frac{\langle \mathbf{b}, \mathbf{u}_1 \rangle}{\langle \mathbf{u}_1, \mathbf{u}_1 \rangle} \mathbf{u}_1 + \dots + \frac{\langle \mathbf{b}, \mathbf{u}_k \rangle}{\langle \mathbf{u}_k, \mathbf{u}_k \rangle} \mathbf{u}_k$$

for every vector $\mathbf{b}$ in $\mathscr{L}$. This means that, if $\{\mathbf{u}_1, \dots, \mathbf{u}_k\}$ is a set of orthogonal nonzero vectors, then the best approximation to the vector $\mathbf{b}$ by vectors from the subspace $\operatorname{Span}\{\mathbf{u}_1, \dots, \mathbf{u}_k\}$ is the vector

$$\frac{\langle \mathbf{b}, \mathbf{u}_1 \rangle}{\langle \mathbf{u}_1, \mathbf{u}_1 \rangle} \mathbf{u}_1 + \dots + \frac{\langle \mathbf{b}, \mathbf{u}_k \rangle}{\langle \mathbf{u}_k, \mathbf{u}_k \rangle} \mathbf{u}_k.$$

Example 10.51. If f is a function in the inner product space $\mathscr{C}_{[0,2\pi]}$ with the inner product $\langle f, g \rangle = \int_0^{2\pi} f(t)g(t)\,dt$, then the best approximation of f by functions from

$$\operatorname{Span}\{1, \sin t, \dots, \sin nt, \cos t, \dots, \cos nt\}$$

is the function

$$\frac{1}{2\pi} \int_0^{2\pi} f(t)\,dt + \frac{1}{\pi} \sum_{k=1}^{n} \left(\int_0^{2\pi} f(t) \sin kt\,dt \sin kt + \int_0^{2\pi} f(t) \cos kt\,dt \cos kt \right).$$

For example, the best approximation of the function $f(t) = t^2$ by functions from $\operatorname{Span}\{1, \sin t, \sin 2t, \cos t, \cos 2t\}$ is the function

$$\frac{4}{3}\pi^2 - 4\pi \sin t + 4\cos t - 2\pi \sin 2t + \cos 2t.$$

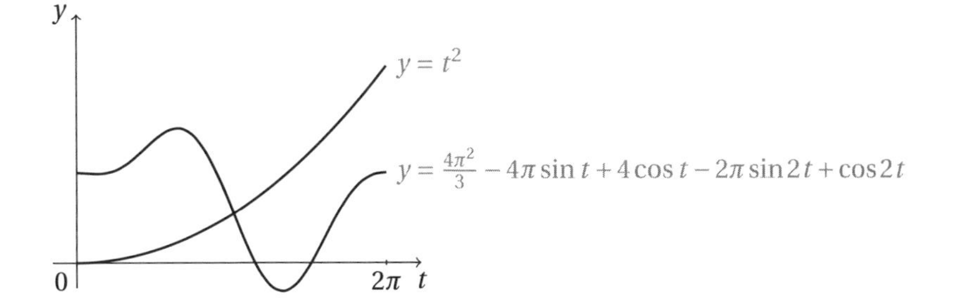

Figure 10.2: The best approximation of the function $f(t) = t^2$ by functions from $\operatorname{Span}\{1, \sin t, \sin 2t, \cos t, \cos 2t\}$.

It is not always intuitive why a function from a subspace is the best approximation of a given function. In Figure 10.1 the function $y = 4e - 10 + (18 - 6e)t$ looks like a "good" approximation of the function $y = e^t$. On the other hand, the function $y = \frac{4\pi^2}{3} - 4\pi \sin t + 4\cos t - 2\pi \sin 2t + \cos 2t$ in Figure 10.2 looks like a rather "poor" approximation of the function $y = t^2$. We need to remember that we are looking for a function $y = g(t)$ that minimizes the integral $\int_0^{2\pi} (f(t) - t^2)^2\, dt$ and not the distance between the graphs of the functions. Moreover, a subspace may not be a good choice for approximating a given function. We are facing the same situation on the plane. If a point on the plane is far from a line, then the best approximation by points on the line is not going to be a good approximation.

The Gram-Schmidt process in inner product spaces

The Gram-Schmidt orthogonalization process is a very useful tool in $\mathbb{R}^n$. As we will see below, the method is working well in every inner product space and the only difference is that we are using the inner product instead of the dot product. The following theorem, which is the basis for the Gram-Schmidt orthogonalization process, is essentially a copy of Theorem 3.36 with all dot products replaced by inner products. We can also easily adapt the proof of Theorem 3.36 to the setting of the general inner product space.

Theorem 10.52. *Let* $\{\mathbf{u}_1, \dots, \mathbf{u}_m\}$ *be an orthogonal set of nonzero vectors in an inner product space* $\mathscr{L}$ *and let* $\mathscr{V} = \operatorname{Span}\{\mathbf{u}_1, \dots, \mathbf{u}_m\}$.

(a) *The vector* $\mathbf{v}$ *is in* $\mathscr{V}$ *if and only if*

$$\mathbf{v} = \operatorname{proj}_{\mathscr{V}} \mathbf{v} = \frac{\langle \mathbf{u}_1, \mathbf{v} \rangle}{\langle \mathbf{u}_1, \mathbf{u}_1 \rangle} \mathbf{u}_1 + \frac{\langle \mathbf{u}_2, \mathbf{v} \rangle}{\langle \mathbf{u}_2, \mathbf{u}_2 \rangle} \mathbf{u}_2 + \cdots + \frac{\langle \mathbf{u}_m, \mathbf{v} \rangle}{\langle \mathbf{u}_m, \mathbf{u}_m \rangle} \mathbf{u}_m.$$

In this case we have

$$\operatorname{Span}\{\mathbf{u}_1, \dots, \mathbf{u}_m, \mathbf{v}\} = \operatorname{Span}\{\mathbf{u}_1, \dots, \mathbf{u}_m\}.$$

(b) *The vector* $\mathbf{v}$ *is not in* $\mathscr{V}$ *if and only if the vector*

$$\mathbf{u}_{m+1} = \mathbf{v} - \frac{\langle \mathbf{u}_1, \mathbf{v} \rangle}{\langle \mathbf{u}_1, \mathbf{u}_1 \rangle} \mathbf{u}_1 - \frac{\langle \mathbf{u}_2, \mathbf{v} \rangle}{\langle \mathbf{u}_2, \mathbf{u}_2 \rangle} \mathbf{u}_2 - \cdots - \frac{\langle \mathbf{u}_m, \mathbf{v} \rangle}{\langle \mathbf{u}_m, \mathbf{u}_m \rangle} \mathbf{u}_m$$

is a nonzero vector. In this case we have

$$\langle \mathbf{u}_{m+1}, \mathbf{u}_1 \rangle = \langle \mathbf{u}_{m+1}, \mathbf{u}_2 \rangle = \cdots = \langle \mathbf{u}_{m+1}, \mathbf{u}_m \rangle = 0$$

and

$$\operatorname{Span}\{\mathbf{u}_1, \dots, \mathbf{u}_m, \mathbf{v}\} = \operatorname{Span}\{\mathbf{u}_1, \dots, \mathbf{u}_m, \mathbf{u}_{m+1}\}.$$

Example 10.53. We consider the polynomials $1, t, t^2$ as functions in the inner product space $\mathscr{C}_{[-1,1]}$ with the inner product

$$\langle f, g\rangle = \int_a^b f(t)g(t)dt.$$

Find polynomials q_1 and q_2 in $\mathscr{C}_{[-1,1]}$ such that $\{1, q_1, q_2\}$ is an orthogonal set, $\operatorname{Span}\{1, q_1\} = \operatorname{Span}\{1, t\}$, and $\operatorname{Span}\{1, q_1, q_2\} = \operatorname{Span}\{1, t, t^2\}$.

Solution. We use the Gram-Schmidt orthogonalization process:

$$q_1 = t - \frac{\int_{-1}^1 t\,dt}{\int_{-1}^1 dt} = t$$

and

$$q_2 = t^2 - \frac{\int_{-1}^1 t^2 dt}{\int_{-1}^1 dt} - \frac{\int_{-1}^1 t^3 dt}{\int_{-1}^1 t^2 dt} t = t^2 - \frac{1}{3}.$$

It is easy to verify that $\langle 1, t\rangle = 0$, $\langle 1, t^2 - \frac{1}{3}\rangle = 0$, and $\langle t, t^2 - \frac{1}{3}\rangle = 0$. □

Example 10.54. Let $\mathbf{v}_1, \mathbf{v}_2, \mathbf{v}_3, \mathbf{v}_4$ be linearly independent vectors in an inner product space $\mathscr{L}$. Find an orthogonal set $\{\mathbf{u}_1, \mathbf{u}_2, \mathbf{u}_3, \mathbf{u}_4\}$ such that

$$\begin{aligned}
\mathbf{v}_1 &= \mathbf{u}_1, \\
\operatorname{Span}\{\mathbf{v}_1, \mathbf{v}_2\} &= \operatorname{Span}\{\mathbf{u}_1, \mathbf{u}_2\}, \\
\operatorname{Span}\{\mathbf{v}_1, \mathbf{v}_2, \mathbf{v}_3\} &= \operatorname{Span}\{\mathbf{u}_1, \mathbf{u}_2, \mathbf{u}_3\},
\end{aligned}$$

and

$$\operatorname{Span}\{\mathbf{v}_1, \mathbf{v}_2, \mathbf{v}_3, \mathbf{v}_4\} = \operatorname{Span}\{\mathbf{u}_1, \mathbf{u}_2, \mathbf{u}_3, \mathbf{u}_4\}.$$

Solution. We take

$$\begin{aligned}
\mathbf{u}_1 &= \mathbf{v}_1, \\
\mathbf{u}_2 &= \mathbf{v}_2 - \frac{\langle \mathbf{u}_1, \mathbf{v}_2\rangle}{\langle \mathbf{u}_1, \mathbf{u}_1\rangle}\mathbf{u}_1, \\
\mathbf{u}_3 &= \mathbf{v}_3 - \frac{\langle \mathbf{u}_1, \mathbf{v}_3\rangle}{\langle \mathbf{u}_1, \mathbf{u}_1\rangle}\mathbf{u}_1 - \frac{\langle \mathbf{u}_2, \mathbf{v}_3\rangle}{\langle \mathbf{u}_2, \mathbf{u}_2\rangle}\mathbf{u}_2,
\end{aligned}$$

and

$$\mathbf{u}_4 = \mathbf{v}_4 - \frac{\langle \mathbf{u}_1, \mathbf{v}_4\rangle}{\langle \mathbf{u}_1, \mathbf{u}_1\rangle}\mathbf{u}_1 - \frac{\langle \mathbf{u}_2, \mathbf{v}_4\rangle}{\langle \mathbf{u}_2, \mathbf{u}_2\rangle}\mathbf{u}_2 - \frac{\langle \mathbf{u}_3, \mathbf{v}_4\rangle}{\langle \mathbf{u}_3, \mathbf{u}_3\rangle}\mathbf{u}_3.$$

□

Example 10.54 describes a particular form of the Gram-Schmidt process. We give the general result in the following theorem.

Theorem 10.55. *For linearly independent vectors* $\mathbf{v}_1, \dots, \mathbf{v}_m$ *in an inner product space* $\mathscr{L}$ *there are orthogonal vectors* $\mathbf{u}_1, \dots, \mathbf{u}_m$ *in* $\mathscr{L}$ *such that*

$$\text{Span}\{\mathbf{v}_1, \dots, \mathbf{v}_k\} = \text{Span}\{\mathbf{u}_1, \dots, \mathbf{u}_k\}$$

for every integer $k = 1, 2, \dots, m$.

Exercises 10.2

1. Consider the vector space $\mathscr{M}_{2\times 2}$. Show that

$$\left\langle \begin{bmatrix} u_1 & u_2 \\ u_3 & u_4 \end{bmatrix}, \begin{bmatrix} v_1 & v_2 \\ v_3 & v_4 \end{bmatrix} \right\rangle = u_1 v_1 + u_2 v_2 + u_3 v_3 + u_4 v_4$$

defines an inner product on the vector space $\mathscr{M}_{2\times 2}$.

2. Show that

$$\left\langle \begin{bmatrix} u_1 \\ u_2 \\ u_3 \end{bmatrix}, \begin{bmatrix} v_1 \\ v_2 \\ v_3 \end{bmatrix} \right\rangle = 5 u_1 v_1 + \frac{1}{2} u_2 v_2 + \sqrt{7} u_3 v_3$$

defines an inner product on the vector space $\mathbb{R}^3$.

3. Show that $\langle f, g\rangle = \int_0^1 f(t) g(t)\, t\, dt$ defines an inner product on $\mathscr{C}_{[0,1]}$.

4. Show that $\langle f, g\rangle = \int_{-1}^1 f(t) g(t) t^2\, dt$ defines an inner product on $\mathscr{C}_{[-1,1]}$.

5. Let $\mathbf{u}$, $\mathbf{v}$, and $\mathbf{w}$ be vectors in an inner product space $\mathscr{L}$. Show that

$$\langle \mathbf{u} + \mathbf{v}, \mathbf{w}\rangle = \langle \mathbf{u}, \mathbf{w}\rangle + \langle \mathbf{v}, \mathbf{w}\rangle.$$

6. Let $\mathbf{v}$ be a vector in an inner product space $\mathscr{L}$. Show that $\langle \mathbf{0}, \mathbf{v}\rangle = 0$.

7. Let $\mathbf{v}$ be a vector in an inner product space $\mathscr{L}$ and let c be a real number. Show that $\|c\mathbf{v}\| = |c|\|\mathbf{v}\|$.

8. Let $\mathbf{u}$ and $\mathbf{v}$ be vectors in an inner product space $\mathscr{L}$ and let c be a real number. Show that $\langle \mathbf{u}, c\mathbf{v}\rangle = c\langle \mathbf{u}, \mathbf{v}\rangle$.

9. Consider the inner product space $\mathscr{C}_{[0,1]}$ with the inner product defined by $\langle f, g\rangle = \int_0^1 f(t) g(t) e^t\, dt$. Calculate $\langle t, e^t\rangle$.

10. Consider the inner product space $\mathscr{C}_{[0,1]}$ with the inner product defined by $\langle f,g\rangle = \int_0^1 f(t)g(t)t\,dt$. Calculate $\langle t^2, t^{31}\rangle$.

11. Consider the inner product space $\mathscr{C}_{[-1,1]}$ with the inner product defined by $\langle f,g\rangle = \int_{-1}^1 f(t)g(t)t^2\,dt$. Calculate $\|t\|$.

12. Consider the inner product space $\mathscr{C}_{[0,1]}$ with the inner product defined by $\langle f,g\rangle = \int_0^1 f(t)g(t)e^{3t}\,dt$. Calculate $\|e^{5t}\|$.

13. Let $\mathbf{u}$ and $\mathbf{v}$ be vectors in an inner product space $\mathscr{L}$ and let $\mathbf{u} \neq \mathbf{0}$. Show that

$$\left\langle \mathbf{u}, \mathbf{v} - \frac{\langle \mathbf{v},\mathbf{u}\rangle}{\langle \mathbf{u},\mathbf{u}\rangle}\mathbf{u} \right\rangle = 0.$$

14. Let $\mathbf{u}$ and $\mathbf{v}$ be vectors in an inner product space $\mathscr{L}$ and let $\mathbf{u} \neq \mathbf{0}$. Show that

$$\|\mathbf{v}\|^2 = \left\| \mathbf{v} - \frac{\langle \mathbf{v},\mathbf{u}\rangle}{\langle \mathbf{u},\mathbf{u}\rangle}\mathbf{u} \right\|^2 + \left\| \frac{\langle \mathbf{v},\mathbf{u}\rangle}{\langle \mathbf{u},\mathbf{u}\rangle}\mathbf{u} \right\|^2.$$

15. Let $\mathbf{u}$ and $\mathbf{v}$ be vectors in an inner product space $\mathscr{L}$ and let $\mathbf{u} \neq \mathbf{0}$. Show that

$$\left\| \frac{\langle \mathbf{v},\mathbf{u}\rangle}{\langle \mathbf{u},\mathbf{u}\rangle}\mathbf{u} \right\| = \frac{|\langle \mathbf{v},\mathbf{u}\rangle|}{\|\mathbf{u}\|}.$$

16. Let $\mathbf{u}$ and $\mathbf{v}$ be vectors in an inner product space $\mathscr{L}$. Show that

$$|\langle \mathbf{v},\mathbf{u}\rangle| \le \|\mathbf{u}\|\,\|\mathbf{v}\|.$$

17. Let $\mathscr{L}$ be an inner product space. Show that for all vectors $\mathbf{u}$ and $\mathbf{v}$ in $\mathscr{L}$ we have

$$\|\mathbf{u}+\mathbf{v}\| \le \|\mathbf{u}\| + \|\mathbf{v}\|.$$

18. Consider the inner product space $\mathscr{C}_{[-\pi,\pi]}$ with the inner product defined by $\langle f,g\rangle = \int_{-\pi}^{\pi} f(t)g(t)\,dt$. Show that the set of functions $\{1, \cos t, \dots, \cos nt, \sin t, \dots, \sin nt\}$ is an orthogonal set.

19. Consider the vector space $\mathscr{C}_{[0,1]}$ with the inner product $\langle f,g\rangle = \int_0^1 f(t)g(t)\,dt$. Calculate the best approximation of the function t^α, where α is a real number greater than -1, by functions from $\operatorname{Span}\{1, t\}$.

20. Consider the vector space $\mathscr{C}_{[0,1]}$ with the inner product $\langle f,g\rangle = \int_0^1 f(t)g(t)\,dt$. Calculate the best approximation of the function t^α,where α is a real number greater than 1, by functions from $\operatorname{Span}\{1, t, t^2\}$.

21. Consider the inner product space $\mathscr{C}_{[0,1]}$ with the inner product defined by $\langle f, g\rangle = \int_0^1 f(t)g(t)dt$. Show that the polynomial $x_0 + x_1 t + \cdots + x_k t^k$ is the best approximation of the function f by the elements of $\operatorname{Span}\{1,\dots,t^k\}$ if and only if the numbers $x_1, x_2, \dots, x_k$ are solutions of the system

$$\begin{cases} x_0 & + & \frac{x_1}{2} & + & \dots & + & \frac{x_k}{k+1} & = & \int_0^1 f(t)dt \\ \frac{x_0}{2} & + & \frac{x_1}{3} & + & \dots & + & \frac{x_k}{k+2} & = & \int_0^1 tf(t)dt \\ & \vdots & & \vdots & & \vdots & & & \vdots \\ \frac{x_0}{k+1} & + & \frac{x_1}{k+2} & + & \dots & + & \frac{x_k}{2k+1} & = & \int_0^1 t^k f(t)dt \end{cases}.$$

22. Consider the inner product space $\mathscr{C}_{[-1,1]}$ with the inner product defined by $\langle f, g\rangle = \int_{-1}^1 f(t)g(t)dt$ and suppose that the function $x_0+x_1t+x_2t^2+x_3t^3$ is the best approximation of the function f by the elements of $\operatorname{Span}\{1,t,t^2,t^3\}$. Determine real numbers a_j, b_j, c_j, d_j for $1 \le j \le 4$ such that the following equalities hold

$$\begin{cases} a_0x_0 & + & a_1x_1 & + & a_2x_2 & + & a_3x_3 & = & \int_{-1}^1 f(t)dt \\ b_0x_0 & + & b_1x_1 & + & b_2x_2 & + & b_3x_3 & = & \int_{-1}^1 tf(t)dt \\ c_0x_0 & + & c_1x_1 & + & c_2x_2 & + & c_3x_3 & = & \int_{-1}^1 t^2f(t)dt \\ d_0x_0 & + & d_1x_1 & + & d_2x_2 & + & d_3x_3 & = & \int_{-1}^1 t^3f(t)dt \end{cases}.$$

23. Consider the vector space $\mathscr{C}_{[-\pi,\pi]}$ with the inner product defined by $\langle f, g\rangle = \int_{-\pi}^{\pi} f(t)g(t)dt$. Calculate the best approximation of the function $\cos^3 t$ by functions from $\operatorname{Span}\{1, \cos t, \cos 2t, \cos 3t, \sin t, \sin 2t, \sin 3t\}$.

24. Consider the vector space $\mathscr{C}_{[-\pi,\pi]}$ with the inner product defined by $\langle f, g\rangle = \int_{-\pi}^{\pi} f(t)g(t)dt$. Calculate the best approximation of the function $\sin^3 t$ by functions from $\operatorname{Span}\{1, \cos t, \cos 2t, \cos 3t, \sin t, \sin 2t, \sin 3t\}$

25. Consider the vector space $\mathscr{C}_{[-\pi,\pi]}$ with the inner product defined by $\langle f, g\rangle = \int_{-\pi}^{\pi} f(t)g(t)dt$. Calculate the best approximation of the function t^2 by functions from $\operatorname{Span}\{1, \cos t, \cos 2t, \sin t, \sin 2t\}$.

26. Consider the vector space $\mathscr{C}_{[-\pi,\pi]}$ with the inner product defined by $\langle f, g\rangle = \int_{-\pi}^{\pi} f(t)g(t)dt$. Calculate the best approximation of the function t by functions from $\operatorname{Span}\{1, \cos t, \cos 2t, \cos 3t \cos 4t, \sin t, \sin 2t, \sin 3t, \sin 4t\}$.

27. Consider the vector space $\mathscr{C}_{[0,2\pi]}$ with the inner product defined by $\langle f, g\rangle = \int_0^{2\pi} f(t)g(t)dt$. Calculate the best approximation of the continuous function f defined by $f(t) = \pi$ for $0 \le t \le \pi$ and $f(t) = t$ for $\pi \le t \le 2\pi$ by functions from $\operatorname{Span}\{1, \cos t, \cos 2t, \cos 3t, \sin t, \sin 2t, \sin 3t\}$.

28. Consider the vector space $\mathscr{C}_{[0,2\pi]}$ with the inner product defined by $\langle f,g\rangle = \int_0^{2\pi} f(t)g(t)dt$. Calculate the best approximation of the continuous function f defined by $f(t) = t$ for $0 \le t \le \pi$ and $f(t) = \pi$ for $\pi \le t \le 2\pi$ by functions from $\operatorname{Span}\{1, \cos t, \cos 2t, \cos 3t, \sin t, \sin 2t, \sin 3t\}$.

29. Consider the vector space $\mathscr{C}_{[0,2\pi]}$ with the inner product defined by $\langle f,g\rangle = \int_0^{2\pi} f(t)g(t)dt$. Calculate the best approximation of the continuous function e^t by functions from $\operatorname{Span}\{1, \cos t, \sin t\}$.

30. Consider the vector space $\mathscr{C}_{[0,2\pi]}$ with the inner product defined by $\langle f,g\rangle = \int_0^{2\pi} f(t)g(t)dt$. Calculate the best approximation of the continuous function f defined by $f(t) = 0$ for $0 \le t \le \frac{3\pi}{2}$ and $f(t) = t - \frac{3\pi}{2}$ for $\frac{3\pi}{2} \le t \le 2\pi$ by functions from $\operatorname{Span}\{1, \cos t, \cos 2t, \sin t, \sin 2t\}$.

31. Consider the polynomials 1, t, and t^2 as functions in the inner product space $\mathscr{C}_{[0,1]}$ with the inner product defined by $\langle f,g\rangle = \int_0^1 f(t)g(t)dt$. Find polynomials q_1 and q_2 in $\mathscr{C}_{[0,1]}$ such that $\{1, q_1, q_2\}$ is an orthogonal set, $\operatorname{Span}\{1, q_1\} = \operatorname{Span}\{1, t\}$, and $\operatorname{Span}\{1, q_1, q_2\} = \operatorname{Span}\{1, t, t^2\}$.

32. Consider the polynomials 1, t, and t^2 as functions in the inner product space $\mathscr{C}_{[-1,1]}$ with the inner product defined by $\langle f,g\rangle = \int_{-1}^1 f(t)g(t)dt$. Find polynomials q_1 and q_2 in $\mathscr{C}_{[-1,1]}$ such that $\{1, q_1, q_2\}$ is an orthogonal set, $\operatorname{Span}\{1, q_1\} = \operatorname{Span}\{1, t\}$, and $\operatorname{Span}\{1, q_1, q_2\} = \operatorname{Span}\{1, t, t^2\}$.

Chapter 11

Examples of solutions with the aid of a computer algebra system

In this chapter we discuss some problems that are designed to be solved with the aid of a computer algebra system. We present solutions using Maple, but other systems can be used.

The purpose of this chapter is not only to explain how to facilitate calculations using a computer algebra system, but also to improve understanding of the standard questions and methods of linear algebra. In order to accomplish this goal we present solutions which are more relevant for understanding of the problem, but not necessarily the shortest solutions. We explain every step of the solution and we solve more problems for more difficult topics, like the subspaces of $\mathbb{R}^n$ or Jordan forms. For some problems we give two solutions. For most problems we also verify the result, which emphasizes what has been accomplished by the solution.

The reader does not need to have any previous experience with Maple. Studying our solutions should be enough to understand how Maple works and to try it on different problems.

In our solutions we made an attempt to illustrate the Maple code as it appears in Maple. The reader may notice some small differences due to differences between different versions of Maple and different operating systems. There is one deliberate change in our presentation. When multiple vectors are defined, in Maple they are displayed one vector per line. For example, if we type

$> p := \langle 7,4,11,0\rangle; q := \langle 1,2,3,10\rangle; r := \langle 1,1,2,3\rangle; s := \langle 2,1,3,-1\rangle$

then in Maple we get

$$p := \begin{bmatrix} 7 \\ 4 \\ 11 \\ 0 \end{bmatrix}$$

$$q := \begin{bmatrix} 1 \\ 2 \\ 3 \\ 10 \end{bmatrix}$$

$$r := \begin{bmatrix} 1 \\ 1 \\ 2 \\ 3 \end{bmatrix}$$

$$s := \begin{bmatrix} 2 \\ 1 \\ 3 \\ -1 \end{bmatrix}.$$

In our presentation, to save space, we list all vectors in a single line. In the example above we would have

$$p := \begin{bmatrix} 7 \\ 4 \\ 11 \\ 0 \end{bmatrix}, q := \begin{bmatrix} 1 \\ 2 \\ 3 \\ 10 \end{bmatrix}, r := \begin{bmatrix} 1 \\ 1 \\ 2 \\ 3 \end{bmatrix}, s := \begin{bmatrix} 2 \\ 1 \\ 3 \\ -1 \end{bmatrix}.$$

Since the tools used in the presented solutions require the LinearAlgebra package, we always include

> *with*(*LinearAlgebra*) :

at the top of our Maple worksheet.

Matrices

Problem 11.1. Solve the equation $\begin{bmatrix} 1 & 1 & 2 \\ 1 & 2 & 1 \\ 3 & 1 & 1 \\ 1 & 2 & 3 \end{bmatrix} X = \begin{bmatrix} 4 & 12 \\ 11 & 7 \\ 12 & 18 \\ 7 & 15 \end{bmatrix}$ and verify the result.

Solution.

> $with(LinearAlgebra):$

First we define the matrices.

> $A := \langle\langle 1,1,3,1\rangle|\langle 1,2,1,2\rangle|\langle 2,1,1,3\rangle\rangle; B := \langle\langle 4,11,12,7\rangle|\langle 12,7,18,15\rangle\rangle$

$$A := \begin{bmatrix} 1 & 1 & 2 \\ 1 & 2 & 1 \\ 3 & 1 & 1 \\ 1 & 2 & 3 \end{bmatrix}, B := \begin{bmatrix} 4 & 12 \\ 11 & 7 \\ 12 & 18 \\ 7 & 15 \end{bmatrix}.$$

Next we determine the matrix X.

> $X := LinearSolve(A, B)$

$$X := \begin{bmatrix} 3 & 5 \\ 5 & -1 \\ -2 & 4 \end{bmatrix}.$$

Now we verify the result.

> $A.X$

$$\begin{bmatrix} 4 & 12 \\ 11 & 7 \\ 12 & 18 \\ 7 & 15 \end{bmatrix}.$$

□

Problem 11.2. We consider the vectors $\mathbf{a} = \begin{bmatrix} 2 \\ 1 \\ 1 \\ 3 \\ -1 \end{bmatrix}$, $\mathbf{b} = \begin{bmatrix} 1 \\ 3 \\ 1 \\ 1 \\ 1 \end{bmatrix}$, $\mathbf{c} = \begin{bmatrix} 1 \\ 13 \\ 3 \\ -1 \\ 7 \end{bmatrix}$, $\mathbf{d} = \begin{bmatrix} 4 \\ 17 \\ 5 \\ 3 \\ 7 \end{bmatrix}$, and $\mathbf{e} = \begin{bmatrix} 5 \\ 5 \\ 3 \\ 7 \\ -1 \end{bmatrix}$. Solve the equation $x\mathbf{a} + y\mathbf{b} + z\mathbf{c} + w\mathbf{d} = \mathbf{e}$.

Solution.

> $with(LinearAlgebra):$

First we need to define the vectors.

> $a := \langle 2,1,1,3,-1\rangle; b := \langle 1,3,1,1,1\rangle; c := \langle 1,13,3,-1,7\rangle;$
$d := \langle 4,17,5,3,7\rangle; e := \langle 5,5,3,7,-1\rangle$

$$a := \begin{bmatrix} 2 \\ 1 \\ 1 \\ 3 \\ -1 \end{bmatrix}, b := \begin{bmatrix} 1 \\ 3 \\ 1 \\ 1 \\ 1 \end{bmatrix}, c := \begin{bmatrix} 1 \\ 13 \\ 3 \\ -1 \\ 7 \end{bmatrix}, d := \begin{bmatrix} 4 \\ 17 \\ 5 \\ 3 \\ 7 \end{bmatrix}, e := \begin{bmatrix} 5 \\ 5 \\ 3 \\ 7 \\ -1 \end{bmatrix}.$$

Now we solve the equation.

> $LinearSolve(\langle a,b,c,d\rangle, e)$

$$\begin{bmatrix} 2+2\_t_3+\_t_4 \\ 1-5\_t_3-6\_t_4 \\ \_t_3 \\ \_t_4 \end{bmatrix}.$$

This means that the general solution is

$$\begin{bmatrix} x \\ y \\ z \\ w \end{bmatrix} = \begin{bmatrix} 2 \\ 1 \\ 0 \\ 0 \end{bmatrix} + s\begin{bmatrix} 2 \\ -5 \\ 1 \\ 0 \end{bmatrix} + t\begin{bmatrix} 1 \\ -6 \\ 0 \\ 1 \end{bmatrix}.$$

□

The vector space $\mathbb{R}^n$

Problem 11.3. We consider the vectors $\mathbf{a} = \begin{bmatrix} 2 \\ 1 \\ 0 \\ 1 \end{bmatrix}$, $\mathbf{b} = \begin{bmatrix} 1 \\ 2 \\ 1 \\ 1 \end{bmatrix}$, $\mathbf{c} = \begin{bmatrix} 5 \\ 7 \\ 3 \\ 4 \end{bmatrix}$, and $\mathbf{d} = \begin{bmatrix} 12 \\ 9 \\ 2 \\ 7 \end{bmatrix}$. Show that $\text{Span}\{\mathbf{a},\mathbf{b}\} = \text{Span}\{\mathbf{c},\mathbf{d}\}$ and then find the change of basis matrix from the basis $\{\mathbf{a},\mathbf{b}\}$ to the basis $\{\mathbf{c},\mathbf{d}\}$. Find the coordinates of the vector $\mathbf{v} = 7\mathbf{a} - 3\mathbf{b}$ in the basis $\{\mathbf{c},\mathbf{d}\}$.

Solution.

> $with(LinearAlgebra):$

We start by defining the vectors.

> $a := \langle 2,1,0,1\rangle; b := \langle 1,2,1,1\rangle; c := \langle 5,7,3,4\rangle; d := \langle 12,9,2,7\rangle$

$$a := \begin{bmatrix} 2 \\ 1 \\ 0 \\ 1 \end{bmatrix}, b := \begin{bmatrix} 1 \\ 2 \\ 1 \\ 1 \end{bmatrix}, c := \begin{bmatrix} 5 \\ 7 \\ 3 \\ 4 \end{bmatrix}, d := \begin{bmatrix} 12 \\ 9 \\ 2 \\ 7 \end{bmatrix}.$$

In order to show that $\text{Span}\{\mathbf{a},\mathbf{b}\} = \text{Span}\{\mathbf{c},\mathbf{d}\}$ we show that the vectors **c** and **d** are in $\text{Span}\{\mathbf{a},\mathbf{b}\}$ and that the vectors **a** and **b** are in $\text{Span}\{\mathbf{c},\mathbf{d}\}$.

> $ReducedRowEchelonForm(\langle a|b|c|d\rangle)$

$$\begin{bmatrix} 1 & 0 & 1 & 5 \\ 0 & 1 & 3 & 2 \\ 0 & 0 & 0 & 0 \\ 0 & 0 & 0 & 0 \end{bmatrix}.$$

The above implies that the vectors **c** and **d** are in $\text{Span}\{\mathbf{a},\mathbf{b}\}$.

> *ReducedRowEchelonForm*$(\langle c|d|a|b\rangle)$

$$\begin{bmatrix} 1 & 0 & -\frac{2}{13} & \frac{5}{13} \\ 0 & 1 & \frac{3}{13} & -\frac{1}{13} \\ 0 & 0 & 0 & 0 \\ 0 & 0 & 0 & 0 \end{bmatrix}.$$

The above implies that the vectors **a** and **b** are in Span$\{\mathbf{c},\mathbf{d}\}$. Moreover, the change of basis matrix from the basis $\{\mathbf{a},\mathbf{b}\}$ to the basis $\{\mathbf{c},\mathbf{d}\}$ is $\begin{bmatrix} -\frac{2}{13} & \frac{5}{13} \\ \frac{3}{13} & -\frac{1}{13} \end{bmatrix}$.

Note that the same matrix can be obtained as follows.

> *LinearSolve*$(\langle c,d\rangle,\langle a,b\rangle)$

$$C := \begin{bmatrix} -\frac{2}{13} & \frac{5}{13} \\ \frac{3}{13} & -\frac{1}{13} \end{bmatrix}.$$

Now we find the coordinates of the vector $\mathbf{v} = 7\mathbf{a} - 3\mathbf{b}$ in the basis $\{\mathbf{c},\mathbf{d}\}$.

> *LinearSolve*$(\langle c,d\rangle, 7a-3b)$

$$\begin{bmatrix} -\frac{29}{13} \\ \frac{24}{13} \end{bmatrix}.$$

This means that $\mathbf{v} = 7\mathbf{a} - 3\mathbf{b} = -\frac{29}{13}\mathbf{c} + \frac{24}{13}\mathbf{d}$. The same result can be obtained by multiplying the vector **v** by the matrix C.

> $C.\langle 7,-3\rangle$

$$\begin{bmatrix} -\frac{29}{13} \\ \frac{24}{13} \end{bmatrix}.$$

□

Problem 11.4. For the vectors

$$\mathbf{p} = \begin{bmatrix} 7 \\ 4 \\ 11 \\ 0 \end{bmatrix}, \quad \mathbf{q} = \begin{bmatrix} 1 \\ 2 \\ 3 \\ 10 \end{bmatrix}, \quad \mathbf{r} = \begin{bmatrix} 1 \\ 1 \\ 2 \\ 3 \end{bmatrix}, \quad \mathbf{s} = \begin{bmatrix} 2 \\ 1 \\ 3 \\ -1 \end{bmatrix} \quad \text{and} \quad \mathbf{t} = \begin{bmatrix} 4 \\ 3 \\ 7 \\ 5 \end{bmatrix},$$

show that $\{\mathbf{p}, \mathbf{q}\}$ is a basis in $\operatorname{Span}\{\mathbf{r}, \mathbf{s}, \mathbf{t}\}$

Solution.

> $with(LinearAlgebra):$

We start by defining the vectors.

> $p := \langle 7,4,11,0\rangle; q := \langle 1,2,3,10\rangle; r := \langle 1,1,2,3\rangle;$
$s := \langle 2,1,3,-1\rangle; t := \langle 4,3,7,5\rangle$

$$p := \begin{bmatrix} 7 \\ 4 \\ 11 \\ 0 \end{bmatrix}, \; q := \begin{bmatrix} 1 \\ 2 \\ 3 \\ 10 \end{bmatrix}, \; r := \begin{bmatrix} 1 \\ 1 \\ 2 \\ 3 \end{bmatrix}, \; s := \begin{bmatrix} 2 \\ 1 \\ 3 \\ -1 \end{bmatrix}, \; t := \begin{bmatrix} 4 \\ 3 \\ 7 \\ 5 \end{bmatrix}.$$

Now we find the reduced row echelon form of the matrix with columns **p**, **q**, **r**, **s**, and **t**.

> $ReducedRowEchelonForm(\langle p, q, r, s, t\rangle)$

$$\begin{bmatrix} 1 & 0 & \frac{1}{10} & \frac{3}{10} & \frac{1}{2} \\ 0 & 1 & \frac{3}{10} & -\frac{1}{10} & \frac{1}{2} \\ 0 & 0 & 0 & 0 & 0 \\ 0 & 0 & 0 & 0 & 0 \end{bmatrix}.$$

Since the columns corresponding to the vectors **p** and **q** are the only pivot columns and $\dim(\operatorname{Span}\{\mathbf{r}, \mathbf{s}, \mathbf{t}\}) \geq 2$, the set $\{\mathbf{p}, \mathbf{q}\}$ is a basis in $\operatorname{Span}\{\mathbf{p}, \mathbf{q}, \mathbf{r}, \mathbf{s}, \mathbf{t}\} = \operatorname{Span}\{\mathbf{r}, \mathbf{s}, \mathbf{t}\}$.

□

Problem 11.5. Find real numbers p and q such the vector $\begin{bmatrix} 2 \\ 5 \\ p \\ q \end{bmatrix}$ is in Span$\{\mathbf{u}, \mathbf{v}\}$ where $\mathbf{u} = \begin{bmatrix} 4 \\ 2 \\ 1 \\ 3 \end{bmatrix}$ and $\mathbf{v} = \begin{bmatrix} 1 \\ 3 \\ 3 \\ 2 \end{bmatrix}$. Verify the result.

Solution.

> $with(LinearAlgebra):$

If $\begin{bmatrix} 2 \\ 5 \\ p \\ q \end{bmatrix} = s\begin{bmatrix} 4 \\ 2 \\ 1 \\ 3 \end{bmatrix} + t\begin{bmatrix} 1 \\ 3 \\ 3 \\ 2 \end{bmatrix}$ for some numbers s and t, then we must have $\begin{bmatrix} 2 \\ 5 \end{bmatrix} = s\begin{bmatrix} 4 \\ 2 \end{bmatrix} + t\begin{bmatrix} 1 \\ 3 \end{bmatrix}$ for the same numbers s and t. First we solve the second equation for s and t.

> $A := Matrix([[4,1],[2,3]])$

$$A := \begin{bmatrix} 4 & 1 \\ 2 & 3 \end{bmatrix}.$$

> $LinearSolve(A, Vector([2,5]))$

$$\begin{bmatrix} \frac{1}{10} \\ \frac{8}{5} \end{bmatrix}.$$

This means that $s = \frac{1}{10}$ and $t = \frac{8}{5}$. Now we can find numbers p and q.

> $u := Vector([4,2,1,3]);\ v := Vector([1,3,3,2])$

$$u := \begin{bmatrix} 4 \\ 2 \\ 1 \\ 3 \end{bmatrix},\ v := \begin{bmatrix} 1 \\ 3 \\ 3 \\ 2 \end{bmatrix}.$$

> $w := \frac{1}{10}u + \frac{8}{5}v$

$$w := \begin{bmatrix} 2 \\ 5 \\ \frac{49}{10} \\ \frac{7}{2} \end{bmatrix}.$$

Now we verify the result.

> *ReducedRowEchelonForm*$(\langle u, v, w\rangle)$

$$\begin{bmatrix} 1 & 0 & \frac{1}{10} \\ 0 & 1 & \frac{8}{5} \\ 0 & 0 & 0 \\ 0 & 0 & 0 \end{bmatrix}.$$

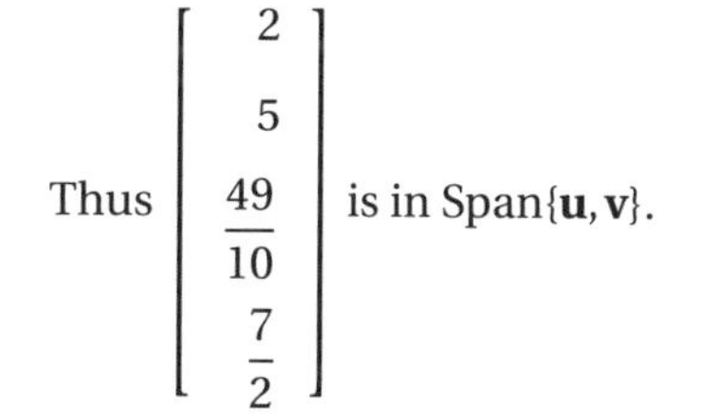

Thus $\begin{bmatrix} 2 \\ 5 \\ \frac{49}{10} \\ \frac{7}{2} \end{bmatrix}$ is in Span$\{\mathbf{u}, \mathbf{v}\}$.

□

Problem 11.6. We consider the vectors $\mathbf{a} = \begin{bmatrix} 2 \\ 3 \\ 1 \\ 1 \end{bmatrix}$, $\mathbf{b} = \begin{bmatrix} 1 \\ 1 \\ 3 \\ 1 \end{bmatrix}$, $\mathbf{c} = \begin{bmatrix} 8 \\ 13 \\ -1 \\ 3 \end{bmatrix}$, and $\mathbf{d} = \begin{bmatrix} 5 \\ 6 \\ 10 \\ 4 \end{bmatrix}$. Show that Span$\{\mathbf{a}, \mathbf{b}\}$ = Span$\{\mathbf{c}, \mathbf{d}\}$ by writing every linear combination of the vectors **a** and **b** as a linear combination of the vectors **c** and **d** and every linear combination of the vectors **c** and **d** as a linear combination of the vectors **a** and **b**.

Solution.

> *with*(*LinearAlgebra*) :

We start by defining the vectors.

$> a := \langle 2,3,1,1\rangle; b := \langle 1,1,3,1\rangle; c := \langle 8,13,-1,3\rangle; d := \langle 5,6,10,4\rangle$

$$a := \begin{bmatrix} 2 \\ 3 \\ 1 \\ 1 \end{bmatrix}, \ b := \begin{bmatrix} 1 \\ 1 \\ 3 \\ 1 \end{bmatrix}, \ c := \begin{bmatrix} 8 \\ 13 \\ -1 \\ 3 \end{bmatrix}, \ d := \begin{bmatrix} 5 \\ 6 \\ 10 \\ 4 \end{bmatrix}.$$

First we show that $x\mathbf{a} + y\mathbf{b}$ as a linear combination of the vectors $\mathbf{c}$ and $\mathbf{d}$ for arbitrary numbers x and y.

$> LinearSolve(\langle c|d\rangle, x \cdot a + y \cdot b)$

$$\begin{bmatrix} -\frac{1}{17}y + \frac{3}{17}x \\ \frac{5}{17}y + \frac{2}{17}x \end{bmatrix}.$$

Next we show that $p\mathbf{c} + q\mathbf{d}$ as a linear combination of the vectors $\mathbf{a}$ and $\mathbf{b}$ for arbitrary numbers p and q.

$> LinearSolve(\langle a|b\rangle, p \cdot c + q \cdot d)$

$$\begin{bmatrix} q + 5p \\ 3q - 2p \end{bmatrix}.$$

The above shows that

$$x\mathbf{a} + y\mathbf{b} = \left(\frac{3}{17}x - \frac{1}{17}y\right)\mathbf{c} + \left(\frac{2}{17}x + \frac{5}{17}y\right)\mathbf{d}$$

and

$$p\mathbf{c} + q\mathbf{d} = (5p + q)\mathbf{a} + (-2p + 3q)\mathbf{b}.$$

□

Problem 11.7. Consider the vectors $\mathbf{a} = \begin{bmatrix} 1 \\ 4 \\ -3 \\ 2 \\ -1 \end{bmatrix}, \mathbf{b} = \begin{bmatrix} 1 \\ 1 \\ 2 \\ 1 \\ 2 \end{bmatrix}, \mathbf{c} = \begin{bmatrix} 2 \\ -1 \\ -2 \\ 1 \\ 2 \end{bmatrix}, \mathbf{d} = \begin{bmatrix} 1 \\ 13 \\ -7 \\ 5 \\ -5 \end{bmatrix},$

and $\mathbf{e} = \begin{bmatrix} 7 \\ 1 \\ -9 \\ 5 \\ 5 \end{bmatrix}$. Show that the vectors **d** and **e** are in Span$\{\mathbf{a}, \mathbf{b}, \mathbf{c}\}$ and extend the set $\{\mathbf{d}, \mathbf{e}\}$ to a basis of Span$\{\mathbf{a}, \mathbf{b}, \mathbf{c}\}$.

Solution.

> $with(LinearAlgebra):$

First we define the vectors.

> $a := \langle 1, 4, -3, 2, -1 \rangle; b := \langle 1, 1, 2, 1, 2 \rangle; c := \langle 2, -1, -2, 1, 2 \rangle;$
$d := \langle 1, 13, -7, 5, -5 \rangle; e := \langle 7, 1, -9, 5, 5 \rangle$

$$a := \begin{bmatrix} 1 \\ 4 \\ -3 \\ 2 \\ -1 \end{bmatrix}, b := \begin{bmatrix} 1 \\ 1 \\ 2 \\ 1 \\ 2 \end{bmatrix}, c := \begin{bmatrix} 2 \\ -1 \\ -2 \\ 1 \\ 2 \end{bmatrix}, d := \begin{bmatrix} 1 \\ 13 \\ -7 \\ 5 \\ -5 \end{bmatrix}, e := \begin{bmatrix} 7 \\ 1 \\ -9 \\ 5 \\ 5 \end{bmatrix}.$$

To show that the vectors **d** and **e** are in Span$\{\mathbf{a}, \mathbf{b}, \mathbf{c}\}$ we find the reduced row echelon form of the matrix $\begin{bmatrix} \mathbf{a} & \mathbf{b} & \mathbf{c} & \mathbf{d} & \mathbf{e} \end{bmatrix}$.

> $ReducedRowEchelonForm(\langle a|b|c|d|e \rangle)$

$$\begin{bmatrix} 1 & 0 & 0 & 3 & 1 \\ 0 & 1 & 0 & 0 & 0 \\ 0 & 0 & 1 & -1 & 3 \\ 0 & 0 & 0 & 0 & 0 \\ 0 & 0 & 0 & 0 & 0 \end{bmatrix}.$$

Finally, to find a basis of Span$\{\mathbf{a}, \mathbf{b}, \mathbf{c}\}$ that contains vectors **d** and **e** we find the reduced row echelon form of the matrix $\begin{bmatrix} \mathbf{d} & \mathbf{e} & \mathbf{a} & \mathbf{b} & \mathbf{c} \end{bmatrix}$.

> *ReducedRowEchelonForm*$(\langle d|e|a|b|c\rangle)$

$$\begin{bmatrix} 1 & 0 & \frac{3}{10} & 0 & -\frac{1}{10} \\ 0 & 1 & \frac{1}{10} & 0 & \frac{3}{10} \\ 0 & 0 & 0 & 1 & 0 \\ 0 & 0 & 0 & 0 & 0 \\ 0 & 0 & 0 & 0 & 0 \end{bmatrix}.$$

Therefore we can take $\{\mathbf{d}, \mathbf{e}, \mathbf{b}\}$.

□

Problem 11.8. We consider the vectors $\mathbf{a} = \begin{bmatrix} 1 \\ 2 \\ 1 \\ 1 \end{bmatrix}$, $\mathbf{b} = \begin{bmatrix} 2 \\ 1 \\ 2 \\ 3 \end{bmatrix}$, $\mathbf{c} = \begin{bmatrix} 1 \\ -4 \\ 1 \\ 3 \end{bmatrix}$, and $\mathbf{d} = \begin{bmatrix} 1 \\ 1 \\ 1 \\ 2 \end{bmatrix}$. Show that Span$\{\mathbf{a}, \mathbf{b}, \mathbf{c}, \mathbf{d}\}$ has dimension 3 and find all subsets of 3 vectors from the set $\{\mathbf{a}, \mathbf{b}, \mathbf{c}, \mathbf{d}\}$ which are bases in Span$\{\mathbf{a}, \mathbf{b}, \mathbf{c}, \mathbf{d}\}$.

Solution.

> *with*(*LinearAlgebra*) :

First we define the vectors.

> $a := \langle 1,2,1,1\rangle; b := \langle 12,1,2,3\rangle; c := \langle 1,-4,1,3\rangle; d := \langle 1,1,1,2\rangle$

$$a := \begin{bmatrix} 1 \\ 2 \\ 1 \\ 1 \end{bmatrix}, b := \begin{bmatrix} 2 \\ 1 \\ 2 \\ 3 \end{bmatrix}, c := \begin{bmatrix} 1 \\ -4 \\ 1 \\ 3 \end{bmatrix}, d := \begin{bmatrix} 1 \\ 1 \\ 1 \\ 2 \end{bmatrix}.$$

Next we find the reduced row echelon form of the matrix $\begin{bmatrix} \mathbf{a} & \mathbf{b} & \mathbf{c} & \mathbf{d} \end{bmatrix}$.

> $ReducedRowEchelonForm(\langle a|b|c|d\rangle)$

$$\begin{bmatrix} 1 & 0 & -3 & 0 \\ 0 & 1 & 2 & 0 \\ 0 & 0 & 0 & 1 \\ 0 & 0 & 0 & 0 \end{bmatrix}.$$

This shows that Span$\{\mathbf{a}, \mathbf{b}, \mathbf{c}, \mathbf{d}\}$ has dimension 3.

Now we find the reduced row echelon form of each the matrices $\begin{bmatrix} \mathbf{a} & \mathbf{b} & \mathbf{c} \end{bmatrix}$, $\begin{bmatrix} \mathbf{a} & \mathbf{c} & \mathbf{d} \end{bmatrix}$, $\begin{bmatrix} \mathbf{b} & \mathbf{c} & \mathbf{d} \end{bmatrix}$, and $\begin{bmatrix} \mathbf{a} & \mathbf{b} & \mathbf{d} \end{bmatrix}$ to find out which combination of three vectors from the set $\{\mathbf{a}, \mathbf{b}, \mathbf{c}, \mathbf{d}\}$ spans a subspace of dimension 3.

> $ReducedRowEchelonForm(\langle a|b|c\rangle); ReducedRowEchelonForm(\langle a|c|d\rangle);$
$ReducedRowEchelonForm(\langle b|c|d\rangle); ReducedRowEchelonForm(\langle a|b|d\rangle)$

$$\begin{bmatrix} 1 & 0 & -3 \\ 0 & 1 & 2 \\ 0 & 0 & 0 \\ 0 & 0 & 0 \end{bmatrix}, \begin{bmatrix} 1 & 0 & 0 \\ 0 & 1 & 0 \\ 0 & 0 & 1 \\ 0 & 0 & 0 \end{bmatrix}, \begin{bmatrix} 1 & 0 & 0 \\ 0 & 1 & 0 \\ 0 & 0 & 1 \\ 0 & 0 & 0 \end{bmatrix}, \begin{bmatrix} 1 & 0 & 0 \\ 0 & 1 & 0 \\ 0 & 0 & 1 \\ 0 & 0 & 0 \end{bmatrix}.$$

Consequently, $\{\mathbf{a}, \mathbf{c}, \mathbf{d}\}$, $\{\mathbf{b}, \mathbf{c}, \mathbf{d}\}$, $\{\mathbf{a}, \mathbf{b}, \mathbf{d}\}$ are bases in Span$\{\mathbf{a}, \mathbf{b}, \mathbf{c}, \mathbf{d}\}$, but $\{\mathbf{a}, \mathbf{b}, \mathbf{c}\}$ is not.

□

Problem 11.9. Consider the vectors $\mathbf{a} = \begin{bmatrix} 1 \\ 3 \\ 1 \\ 1 \\ 2 \end{bmatrix}$, $\mathbf{b} = \begin{bmatrix} 1 \\ -7 \\ 2 \\ 3 \\ 1 \end{bmatrix}$, $\mathbf{c} = \begin{bmatrix} 1 \\ 2 \\ -1 \\ -3 \\ -5 \end{bmatrix}$, $\mathbf{d} = \begin{bmatrix} 2 \\ -1 \\ 2 \\ 2 \\ 1 \end{bmatrix}$, and $\mathbf{e} = \begin{bmatrix} 1 \\ 0 \\ 2 \\ 3 \\ 4 \end{bmatrix}$. Find a basis in Span$\{\mathbf{a}, \mathbf{b}, \mathbf{c}, \mathbf{d}, \mathbf{e}\}$ by using elementary row operations on the transpose of the matrix $\begin{bmatrix} 1 & 1 & 1 & 2 & 1 \\ 3 & -7 & 2 & -1 & 0 \\ 1 & 2 & -1 & 2 & 2 \\ 1 & 3 & -3 & 2 & 3 \\ 2 & 1 & -5 & 1 & 4 \end{bmatrix}$ and determine the coordinates of the vectors **a**, **b**, **c**, **d**, and **e** in that basis.

Solution.

> $with(LinearAlgebra):$

First we define the vectors, the matrix, and its transpose.

> $a := \langle 1,3,1,1,2\rangle; b := \langle 1,-7,2,3,1\rangle; c := \langle 1,2,-1,-3,-5\rangle;$
$d := \langle 2,-1,2,2,1\rangle; e := \langle 1,0,2,3,4\rangle$

$$a := \begin{bmatrix} 1 \\ 3 \\ 1 \\ 1 \\ 2 \end{bmatrix}, b := \begin{bmatrix} 1 \\ -7 \\ 2 \\ 3 \\ 1 \end{bmatrix}, c := \begin{bmatrix} 1 \\ 2 \\ -1 \\ -3 \\ -5 \end{bmatrix}, d := \begin{bmatrix} 1 \\ 13 \\ -7 \\ 5 \\ -5 \end{bmatrix}, e := \begin{bmatrix} 1 \\ 0 \\ 2 \\ 3 \\ 4 \end{bmatrix}.$$

> $M := \langle a|b|c|d|e\rangle$

$$M := \begin{bmatrix} 1 & 1 & 1 & 2 & 1 \\ 3 & -7 & 2 & -1 & 0 \\ 1 & 2 & -1 & 2 & 2 \\ 1 & 3 & -3 & 2 & 3 \\ 2 & 1 & -5 & 1 & 4 \end{bmatrix}.$$

> $P := Transpose(M)$

$$P := \begin{bmatrix} 1 & 3 & 1 & 1 & 2 \\ 1 & -7 & 2 & 3 & 1 \\ 1 & 2 & -1 & -3 & -5 \\ 2 & -1 & 2 & 2 & 1 \\ 1 & 0 & 2 & 3 & 4 \end{bmatrix}.$$

Now we find the reduced row echelon form of the matrix P.

> $Q := ReducedRowEchelonForm(P)$

$$Q := \begin{bmatrix} 1 & 0 & 0 & -1 & -\frac{18}{7} \\ 0 & 1 & 0 & 0 & \frac{3}{7} \\ 0 & 0 & 1 & 2 & \frac{23}{7} \\ 0 & 0 & 0 & 0 & 0 \\ 0 & 0 & 0 & 0 & 0 \end{bmatrix}.$$

The dimension of the subspace Span$\{\mathbf{a},\mathbf{b},\mathbf{c},\mathbf{d},\mathbf{e}\}$ is 3. We can use the transposed rows of the matrix Q to obtain a basis of Span$\{\mathbf{a},\mathbf{b},\mathbf{c},\mathbf{d},\mathbf{e}\}$. First we define the vectors:

> $f := 7 \cdot Column(Transpose(Q), [1]); g := 7 \cdot Column(Transpose(Q), [2]); h := 7 \cdot Column(Transpose(Q), [3])$

$$f := \begin{bmatrix} 7 \\ 0 \\ 0 \\ -7 \\ -18 \end{bmatrix}, g := \begin{bmatrix} 0 \\ 7 \\ 0 \\ 0 \\ 3 \end{bmatrix}, h := \begin{bmatrix} 0 \\ 0 \\ 7 \\ 14 \\ 23 \end{bmatrix}.$$

Span$\{\mathbf{f},\mathbf{g},\mathbf{h}\}$ is a basis of Span$\{\mathbf{a},\mathbf{b},\mathbf{c},\mathbf{d},\mathbf{e}\}$. Now we find the coordinates of vectors **a**, **b**, **c**, **d**, and **e** in this basis.

> $LinearSolve(\langle f,g,h\rangle\rangle, a), LinearSolve(\langle f,g,h\rangle\rangle, b), LinearSolve(\langle f,g,h\rangle\rangle, c), LinearSolve(\langle f,g,h\rangle\rangle, d), LinearSolve(\langle f,g,h\rangle\rangle, e)$

$$\begin{bmatrix} \frac{1}{7} \\ \frac{3}{7} \\ \frac{1}{7} \end{bmatrix}, \begin{bmatrix} \frac{1}{7} \\ -1 \\ \frac{2}{7} \end{bmatrix}, \begin{bmatrix} \frac{1}{7} \\ \frac{2}{7} \\ -\frac{1}{7} \end{bmatrix}, \begin{bmatrix} \frac{2}{7} \\ -\frac{1}{7} \\ \frac{2}{7} \end{bmatrix}, \begin{bmatrix} \frac{1}{7} \\ 0 \\ \frac{2}{7} \end{bmatrix}.$$

The column vectors give us the desired coordinates. For example, we have $\mathbf{a} = \frac{1}{7}\mathbf{f} + \frac{3}{7}\mathbf{g} + \frac{1}{7}\mathbf{h}$.

□

Problem 11.10. Consider the vectors $\mathbf{a} = \begin{bmatrix} 2 \\ -1 \\ 3 \end{bmatrix}$, $\mathbf{b} = \begin{bmatrix} 1 \\ 1 \\ 1 \end{bmatrix}$, $\mathbf{c} = \begin{bmatrix} 1 \\ 2 \\ 2 \end{bmatrix}$, $\mathbf{d} = \begin{bmatrix} 3 \\ 1 \\ 1 \end{bmatrix}$, $\mathbf{e} = \begin{bmatrix} 5 \\ 0 \\ 1 \end{bmatrix}$, $\mathbf{f} = \begin{bmatrix} 1 \\ -1 \\ 4 \end{bmatrix}$. Show that the sets $\{\mathbf{a}, \mathbf{b}, \mathbf{c}\}$ and $\{\mathbf{d}, \mathbf{e}, \mathbf{f}\}$ are bases of $\mathbb{R}^3$ and find the change of basis matrix from the basis $\{\mathbf{a}, \mathbf{b}, \mathbf{c}\}$ to the basis $\{\mathbf{d}, \mathbf{e}, \mathbf{f}\}$. Find the coordinates of the vector $x\mathbf{a} + y\mathbf{b} + z\mathbf{c}$ in the basis $\{\mathbf{d}, \mathbf{e}, \mathbf{f}\}$ and verify the result using change of basis matrix from the basis $\{\mathbf{a}, \mathbf{b}, \mathbf{c}\}$ to the basis $\{\mathbf{d}, \mathbf{e}, \mathbf{f}\}$.

Solution.

> $with(LinearAlgebra):$

First we define the vectors.

> $a := \langle 2, -1, 3\rangle; b := \langle 1, 1, 1\rangle; c := \langle 1, 2, 2\rangle;$
> $d := \langle 3, 1, 1\rangle; e := \langle 5, 0, 1\rangle; f := \langle 1, -1, 4\rangle$

$$a := \begin{bmatrix} 2 \\ -1 \\ 3 \end{bmatrix}, \ b := \begin{bmatrix} 1 \\ 1 \\ 1 \end{bmatrix}, \ c := \begin{bmatrix} 1 \\ 2 \\ 2 \end{bmatrix},$$

$$d := \begin{bmatrix} 3 \\ 1 \\ 1 \end{bmatrix}, \ e := \begin{bmatrix} 5 \\ 0 \\ 1 \end{bmatrix}, \ f := \begin{bmatrix} 1 \\ -1 \\ 4 \end{bmatrix}.$$

Now we find the reduced row echelon form of the matrices $\begin{bmatrix} \mathbf{a} & \mathbf{b} & \mathbf{c} \end{bmatrix}$ and $\begin{bmatrix} \mathbf{d} & \mathbf{e} & \mathbf{f} \end{bmatrix}$ to determine if the vectors are linearly independent.

> $ReducedRowEchelonForm(\langle a|b|c\rangle); ReducedRowEchelonForm(\langle d|e|f\rangle)$

$$\begin{bmatrix} 1 & 0 & 0 \\ 0 & 1 & 0 \\ 0 & 0 & 1 \end{bmatrix}, \begin{bmatrix} 1 & 0 & 0 \\ 0 & 1 & 0 \\ 0 & 0 & 1 \end{bmatrix}.$$

Thus the sets $\{\mathbf{a}, \mathbf{b}, \mathbf{c}\}$ and $\{\mathbf{d}, \mathbf{e}, \mathbf{f}\}$ are bases of $\mathbb{R}^3$.

Next we find the change of basis matrix from the basis $\{\mathbf{a}, \mathbf{b}, \mathbf{c}\}$ to the basis $\{\mathbf{d}, \mathbf{e}, \mathbf{f}\}$.

> $C := LinearSolve(\langle d, e, f\rangle, \langle a, b, c\rangle)$

$$\begin{bmatrix} -\frac{2}{7} & \frac{23}{21} & \frac{47}{21} \\ \frac{3}{7} & -\frac{10}{21} & -\frac{25}{21} \\ \frac{5}{7} & 2/21 & \frac{5}{21} \end{bmatrix}.$$

Now we find the coordinates of the vector $x\mathbf{a} + y\mathbf{b} + z\mathbf{c}$ in the basis $\{\mathbf{d}, \mathbf{e}, \mathbf{f}\}$.

> $LinearSolve(\langle d, e, f\rangle, x \cdot a + y \cdot b + z \cdot c)$

$$\begin{bmatrix} \frac{47}{21}z + \frac{23}{21}y - \frac{2}{7}x \\ -\frac{25}{21}z - \frac{10}{21}y + \frac{3}{7}x \\ \frac{5}{21}z + \frac{2}{21}y + \frac{5}{7}x \end{bmatrix}.$$

This means that

$$x\mathbf{a}+y\mathbf{b}+z\mathbf{c} = \left(\frac{47}{21}z + \frac{23}{21}y - \frac{2}{7}x\right)\mathbf{d} + \left(-\frac{25}{21}z - \frac{10}{21}y + \frac{3}{7}x\right)\mathbf{e} + \left(\frac{5}{21}z + \frac{2}{21}y + \frac{5}{7}x\right)\mathbf{f}.$$

We verify this result by calculating the product $C\begin{bmatrix} x \\ y \\ z \end{bmatrix}$.

> $C.\langle x, y, z\rangle$

$$\begin{bmatrix} \frac{47}{21}z + \frac{23}{21}y - \frac{2}{7}x \\ -\frac{25}{21}z - \frac{10}{21}y + \frac{3}{7}x \\ \frac{5}{21}z + \frac{2}{21}y + \frac{5}{7}x \end{bmatrix}.$$

□

Problem 11.11. Find a basis of the vector space $N(A)$ for $A = \begin{bmatrix} 1 & 3 & 4 & 1 \\ 2 & 1 & 3 & 4 \end{bmatrix}$.

Solution.

> $with(LinearAlgebra):$

First we define the matrix A.

> $A := Matrix([[1,3,4,1],[2,1,3,4]])$

$$A := \begin{bmatrix} 1 & 3 & 4 & 1 \\ 2 & 1 & 3 & 4 \end{bmatrix}.$$

Now we solve the equation $AX = \begin{bmatrix} 0 \\ 0 \end{bmatrix}$.

> $LinearSolve(A, \langle 0,0 \rangle)$

$$\begin{bmatrix} -\frac{11}{2}\_t_2 - \frac{13}{2}\_t_3 \\ \_t_2 \\ \_t_3 \\ \frac{5}{2}\_t_2 + \frac{5}{2}\_t_3 \end{bmatrix}.$$

This means that $N(A) = \operatorname{Span}\left\{ \begin{bmatrix} -\frac{11}{2} \\ 1 \\ 0 \\ \frac{5}{2} \end{bmatrix}, \begin{bmatrix} -\frac{13}{2} \\ 0 \\ 1 \\ \frac{5}{2} \end{bmatrix} \right\}$. Note that we can also write $N(A) = \operatorname{Span}\left\{ \begin{bmatrix} 11 \\ -2 \\ 0 \\ -5 \end{bmatrix}, \begin{bmatrix} 13 \\ 0 \\ -2 \\ -5 \end{bmatrix} \right\}$.

□

Problem 11.12. Show that the vectors $\mathbf{a} = \begin{bmatrix} 1 \\ 1 \\ 2 \\ 3 \end{bmatrix}$, $\mathbf{b} = \begin{bmatrix} 2 \\ 1 \\ 2 \\ 3 \end{bmatrix}$, and $\mathbf{c} = \begin{bmatrix} 1 \\ 2 \\ 2 \\ 3 \end{bmatrix}$ are linearly independent and find a vector $\mathbf{d}$ such that the set $\{\mathbf{a}, \mathbf{b}, \mathbf{c}, \mathbf{d}\}$ is a basis in $\mathbb{R}^4$.

Solution.

> $with(LinearAlgebra):$

First we define the vectors.

> $a := \langle 1,1,2,3\rangle; b := \langle 2,1,2,3\rangle; c := \langle 1,2,2,3\rangle$

$$a := \begin{bmatrix} 1 \\ 1 \\ 2 \\ 3 \end{bmatrix}, b := \begin{bmatrix} 2 \\ 1 \\ 2 \\ 3 \end{bmatrix}, c := \begin{bmatrix} 1 \\ 2 \\ 2 \\ 3 \end{bmatrix}.$$

Now we find the reduced row echelon form of the matrix $\begin{bmatrix} \mathbf{a} & \mathbf{b} & \mathbf{c} \end{bmatrix}$.

> $ReducedRowEchelonForm(\langle a, b, c\rangle)$

$$\begin{bmatrix} 1 & 0 & 0 \\ 0 & 1 & 0 \\ 0 & 0 & 1 \\ 0 & 0 & 0 \end{bmatrix}.$$

This proves that the vectors **a**, **b**, and **c** are linearly independent. To find the vector **d** we find the reduced row echelon form of the matrix $\begin{bmatrix} \mathbf{a} & \mathbf{b} & \mathbf{c} & \mathbf{e}_1 & \mathbf{e}_2 & \mathbf{e}_3 & \mathbf{e}_4 \end{bmatrix}$.

> $ReducedRowEchelonForm(\langle a|b|c|\langle 1,0,0,0\rangle|\langle 0,1,0,0\rangle|\langle 0,0,1,0\rangle|\langle 0,0,0,1\rangle\rangle)$

$$\begin{bmatrix} 1 & 0 & 0 & -1 & -1 & 0 & 1 \\ 0 & 1 & 0 & 1 & 0 & 0 & -\frac{1}{3} \\ 0 & 0 & 1 & 0 & 1 & 0 & -\frac{1}{3} \\ 0 & 0 & 0 & 0 & 0 & 1 & -\frac{2}{3} \end{bmatrix}.$$

Consequently, we can take $\mathbf{d} = \mathbf{e}_3 = \begin{bmatrix} 0 \\ 0 \\ 1 \\ 0 \end{bmatrix}$.

□

Orthogonality in $\mathbb{R}^n$

Problem 11.13. Show that the vectors $\mathbf{a} = \begin{bmatrix} 5 \\ 2 \\ 7 \\ 1 \end{bmatrix}$ and $\mathbf{b} = \begin{bmatrix} 2 \\ 1 \\ -1 \\ -5 \end{bmatrix}$ are orthogonal and find vectors $\mathbf{c}$ and $\mathbf{d}$ such that the set $\{\mathbf{a}, \mathbf{b}, \mathbf{c}, \mathbf{d}\}$ is an orthogonal basis in $\mathbb{R}^4$. Verify the result.

Solution.

> $with(LinearAlgebra):$

First we define vectors **a** and **b**.

> $a := \langle 5, 2, 7, 1 \rangle; b := \langle 2, 1, -1, -5 \rangle$

$$a := \begin{bmatrix} 5 \\ 2 \\ 7 \\ 1 \end{bmatrix}, \; b := \begin{bmatrix} 2 \\ 1 \\ -1 \\ -5 \end{bmatrix}.$$

To check that the vectors are orthogonal we use the dot product.

> $DotProduct(a, b)$

$$0.$$

To find vectors **c** and **d** we use the Gram-Schmidt method.

> $Gram\text{-}Schmidt([a, b, \langle 1, 0, 0, 0 \rangle, \langle 0, 1, 0, 0 \rangle, \langle 0, 0, 1, 0 \rangle, \langle 0, 0, 0, 1 \rangle])$

$$\left[\begin{bmatrix} 5 \\ 2 \\ 7 \\ 1 \end{bmatrix}, \begin{bmatrix} 2 \\ 1 \\ -1 \\ -5 \end{bmatrix}, \begin{bmatrix} \frac{1358}{2449} \\ -\frac{468}{2449} \\ -\frac{927}{2449} \\ \frac{635}{2449} \end{bmatrix}, \begin{bmatrix} 0 \\ \frac{578}{679} \\ -\frac{187}{679} \\ \frac{153}{679} \end{bmatrix} \right].$$

Consequently, we can use $\mathbf{c} = \begin{bmatrix} 1358 \\ -468 \\ -927 \\ 635 \end{bmatrix}$ and $\mathbf{d} = \begin{bmatrix} 0 \\ 578 \\ -187 \\ 153 \end{bmatrix}$. To verify the result we first define vectors **c** and **d**.

> $c := \langle 1358, -468, -927, 635 \rangle; d := \langle 0, 578, -187, 153 \rangle$

$$c := \begin{bmatrix} 1358 \\ -468 \\ -927 \\ 635 \end{bmatrix}, d := \begin{bmatrix} 0 \\ 578 \\ -187 \\ 153 \end{bmatrix}.$$

To check that the vectors are orthogonal we use the dot product.

> $DotProduct(a, c), DotProduct(a, d), DotProduct(b, c), DotProduct(b, d),$
> $DotProduct(c, d)$

$$0, 0, 0, 0, 0.$$

□

Problem 11.14. Verify that $(\mathbf{N}(A))^{\perp} = \mathbf{C}(A^T)$ for the matrix $A = \begin{bmatrix} 4 & 1 & 2 & -1 \\ 7 & 4 & 5 & 2 \\ 1 & 1 & 1 & 1 \end{bmatrix}$.

Solution.

> $with(LinearAlgebra):$

First we define the matrix A.

> $A := Matrix([[4, 1, 2, -1], [7, 4, 5, 2], [1, 1, 1, 1]])$

$$A := \begin{bmatrix} 4 & 1 & 2 & -1 \\ 7 & 4 & 5 & 2 \\ 1 & 1 & 1 & 1 \end{bmatrix}.$$

To find $(\mathbf{N}(A))^{\perp}$, we first find $\mathbf{N}(A)$.

> $LinearSolve(A, Vector([0, 0, 0]))$

$$\begin{bmatrix} \frac{\_t0_2}{2} + \frac{3\_t0_4}{2} \\ \_t0_2 \\ -\frac{3\_t0_2}{2} - \frac{5\_t0_4}{2} \\ \_t0_4 \end{bmatrix}.$$

This means that $\mathbf{N}(A) = \text{Span}\left\{\begin{bmatrix} 1 \\ 2 \\ -3 \\ 0 \end{bmatrix}, \begin{bmatrix} 3 \\ 0 \\ -5 \\ 2 \end{bmatrix}\right\}$. For the purpose of further calculations we define a matrix B that has these two vectors as columns.

> $B := Matrix([[1,3],[2,0],[-3,-5],[0,2]])$

$$B := \begin{bmatrix} 1 & 3 \\ 2 & 0 \\ -3 & -5 \\ 0 & 2 \end{bmatrix}.$$

Now we find $(\mathbf{N}(A))^{\perp}$.

> $LinearSolve(Transpose(B), Vector([0,0]))$

$$\begin{bmatrix} -2\_t1_2 + 3\_t1_3 \\ \_t1_2 \\ \_t1_3 \\ 3\_t1_2 - 2\_t1_3 \end{bmatrix}.$$

This means that $(\mathbf{N}(A))^{\perp} = \text{Span}\left\{\begin{bmatrix} -2 \\ 1 \\ 0 \\ 3 \end{bmatrix}, \begin{bmatrix} 3 \\ 0 \\ 1 \\ -2 \end{bmatrix}\right\}$. For the purpose of further calculations we define a matrix C that has these two vectors as columns.

> $C := Matrix([[-2,3],[1,0],[0,1],[3,-2]])$

$$C := \begin{bmatrix} -2 & 3 \\ 1 & 0 \\ 0 & 1 \\ 3 & -2 \end{bmatrix}.$$

Now we show that $(\mathbf{N}(A))^{\perp} = \mathbf{C}(A^T)$.

> $E := Transpose(A)$

$$E := \begin{bmatrix} 4 & 7 & 1 \\ 1 & 4 & 1 \\ 2 & 5 & 1 \\ -1 & 2 & 1 \end{bmatrix}.$$

> $ReducedRowEchelonForm(\langle C|E\rangle)$

$$\begin{bmatrix} 1 & 0 & 1 & 4 & 1 \\ 0 & 1 & 2 & 5 & 1 \\ 0 & 0 & 0 & 0 & 0 \\ 0 & 0 & 0 & 0 & 0 \end{bmatrix}.$$

This means that $\mathbf{C}(A^T)$ is a subspace of $(\mathbf{N}(A))^{\perp}$.

> $ReducedRowEchelonForm(\langle E|C\rangle)$

$$\begin{bmatrix} 1 & 0 & -\frac{1}{3} & -\frac{5}{3} & \frac{4}{3} \\ 0 & 1 & \frac{1}{3} & \frac{2}{3} & -\frac{1}{3} \\ 0 & 0 & 0 & 0 & 0 \\ 0 & 0 & 0 & 0 & 0 \end{bmatrix}.$$

This means that $(\mathbf{N}(A))^{\perp}$ is a subspace of $\mathbf{C}(A^T)$.

□

Problem 11.15. Let $\mathbf{a} = \begin{bmatrix} 1 \\ 1 \\ 2 \\ -1 \end{bmatrix}$, $\mathbf{b} = \begin{bmatrix} 2 \\ 1 \\ 0 \\ 1 \end{bmatrix}$, and $\mathbf{c} = \begin{bmatrix} 1 \\ 2 \\ 1 \\ 1 \end{bmatrix}$. Find an orthonormal basis in Span$\{\mathbf{a}, \mathbf{b}, \mathbf{c}\}$ and then use it to find the QR decomposition of the matrix $A = \begin{bmatrix} \mathbf{a} & \mathbf{b} & \mathbf{c} \end{bmatrix}$. Verify the result.

Solution.

> $with(LinearAlgebra):$

First we define the vectors **a**, **b**, and **c**, and then use the Gram-Schmidt method to find an orthogonal basis in Span$\{\mathbf{a}, \mathbf{b}, \mathbf{c}\}$.

> $a := \langle 1,1,2,-1\rangle; b := \langle 2,1,0,1\rangle; c := \langle 1,2,1,1\rangle$

$$a := \begin{bmatrix} 1 \\ 1 \\ 2 \\ -1 \end{bmatrix}, b := \begin{bmatrix} 2 \\ 1 \\ 0 \\ 1 \end{bmatrix}, c := \begin{bmatrix} 1 \\ 2 \\ 1 \\ 1 \end{bmatrix}.$$

> $Gram\text{-}Schmidt([a, b, c])$

$$\left[\begin{bmatrix} 1 \\ 1 \\ 2 \\ -1 \end{bmatrix}, \begin{bmatrix} \frac{12}{7} \\ \frac{5}{7} \\ -\frac{4}{7} \\ \frac{9}{7} \end{bmatrix}, \begin{bmatrix} -\frac{15}{19} \\ \frac{35}{38} \\ \frac{5}{19} \\ \frac{25}{38} \end{bmatrix}\right].$$

Now we normalize this orthogonal basis.

> $B := \langle Normalize(a, 2)|Normalize(\langle 12, 5, -4, 9\rangle, 2)|Normalize(\langle -30, 35, 10, 25\rangle, 2)\rangle$

$$B := \begin{bmatrix} \frac{1}{7}\sqrt{7} & \frac{6}{133}\sqrt{266} & -\frac{1}{19}\sqrt{114} \\ \frac{1}{7}\sqrt{7} & \frac{5}{266}\sqrt{266} & \frac{7}{114}\sqrt{114} \\ \frac{2}{7}\sqrt{7} & -\frac{2}{133}\sqrt{266} & \frac{1}{57}\sqrt{114} \\ -\frac{1}{7}\sqrt{7} & \frac{9}{266}\sqrt{266} & \frac{5}{114}\sqrt{114} \end{bmatrix}.$$

To obtain the QR decomposition we solve the equation $B\mathbf{x} = A$.

> $X := LinearSolve(B, \langle a|b|c\rangle)$

$$\begin{bmatrix} \sqrt{7} & \frac{2}{7}\sqrt{7} & \frac{4}{7}\sqrt{7} \\ 0 & \frac{1}{7}\sqrt{266} & \frac{27}{266}\sqrt{266} \\ 0 & 0 & \frac{5}{38}\sqrt{114} \end{bmatrix}.$$

Finally we verify the result.

> $QRDecomposition(\langle a|b|c\rangle)$

$$\begin{bmatrix} \frac{1}{7}\sqrt{7} & \frac{6}{133}\sqrt{266} & -\frac{1}{19}\sqrt{114} \\ \frac{1}{7}\sqrt{7} & \frac{5}{266}\sqrt{266} & \frac{7}{114}\sqrt{114} \\ \frac{2}{7}\sqrt{7} & -\frac{2}{133}\sqrt{266} & \frac{1}{57}\sqrt{114} \\ -\frac{1}{7}\sqrt{7} & \frac{9}{266}\sqrt{266} & \frac{5}{114}\sqrt{114} \end{bmatrix}, \begin{bmatrix} \sqrt{7} & \frac{2}{7}\sqrt{7} & \frac{4}{7}\sqrt{7} \\ 0 & \frac{1}{7}\sqrt{266} & \frac{27}{266}\sqrt{266} \\ 0 & 0 & \frac{5}{38}\sqrt{114} \end{bmatrix}.$$

□

Problem 11.16. Find the projection **p** of the vector **z** on Span{**x**, **y**} and the coordinates of **p** in the basis {**x**, **y**}, where

$$\mathbf{x} = \begin{bmatrix} 3 \\ 1 \\ 5 \\ 2 \end{bmatrix}, \quad \mathbf{y} = \begin{bmatrix} 2 \\ 3 \\ 1 \\ 3 \end{bmatrix}, \quad \text{and} \quad \mathbf{z} = \begin{bmatrix} 5 \\ -1 \\ 0 \\ 1 \end{bmatrix}.$$

Verify that $(\mathbf{z}-\mathbf{p})\cdot\mathbf{x} = 0$ and $(\mathbf{z}-\mathbf{p})\cdot\mathbf{y} = 0$.

Solution.

> $with(LinearAlgebra):$

First we define the vectors **x**, **y**, and **z**.

> $x := \langle 3,1,5,2\rangle; y := \langle 2,3,1,3\rangle; z := \langle 5,-1,0,1\rangle$

$$x := \begin{bmatrix} 3 \\ 1 \\ 5 \\ 2 \end{bmatrix}, y := \begin{bmatrix} 2 \\ 3 \\ 1 \\ 3 \end{bmatrix}, z := \begin{bmatrix} 5 \\ -1 \\ 0 \\ 1 \end{bmatrix}.$$

Now we find the projection matrix on Span{**x**, **y**}.

> $PM := ProjectionMatrix([x, y])$

$$PM := \begin{bmatrix} \frac{123}{497} & \frac{83}{497} & \frac{163}{497} & \frac{16}{71} \\ \frac{83}{497} & \frac{254}{497} & -\frac{88}{497} & \frac{31}{71} \\ \frac{163}{497} & -\frac{88}{497} & \frac{414}{497} & \frac{1}{71} \\ \frac{16}{71} & \frac{31}{71} & \frac{1}{71} & \frac{29}{71} \end{bmatrix}.$$

Then we find the projection $\mathbf{p}$ by multiplying the vector $\mathbf{z}$ by the projection matrix PM.

> $PM.z$

$$\begin{bmatrix} \frac{92}{71} \\ \frac{54}{71} \\ \frac{130}{71} \\ \frac{78}{71} \end{bmatrix}.$$

To find the coordinates of $\mathbf{p}$ in the basis $\{\mathbf{x}, \mathbf{y}\}$ we solve the equation $\begin{bmatrix} \mathbf{x} & \mathbf{y} \end{bmatrix} \begin{bmatrix} a \\ b \end{bmatrix} = \mathbf{p}$.

> $LinearSolve(\langle x|y\rangle, PM.z)$

$$\begin{bmatrix} \frac{24}{71} \\ \frac{10}{71} \end{bmatrix}.$$

This means that $\mathbf{p} = \frac{24}{71}\mathbf{x} + \frac{10}{71}\mathbf{y}$.

Finally we verify that $(\mathbf{z}-\mathbf{p})\cdot\mathbf{x} = (\mathbf{z}-\mathbf{p})^T\mathbf{x}$ and $(\mathbf{z}-\mathbf{p})\cdot\mathbf{y} = (\mathbf{z}-\mathbf{p})^T\mathbf{y}$.

> $Transpose(z - PM.z).x$

$$0.$$

> $Transpose(z - PM.z).y$

$$0.$$

□

Problem 11.17. We consider the vectors $\mathbf{a} = \begin{bmatrix} 1 \\ 2 \\ 1 \\ 1 \end{bmatrix}$, $\mathbf{b} = \begin{bmatrix} 7 \\ 4 \\ 11 \\ 0 \end{bmatrix}$, $\mathbf{c} = \begin{bmatrix} 1 \\ 2 \\ 3 \\ 10 \end{bmatrix}$, and

$\mathbf{d} = \begin{bmatrix} 0 \\ 20 \\ -6 \\ 23 \end{bmatrix}$. Show, using the Gram-Schmidt process, that the vector **d** is in Span{**a**, **b**, **c**}. Verify the result by solving the equation $x\mathbf{a} + y\mathbf{b} + z\mathbf{c} = \mathbf{d}$.

Solution.

> $with(LinearAlgebra):$

First we need to define the vectors.

> $a := \langle 1,2,1,1\rangle; b := \langle 7,4,11,0\rangle; c := \langle 1,2,3,10\rangle; d := \langle 0,20,-6,23\rangle$

$$a := \begin{bmatrix} 1 \\ 2 \\ 1 \\ 1 \end{bmatrix}, \; b := \begin{bmatrix} 7 \\ 4 \\ 11 \\ 0 \end{bmatrix}, \; c := \begin{bmatrix} 1 \\ 2 \\ 3 \\ 10 \end{bmatrix}, \; d := \begin{bmatrix} 0 \\ 20 \\ -6 \\ 23 \end{bmatrix}.$$

Now we apply the Gram-Schmidt process to the vectors **a**, **b**, and **c**.

> $Gram\text{-}Schmidt([a,b,c])$

$$\left[\begin{bmatrix} 1 \\ 2 \\ 1 \\ 1 \end{bmatrix}, \begin{bmatrix} \frac{23}{7} \\ -\frac{24}{7} \\ \frac{51}{7} \\ -\frac{26}{7} \end{bmatrix}, \begin{bmatrix} -\frac{275}{313} \\ -\frac{1210}{313} \\ \frac{615}{313} \\ \frac{2080}{313} \end{bmatrix} \right].$$

Next we apply the Gram-Schmidt process to the vectors **a**, **b**, **c**, and **d**.

> $Gram\text{-}Schmidt([a,b,c,d])$

$$\left[\begin{bmatrix} 1 \\ 2 \\ 1 \\ 1 \end{bmatrix}, \begin{bmatrix} \frac{23}{7} \\ -\frac{24}{7} \\ \frac{51}{7} \\ -\frac{26}{7} \end{bmatrix}, \begin{bmatrix} -\frac{275}{313} \\ -\frac{1210}{313} \\ \frac{615}{313} \\ \frac{2080}{313} \end{bmatrix} \right].$$

Since the result is the same, we know that the vector $\mathbf{d}$ must be in Span$\{\mathbf{a},\mathbf{b},\mathbf{c}\}$.

Now we verify the result by solving the equation $x\mathbf{a}+y\mathbf{b}+z\mathbf{c}=\mathbf{d}$.

> $LinearSolve(\langle a,b,c\rangle,d)$

$$\begin{bmatrix} 13 \\ -2 \\ 1 \end{bmatrix}.$$

This means that $\mathbf{d}=13\mathbf{a}-2\mathbf{b}+\mathbf{c}$.

□

Problem 11.18. We consider the vectors $\mathbf{a}=\begin{bmatrix} 1 \\ 1 \\ 3 \\ 1 \end{bmatrix}$, $\mathbf{b}=\begin{bmatrix} 1 \\ 2 \\ 1 \\ 2 \end{bmatrix}$, and $\mathbf{c}=\begin{bmatrix} 2 \\ 1 \\ 1 \\ 3 \end{bmatrix}$. Use normal equations to find the coordinates of the projection of $\mathbf{c}$ on Span$\{\mathbf{a},\mathbf{b}\}$ in the basis $\{\mathbf{a},\mathbf{b}\}$. Find the projection and verify the result.

Solution.

> $with(LinearAlgebra):$

First we define the vectors.

> $a:=\langle 1,1,3,1\rangle; b:=\langle 1,2,1,2\rangle; c:=\langle 2,1,1,3\rangle$

$$a:=\begin{bmatrix} 1 \\ 1 \\ 3 \\ 1 \end{bmatrix},\ b:=\begin{bmatrix} 1 \\ 2 \\ 1 \\ 2 \end{bmatrix},\ c:=\begin{bmatrix} 2 \\ 1 \\ 1 \\ 3 \end{bmatrix}.$$

Next we solve the normal equation.

> $LinearSolve(Transpose(\langle a|b\rangle).\langle a|b\rangle, Transpose(\langle a|b\rangle).c)$

$$\begin{bmatrix} \frac{1}{28} \\ \frac{15}{14} \end{bmatrix}.$$

The projection of $\mathbf{c}$ on Span$\{\mathbf{a},\mathbf{b}\}$ is

> $p := \frac{1}{28}a + \frac{15}{14}b$

$$p := \begin{bmatrix} \frac{31}{28} \\ \frac{61}{28} \\ \frac{33}{28} \\ \frac{61}{28} \end{bmatrix}.$$

In order to verify this result we calculate the projection of **c** on Span{**a**, **b**} using the projection matrix on Span{**a**, **b**}:

> *ProjectionMatrix*({a, b}).c

$$\begin{bmatrix} \frac{31}{28} \\ \frac{61}{28} \\ \frac{33}{28} \\ \frac{61}{28} \end{bmatrix}.$$

□

Problem 11.19. We consider the vectors $\mathbf{a} = \begin{bmatrix} 2 \\ 1 \\ 1 \\ 3 \\ -1 \end{bmatrix}$, $\mathbf{b} = \begin{bmatrix} 1 \\ 3 \\ 1 \\ 1 \\ 1 \end{bmatrix}$, $\mathbf{c} = \begin{bmatrix} 1 \\ 13 \\ 3 \\ -1 \\ 7 \end{bmatrix}$, $\mathbf{d} = \begin{bmatrix} 4 \\ 17 \\ 5 \\ 3 \\ 7 \end{bmatrix}$, and $\mathbf{e} = \begin{bmatrix} 1 \\ 1 \\ 3 \\ 2 \\ 1 \end{bmatrix}$. Show that the equation $x\mathbf{a} + y\mathbf{b} + z\mathbf{c} + w\mathbf{d} = \mathbf{e}$ is inconsistent and find the least squares solution of this equation.

Solution.

> *with*(*LinearAlgebra*) :

First we define the vectors.

> $a := \langle 2,1,1,3,-1\rangle; b := \langle 1,3,1,1,1\rangle; c := \langle 1,13,3,-1,7\rangle;$
$d := \langle 4,17,5,3,7\rangle; e := \langle 1,1,3,2,1\rangle$

$$a := \begin{bmatrix} 2 \\ 1 \\ 1 \\ 3 \\ -1 \end{bmatrix}, b := \begin{bmatrix} 1 \\ 3 \\ 1 \\ 1 \\ 1 \end{bmatrix}, c := \begin{bmatrix} 1 \\ 13 \\ 3 \\ -1 \\ 7 \end{bmatrix}, d := \begin{bmatrix} 4 \\ 17 \\ 5 \\ 3 \\ 7 \end{bmatrix}, e := \begin{bmatrix} 1 \\ 1 \\ 3 \\ 2 \\ 1 \end{bmatrix}.$$

Now we find the reduced row echelon form of the matrix $\begin{bmatrix} \mathbf{a} & \mathbf{b} & \mathbf{c} & \mathbf{d} & \mathbf{e} \end{bmatrix}$.

> $ReducedRowEchelonForm(\langle a|b|c|d|e\rangle)$

$$\begin{bmatrix} 1 & 0 & -2 & -1 & 0 \\ 0 & 1 & 5 & 6 & 0 \\ 0 & 0 & 0 & 0 & 1 \\ 0 & 0 & 0 & 0 & 0 \\ 0 & 0 & 0 & 0 & 0 \end{bmatrix}.$$

Consequently, the equation $x\mathbf{a} + y\mathbf{b} + z\mathbf{c} + w\mathbf{d} = \mathbf{e}$ is inconsistent.

To find the least square solution of the equation $x\mathbf{a} + y\mathbf{b} + z\mathbf{c} + w\mathbf{d} = \mathbf{e}$ we solve the normal equations.

> $LinearSolve(Transpose(\langle a|b|c|d\rangle).\langle a|b|c|d\rangle, Transpose(\langle a|b|c|d\rangle).e)$

$$\begin{bmatrix} \frac{7}{16} + 2\_t_3 + \_t_4 \\ \frac{1}{2} - 5\_t_3 - 6\_t_4 \\ \_t_3 \\ \_t_4 \end{bmatrix}.$$

This means that every least squares solution is of the form

$$\begin{bmatrix} x \\ y \\ z \\ w \end{bmatrix} = \begin{bmatrix} \frac{7}{16} \\ \frac{1}{2} \\ 0 \\ 0 \end{bmatrix} + s \begin{bmatrix} 2 \\ -5 \\ 1 \\ 0 \end{bmatrix} + t \begin{bmatrix} 1 \\ -6 \\ 0 \\ 1 \end{bmatrix}.$$

□

Determinants

Problem 11.20. Solve the equation $A\mathbf{x} = \mathbf{b}$ where $A = \begin{bmatrix} 3 & 4 & -2 \\ 2 & 1 & 3 \\ 5 & s & 1 \end{bmatrix}$ and $\mathbf{b} = \begin{bmatrix} 1 \\ t \\ 2 \end{bmatrix}$.

Solution.

> $with(LinearAlgebra):$

First we define the matrix A and the vector **b**.

> $A := Matrix([[3,4,-2],[2,1,3],[5,s,1]]); b := \langle 1,t,2 \rangle$

$$A := \begin{bmatrix} 3 & 4 & -2 \\ 2 & 1 & 3 \\ 5 & s & 1 \end{bmatrix}, \mathbf{b} = \begin{bmatrix} 1 \\ t \\ 2 \end{bmatrix}.$$

If the matrix A is invertible, then the solution is unique and we have $X = A^{-1}\mathbf{b}$. The matrix A is invertible if and only if the determinant of A is different from 0. We calculate this determinant.

> $Determinant(A)$

$$-13s + 65.$$

This means that the determinant is different from 0 if and only if s is different from 5. In this case the solution is

> $X := MatrixInverse(A).b$

$$X := \begin{bmatrix} \frac{1}{13}\frac{3s-1}{s-5} + \frac{2}{13}\frac{(s+2)\,t}{s-5} - \frac{1}{13(s-5)} \\ \frac{28}{s-5} - \frac{t}{s-5} \\ -\frac{1}{13}\frac{2s-5}{s-5} + \frac{1}{13}\frac{(3s-20)\,t}{s-5} + \frac{10}{13(s-5)} \end{bmatrix}.$$

We can verify the result.

> $simplify(A.X)$

$$\begin{bmatrix} 1 \\ t \\ 2 \end{bmatrix}.$$

If the matrix A is not invertible, that is when $s = 5$, then we have

> *ReducedRowEchelonForm*(*Matrix*([[3,4,−2],[2,1,3],[5,5,1]]))

$$\begin{bmatrix} 1 & 0 & \frac{14}{5} \\ 0 & 1 & -\frac{13}{5} \\ 0 & 0 & 0 \end{bmatrix}.$$

Thus the equation $A\mathbf{x} = \mathbf{b}$ is consistent if and only if the vector $\mathbf{b}$ is in $\text{Span}\left\{\begin{bmatrix} 3 \\ 2 \\ 5 \end{bmatrix}, \begin{bmatrix} 4 \\ 1 \\ 5 \end{bmatrix}\right\}$. To find the value of t when this happens we use the determinant.

> *Determinant*(⟨⟨3,2,5⟩|⟨4,1,5⟩|*b*⟩)

$$5t - 5.$$

We conclude that, when $s = 5$ and $t = 1$, then the solution is

> *LinearSolve*(*Matrix*([[3,4,−2],[2,1,3],[5,5,1]]),⟨1,1,2⟩)

$$\begin{bmatrix} \frac{3}{5} - \frac{14}{5}\_t_1 \\ -\frac{1}{5} + \frac{13}{5}\_t_1 \\ \_t_1 \end{bmatrix}.$$

In other words, the general solution is $\mathbf{x} = \begin{bmatrix} \frac{3}{5} \\ -\frac{1}{5} \\ 0 \end{bmatrix} + u \begin{bmatrix} -14 \\ 13 \\ 5 \end{bmatrix}$, where u is an arbitrary real number.

□

Problem 11.21. Consider the vectors $\mathbf{a} = \begin{bmatrix} 2 \\ 5 \\ 4 \end{bmatrix}$, $\mathbf{b} = \begin{bmatrix} 1 \\ 3 \\ 1 \end{bmatrix}$, $\mathbf{c} = \begin{bmatrix} 1 \\ 7 \\ 3 \end{bmatrix}$, and $\mathbf{d} = \begin{bmatrix} 5 \\ 9 \\ 8 \end{bmatrix}$. Use Cramer's Rule to solve the equation $x\mathbf{a} + y\mathbf{b} + z\mathbf{c} = \mathbf{d}$ and verify the result.

Solution.

$> with(LinearAlgebra):$

First we define the vectors.

$> a := \langle 2,5,4\rangle; b := \langle 1,3,1\rangle; c := \langle 1,7,3\rangle; d := \langle 5,9,8\rangle$

$$a := \begin{bmatrix} 2 \\ 5 \\ 4 \end{bmatrix}, b := \begin{bmatrix} 1 \\ 3 \\ 1 \end{bmatrix}, c := \begin{bmatrix} 1 \\ 7 \\ 3 \end{bmatrix}, d := \begin{bmatrix} 5 \\ 9 \\ 8 \end{bmatrix}.$$

Next we use Cramer's Rule to solve the equation.

$> x := \dfrac{Determinant(\langle d|b|c\rangle)}{Determinant(\langle a|b|c\rangle)}$

$$x := \frac{12}{5}.$$

$> y := \dfrac{Determinant(\langle a|d|c\rangle)}{Determinant(\langle a|b|c\rangle)}$

$$y := \frac{11}{10}.$$

$> z := \dfrac{Determinant(\langle a|b|d\rangle)}{Determinant(\langle a|b|c\rangle)}$

$$z := -\frac{9}{10}.$$

Now we verify the result.

$> x \cdot a + y \cdot b + z \cdot c$

$$\begin{bmatrix} 5 \\ 9 \\ 8 \end{bmatrix}.$$

□

Eigenvalues and eigenvectors

Problem 11.22. Determine a symmetric matrix A with an eigenvector $\begin{bmatrix} 1 \\ 1 \\ 1 \\ 1 \end{bmatrix}$ corresponding to the eigenvalue 2, an eigenvector $\begin{bmatrix} -1 \\ 0 \\ 1 \\ 0 \end{bmatrix}$ corresponding to the eigenvalue 3, an eigenvector $\begin{bmatrix} 1 \\ 0 \\ 1 \\ -2 \end{bmatrix}$ corresponding to the eigenvalue 5, and an eigenvector $\begin{bmatrix} 1 \\ -3 \\ 1 \\ 1 \end{bmatrix}$ corresponding to the eigenvalue 7. Verify the result.

Solution.

> $with(LinearAlgebra):$

We note that the given vectors are orthogonal and consequently we have

> $A := \frac{2}{4} * OuterProductMatrix(\langle 1,1,1,1\rangle, \langle 1,1,1,1\rangle) +$
$\frac{3}{2} * OuterProductMatrix(\langle 1,0,-1,0\rangle, \langle 1,0,-1,0\rangle) +$
$\frac{5}{6} * OuterProductMatrix(\langle 1,0,1,-2\rangle, \langle 1,0,1,-2\rangle) +$
$\frac{7}{12} * OuterProductMatrix(\langle 1,-3,1,1\rangle, \langle 1,-3,1,1\rangle)$

$$A := \begin{bmatrix} \frac{41}{12} & -\frac{5}{4} & \frac{5}{12} & -\frac{7}{12} \\ -\frac{5}{4} & \frac{23}{4} & -\frac{5}{4} & -\frac{5}{4} \\ \frac{5}{12} & -\frac{5}{4} & \frac{41}{12} & -\frac{7}{12} \\ -\frac{7}{12} & -\frac{5}{4} & -\frac{7}{12} & \frac{53}{12} \end{bmatrix}.$$

Now we verify the result.

> *Eigenvectors*(*A*)

$$\begin{bmatrix} 5 \\ 2 \\ 3 \\ 7 \end{bmatrix}, \begin{bmatrix} -\frac{1}{2} & 1 & -1 & 1 \\ 0 & 1 & 0 & -3 \\ -\frac{1}{2} & 1 & 1 & 1 \\ 1 & 1 & 0 & 1 \end{bmatrix}.$$

Note that $\begin{bmatrix} 1 \\ 0 \\ 1 \\ -2 \end{bmatrix} = -2 \begin{bmatrix} -\frac{1}{2} \\ 0 \\ -\frac{1}{2} \\ 1 \end{bmatrix}$.

□

Problem 11.23. Show that the matrix $A = \begin{bmatrix} 7 & 2 & 4 \\ -12 & 17 & 8 \\ -6 & 2 & 17 \end{bmatrix}$ can be diagonalized. Verify the result.

Solution.

> *with*(*LinearAlgebra*) :

We start by defining the matrix *A*.

> *A* := *Matrix*([[7,2,4],[−12,17,8],[−6,2,17]])

$$A := \begin{bmatrix} 7 & 2 & 4 \\ -12 & 17 & 8 \\ -6 & 2 & 17 \end{bmatrix}.$$

Next we find the eigenvalues and eigenvectors of the matrix *A*.

> *Eigenvectors*(*A*)

$$\begin{bmatrix} 15 \\ 13 \\ 13 \end{bmatrix}, \begin{bmatrix} 1 & \frac{2}{3} & \frac{1}{3} \\ 2 & 0 & 1 \\ 1 & 1 & 0 \end{bmatrix}.$$

The matrix can be diagonalized because it has two independent eigenvectors corresponding to the eigenvalue 13.

To verify the result we first define a matrix *P*.

> $P := \langle\langle 1,2,1\rangle|\langle 2,0,3\rangle|\langle 1,3,0\rangle\rangle$

$$P := \begin{bmatrix} 1 & 2 & 1 \\ 2 & 0 & 3 \\ 1 & 3 & 0 \end{bmatrix}.$$

Now we calculate the product $P\begin{bmatrix} 15 & 0 & 0 \\ 0 & 13 & 0 \\ 0 & 0 & 13 \end{bmatrix}P^{-1}$.

> $P.\langle\langle 15,0,0\rangle|\langle 0,13,0\rangle|\langle 0,0,13\rangle\rangle.MatrixInverse(P)$

$$\begin{bmatrix} 7 & 2 & 4 \\ -12 & 17 & 8 \\ -6 & 2 & 17 \end{bmatrix}.$$

□

Problem 11.24. Orthogonal diagonalize the matrix $A = \begin{bmatrix} 2 & 1 & 2 \\ 1 & 2 & -2 \\ 2 & -2 & -1 \end{bmatrix}$.

Solution.

> *with*(*LinearAlgebra*) :

First we define the matrix A.

> $A := Matrix([[2,1,2],[1,2,-2],[2,-2,-1]]);$

$$A := \begin{bmatrix} 2 & 1 & 2 \\ 1 & 2 & -2 \\ 2 & -2 & -1 \end{bmatrix}.$$

Then we find the eigenvalues and eigenvectors of A.

> *Eigenvectors*(A);

$$\begin{bmatrix} -3 \\ 3 \\ 3 \end{bmatrix}, \begin{bmatrix} -\frac{1}{2} & 2 & 1 \\ \frac{1}{2} & 0 & 1 \\ 1 & 1 & 0 \end{bmatrix}.$$

We see that the matrix A has two eigenvalues: -3 and 3. The vector $\begin{bmatrix} -\frac{1}{2} \\ \frac{1}{2} \\ 1 \end{bmatrix}$ is an eigenvector corresponding to the eigenvalue -3 and the vectors $\begin{bmatrix} 2 \\ 0 \\ 1 \end{bmatrix}$ and $\begin{bmatrix} 1 \\ 1 \\ 0 \end{bmatrix}$ are eigenvectors corresponding to the eigenvalue 3. The vector $\begin{bmatrix} -\frac{1}{2} \\ \frac{1}{2} \\ 1 \end{bmatrix}$ is orthogonal to both vectors $\begin{bmatrix} 2 \\ 0 \\ 1 \end{bmatrix}$ and $\begin{bmatrix} 1 \\ 1 \\ 0 \end{bmatrix}$, but these last two vectors are not orthogonal to each other. To fix this problem we use the Gram-Schmidt method.

> $\mathit{Gram\text{-}Schmidt}([\langle 2,0,1\rangle,\langle 1,1,0\rangle]);$

$$\left[\begin{bmatrix} 2 \\ 0 \\ 1 \end{bmatrix}, \begin{bmatrix} \frac{1}{5} \\ 1 \\ -\frac{2}{5} \end{bmatrix}\right].$$

The vectors $\begin{bmatrix} 2 \\ 0 \\ 1 \end{bmatrix}$ and $\begin{bmatrix} \frac{1}{5} \\ 1 \\ -\frac{2}{5} \end{bmatrix}$ are eigenvectors corresponding to the eigenvalue 3 and the vectors $\begin{bmatrix} -\frac{1}{2} \\ \frac{1}{2} \\ 1 \end{bmatrix}$, $\begin{bmatrix} 2 \\ 0 \\ 1 \end{bmatrix}$, and $\begin{bmatrix} \frac{1}{5} \\ 1 \\ -\frac{2}{5} \end{bmatrix}$ are orthogonal. To simplify the outcome we multiply the first vector by 2 and the third one by 5 and then we normalize the vectors.

> $P := \langle \mathit{Normalize}(\langle -1,1,2\rangle,2)\,|\,\mathit{Normalize}(\langle 2,0,1\rangle,2)\,|\,\mathit{Normalize}(\langle 1,5,-2\rangle,2)\rangle;$

$$P := \begin{bmatrix} -\frac{\sqrt{6}}{6} & \frac{2\sqrt{5}}{5} & \frac{\sqrt{30}}{30} \\ \frac{\sqrt{6}}{6} & 0 & \frac{\sqrt{30}}{6} \\ \frac{\sqrt{6}}{3} & \frac{\sqrt{5}}{5} & -\frac{\sqrt{30}}{15} \end{bmatrix}.$$

Finally we verify the result.

> $P.Matrix([[-3,0,0],[0,3,0],[0,0,3]]).Transpose(P);$

$$\begin{bmatrix} 2 & 1 & 2 \\ 1 & 2 & -2 \\ 2 & -2 & -1 \end{bmatrix}.$$

□

Jordan forms

Problem 11.25. Let $A = \begin{bmatrix} -3 & 7 & -1 \\ -12 & 17 & -2 \\ -13 & 7 & 9 \end{bmatrix}$. Find an invertible matrix P and a Jordan canonical form J such that $A = PJP^{-1}$. Verify the result.

Solution.

> *with(LinearAlgebra)* :

First we define the matrix A and then find its eigenvalues and eigenvectors.

> $A := Matrix([[-3,7,-1],[-12,17,-2],[-13,7,9]])$

$$A := \begin{bmatrix} -3 & 7 & -1 \\ -12 & 17 & -2 \\ -13 & 7 & 9 \end{bmatrix}.$$

> *Eigenvectors*(A)

$$\begin{bmatrix} 10 \\ 10 \\ 3 \end{bmatrix}, \begin{bmatrix} 1 & 0 & 1 \\ 2 & 0 & 1 \\ 1 & 0 & 1 \end{bmatrix}.$$

Note that the matrix A can not be diagonalized because the eigenspace coresponding to the eigenvalue 10 has dimension 1.

Next we define the matrix $B = A - 10I_3$ and then solve the equation $B^2X = \mathbf{0}$.

> $B := A - 10.\langle\langle 1,0,0\rangle|\langle 0,1,0\rangle|\langle 0,0,1\rangle\rangle$

$$\begin{bmatrix} -13 & 7 & -1 \\ -12 & 7 & -2 \\ -13 & 7 & -1 \end{bmatrix}.$$

> $LinearSolve(B\hat{\ }2, \langle 0,0,0\rangle)$

$$\begin{bmatrix} \_t_1 \\ 2\_t_1 \\ \_t_3 \end{bmatrix}.$$

This means that $N(B^2) = \text{Span}\left\{\begin{bmatrix} 1 \\ 2 \\ 0 \end{bmatrix}, \begin{bmatrix} 0 \\ 0 \\ 1 \end{bmatrix}\right\}$. Since the vector $\begin{bmatrix} 0 \\ 0 \\ 1 \end{bmatrix}$ is not an eigenvector corresponding to the eigenvalue 10, we can take

> $P := \langle B.\langle 0,0,1\rangle|\langle 0,0,1\rangle|\langle 1,1,1\rangle\rangle$

$$P := \begin{bmatrix} -1 & 0 & 1 \\ -2 & 0 & 1 \\ -1 & 1 & 1 \end{bmatrix}$$

and

> $J := \langle\langle 10,0,0\rangle|\langle 1,10,0\rangle|\langle 0,0,3\rangle\rangle$

$$J := \begin{bmatrix} 10 & 1 & 0 \\ 0 & 10 & 0 \\ 0 & 0 & 3 \end{bmatrix}.$$

Now we verify our result. We need to calculate the matrix PJP^{-1}.

> $P.J.MatrixInverse(P)$

$$\begin{bmatrix} -3 & 7 & -1 \\ -12 & 17 & -2 \\ -13 & 7 & 9 \end{bmatrix}.$$

□

Problem 11.26. Let $A = \begin{bmatrix} 3 & -4 & 4 & -1 \\ -3 & 7 & -7 & 4 \\ -4 & 4 & -5 & 5 \\ -4 & 0 & -4 & 7 \end{bmatrix}$. Find an invertible matrix P and a Jordan canonical form J such that $A = PJP^{-1}$. Verify the result.

Solution.

> $with(LinearAlgebra):$

First we define the matrix A and then find its eigenvalues and eigenvectors.

> $A := Matrix([[3,-4,4,-1],[-3,7,-7,4],[-4,4,-5,5],[-4,0,-4,7]])$

$$A := \begin{bmatrix} 3 & -4 & 4 & -1 \\ -3 & 7 & -7 & 4 \\ -4 & 4 & -5 & 5 \\ -4 & 0 & -4 & 7 \end{bmatrix}.$$

> $Eigenvectors(A)$

$$\begin{bmatrix} 3 \\ 3 \\ 3 \\ 3 \end{bmatrix}, \begin{bmatrix} 1 & -1 & 0 & 0 \\ -\frac{1}{4} & 1 & 0 & 0 \\ 0 & 1 & 0 & 0 \\ 1 & 0 & 0 & 0 \end{bmatrix}.$$

Note that matrix A cannot be diagonalized because the eigenspace corresponding to the eigenvalue 3 has dimension that is less than 4.

Now we define the matrix $B = A - 3I_4$ and then solve the equation $B^2X = \mathbf{0}$.

> $B := A - 3.\langle\langle 1,0,0,0\rangle|\langle 0,1,0,0\rangle|\langle 0,0,1,0\rangle|\langle 0,0,0,1\rangle\rangle$

$$B := \begin{bmatrix} 0 & -4 & 4 & -1 \\ -3 & 4 & -7 & 4 \\ -4 & 4 & -8 & 5 \\ -4 & 0 & -4 & 4 \end{bmatrix}.$$

> *LinearSolve*(B^2, ⟨0, 0, 0, 0⟩)

$$\begin{bmatrix} -t_1 \\ -t_2 \\ -t_3 \\ -t_4 \end{bmatrix}.$$

This means that the vectors $\begin{bmatrix} 1 \\ 0 \\ 0 \\ 0 \end{bmatrix}$, $\begin{bmatrix} 0 \\ 1 \\ 0 \\ 0 \end{bmatrix}$, $\begin{bmatrix} 0 \\ 0 \\ 1 \\ 0 \end{bmatrix}$, and $\begin{bmatrix} 0 \\ 0 \\ 0 \\ 1 \end{bmatrix}$ are in $N(B^2)$ and consequently $N(B^2) = \mathbb{R}^4$. Now we extend a basis of eigenvectors corresponding to the eigenvalue 3 to a basis of $\mathbb{R}^4$.

> *ReducedRowEchelonForm*(⟨⟨4, −1, 0, 4⟩|⟨−1, 1, 1, 0⟩|⟨1, 0, 0, 0⟩|⟨0, 1, 0, 0⟩|⟨0, 0, 1, 0⟩| ⟨0, 0, 0, 1⟩⟩)

$$\begin{bmatrix} 1 & 0 & 0 & 0 & 0 & \frac{1}{4} \\ 0 & 1 & 0 & 0 & 1 & 0 \\ 0 & 0 & 1 & 0 & 1 & -1 \\ 0 & 0 & 0 & 1 & -1 & \frac{1}{4} \end{bmatrix}.$$

This means that the vectors $\begin{bmatrix} 4 \\ -1 \\ 0 \\ 4 \end{bmatrix}$, $\begin{bmatrix} -1 \\ 1 \\ 1 \\ 0 \end{bmatrix}$ $\begin{bmatrix} 1 \\ 0 \\ 0 \\ 0 \end{bmatrix}$, and $\begin{bmatrix} 0 \\ 1 \\ 0 \\ 0 \end{bmatrix}$ are linearly independent and we can define the matrix P.

> P := ⟨B.⟨1, 0, 0, 0⟩|⟨1, 0, 0, 0⟩|B.⟨0, 1, 0, 0⟩|⟨0, 1, 0, 0⟩⟩)

$$P := \begin{bmatrix} 0 & 1 & -4 & 0 \\ -3 & 0 & 4 & 1 \\ -4 & 0 & 4 & 0 \\ -4 & 0 & 0 & 0 \end{bmatrix}.$$

Note that $B\begin{bmatrix} 1 \\ 0 \\ 0 \\ 0 \end{bmatrix}$ and $B\begin{bmatrix} 0 \\ 1 \\ 0 \\ 0 \end{bmatrix}$ are eigenvectors of A corresponding to the eigenvalue 3. Now we define the Jordan canonical form J.

> $J := \langle\langle 3,0,0,0\rangle | \langle 1,3,0,0\rangle | \langle 0,0,3,0\rangle | \langle 0,0,1,3\rangle\rangle$

$$\begin{bmatrix} 3 & 1 & 0 & 0 \\ 0 & 3 & 0 & 0 \\ 0 & 0 & 3 & 1 \\ 0 & 0 & 0 & 3 \end{bmatrix}.$$

Finally we verify the result.

> $P.J.MatrixInverse(P)$

$$\begin{bmatrix} 3 & -4 & 4 & -1 \\ -3 & 7 & -7 & 4 \\ -4 & 4 & -5 & 5 \\ -4 & 0 & -4 & 7 \end{bmatrix}.$$

□

Problem 11.27. Let $A = \begin{bmatrix} 0 & 11 & 7 & -17 \\ -2 & 10 & 4 & -10 \\ -2 & 5 & 7 & -7 \\ -1 & 3 & 2 & -1 \end{bmatrix}$. Find an invertible matrix P and a Jordan canonical form J such that $A = PJP^{-1}$. Verify the result.

Solution.

> $with(LinearAlgebra):$

First we define the matrix A and then find its eigenvalues and eigenvectors.

> $A := Matrix([[0,11,7,-17],[-2,10,4,-10],[-2,5,7,-7],[-1,3,2,-1]])$

$$A := \begin{bmatrix} 0 & 11 & 7 & -17 \\ -2 & 10 & 4 & -10 \\ -2 & 5 & 7 & -7 \\ -1 & 3 & 2 & -1 \end{bmatrix}.$$

> *Eigenvectors*(A)

$$\begin{bmatrix} 4 \\ 4 \\ 4 \\ 4 \end{bmatrix}, \begin{bmatrix} -1 & 4 & 0 & 0 \\ -1 & 3 & 0 & 0 \\ 1 & 0 & 0 & 0 \\ 0 & 1 & 0 & 0 \end{bmatrix}.$$

This matrix cannot be diagonalized because the eigenspace corresponding to the eigenvalue 4 has dimension 2.

Now we define the matrix $B = A - 4I_4$ and then solve the equation $B^2X = \mathbf{0}$ and $B^3X = \mathbf{0}$ to find the subspaces $N(B^2)$ and $N(B^3)$.

> $B := A - 4.\langle\langle 1,0,0,0\rangle|\langle 0,1,0,0\rangle|\langle 0,0,1,0\rangle|\langle 0,0,0,1\rangle\rangle$

$$B := \begin{bmatrix} -4 & 11 & 7 & -17 \\ -2 & 6 & 4 & -10 \\ -2 & 5 & 3 & -7 \\ -1 & 3 & 2 & -5 \end{bmatrix}.$$

> *LinearSolve*$(B\hat{}2, \langle 0,0,0,0\rangle)$

$$\begin{bmatrix} 2\_t_2 + \_t_3 - 2\_t_4 \\ \_t_2 \\ \_t_3 \\ \_t_4 \end{bmatrix}.$$

> *LinearSolve*$(B\hat{}3, \langle 0,0,0,0\rangle)$

$$\begin{bmatrix} \_t_2 \\ \_t_2 \\ \_t_3 \\ \_t_4 \end{bmatrix}.$$

This means that $\left\{ \begin{bmatrix} 2 \\ 1 \\ 0 \\ 0 \end{bmatrix}, \begin{bmatrix} 1 \\ 0 \\ 1 \\ 0 \end{bmatrix}, \begin{bmatrix} -2 \\ 0 \\ 0 \\ 1 \end{bmatrix} \right\}$ is a basis in $N(B^2)$ and $N(B^3) = \mathbb{R}^4$. Now we extend the basis in $N(B^2)$ to a basis of $\mathbb{R}^4$.

> *ReducedRowEchelonForm*$(\langle\langle 2,1,0,0\rangle|\langle 1,0,1,0\rangle|\langle -2,0,0,1\rangle|\langle 1,0,0,0\rangle|$
$\langle 0,1,0,0\rangle|\langle 0,0,1,0\rangle|\langle 0,0,0,1\rangle\rangle)$

$$\begin{bmatrix} 1 & 0 & 0 & 0 & 1 & 0 & 0 \\ 0 & 1 & 0 & 0 & 0 & 1 & 0 \\ 0 & 0 & 1 & 0 & 0 & 0 & 1 \\ 0 & 0 & 0 & 1 & -2 & -1 & 2 \end{bmatrix}.$$

Thus $\left\{\begin{bmatrix} 2 \\ 1 \\ 0 \\ 0 \end{bmatrix}, \begin{bmatrix} 1 \\ 0 \\ 1 \\ 0 \end{bmatrix}, \begin{bmatrix} -2 \\ 0 \\ 0 \\ 1 \end{bmatrix}, \begin{bmatrix} 1 \\ 0 \\ 0 \\ 0 \end{bmatrix}\right\}$ is a basis in $\mathbb{R}^4$.

Since

> *ReducedRowEchelonForm*$(\langle B.B.\langle 1,0,0,0\rangle|\langle 4,3,0,01\rangle|\langle 1,1,-1,0\rangle\rangle)$

$$\begin{bmatrix} 1 & 0 & 1 \\ 0 & 1 & 1 \\ 0 & 0 & 0 \\ 0 & 0 & 0 \end{bmatrix}.$$

$\left\{B^2\begin{bmatrix} 1 \\ 0 \\ 0 \\ 0 \end{bmatrix}, \begin{bmatrix} 4 \\ 3 \\ 0 \\ 1 \end{bmatrix},\right\}$ and $\left\{B^2\begin{bmatrix} 1 \\ 0 \\ 0 \\ 0 \end{bmatrix}, \begin{bmatrix} 1 \\ 1 \\ -1 \\ 0 \end{bmatrix}\right\}$ are bases of the eigenspace corresponding to the eigenvalue 4. We will use the basis $\left\{B^2\begin{bmatrix} 1 \\ 0 \\ 0 \\ 0 \end{bmatrix}, \begin{bmatrix} 4 \\ 3 \\ 0 \\ 1 \end{bmatrix},\right\}$ to define the matrix P.

> $P := \langle B.B.\langle 1,0,0,0\rangle|B.\langle 1,0,0,0\rangle|\langle 1,0,0,0\rangle|\langle 4,3,0,1\rangle\rangle$

$$P := \begin{bmatrix} -3 & -4 & 1 & 4 \\ -2 & -2 & 0 & 3 \\ -1 & -2 & 0 & 0 \\ -1 & -1 & 0 & 1 \end{bmatrix}.$$

> $J := \langle\langle 4,0,0,0\rangle|\langle 1,4,0,0\rangle|\langle 0,1,4,0\rangle|\langle 0,0,0,4\rangle\rangle$

$$J := \begin{bmatrix} 4 & 1 & 0 & 0 \\ 0 & 4 & 1 & 0 \\ 0 & 0 & 4 & 0 \\ 0 & 0 & 0 & 4 \end{bmatrix}.$$

Now we verify the result.

> $P.J.MatrixInverse(P)$

$$\begin{bmatrix} 0 & 11 & 7 & -17 \\ -2 & 10 & 4 & -10 \\ -2 & 5 & 7 & -7 \\ -1 & 3 & 2 & -1 \end{bmatrix}.$$

□

Problem 11.28. Let $A = \begin{bmatrix} 3 & 7 & 8 & -11 & -14 \\ -1 & 3 & 0 & 0 & 1 \\ -1 & 1 & 2 & -1 & 2 \\ -1 & 0 & -1 & 3 & 3 \\ 0 & 2 & 2 & -3 & -1 \end{bmatrix}$. Find an invertible matrix P and a matrix $J = \begin{bmatrix} \lambda & 1 & 0 & 0 & 0 \\ 0 & \lambda & 1 & 0 & 0 \\ 0 & 0 & \lambda & 0 & 0 \\ 0 & 0 & 0 & \lambda & 1 \\ 0 & 0 & 0 & 0 & \lambda \end{bmatrix}$ such that $A = PJP^{-1}$. (A matrix of this form is an example of a 5×5 standard Jordan canonical form.) Verify the result.

Solution.

> $with(LinearAlgebra):$

First we define the matrix A and then find its eigenvalues and eigenvectors.

> $A := Matrix([[3,7,8,-11,-14],[-1,3,0,0,1],[-1,1,2,-1,2],[-1,0,-1,3,3],[0,2,$

2, −3, −1]])

$$A := \begin{bmatrix} 3 & 7 & 8 & -11 & -14 \\ -1 & 3 & 0 & 0 & 1 \\ -1 & 1 & 2 & -1 & 2 \\ -1 & 0 & -1 & 3 & 3 \\ 0 & 2 & 2 & -3 & -1 \end{bmatrix}.$$

> *Eigenvectors*(A)

$$\begin{bmatrix} 2 \\ 2 \\ 2 \\ 2 \\ 2 \end{bmatrix}, \begin{bmatrix} 4 & -1 & 0 & 0 & 0 \\ 3 & -1 & 0 & 0 & 0 \\ 0 & 1 & 0 & 0 & 0 \\ 1 & 0 & 0 & 0 & 0 \\ 1 & 0 & 0 & 0 & 0 \end{bmatrix}.$$

This matrix cannot be diagonalized because the eigenspace corresponding to the eigenvalue 2 has dimension 2. The vectors $\begin{bmatrix} 4 \\ 3 \\ 0 \\ 1 \\ 1 \end{bmatrix}$ and $\begin{bmatrix} -1 \\ -1 \\ 1 \\ 0 \\ 0 \end{bmatrix}$ are linearly independent eigenvectors corresponding to the eigenvalue 2.

Now we define the matrix $B = A - 2I_5$ and then solve the equations $B^2X = \mathbf{0}$ and $B^3X = \mathbf{0}$ to find the subspaces $N(B^2)$ and $N(B^3)$.

> $B := A - 2.\langle\langle 1,0,0,0,0\rangle|\langle 0,1,0,0,0\rangle|\langle 0,0,1,0,0\rangle|\langle 0,0,0,1,0\rangle|\langle 0,0,0,0,1\rangle\rangle$

$$B := \begin{bmatrix} 1 & 7 & 8 & -11 & -14 \\ -1 & 1 & 0 & 0 & 1 \\ -1 & 1 & 0 & -1 & 2 \\ -1 & 0 & -1 & 1 & 3 \\ 0 & 2 & 2 & -3 & -3 \end{bmatrix}.$$

> *LinearSolve*$(B^\wedge 2, \langle 0,0,0,0,0 \rangle)$

$$\begin{bmatrix} 2\_t_2 - 3\_t_3 + 4\_t_4 + 6\_t_5 \\ \_t_2 \\ \_t_3 \\ \_t_4 \\ \_t_5 \end{bmatrix}.$$

> *LinearSolve*$(B^\wedge 3, \langle 0,0,0,0,0 \rangle)$

$$\begin{bmatrix} \_t_1 \\ \_t_2 \\ \_t_3 \\ \_t_4 \\ \_t_5 \end{bmatrix}.$$

This means that $\left\{ \begin{bmatrix} -2 \\ 1 \\ 0 \\ 0 \\ 0 \end{bmatrix}, \begin{bmatrix} -3 \\ 0 \\ 1 \\ 0 \\ 0 \end{bmatrix}, \begin{bmatrix} 4 \\ 0 \\ 0 \\ 1 \\ 0 \end{bmatrix}, \begin{bmatrix} 6 \\ 0 \\ 0 \\ 0 \\ 1 \end{bmatrix} \right\}$ is a basis in $N(B^2)$ and $N(B^3) = \mathbb{R}^5$. Because

> *ReducedRowEchelonForm*$(\langle\langle -2,1,0,0,0\rangle | \langle -3,0,1,0,0\rangle | \langle 4,0,0,1,0\rangle | \langle 6,0,0,0,1\rangle | \langle 1,0,0,0,0\rangle | \langle 0,1,0,0,0\rangle | \langle 0,0,1,0,0\rangle | \langle 0,0,0,1,0\rangle | \langle 0,0,0,0,1\rangle\rangle)$

$$\begin{bmatrix} 1 & 0 & 0 & 0 & 0 & 1 & 0 & 0 & 0 \\ 0 & 1 & 0 & 0 & 0 & 0 & 1 & 0 & 0 \\ 0 & 0 & 1 & 0 & 0 & 0 & 0 & 1 & 0 \\ 0 & 0 & 0 & 1 & 0 & 0 & 0 & 0 & 1 \\ 0 & 0 & 0 & 0 & 1 & 2 & 3 & -4 & -6 \end{bmatrix}$$

the vector $\begin{bmatrix} 1 \\ 0 \\ 0 \\ 0 \\ 0 \end{bmatrix}$ is not in $N(B^2)$.

Now since

```
> ReducedRowEchelonForm(⟨⟨4,3,0,1,1⟩|⟨−1,−1,1,0,0⟩|B.⟨1,0,0,0,0⟩|
⟨−2,1,0,0,0⟩|⟨−3,0,1,0,0⟩|⟨4,0,0,1,0⟩|⟨6,0,0,0,1⟩⟩)
```

$$\begin{bmatrix} 1 & 0 & 0 & 0 & 0 & 0 & 1 \\ 0 & 1 & 0 & 0 & 1 & -1 & 1 \\ 0 & 0 & 1 & 0 & 0 & -1 & 1 \\ 0 & 0 & 0 & 1 & 1 & -2 & -1 \\ 0 & 0 & 0 & 0 & 0 & 0 & 0 \end{bmatrix}$$

if $\mathbf{v}$ is an eigenvector corresponding to the eigenvalue 2, then the vectors $\mathbf{v}$, $B\begin{bmatrix} 1 \\ 0 \\ 0 \\ 0 \\ 0 \end{bmatrix}$, and $\begin{bmatrix} -2 \\ 1 \\ 0 \\ 0 \\ 0 \end{bmatrix}$ are linearly independent.

We are ready to define matrices P and J.

```
> P := ⟨B.B.⟨1,0,0,0,0⟩|B.⟨1,0,0,0,0⟩|⟨1,0,0,0,0⟩|B.⟨−2,1,0,0,0⟩|⟨−2,1,0,0,0⟩⟩
```

$$P := \begin{bmatrix} -3 & 1 & 1 & 5 & -2 \\ -2 & -1 & 0 & 3 & 1 \\ -1 & -1 & 0 & 3 & 0 \\ -1 & -1 & 0 & 2 & 0 \\ -1 & 0 & 0 & 2 & 0 \end{bmatrix}.$$

```
> J := Matrix([[2,1,0,0,0],[0,2,1,0,0],[0,0,2,0,0],[0,0,0,2,1],[0,0,0,0,2]])
```

$$J := \begin{bmatrix} 2 & 1 & 0 & 0 & 0 \\ 0 & 2 & 1 & 0 & 0 \\ 0 & 0 & 2 & 0 & 0 \\ 0 & 0 & 0 & 2 & 1 \\ 0 & 0 & 0 & 0 & 2 \end{bmatrix}.$$

Now we verify the result.

> $P.J.MatrixInverse(P)$

$$\begin{bmatrix} 3 & 7 & 8 & -11 & -14 \\ -1 & 3 & 0 & 0 & 1 \\ -1 & 1 & 2 & -1 & 2 \\ -1 & 0 & -1 & 3 & 3 \\ 0 & 2 & 2 & -3 & -1 \end{bmatrix}.$$

□

Problem 11.29. Let $A = \begin{bmatrix} -11 & -30 & -30 & 49 & 70 \\ -15 & -31 & -33 & 54 & 78 \\ -16 & -31 & -29 & 51 & 76 \\ -11 & -20 & -20 & 35 & 50 \\ -8 & -19 & -19 & 31 & 46 \end{bmatrix}$. Find an invertible matrix P and a matrix $J = \begin{bmatrix} \lambda & 1 & 0 & 0 & 0 \\ 0 & \lambda & 0 & 0 & 0 \\ 0 & 0 & \lambda & 1 & 0 \\ 0 & 0 & 0 & \lambda & 0 \\ 0 & 0 & 0 & 0 & \lambda \end{bmatrix}$ such that $A = PJP^{-1}$. (A matrix of this form is an example of a 5×5 standard Jordan canonical form.) Verify the result.

Solution.

> $with(LinearAlgebra):$

First we define the matrix A and then find its eigenvalues and eigenvectors.

> $A := Matrix([[-11,-30,-30,49,70],[-15,-31,-33,54,78],$
$[-16,-31,-29,51,76],[-11,-20,-20,35,50],[-8,-19,-19,31,46]])$

$$A := \begin{bmatrix} -11 & -30 & -30 & 49 & 70 \\ -15 & -31 & -33 & 54 & 78 \\ -16 & -31 & -29 & 51 & 76 \\ -11 & -20 & -20 & 35 & 50 \\ -8 & -19 & -19 & 31 & 46 \end{bmatrix}.$$

> *Eigenvectors*(A)

$$\begin{bmatrix} 2 \\ 2 \\ 2 \\ 2 \\ 2 \end{bmatrix}, \begin{bmatrix} \frac{10}{7} & \frac{1}{7} & 0 & 0 & 0 \\ \frac{12}{7} & \frac{11}{7} & -1 & 0 & 0 \\ 0 & 0 & 1 & 0 & 0 \\ 0 & 1 & 0 & 0 & 0 \\ 1 & 0 & 0 & 0 & 0 \end{bmatrix}.$$

This matrix cannot be diagonalized because the eigenspace corresponding to the eigenvalue 2 has dimension 3. We note that the vectors $\begin{bmatrix} 10 \\ 12 \\ 0 \\ 0 \\ 7 \end{bmatrix}, \begin{bmatrix} 1 \\ 11 \\ 0 \\ 7 \\ 0 \end{bmatrix}$, and $\begin{bmatrix} 0 \\ -1 \\ 1 \\ 0 \\ 0 \end{bmatrix}$ are linearly independent eigenvectors corresponding to the eigenvalue 2.

Now we define the matrix $B = A - 2I_5$ and then solve the equation $B^2X = \mathbf{0}$ to find the subspace $N(B^2)$.

> $B := A - 2\langle\langle 1,0,0,0,0\rangle|\langle 0,1,0,0,0\rangle|\langle 0,0,1,0,0\rangle|\langle 0,0,0,1,0\rangle|\langle 0,0,0,0,1\rangle\rangle$

$$B := \begin{bmatrix} -13 & -30 & -30 & 49 & 70 \\ -15 & -33 & -33 & 54 & 78 \\ -16 & -31 & -31 & 51 & 76 \\ -11 & -20 & -20 & 33 & 50 \\ -8 & -19 & -19 & 31 & 44 \end{bmatrix}.$$

> $LinearSolve(B^\wedge 2, \langle 0,0,0,0,0\rangle)$

$$\begin{bmatrix} -t_1 \\ -t_2 \\ -t_3 \\ -t_4 \\ -t_5 \end{bmatrix}.$$

This means that $N(B^2) = \mathbb{R}^5$. Since

> $ReducedRowEchelonForm(\langle\langle 10,12,0,0,7\rangle|\langle 1,0,11,2,0\rangle|\langle 0,-1,1,0,0\rangle|$
$\langle 1,0,0,0,0\rangle|\langle 0,1,0,0,0\rangle|\langle 0,0,1,0,0\rangle|\langle 0,0,0,1,0\rangle|\langle 0,0,0,0,1\rangle\rangle)$

$$\begin{bmatrix} 1 & 0 & 0 & 0 & 0 & 0 & 0 & \frac{1}{7} \\ 0 & 1 & 0 & 0 & 0 & 0 & \frac{1}{2} & 0 \\ 0 & 0 & 1 & 0 & 0 & 1 & -\frac{11}{2} & 0 \\ 0 & 0 & 0 & 1 & 0 & 0 & -\frac{1}{2} & -\frac{10}{7} \\ 0 & 0 & 0 & 0 & 1 & 1 & -\frac{11}{2} & -\frac{12}{7} \end{bmatrix}$$

we get that if $\mathbf{v}$ is an eigenvector corresponding to the eigenvalue 2, then the vectors $\mathbf{v}$, $\begin{bmatrix} 1 \\ 0 \\ 0 \\ 0 \\ 0 \end{bmatrix}$, and $\begin{bmatrix} 0 \\ 1 \\ 0 \\ 0 \\ 0 \end{bmatrix}$ are linearly independent.

Moreover, since

> $ReducedRowEchelonForm(\langle B.\langle 1,0,0,0,0\rangle|B.\langle 0,1,0,0,0\rangle|\langle 10,12,0,0,7\rangle|$
$\langle 1,0,11,2,0\rangle|\langle 0,-1,1,0,0\rangle\rangle)$

$$\begin{bmatrix} 1 & 0 & 0 & 0 & \frac{20}{21} \\ 0 & 1 & 0 & 0 & -\frac{11}{21} \\ 0 & 0 & 1 & 0 & -\frac{1}{3} \\ 0 & 0 & 0 & 1 & 0 \\ 0 & 0 & 0 & 0 & 0 \end{bmatrix}$$

the eigenvectors $\begin{bmatrix} 10 \\ 12 \\ 0 \\ 0 \\ 7 \end{bmatrix}$, $B\begin{bmatrix} 1 \\ 0 \\ 0 \\ 0 \\ 0 \end{bmatrix}$ and $B\begin{bmatrix} 0 \\ 1 \\ 0 \\ 0 \\ 0 \end{bmatrix}$ are linearly independent and therefore we can take

> $P := \langle B.\langle 1,0,0,0,0\rangle | \langle 1,0,0,0,0\rangle | B.\langle 0,1,0,0,0\rangle | \langle 0,1,0,0,0\rangle | \langle 10,12,0,0,7\rangle\rangle$

$$P := \begin{bmatrix} -13 & 1 & -30 & 0 & 10 \\ -15 & 0 & -33 & 1 & 12 \\ -16 & 0 & -31 & 0 & 0 \\ -11 & 0 & -20 & 0 & 0 \\ -8 & 0 & -19 & 0 & 7 \end{bmatrix}$$

and

> $J := \langle\langle 2,0,0,0,0\rangle | \langle 1,2,0,0,0\rangle | \langle 0,0,2,0,0\rangle | \langle 0,0,1,2,0\rangle | \langle 0,0,0,0,2\rangle\rangle$

$$J := \begin{bmatrix} 2 & 1 & 0 & 0 & 0 \\ 0 & 2 & 0 & 0 & 0 \\ 0 & 0 & 2 & 1 & 0 \\ 0 & 0 & 0 & 2 & 0 \\ 0 & 0 & 0 & 0 & 2 \end{bmatrix}.$$

Finally we verify the result.

> $P.J.MatrixInverse(P)$

$$\begin{bmatrix} -11 & -30 & -30 & 49 & 70 \\ -15 & -31 & -33 & 54 & 78 \\ -16 & -31 & -29 & 51 & 76 \\ -11 & -20 & -20 & 35 & 50 \\ -8 & -19 & -19 & 31 & 46 \end{bmatrix}.$$

□

Singular value decomposition

Problem 11.30. Determine the outer product expansion of the matrix

$$A = \begin{bmatrix} 5 & 5 \\ -1 & 2 \\ -3 & 1 \\ -5 & 0 \end{bmatrix}$$

and verify the result.

Solution.

> $with(LinearAlgebra):$

First we define the matrix A.

> $A := Matrix([[5,5],[-1,2],[-3,1],[-5,0]])$

$$A := \begin{bmatrix} 5 & 5 \\ -1 & 2 \\ -3 & 1 \\ -5 & 0 \end{bmatrix}.$$

Next we find the matrix $A^T A$.

> $B := Transpose(A).A$

$$B := \begin{bmatrix} 60 & 20 \\ 20 & 30 \end{bmatrix}.$$

Now we find the eigenvalues and the eigenvectors of B.

> $Eigenvectors(B)$

$$\begin{bmatrix} 20 \\ 70 \end{bmatrix}, \begin{bmatrix} -\frac{1}{2} & 2 \\ 1 & 1 \end{bmatrix}.$$

Next we define the column vectors using the second one first since it corresponds to the larger eigenvalue. We also multiply the first column by 2 to simplify the outcome.

> $V1 := Vector([2,1]); V2 := Vector([-1,2])$

$$V1 := \begin{bmatrix} 2 \\ 1 \end{bmatrix}, V2 := \begin{bmatrix} -1 \\ 2 \end{bmatrix}.$$

Now we need to multiply these column vectors by A.

> $U1 := A.V1; U2 := A.V2$

$$U1 := \begin{bmatrix} 15 \\ 0 \\ -5 \\ -10 \end{bmatrix}, \; U2 := \begin{bmatrix} 5 \\ 5 \\ 5 \\ 5 \end{bmatrix}.$$

Next we normalize vectors $V1$ and $V2$.

> $v1 := \frac{1}{\text{sqrt}(5)} V1; v2 := \frac{1}{\text{sqrt}(5)} V2$

$$v1 := \begin{bmatrix} \frac{2\sqrt{5}}{5} \\ \frac{\sqrt{5}}{5} \end{bmatrix}, \; v2 := \begin{bmatrix} -\frac{\sqrt{5}}{5} \\ \frac{2\sqrt{5}}{5} \end{bmatrix}.$$

Finally we find vectors $\mathbf{u}_1$ and $\mathbf{u}_2$.

> $u1 := \frac{1}{\text{sqrt}(70)\text{sqrt}(5)} U1; u2 := \frac{1}{\text{sqrt}(20)\text{sqrt}(5)} U2$

$$u1 := \begin{bmatrix} \frac{3\sqrt{70}\sqrt{5}}{70} \\ 0 \\ -\frac{\sqrt{70}\sqrt{5}}{70} \\ -\frac{\sqrt{70}\sqrt{5}}{35} \end{bmatrix}, \; u2 := \begin{bmatrix} \frac{1}{2} \\ \frac{1}{2} \\ \frac{1}{2} \\ \frac{1}{2} \end{bmatrix}.$$

Now we verify our calculations.

> $\text{sqrt}(70)\, OuterProductMatrix(u1, v1) + \text{sqrt}(20)\, OuterProductMatrix(u2, v2)$

$$\begin{bmatrix} 5 & 5 \\ -1 & 2 \\ -3 & 1 \\ -5 & 0 \end{bmatrix}.$$

□

Problem 11.31. Determine the outer product expansion of the matrix $A = \begin{bmatrix} -5 & 11 & 7 \\ -10 & -5 & 8 \\ -12 & 3 & -6 \end{bmatrix}$.

Solution.

> $with(LinearAlgebra):$

First we define the matrix A.

> $A := \langle\langle -5, -10, -12\rangle | \langle 11, -5, 3\rangle | \langle 7, 8, -6\rangle\rangle$

$$A := \begin{bmatrix} -5 & 11 & 7 \\ -10 & -5 & 8 \\ -12 & 3 & -6 \end{bmatrix}.$$

Next we calculate the eigenvalues and eigenvectors of the matrix $B = A^T A$.

> $B := Transpose(A).A$

$$B := \begin{bmatrix} 269 & -41 & -43 \\ -41 & 155 & 19 \\ -43 & 19 & 149 \end{bmatrix}.$$

> $Eigenvectors(B)$

$$\begin{bmatrix} 144 \\ 297 \\ 132 \end{bmatrix}, \begin{bmatrix} 1 & -3 & \frac{1}{7} \\ 2 & 1 & -\frac{4}{7} \\ 1 & 1 & 1 \end{bmatrix}.$$

Now we define the vectors $\mathbf{V}_1$, $\mathbf{V}_2$, and $\mathbf{V}_3$ in the order of decreasing eigenvalues. Note that we use $\begin{bmatrix} 1 \\ -4 \\ 7 \end{bmatrix}$ instead of $\begin{bmatrix} \frac{1}{7} \\ -\frac{4}{7} \\ 1 \end{bmatrix}$.

> $V1 := \langle -3, 1, 1\rangle;\ V2 := \langle 1, 2, 1\rangle;\ V3 := \langle 1, -4, 7\rangle$

$$V1 := \begin{bmatrix} -3 \\ 1 \\ 1 \end{bmatrix},\ V2 := \begin{bmatrix} 1 \\ 2 \\ 1 \end{bmatrix},\ V3 := \begin{bmatrix} 1 \\ -4 \\ 7 \end{bmatrix}.$$

Next we find the vectors $\mathbf{U}_1$, $\mathbf{U}_2$, and $\mathbf{U}_3$.

> $U1 := A.V1; U2 := A.V2; U3 := A.V3$

$$U1 := \begin{bmatrix} 33 \\ 33 \\ 33 \end{bmatrix}, U2 := \begin{bmatrix} 24 \\ -12 \\ -12 \end{bmatrix}, U3 := \begin{bmatrix} 0 \\ 66 \\ -66 \end{bmatrix}.$$

In order to determine our outer product expansion we need to normalize these vectors.

> $v1 := \frac{1}{\text{sqrt}(11)}\langle -3,1,1\rangle; v2 := \frac{1}{\text{sqrt}(6)}\langle 1,2,1\rangle; v3 := \frac{1}{\text{sqrt}(66)}\langle 1,-4,7\rangle$

$$v1 := \begin{bmatrix} -\frac{3}{11}\sqrt{11} \\ \frac{1}{11}\sqrt{11} \\ \frac{1}{11}\sqrt{11} \end{bmatrix}, v2 := \begin{bmatrix} \frac{1}{6}\sqrt{6} \\ \frac{1}{3}\sqrt{6} \\ \frac{1}{6}\sqrt{6} \end{bmatrix}, v3 := \begin{bmatrix} \frac{1}{66}\sqrt{66} \\ -\frac{2}{33}\sqrt{66} \\ \frac{7}{66}\sqrt{66} \end{bmatrix}.$$

Now we find the vectors $\mathbf{u}_1$, $\mathbf{u}_2$, and $\mathbf{u}_3$.

> $u1 := \frac{1}{\text{sqrt}(297)\ \text{sqrt}(11)}U1$

$$u1 := \begin{bmatrix} \frac{1}{33}\sqrt{33}\sqrt{11} \\ \frac{1}{33}\sqrt{33}\sqrt{11} \\ \frac{1}{33}\sqrt{33}\sqrt{11} \end{bmatrix}.$$

> $u2 := \frac{1}{\text{sqrt}(144)\ \text{sqrt}(6)}U2$

$$u2 := \begin{bmatrix} \frac{1}{3}\sqrt{6} \\ -\frac{1}{6}\sqrt{6} \\ -\frac{1}{6}\sqrt{6} \end{bmatrix}.$$

> $u3 := \dfrac{1}{\text{sqrt}(132)\ \text{sqrt}(66)}\ U3$

$$u3 := \begin{bmatrix} 0 \\ \frac{1}{66}\sqrt{33}\sqrt{66} \\ -\frac{1}{66}\sqrt{33}\sqrt{66} \end{bmatrix}.$$

The outer product expansion is $\sqrt{297}\ u1\ (v1)^T + \sqrt{144}\ u1\ (v1)^T + \sqrt{132}\ u3\ (v3)^T$. Now we verify the result.

> sqrt(297) *OuterProductMatrix*($u1, v1$)+sqrt(144) *OuterProductMatrix*($u2, v2$)+ sqrt(132) *OuterProductMatrix*($u3, v3$)

$$\begin{bmatrix} -5 & 11 & 7 \\ -10 & -5 & 8 \\ -12 & 3 & -6 \end{bmatrix}.$$

□

Problem 11.32. We consider the matrix $A := \begin{bmatrix} 1 & 2 & 3 \\ 2 & 4 & 6 \\ 3 & 1 & 19 \\ 1 & 3 & 1 \end{bmatrix}$. Find the pseudoinverse of the matrix A and the least squares solution of minimal length of the equation $A\mathbf{x} = \begin{bmatrix} 1 \\ 1 \\ 1 \\ 2 \end{bmatrix}$. Show that this solution verifies the normal equations and use this solution to calculate the projection of the vector $\begin{bmatrix} 1 \\ 1 \\ 1 \\ 2 \end{bmatrix}$ on the subspace $C(A)$.

Solution.

> *with*(*LinearAlgebra*) :

First we define the matrix A and the vector $\mathbf{b} := \begin{bmatrix} 1 \\ 1 \\ 1 \\ 2 \end{bmatrix}$.

> $A := \langle\langle 1,2,3,1\rangle|\langle 2,4,1,3\rangle|\langle 3,6,19,1\rangle\rangle; b := \langle 1,1,1,2\rangle$

$$A := \begin{bmatrix} 1 & 2 & 3 \\ 2 & 4 & 6 \\ 3 & 1 & 19 \\ 1 & 3 & 1 \end{bmatrix}, b := \begin{bmatrix} 1 \\ 1 \\ 1 \\ 2 \end{bmatrix}.$$

We find the pseudoinverse of the matrix A using an appropriate Maple tool, namely "MatrixInverse" with the "method = pseudo" parameter.

> $P := MatrixInverse(\langle\langle 1,2,3,1\rangle|\langle 2,4,1,3\rangle|\langle 3,6,19,1\rangle\rangle), method = pseudo)$

$$P := \begin{bmatrix} \frac{31}{1746} & \frac{31}{873} & -\frac{23}{2619} & \frac{79}{2619} \\ \frac{56}{873} & \frac{112}{873} & -\frac{307}{5238} & \frac{599}{5238} \\ -\frac{7}{1746} & -\frac{7}{873} & \frac{146}{2619} & -\frac{46}{2619} \end{bmatrix}.$$

Now we can find the least square solution of minimal length of the equation $A\mathbf{x} = \mathbf{b}$ by multiplying the vector $\mathbf{b}$ by the matrix P.

> $X := P.b$

$$X := \begin{bmatrix} \frac{61}{582} \\ \frac{211}{582} \\ \frac{5}{582} \end{bmatrix}.$$

Next we verify that this $\mathbf{x}$ satisfies the normal equation.

> $Transpose(A).A.X; Transpose(A).b$

$$\begin{bmatrix} 8 \\ 13 \\ 30 \end{bmatrix}, \begin{bmatrix} 8 \\ 13 \\ 30 \end{bmatrix}.$$

Now we find the projection of **b** on the subspace $C(A)$ and then verify the result.

> $p := A.X$

$$\begin{bmatrix} \frac{83}{97} \\ \frac{166}{97} \\ \frac{163}{194} \\ \frac{233}{194} \end{bmatrix}.$$

> $DotProduct(Column(A, [1]), p - b)$

$$0.$$

> $DotProduct(Column(A, [2]), p - b)$

$$0.$$

> $DotProduct(Column(A, [3]), p - b)$

$$0.$$

□

Quadratic forms and positive definite matrices

Problem 11.33. Show that the matrix $A = \begin{bmatrix} 21 & 25 & 51 \\ 25 & 91 & 75 \\ 51 & 75 & 131 \end{bmatrix}$ is positive definite and find its Cholesky decomposition. Verify the result.

Solution.

> $with(LinearAlgebra):$

We start by defining the matrix.

> $A := Matrix([[21, 25, 51], [25, 91, 75], [51, 75, 131]])$

$$A := \begin{bmatrix} 21 & 25 & 51 \\ 25 & 91 & 75 \\ 51 & 75 & 131 \end{bmatrix}.$$

Now we check that the matrix A is positive definite.

> *Determinant*(*Matrix*([[21,25],[25,91]])); *Determinant*(A)

$$1286$$

$$4900.$$

Next we find the Cholesky decomposition of A.

> C:= *LUDecomposition*(A,*method* = 'Cholesky')

$$C := \begin{bmatrix} \sqrt{21} & 0 & 0 \\ \frac{25}{21}\sqrt{21} & \frac{1}{21}\sqrt{27006} & 0 \\ \frac{17}{7}\sqrt{21} & \frac{50}{4501}\sqrt{27006} & \frac{35}{643}\sqrt{1286} \end{bmatrix}.$$

Finally we verify the result.

> C.*Transpose*(C)

$$\begin{bmatrix} 21 & 25 & 51 \\ 25 & 91 & 75 \\ 51 & 75 & 131 \end{bmatrix}.$$

□

Chapter 12

Answers to selected exercises

Section 1.1

1. $\begin{bmatrix} a+d & b+e & c+f \end{bmatrix}$

3. $\begin{bmatrix} a_{11}+b_{11} & a_{12}+b_{12} \\ a_{21}+b_{21} & a_{22}+b_{22} \end{bmatrix}$

5. $\begin{bmatrix} 2a+5d & 2b+5e & 2c+5f \end{bmatrix}$

7. $\begin{bmatrix} 2s+t & 0 & 2s+4t \\ t & s+t & 3s \\ 5s & 2t & -t \end{bmatrix}$

9. $\begin{bmatrix} 3 & 1 & 3 \\ 0 & 1 & 1 \end{bmatrix}$

11. $\begin{bmatrix} ac & ad \\ bc & bd \end{bmatrix}$

13. $\begin{bmatrix} 28 & 21 \\ 16 & 12 \\ 20 & 15 \end{bmatrix}$

15. $\begin{bmatrix} a_1b_{11}+a_2b_{21} & a_1b_{12}+a_2b_{22} & a_1b_{13}+a_2b_{23} \end{bmatrix}$

17. The number of columns in $\begin{bmatrix} a & b & c \end{bmatrix}$ is different from the number of rows in $\begin{bmatrix} d \\ e \\ f \\ g \end{bmatrix}$.

19. The number of columns $\begin{bmatrix} 1 & 1 & 3 & 1 & 1 \\ 7 & 3 & 1 & 1 & 3 \end{bmatrix}$ is different from the number of rows in $\begin{bmatrix} 2 & 1 & 3 \\ 1 & 5 & 1 \\ 4 & 1 & 2 \end{bmatrix}$.

21.

23. $A^T = \begin{bmatrix} 5 & 1 & 2 \\ 7 & 2 & 4 \\ 9 & 3 & 8 \end{bmatrix}$

25. $A^T = \begin{bmatrix} a \\ b \\ c \end{bmatrix}$

27. $A^T = \begin{bmatrix} a & c & e \\ b & d & f \end{bmatrix}$

29. $A^T = \begin{bmatrix} 2 & 7 & 9 \\ 3 & 1 & 8 \\ 1 & 5 & 3 \\ 4 & 2 & 3 \end{bmatrix}$

31. $\begin{bmatrix} \mathbf{a}_1 & \mathbf{a}_2 \end{bmatrix} \begin{bmatrix} x_1 \\ x_2 \end{bmatrix} = \begin{bmatrix} \mathbf{b}_1 & \mathbf{b}_2 \end{bmatrix} \begin{bmatrix} x_1 \\ x_2 \end{bmatrix}$ can be written as $x_1\mathbf{a}_1 + x_2\mathbf{a}_2 = x_1\mathbf{b}_1 + x_2\mathbf{b}_2$. If we take $x_1 = 1$ and $x_2 = 0$, then we get $\mathbf{a}_1 = \mathbf{b}_1$. If we take $x_1 = 0$ and $x_2 = 1$, then we get $\mathbf{a}_2 = \mathbf{b}_2$. Thus $A = B$.

33. From Theorem 1.15 we get $(ABC)D =$

$(AB)(CD)$ and $(ABC)D = A(BCD)$.

35. $(ABC)^T = C^T(AB)^T = C^T B^T A^T$

37. $(BAA^TB^T)^T = (B^T)^T(A^T)^T A^T B^T = BAA^TB^T$

39.

$$\begin{bmatrix} a_{11} \\ a_{21} \end{bmatrix} \begin{bmatrix} b_{11} & b_{12} \end{bmatrix} + \begin{bmatrix} a_{12} \\ a_{22} \end{bmatrix} \begin{bmatrix} b_{21} & b_{22} \end{bmatrix}$$

$$= \begin{bmatrix} a_{11}b_{11} & a_{11}b_{12} \\ a_{21}b_{11} & a_{21}b_{12} \end{bmatrix} + \begin{bmatrix} a_{12}b_{21} & a_{12}b_{22} \\ a_{22}b_{21} & a_{22}b_{22} \end{bmatrix}$$

$$= \begin{bmatrix} a_{11}b_{11}+a_{12}b_{21} & a_{11}b_{12}+a_{12}b_{22} \\ a_{21}b_{11}+a_{22}b_{21} & a_{21}b_{12}+a_{22}b_{22} \end{bmatrix}$$

Section 1.2

1. $\begin{bmatrix} a_1 & b_1 & c_1 & d_1 \\ 9a_2 & 9b_2 & 9c_2 & 9d_2 \\ a_3 & b_3 & c_3 & d_3 \end{bmatrix}$

3. $\begin{bmatrix} a_3 & b_3 & c_3 & d_3 \\ a_2 & b_2 & c_2 & d_2 \\ a_1 & b_1 & c_1 & d_1 \end{bmatrix}$

5. $\begin{bmatrix} 0 \\ 3 \\ 0 \\ 0 \\ 0 \end{bmatrix}$

7. $\begin{bmatrix} a_1 & b_1 \\ a_2 & b_2 \\ a_3+7a_2 & b_3+7a_2 \\ a_4 & b_4 \end{bmatrix}$

9. $\begin{bmatrix} 3 & 12 & -1 \\ 3 & 13 & 0 \\ -2 & -8 & 1 \end{bmatrix}$

11. $\begin{bmatrix} 17 & 18 & 46 & 29 \\ 2 & -1 & -5 & -1 \\ 5 & 5 & 15 & 10 \end{bmatrix}$

13. $\begin{bmatrix} 1 & 0 & 2 & -1 \\ 0 & 1 & 5 & 1 \\ 0 & 0 & 0 & 0 \end{bmatrix}$

15. $\begin{bmatrix} 1 & 0 & 0 & 1 \\ 0 & 1 & 0 & 2 \\ 0 & 0 & 1 & 3 \\ 0 & 0 & 0 & 0 \end{bmatrix}$

17. If $a \neq 0$, the reduced row echelon form is $\begin{bmatrix} 1 & 0 & 0 & \frac{a+1}{a} \\ 0 & 1 & 0 & -\frac{1}{a} \\ 0 & 0 & 1 & -\frac{1}{a} \\ 0 & 0 & 0 & 0 \end{bmatrix}$.

If $a = 0$, then the matrix becomes $\begin{bmatrix} 1 & 0 & 1 & 1 \\ 0 & -1 & 1 & 1 \\ 1 & 1 & 0 & 0 \\ 1 & 1 & 0 & 1 \end{bmatrix}$ and the reduced row echelon form is $\begin{bmatrix} 1 & 0 & 1 & 0 \\ 0 & 1 & -1 & 0 \\ 0 & 0 & 0 & 1 \\ 0 & 0 & 0 & 0 \end{bmatrix}$.

19. $\begin{bmatrix} 1 & 0 & p & q \\ 0 & 1 & q & p+q \\ 0 & 0 & 0 & 0 \end{bmatrix}$

21. $\begin{bmatrix} 1 & 0 & a & c \\ 0 & 1 & b & d \\ 0 & 0 & 0 & 0 \\ 0 & 0 & 0 & 0 \end{bmatrix}$, $\begin{bmatrix} 1 & a & 0 & b \\ 0 & 0 & 1 & c \\ 0 & 0 & 0 & 1 \\ 0 & 0 & 0 & 0 \end{bmatrix}$, $\begin{bmatrix} 1 & a & b & 0 \\ 0 & 0 & 0 & 1 \\ 0 & 0 & 0 & 0 \\ 0 & 0 & 0 & 0 \end{bmatrix}$, $\begin{bmatrix} 0 & 1 & 0 & a \\ 0 & 0 & 1 & b \\ 0 & 0 & 0 & 0 \\ 0 & 0 & 0 & 0 \end{bmatrix}$, $\begin{bmatrix} 0 & 1 & a & 0 \\ 0 & 1 & 0 & 1 \\ 0 & 0 & 0 & 0 \\ 0 & 0 & 0 & 0 \end{bmatrix}$, $\begin{bmatrix} 0 & 0 & 1 & 0 \\ 0 & 0 & 0 & 1 \\ 0 & 0 & 0 & 0 \\ 0 & 0 & 0 & 0 \end{bmatrix}$.

23. The reduced row echelon form of A is $\begin{bmatrix} 1 & 0 & 4 \\ 0 & 1 & -1 \end{bmatrix}$ and $P = \begin{bmatrix} \frac{5}{17} & -\frac{2}{17} \\ \frac{1}{17} & \frac{3}{17} \end{bmatrix}$.

25. The reduced row echelon form of A is $\begin{bmatrix} 1 \\ 0 \\ 0 \end{bmatrix}$ and $P = \begin{bmatrix} 0 & 0 & \frac{1}{50} \\ 1 & 0 & -\frac{3}{10} \\ 0 & 1 & -\frac{9}{25} \end{bmatrix}$.

27. The reduced row echelon form of A is $\begin{bmatrix} 1 & 0 \\ 0 & 1 \\ 0 & 0 \end{bmatrix}$ and $P = \begin{bmatrix} 0 & -\frac{3}{7} & \frac{5}{7} \\ 0 & \frac{2}{7} & -\frac{1}{7} \\ 1 & -\frac{1}{7} & -\frac{3}{7} \end{bmatrix}$.

29. The reduced row echelon form of A is $\begin{bmatrix} 1 \\ 0 \\ 0 \\ 0 \end{bmatrix}$ and $P = \begin{bmatrix} 0 & 0 & 0 & \frac{1}{5} \\ 1 & 0 & 0 & -\frac{1}{5} \\ 0 & 1 & 0 & -\frac{3}{5} \\ 0 & 0 & 1 & -\frac{2}{5} \end{bmatrix}$.

Section 1.3

1. $A = \begin{bmatrix} -4 & 1 \\ 38 & -10 \end{bmatrix}$ and

$$A^{-1} = \begin{bmatrix} -1 & 0 \\ 0 & 1 \end{bmatrix}\begin{bmatrix} 0 & 1 \\ 1 & 0 \end{bmatrix}\begin{bmatrix} 1 & -4 \\ 0 & 1 \end{bmatrix}\begin{bmatrix} 1 & 0 \\ 5 & 1 \end{bmatrix}\begin{bmatrix} 1 & 0 \\ 0 & \frac{1}{2} \end{bmatrix} = \begin{bmatrix} -5 & -1/2 \\ -19 & -2 \end{bmatrix}.$$

3. $A = \begin{bmatrix} 1 & 3 & 6 \\ 0 & 4 & 9 \\ 0 & 1 & 2 \end{bmatrix}$ and

$$A^{-1} = \begin{bmatrix} 1 & 0 & 0 \\ 0 & 1 & -2 \\ 0 & 0 & 1 \end{bmatrix}\begin{bmatrix} 1 & 0 & 0 \\ 0 & 0 & 1 \\ 0 & 1 & 0 \end{bmatrix}\begin{bmatrix} 1 & 0 & 0 \\ 0 & 1 & -4 \\ 0 & 0 & 1 \end{bmatrix}\begin{bmatrix} 1 & 0 & -3 \\ 0 & 1 & 0 \\ 0 & 0 & 1 \end{bmatrix} = \begin{bmatrix} 1 & 0 & -3 \\ 0 & -2 & 9 \\ 0 & 1 & -4 \end{bmatrix}.$$

5. $A = \begin{bmatrix} 1 & 4 & 0 & 0 \\ 0 & 0 & 0 & 1 \\ 0 & 0 & 7 & 0 \\ 0 & 1 & 0 & 0 \end{bmatrix}$ and

$$A^{-1} = \begin{bmatrix} 1 & 0 & 0 & 0 \\ 0 & 1 & 0 & 0 \\ 0 & 0 & 1/7 & 0 \\ 0 & 0 & 0 & 1 \end{bmatrix}\begin{bmatrix} 1 & 0 & 0 & 0 \\ 0 & 0 & 0 & 1 \\ 0 & 0 & 1 & 0 \\ 0 & 1 & 0 & 0 \end{bmatrix}\begin{bmatrix} 1 & 0 & 0 & -4 \\ 0 & 1 & 0 & 0 \\ 0 & 0 & 1 & 0 \\ 0 & 0 & 0 & 1 \end{bmatrix} = \begin{bmatrix} 1 & 0 & 0 & -4 \\ 0 & 0 & 0 & 1 \\ 0 & 0 & 1/7 & 0 \\ 0 & 1 & 0 & 0 \end{bmatrix}.$$

7. The reduced row echelon form of the augmented matrix is $\begin{bmatrix} 1 & 0 & 0 & -\frac{5}{14} \\ 0 & 1 & 0 & \frac{9}{14} \\ 0 & 0 & 1 & \frac{3}{14} \end{bmatrix}$ and the solution is $x = -\frac{5}{14}$, $y = \frac{9}{14}$, $z = \frac{3}{14}$.

9. The reduced row echelon form of the augmented matrix is $\begin{bmatrix} 1 & 0 & -1 & -\frac{16}{3} \\ 0 & 1 & 1 & \frac{5}{3} \\ 0 & 0 & 0 & 0 \end{bmatrix}$ and the solution is $x = z - \frac{16}{3}$ and $y = z + \frac{5}{3}$, where z is a free variable.

11. The reduced row echelon form of the augmented matrix is $\begin{bmatrix} 1 & 0 & 4 & 0 \\ 0 & 1 & 3 & 0 \\ 0 & 0 & 0 & 1 \end{bmatrix}$, so the system has no solution.

13. If $a \neq \frac{18}{7}$, the solution is $x = \frac{18-3a}{-18+7a}$, $y = \frac{-12+6a}{-18+7a}$, $z = -\frac{8}{-18+7a}$. If $a = \frac{18}{7}$, the reduced row echelon form of the augmented matrix is $\begin{bmatrix} 1 & 0 & \frac{9}{7} & 0 \\ 0 & 1 & \frac{3}{7} & 0 \\ 0 & 0 & 0 & 1 \end{bmatrix}$, so the system has no solution.

15. The reduced row echelon form of the augmented matrix is $\begin{bmatrix} 1 & 0 & 0 & 2 & -1 \\ 0 & 1 & 0 & 5 & 4 \\ 0 & 0 & 1 & 3 & 2 \\ 0 & 0 & 0 & 0 & 0 \end{bmatrix}$ and the solution is $x = -2w - 1$, $y = -5w + 4$, $z = -3w + 2$, where w is a free variable.

17. The reduced row echelon form of the augmented matrix is $\begin{bmatrix} 1 & 0 & 3 & 2 & 0 \\ 0 & 1 & 2 & 5 & 1 \\ 0 & 0 & 0 & 0 & 0 \\ 0 & 0 & 0 & 0 & 0 \end{bmatrix}$ and the solution is $x = -3z - 2w$, $y = -2z - 5w + 1$, where z and w are free variables.

19. The reduced row echelon form of the matrix is $\begin{bmatrix} 1 & 0 & 5/2 \\ 0 & 1 & -7/2 \\ 0 & 0 & 0 \end{bmatrix}$.

21. The reduced row echelon form of the matrix is $\begin{bmatrix} 1 & 0 & 0 & 1 \\ 0 & 1 & 0 & -1 \\ 0 & 0 & 1 & 1 \\ 0 & 0 & 0 & 0 \end{bmatrix}$.

23. The reduced row echelon form of the matrix is $\begin{bmatrix} 1 & 0 & 0 & -1 \\ 0 & 1 & 0 & 2 \\ 0 & 0 & 1 & -1 \\ 0 & 0 & 0 & 0 \end{bmatrix}$.

25. The reduced row echelon form of the matrix $\begin{bmatrix} 1 & 1 & 1 & 1 & 0 & 0 \\ 1 & 1 & -1 & 0 & 1 & 0 \\ 1 & 2 & 3 & 0 & 0 & 1 \end{bmatrix}$ is $\begin{bmatrix} 1 & 0 & 0 & 5/2 & -1/2 & -1 \\ 0 & 1 & 0 & -2 & 1 & 1 \\ 0 & 0 & 1 & 1/2 & -1/2 & 0 \end{bmatrix}$ and the inverse is $\begin{bmatrix} 5/2 & -1/2 & -1 \\ -2 & 1 & 1 \\ 1/2 & -1/2 & 0 \end{bmatrix}$.

27. The reduced row echelon form of the matrix $\begin{bmatrix} 2 & 4 & 3 & 1 & 0 & 0 \\ a & 5 & 1 & 0 & 1 & 0 \\ 1 & 1 & 1 & 0 & 0 & 1 \end{bmatrix}$ is $\begin{bmatrix} 1 & 0 & 0 & -\frac{4}{3+a} & \frac{1}{3+a} & \frac{11}{3+a} \\ 0 & 1 & 0 & \frac{-1+a}{3+a} & \frac{1}{3+a} & -\frac{3a-2}{3+a} \\ 0 & 0 & 1 & -\frac{-5+a}{3+a} & -\frac{2}{3+a} & 2\frac{2a-5}{3+a} \end{bmatrix}$ and the inverse is $\begin{bmatrix} -\frac{4}{3+a} & \frac{1}{3+a} & \frac{11}{3+a} \\ \frac{-1+a}{3+a} & \frac{1}{3+a} & -\frac{3a-2}{3+a} \\ -\frac{-5+a}{3+a} & -\frac{2}{3+a} & 2\frac{2a-5}{3+a} \end{bmatrix}$.

29. The reduced row echelon form of the matrix $\begin{bmatrix} 1 & 1 & 2 & 1 & 1 & 0 & 0 & 0 \\ 2 & 3 & 0 & 2 & 0 & 1 & 0 & 0 \\ 1 & 1 & a & 0 & 0 & 0 & 1 & 0 \\ 1 & 1 & 1 & 1 & 0 & 0 & 0 & 1 \end{bmatrix}$ is $\begin{bmatrix} 1 & 0 & 0 & 0 & -2-a & -1 & 1 & a+4 \\ 0 & 1 & 0 & 0 & 2 & 1 & 0 & -4 \\ 0 & 0 & 1 & 0 & 1 & 0 & 0 & -1 \\ 0 & 0 & 0 & 1 & -1+a & 0 & -1 & 2-a \end{bmatrix}$ and the inverse is $\begin{bmatrix} -2-a & -1 & 1 & a+4 \\ 2 & 1 & 0 & -4 \\ 1 & 0 & 0 & -1 \\ -1+a & 0 & -1 & 2-a \end{bmatrix}$.

31. $A = \begin{bmatrix} 1 & 4 \\ 0 & 1 \end{bmatrix}\begin{bmatrix} 3 & 0 \\ 0 & 1 \end{bmatrix}\begin{bmatrix} 1 & 0 \\ -2 & 1 \end{bmatrix}\begin{bmatrix} 1 & 0 \\ 0 & 5 \end{bmatrix}$, because
$\begin{bmatrix} 1 & 0 \\ 0 & \frac{1}{5} \end{bmatrix}\begin{bmatrix} 1 & 0 \\ 2 & 1 \end{bmatrix}\begin{bmatrix} \frac{1}{3} & 0 \\ 0 & 1 \end{bmatrix}\begin{bmatrix} 1 & -4 \\ 0 & 1 \end{bmatrix}\begin{bmatrix} -5 & 20 \\ -2 & 5 \end{bmatrix} = \begin{bmatrix} 1 & 0 \\ 0 & 1 \end{bmatrix}$, and we have
$A^{-1} = \begin{bmatrix} 1 & 0 \\ 0 & \frac{1}{5} \end{bmatrix}\begin{bmatrix} 1 & 0 \\ 2 & 1 \end{bmatrix}\begin{bmatrix} \frac{1}{3} & 0 \\ 0 & 1 \end{bmatrix}\begin{bmatrix} 1 & -4 \\ 0 & 1 \end{bmatrix} = \begin{bmatrix} \frac{1}{3} & -\frac{4}{3} \\ \frac{2}{15} & -\frac{1}{3} \end{bmatrix}$.

33. $A = \begin{bmatrix} 1 & 0 & 1 \\ 0 & 1 & 0 \\ 0 & 0 & 1 \end{bmatrix}\begin{bmatrix} 1 & 0 & 0 \\ 0 & 1 & 0 \\ 1 & 0 & 1 \end{bmatrix}\begin{bmatrix} 1 & 0 & 0 \\ 0 & 0 & 1 \\ 0 & 1 & 0 \end{bmatrix}\begin{bmatrix} 1 & 0 & 0 \\ 0 & 1 & 0 \\ 0 & -2 & 1 \end{bmatrix}$, because
$\begin{bmatrix} 1 & 0 & 0 \\ 0 & 1 & 0 \\ 0 & 2 & 1 \end{bmatrix}\begin{bmatrix} 1 & 0 & 0 \\ 0 & 0 & 1 \\ 0 & 1 & 0 \end{bmatrix}\begin{bmatrix} 1 & 0 & 0 \\ 0 & 1 & 0 \\ -1 & 0 & 1 \end{bmatrix}\begin{bmatrix} 1 & 0 & -1 \\ 0 & 1 & 0 \\ 0 & 0 & 1 \end{bmatrix}\begin{bmatrix} 2 & 1 & 0 \\ 0 & -2 & 1 \\ 1 & 1 & 0 \end{bmatrix}$
$= \begin{bmatrix} 1 & 0 & 0 \\ 0 & 1 & 0 \\ 0 & 0 & 1 \end{bmatrix}$, and we have $A^{-1} = \begin{bmatrix} 1 & 0 & 0 \\ 0 & 1 & 0 \\ 0 & 2 & 1 \end{bmatrix}\begin{bmatrix} 1 & 0 & 0 \\ 0 & 0 & 1 \\ 0 & 1 & 0 \end{bmatrix}\begin{bmatrix} 1 & 0 & 0 \\ 0 & 1 & 0 \\ -1 & 0 & 1 \end{bmatrix}$
$\begin{bmatrix} 1 & 0 & -1 \\ 0 & 1 & 0 \\ 0 & 0 & 1 \end{bmatrix} = \begin{bmatrix} 1 & 0 & -1 \\ -1 & 0 & 2 \\ -2 & 1 & 4 \end{bmatrix}$.

35. Since $\begin{bmatrix} 2 & 3 & 1 & 1 & -1 \\ 4 & 7 & 2 & 1 & 2 \end{bmatrix} \sim \begin{bmatrix} 1 & 0 & 1/2 & 2 & -13/2 \\ 0 & 1 & 0 & -1 & 4 \end{bmatrix}$, we have
$\begin{bmatrix} p & q & r \\ x & y & z \end{bmatrix} = \begin{bmatrix} 2 & 3 \\ 4 & 7 \end{bmatrix}^{-1}\begin{bmatrix} 1 & 1 & -1 \\ 2 & 1 & 2 \end{bmatrix} = \begin{bmatrix} 1/2 & 2 & -13/2 \\ 0 & -1 & 4 \end{bmatrix}$.

37. Since $\begin{bmatrix} 2 & 3 & 1 & 1 & 1 \\ 1 & 1 & 2 & 1 & 2 \\ 1 & 1 & 3 & 3 & -1 \end{bmatrix} \sim \begin{bmatrix} 1 & 0 & 0 & -8 & 20 \\ 0 & 1 & 0 & 5 & -12 \\ 0 & 0 & 1 & 2 & -3 \end{bmatrix}$ we have $\begin{bmatrix} p & x \\ q & y \\ r & z \end{bmatrix} = \begin{bmatrix} -8 & 20 \\ 5 & -12 \\ 2 & -3 \end{bmatrix}$.

Section 2.1

1. The columns are linearly dependent.
3. The columns are linearly independent.
5. $a = 2$.
7. $a = b = 7$.
9. $\begin{bmatrix} 1 \\ 3 \\ 4 \end{bmatrix} = 5\begin{bmatrix} 1 \\ 1 \\ 2 \end{bmatrix} - 2\begin{bmatrix} 2 \\ 1 \\ 3 \end{bmatrix}$
11. $\begin{bmatrix} 4 \\ -1 \\ 1 \end{bmatrix} = -2\begin{bmatrix} 1 \\ 2 \\ 1 \end{bmatrix} + 3\begin{bmatrix} 2 \\ 1 \\ 1 \end{bmatrix}$,
$\begin{bmatrix} 0 \\ 3 \\ 1 \end{bmatrix} = 2\begin{bmatrix} 1 \\ 2 \\ 1 \end{bmatrix} - \begin{bmatrix} 2 \\ 1 \\ 1 \end{bmatrix}$,
$\begin{bmatrix} 5 \\ -2 \\ 1 \end{bmatrix} = -3\begin{bmatrix} 1 \\ 2 \\ 1 \end{bmatrix} + 4\begin{bmatrix} 2 \\ 1 \\ 1 \end{bmatrix}$.
13. $\begin{bmatrix} 4 \\ 4 \\ 8 \end{bmatrix} = 4\begin{bmatrix} 3 \\ 2 \\ 5 \end{bmatrix} - 4\begin{bmatrix} 2 \\ 1 \\ 3 \end{bmatrix}$,
$\begin{bmatrix} 4 \\ -2 \\ 1 \end{bmatrix} = -9\begin{bmatrix} 3 \\ 2 \\ 5 \end{bmatrix} + 15\begin{bmatrix} 2 \\ 1 \\ 3 \end{bmatrix} + \begin{bmatrix} 1 \\ 1 \\ 1 \end{bmatrix}$.
15. $\begin{bmatrix} 1 & 2 & 3 & 2 & 1 \\ 3 & 1 & 2 & 3 & 1 \\ 2 & -1 & -1 & 1 & 1 \\ 1 & 1 & 2 & 3 & 1 \end{bmatrix} \sim \begin{bmatrix} 1 & 0 & 0 & 0 & 0 \\ 0 & 1 & 0 & -5 & 0 \\ 0 & 0 & 1 & 4 & 0 \\ 0 & 0 & 0 & 0 & 1 \end{bmatrix}$,
$\mathbf{c}_4$ is a nonpivot column, and the equation $x_1\mathbf{c}_1 + x_2\mathbf{c}_2 + x_3\mathbf{c}_3 + x_5\mathbf{c}_5 = \mathbf{0}$ implies that $x_1 = x_2 = x_3 = x_5 = 0$.
17. $x_1\mathbf{c}_1 + x_2\mathbf{c}_2 + x_4\mathbf{c}_4 = \mathbf{0}$ implies $x_1 = x_2 = x_4 = 0$.
$x_1\mathbf{c}_1 + x_2\mathbf{c}_2 + x_5\mathbf{c}_5 = \mathbf{0}$ implies $x_1 = x_2 = x_5 = 0$.

$x_1\mathbf{c}_1+x_3\mathbf{c}_3+x_4\mathbf{c}_4=\mathbf{0}$ implies $x_1=x_3=x_4=0$.
$x_1\mathbf{c}_1+x_3\mathbf{c}_3+x_5\mathbf{c}_5=\mathbf{0}$ implies $x_1=x_3=x_5=0$.

19. $\{\mathbf{c}_1,\mathbf{c}_2,\mathbf{c}_4\}$, $\{\mathbf{c}_1,\mathbf{c}_2,\mathbf{c}_5\}$, $\{\mathbf{c}_1,\mathbf{c}_3,\mathbf{c}_4\}$, $\{\mathbf{c}_1,\mathbf{c}_3,\mathbf{c}_5\}$, $\{\mathbf{c}_2,\mathbf{c}_3,\mathbf{c}_4\}$, $\{\mathbf{c}_2,\mathbf{c}_3,\mathbf{c}_5\}$, $\{\mathbf{c}_1,\mathbf{c}_4,\mathbf{c}_5\}$, and $\{\mathbf{c}_2,\mathbf{c}_4,\mathbf{c}_5\}$.

Section 2.2

1. $\operatorname{Span}\left\{\begin{bmatrix}1\\2\\1\\3\end{bmatrix},\begin{bmatrix}2\\1\\4\\5\end{bmatrix}\right\}=\left\{\begin{bmatrix}a+2b\\2b+a\\a+4b\\3a+5b\end{bmatrix}: a,b \text{ in } \mathbb{R}\right\}$

3. $\begin{bmatrix}a+b\\b+5c\\a-2b\\a\end{bmatrix}=a\begin{bmatrix}1\\0\\1\\1\end{bmatrix}+b\begin{bmatrix}1\\1\\-2\\0\end{bmatrix}+c\begin{bmatrix}0\\5\\0\\0\end{bmatrix}$

5. The vector $\begin{bmatrix}2\\0\\0\end{bmatrix}=\begin{bmatrix}1\\0\\0\end{bmatrix}+\begin{bmatrix}1\\0\\0\end{bmatrix}$ cannot be written in the form $\begin{bmatrix}a+b+1\\a\\b\end{bmatrix}$.

7. The vector $\begin{bmatrix}0\\0\\4\\0\end{bmatrix}=\begin{bmatrix}0\\0\\2\\0\end{bmatrix}+\begin{bmatrix}0\\0\\2\\0\end{bmatrix}$ cannot be written in the form $\begin{bmatrix}a\\a-b\\2\\b\end{bmatrix}$.

9. $\begin{bmatrix}a\\b\\c\\d\\e\end{bmatrix}=\begin{bmatrix}a\\2b\\3a\\5b\\a+3b\end{bmatrix}=a\begin{bmatrix}1\\0\\3\\0\\1\end{bmatrix}+b\begin{bmatrix}0\\2\\0\\5\\3\end{bmatrix}$

11. $\begin{bmatrix}a\\b\\c\\d\end{bmatrix}=\begin{bmatrix}b+c\\b\\c\\-3b-3c\end{bmatrix}=b\begin{bmatrix}1\\1\\0\\-3\end{bmatrix}+c\begin{bmatrix}1\\0\\1\\-3\end{bmatrix}$

13. $\begin{bmatrix}5\\4\\5\\3\end{bmatrix}=\begin{bmatrix}3\\2\\1\\1\end{bmatrix}+2\begin{bmatrix}1\\1\\2\\1\end{bmatrix}$

15. $\begin{bmatrix}0\\0\\0\\1\end{bmatrix}=\begin{bmatrix}1\\2\\1\\1\end{bmatrix}+\begin{bmatrix}1\\1\\0\\1\end{bmatrix}-\begin{bmatrix}3\\2\\1\\1\end{bmatrix}$

17. $\begin{bmatrix}1\\1\\1\\1\end{bmatrix}$ is not in $\operatorname{Span}\left\{\begin{bmatrix}1\\1\\2\\1\end{bmatrix},\begin{bmatrix}1\\2\\1\\1\end{bmatrix}\right\}$.

19. $\begin{bmatrix}1\\0\\1\\0\end{bmatrix}$ is not in $\operatorname{Span}\left\{\begin{bmatrix}4\\1\\0\\5\end{bmatrix},\begin{bmatrix}1\\1\\0\\1\end{bmatrix},\begin{bmatrix}1\\1\\1\\1\end{bmatrix}\right\}$.

21. $\left\{\begin{bmatrix}1\\0\\-1\end{bmatrix},\begin{bmatrix}0\\1\\1\end{bmatrix}\right\}$

23. $\left\{\begin{bmatrix}1\\0\\-2\\-1\end{bmatrix},\begin{bmatrix}0\\1\\2\\2\end{bmatrix}\right\}$

25. $\begin{bmatrix}1\\1\\1\\1\end{bmatrix}=\frac{1}{2}\begin{bmatrix}1\\1\\0\\1\end{bmatrix}$

27. $a=-55$, $b=-17$

29. $\left\{\begin{bmatrix}1\\1\\2\\3\end{bmatrix},\begin{bmatrix}2\\3\\3\\5\end{bmatrix}\right\}$

31. $\left\{\begin{bmatrix}-5\\1\\0\\3\end{bmatrix},\begin{bmatrix}1\\0\\1\\-2\end{bmatrix}\right\}$

33. $\left\{\begin{bmatrix}1\\-1\\-1\\0\end{bmatrix}\right\}$

35. $\left\{\begin{bmatrix}-\frac{4}{3}\\1\\0\\\frac{5}{3}\end{bmatrix},\begin{bmatrix}\frac{1}{3}\\0\\1\\-\frac{5}{3}\end{bmatrix}\right\}$

37. $\left\{\begin{bmatrix}-3\\2\\1\\0\\0\end{bmatrix},\begin{bmatrix}-1\\1\\0\\1\\0\end{bmatrix},\begin{bmatrix}-4\\3\\0\\0\\1\end{bmatrix}\right\}$

Section 2.3

1. $\begin{bmatrix}1&2&1\\2&1&1\\-1&4&1\\3&3&2\end{bmatrix}\sim\begin{bmatrix}1&0&\frac{1}{3}\\0&1&\frac{1}{3}\\0&0&0\\0&0&0\end{bmatrix}$.

All subsets with 2 vectors from the set $\left\{\begin{bmatrix}1\\2\\-1\\3\end{bmatrix},\begin{bmatrix}2\\1\\4\\3\end{bmatrix},\begin{bmatrix}1\\1\\1\\2\end{bmatrix}\right\}$ are bases.

3. $\begin{bmatrix}1&2&2&1\\1&-5&1&-1\\1&3&5&2\\1&3&1&1\end{bmatrix}\sim\begin{bmatrix}1&0&0&0\\0&1&0&\frac{1}{4}\\0&0&1&\frac{1}{4}\\0&0&0&0\end{bmatrix}$.

The bases are $\left\{\begin{bmatrix}1\\1\\1\\1\end{bmatrix},\begin{bmatrix}2\\-5\\3\\3\end{bmatrix},\begin{bmatrix}2\\1\\5\\1\end{bmatrix}\right\}$, $\left\{\begin{bmatrix}1\\1\\1\\1\end{bmatrix},\begin{bmatrix}2\\-5\\3\\3\end{bmatrix},\begin{bmatrix}1\\-1\\2\\1\end{bmatrix}\right\}$, $\left\{\begin{bmatrix}1\\1\\1\\1\end{bmatrix},\begin{bmatrix}2\\1\\5\\1\end{bmatrix},\begin{bmatrix}1\\-1\\2\\1\end{bmatrix}\right\}$.

5. $\begin{bmatrix}1&2&4&1\\2&1&2&1\\2&1&2&1\\2&1&2&1\end{bmatrix}\sim\begin{bmatrix}1&0&0&\frac{1}{3}\\0&1&2&\frac{1}{3}\\0&0&0&0\\0&0&0&0\end{bmatrix}$. The bases are $\left\{\begin{bmatrix}1\\2\\2\\2\end{bmatrix},\begin{bmatrix}2\\1\\1\\1\end{bmatrix}\right\}$, $\left\{\begin{bmatrix}1\\2\\2\\2\end{bmatrix},\begin{bmatrix}4\\2\\2\\2\end{bmatrix}\right\}$, $\left\{\begin{bmatrix}1\\2\\2\\2\end{bmatrix},\begin{bmatrix}1\\1\\1\\1\end{bmatrix}\right\}$, $\left\{\begin{bmatrix}2\\1\\1\\1\end{bmatrix},\begin{bmatrix}1\\1\\1\\1\end{bmatrix}\right\}$, and $\left\{\begin{bmatrix}4\\2\\2\\2\end{bmatrix},\begin{bmatrix}1\\1\\1\\1\end{bmatrix}\right\}$.

7. Since $\begin{bmatrix}1&2&1&4&3\\1&2&2&3&2\\2&4&1&9&7\end{bmatrix}\sim\begin{bmatrix}1&2&0&5&4\\0&0&1&-1&-1\\0&0&0&0&0\end{bmatrix}$, we have $\dim \mathbf{C}(A)=2$ and $\{\mathbf{c}_1,\mathbf{c}_3\}$ is a basis. The other bases are $\{\mathbf{c}_1,\mathbf{c}_4\}$, $\{\mathbf{c}_1,\mathbf{c}_5\}$, $\{\mathbf{c}_2,\mathbf{c}_3\}$, $\{\mathbf{c}_2,\mathbf{c}_4\}$, $\{\mathbf{c}_5,\mathbf{c}_5\}$, $\{\mathbf{c}_3,\mathbf{c}_4\}$, $\{\mathbf{c}_3,\mathbf{c}_5\}$, $\{\mathbf{c}_4,\mathbf{c}_5\}$.

9. $\begin{bmatrix}1&1&1&1\\2&1&1&1\\1&1&2&3\\1&2&1&0\end{bmatrix}\sim\begin{bmatrix}1&0&0&0\\0&1&0&-1\\0&0&1&2\\0&0&0&0\end{bmatrix}$

The sets $\{\mathbf{c}_1,\mathbf{c}_2,\mathbf{c}_3\}$, $\{\mathbf{c}_1,\mathbf{c}_2,\mathbf{c}_4\}$, and $\{\mathbf{c}_1,\mathbf{c}_3,\mathbf{c}_4\}$ are bases of $\mathbf{C}(A)$.

The set $\{\mathbf{c}_2,\mathbf{c}_3,\mathbf{c}_4\}$ is not a basis in $\mathbf{C}(A)$ because the vectors $\mathbf{c}_2,\mathbf{c}_3,\mathbf{c}_4$ are linearly dependent.

11. Since $\begin{bmatrix}1&2&3&2\\3&1&2&3\\5&5&8&7\end{bmatrix}\sim\begin{bmatrix}1&0&\frac{1}{5}&\frac{4}{5}\\0&1&\frac{7}{5}&\frac{3}{5}\\0&0&0&0\end{bmatrix}$, the set $\left\{\begin{bmatrix}1\\3\\5\end{bmatrix},\begin{bmatrix}2\\1\\5\end{bmatrix}\right\}$ is a basis in $\mathbf{C}(A)$, the set $\left\{\begin{bmatrix}1\\0\\\frac{1}{5}\\\frac{4}{5}\end{bmatrix},\begin{bmatrix}0\\1\\\frac{7}{5}\\\frac{3}{5}\end{bmatrix}\right\}$ is a basis in $\mathbf{C}(A^T)$, and $\dim \mathbf{C}(A)=\dim \mathbf{C}(A^T)=2$.

13. Since $\begin{bmatrix}1&2&1\\2&1&3\\1&-1&2\\3&3&4\\5&4&7\end{bmatrix}\sim\begin{bmatrix}1&0&\frac{5}{3}\\0&1&-\frac{1}{3}\\0&0&0\\0&0&0\\0&0&0\end{bmatrix}$, the set $\{\mathbf{c}_1,\mathbf{c}_2\}$ is a basis in $\mathbf{C}(A)$, the set $\left\{\begin{bmatrix}1\\0\\\frac{5}{3}\end{bmatrix},\begin{bmatrix}0\\1\\-\frac{1}{3}\end{bmatrix}\right\}$ is a basis in $\mathbf{C}(A^T)$, and $\dim \mathbf{C}(A)=\dim \mathbf{C}(A^T)=2$.

15. The set $\{\mathbf{c}_1,\mathbf{c}_2,\mathbf{c}_4\}$ is a basis in $\mathbf{C}(A)$.

Since the vector $\begin{bmatrix} x_1 \\ x_2 \\ x_3 \\ x_4 \end{bmatrix}$ is in $\mathbf{N}(A)$ if $\begin{bmatrix} 1 & 0 & 4 & 0 \\ 0 & 1 & 2 & 0 \\ 0 & 0 & 0 & 1 \end{bmatrix} \begin{bmatrix} x_1 \\ x_2 \\ x_3 \\ x_4 \end{bmatrix} = \begin{bmatrix} 0 \\ 0 \\ 0 \end{bmatrix}$, we have $x_1 = -4x_3$, $x_2 = -2x_3$, and $x_4 = 0$. Consequently, $\left\{ \begin{bmatrix} -4 \\ -2 \\ 1 \\ 0 \end{bmatrix} \right\}$ is a basis in $\mathbf{N}(A)$ and $\dim \mathbf{C}(A) + \dim \mathbf{N}(A) = 3 + 1 =$ the number of columns of the matrix A.

17. The set $\{\mathbf{c}_1, \mathbf{c}_2, \mathbf{c}_4\}$ is a basis in $\mathbf{C}(A)$, the set $\left\{ \begin{bmatrix} -3 \\ -7 \\ 1 \\ 0 \\ 0 \end{bmatrix}, \begin{bmatrix} -2 \\ -4 \\ -5 \\ 0 \\ 1 \end{bmatrix} \right\}$ is a basis in $\mathbf{N}(A)$, and $\dim \mathbf{C}(A) + \dim \mathbf{N}(A) = 3 + 2 = 5$.

19. The set $\{\mathbf{c}_1, \mathbf{c}_3, \mathbf{c}_5\}$ is a basis in $\mathbf{C}(A)$, the set $\left\{ \begin{bmatrix} -a \\ 1 \\ 0 \\ 0 \\ 0 \end{bmatrix}, \begin{bmatrix} -b \\ 0 \\ -c \\ 1 \\ 0 \end{bmatrix} \right\}$ is a basis in $\mathbf{N}(A)$, and $\dim \mathbf{C}(A) + \dim \mathbf{N}(A) = 3 + 2 = 5$.

21. The set $\{\mathbf{c}_1, \mathbf{c}_2, \mathbf{c}_4\}$ is a basis in $\mathbf{C}(A)$, the set $\left\{ \begin{bmatrix} -a \\ -b \\ 1 \\ 0 \end{bmatrix} \right\}$ is a basis in $\mathbf{N}(A)$, and $\dim \mathbf{C}(A) + \dim \mathbf{N}(A) = 2 + 1 = 3$.

23. $\dim \mathbf{C}(A) = 3$, $\dim \mathbf{N}(A) = 2$, $\dim \mathbf{C}(A^T) = 3$, $\dim \mathbf{N}(A^T) = 2$.

25. $\dim \mathbf{C}(A) = 5$, $\dim \mathbf{N}(A) = 2$, $\dim \mathbf{C}(A^T) = 5$, $\dim \mathbf{N}(A^T) = 0$.

27. We have $\begin{bmatrix} 3 & 1 & 1 \\ 1 & 1 & 2 \\ 1 & 3 & 7 \end{bmatrix} \sim \begin{bmatrix} 1 & 0 & -\frac{1}{2} \\ 0 & 1 & \frac{5}{2} \\ 0 & 0 & 0 \end{bmatrix}$ and

$$A^T = \begin{bmatrix} 3 & 1 & 1 \\ 1 & 1 & 3 \\ 1 & 2 & 7 \end{bmatrix} \sim \begin{bmatrix} 1 & 0 & -\frac{1}{2} \\ 0 & 1 & \frac{5}{2} \\ 0 & 0 & 0 \end{bmatrix}.$$

The set $\left\{ \begin{bmatrix} 3 \\ 1 \\ 1 \end{bmatrix}, \begin{bmatrix} 1 \\ 1 \\ 3 \end{bmatrix} \right\}$ is a basis in $\mathbf{C}(A)$.

The set $\left\{ \begin{bmatrix} 1 \\ -5 \\ 2 \end{bmatrix} \right\}$ is a basis in $\mathbf{N}(A)$.

The set $\left\{ \begin{bmatrix} 3 \\ 1 \\ 1 \end{bmatrix}, \begin{bmatrix} 1 \\ 1 \\ 2 \end{bmatrix} \right\}$ is a basis in $\mathbf{C}(A^T)$.

The set $\left\{ \begin{bmatrix} 1 \\ -4 \\ 1 \end{bmatrix} \right\}$ is a basis in $\mathbf{N}(A^T)$.

29. We have $\begin{bmatrix} 2 & 1 & 1 & 3 \\ 1 & 2 & 3 & 1 \end{bmatrix} \sim \begin{bmatrix} 1 & 0 & -\frac{1}{3} & \frac{5}{3} \\ 0 & 1 & \frac{5}{3} & -\frac{1}{3} \end{bmatrix}$ and $A^T = \begin{bmatrix} 2 & 1 \\ 1 & 2 \\ 1 & 3 \\ 3 & 1 \end{bmatrix} \sim \begin{bmatrix} 1 & 0 \\ 0 & 1 \\ 0 & 0 \\ 0 & 0 \end{bmatrix}$.

The set $\left\{ \begin{bmatrix} 2 \\ 1 \end{bmatrix}, \begin{bmatrix} 1 \\ 2 \end{bmatrix} \right\}$ is a basis in $\mathbf{C}(A)$.

The set $\left\{ \begin{bmatrix} -1 \\ 5 \\ 3 \\ 0 \end{bmatrix}, \begin{bmatrix} 5 \\ -1 \\ 0 \\ 3 \end{bmatrix} \right\}$ is a basis in $\mathbf{N}(A)$.

The set $\left\{ \begin{bmatrix} 2 \\ 1 \\ 1 \\ 3 \end{bmatrix}, \begin{bmatrix} 1 \\ 2 \\ 3 \\ 1 \end{bmatrix} \right\}$ is a basis in $\mathbf{C}(A^T)$.

$\mathbf{N}(A^T) = \{\mathbf{0}\}$.

31. $\left(A^T \begin{bmatrix} x_1 \\ \vdots \\ x_m \end{bmatrix} \right)^T = \begin{bmatrix} x_1 \\ \vdots \\ x_m \end{bmatrix}^T (A^T)^T = \begin{bmatrix} x_1 & \dots & x_m \end{bmatrix} A.$

33. Since $\begin{bmatrix} 1 & 1 & 3 & 1 \\ 1 & 2 & 1 & 3 \\ 3 & 4 & 7 & 5 \end{bmatrix} \sim \begin{bmatrix} 1 & 0 & 5 & -1 \\ 0 & 1 & -2 & 2 \\ 0 & 0 & 0 & 0 \end{bmatrix}$, the transition matrix from $\mathscr{B}$ to $\mathscr{C}$ is $\begin{bmatrix} 5 & -1 \\ -2 & 2 \end{bmatrix}$ and the transition matrix from $\mathscr{C}$ to $\mathscr{B}$ is $\begin{bmatrix} 1/4 & 1/8 \\ 1/4 & 5/8 \end{bmatrix}$.

35. $\begin{bmatrix} \frac{1}{3} & \frac{1}{3} \\ -\frac{1}{3} & \frac{2}{3} \end{bmatrix} \begin{bmatrix} 1 & 2 \\ 1 & 3 \end{bmatrix} = \begin{bmatrix} \frac{2}{3} & \frac{5}{3} \\ \frac{1}{3} & \frac{4}{3} \end{bmatrix}$

37. $\begin{bmatrix} 3 & 4 \\ 2 & 5 \end{bmatrix}^{-1} = \begin{bmatrix} 5/7 & -4/7 \\ -2/7 & 3/7 \end{bmatrix}$

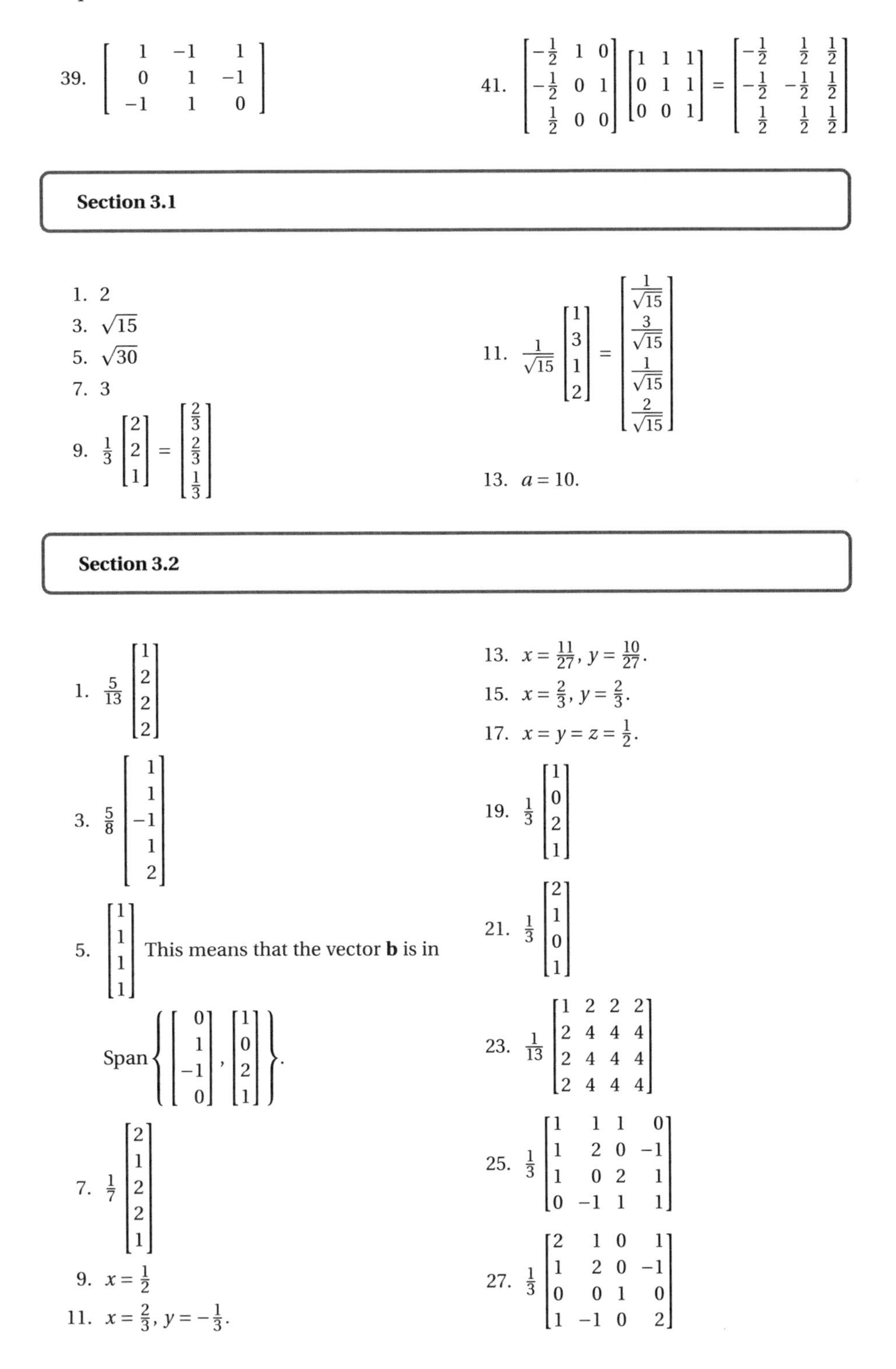

39. $\begin{bmatrix} 1 & -1 & 1 \\ 0 & 1 & -1 \\ -1 & 1 & 0 \end{bmatrix}$

41. $\begin{bmatrix} -\frac{1}{2} & 1 & 0 \\ -\frac{1}{2} & 0 & 1 \\ \frac{1}{2} & 0 & 0 \end{bmatrix}\begin{bmatrix} 1 & 1 & 1 \\ 0 & 1 & 1 \\ 0 & 0 & 1 \end{bmatrix} = \begin{bmatrix} -\frac{1}{2} & \frac{1}{2} & \frac{1}{2} \\ -\frac{1}{2} & -\frac{1}{2} & \frac{1}{2} \\ \frac{1}{2} & \frac{1}{2} & \frac{1}{2} \end{bmatrix}$

Section 3.1

1. 2

3. $\sqrt{15}$

5. $\sqrt{30}$

7. 3

9. $\frac{1}{3}\begin{bmatrix} 2 \\ 2 \\ 1 \end{bmatrix} = \begin{bmatrix} \frac{2}{3} \\ \frac{2}{3} \\ \frac{1}{3} \end{bmatrix}$

11. $\frac{1}{\sqrt{15}}\begin{bmatrix} 1 \\ 3 \\ 1 \\ 2 \end{bmatrix} = \begin{bmatrix} \frac{1}{\sqrt{15}} \\ \frac{3}{\sqrt{15}} \\ \frac{1}{\sqrt{15}} \\ \frac{2}{\sqrt{15}} \end{bmatrix}$

13. $a = 10$.

Section 3.2

1. $\frac{5}{13}\begin{bmatrix} 1 \\ 2 \\ 2 \\ 2 \end{bmatrix}$

3. $\frac{5}{8}\begin{bmatrix} 1 \\ 1 \\ -1 \\ 1 \\ 2 \end{bmatrix}$

5. $\begin{bmatrix} 1 \\ 1 \\ 1 \\ 1 \end{bmatrix}$ This means that the vector **b** is in

$$\text{Span}\left\{\begin{bmatrix} 0 \\ 1 \\ -1 \\ 0 \end{bmatrix}, \begin{bmatrix} 1 \\ 0 \\ 2 \\ 1 \end{bmatrix}\right\}.$$

7. $\frac{1}{7}\begin{bmatrix} 2 \\ 1 \\ 2 \\ 2 \\ 1 \end{bmatrix}$

9. $x = \frac{1}{2}$

11. $x = \frac{2}{3}, y = -\frac{1}{3}$.

13. $x = \frac{11}{27}, y = \frac{10}{27}$.

15. $x = \frac{2}{3}, y = \frac{2}{3}$.

17. $x = y = z = \frac{1}{2}$.

19. $\frac{1}{3}\begin{bmatrix} 1 \\ 0 \\ 2 \\ 1 \end{bmatrix}$

21. $\frac{1}{3}\begin{bmatrix} 2 \\ 1 \\ 0 \\ 1 \end{bmatrix}$

23. $\frac{1}{13}\begin{bmatrix} 1 & 2 & 2 & 2 \\ 2 & 4 & 4 & 4 \\ 2 & 4 & 4 & 4 \\ 2 & 4 & 4 & 4 \end{bmatrix}$

25. $\frac{1}{3}\begin{bmatrix} 1 & 1 & 1 & 0 \\ 1 & 2 & 0 & -1 \\ 1 & 0 & 2 & 1 \\ 0 & -1 & 1 & 1 \end{bmatrix}$

27. $\frac{1}{3}\begin{bmatrix} 2 & 1 & 0 & 1 \\ 1 & 2 & 0 & -1 \\ 0 & 0 & 1 & 0 \\ 1 & -1 & 0 & 2 \end{bmatrix}$

Section 3.3

1. Since $\mathbf{u}_1 \cdot \mathbf{u}_2 = 0$, the projection of the vector $\mathbf{v}$ on Span $\{\mathbf{u}_1, \mathbf{u}_2\}$ is $\frac{\mathbf{u}_1\cdot\mathbf{v}}{\mathbf{u}_1\cdot\mathbf{u}_1}\mathbf{u}_1 + \frac{\mathbf{u}_2\cdot\mathbf{v}}{\mathbf{u}_2\cdot\mathbf{u}_2}\mathbf{u}_2 = \mathbf{v}$.

3. $\mathbf{v}_1 = \begin{bmatrix} 1 \\ 1 \\ 1 \\ 1 \end{bmatrix}$ and $\mathbf{v}_2 = \begin{bmatrix} 3 \\ -1 \\ -1 \\ -1 \end{bmatrix}$

5. $\mathbf{v}_1 = \begin{bmatrix} 1 \\ 1 \\ 0 \\ 1 \\ 0 \end{bmatrix}$ and $\mathbf{v}_2 = \begin{bmatrix} 2 \\ -1 \\ 3 \\ -1 \\ 3 \end{bmatrix}$

7. $\mathbf{u}_3 = \begin{bmatrix} 1 \\ 0 \\ -1 \\ -1 \end{bmatrix}$

9. $\mathbf{u}_2 = \begin{bmatrix} 1 \\ 2 \\ -1 \\ 0 \end{bmatrix}$ and $\mathbf{u}_3 = \begin{bmatrix} 1 \\ -1 \\ -1 \\ 3 \end{bmatrix}$

11. $Q = \frac{\sqrt{2}}{2}\begin{bmatrix} 1 & 1 \\ 1 & -1 \end{bmatrix}$, $R = \frac{\sqrt{2}}{2}\begin{bmatrix} 2 & 3 \\ 0 & 1 \end{bmatrix}$

13. $Q = \begin{bmatrix} \frac{\sqrt{3}}{3} & 0 \\ -\frac{\sqrt{3}}{3} & \frac{\sqrt{2}}{2} \\ \frac{\sqrt{3}}{3} & \frac{\sqrt{2}}{2} \end{bmatrix}$, $R = \begin{bmatrix} \sqrt{3} & \sqrt{3} \\ 0 & 2\sqrt{2} \end{bmatrix}$

15. $Q = \begin{bmatrix} \frac{\sqrt{3}}{3} & -\frac{2\sqrt{33}}{33} \\ 0 & \frac{\sqrt{33}}{11} \\ \frac{\sqrt{3}}{3} & \frac{4\sqrt{33}}{33} \\ -\frac{\sqrt{3}}{3} & \frac{2\sqrt{33}}{33} \end{bmatrix}$, $R = \begin{bmatrix} \sqrt{3} & -\frac{\sqrt{3}}{3} \\ 0 & \frac{\sqrt{33}}{3} \end{bmatrix}$

17. $Q = \begin{bmatrix} \frac{\sqrt{2}}{2} & \frac{\sqrt{2}}{2} & 0 \\ 0 & 0 & 1 \\ \frac{\sqrt{2}}{2} & -\frac{\sqrt{2}}{2} & 0 \end{bmatrix}$,

$R = \begin{bmatrix} \sqrt{2} & \frac{\sqrt{2}}{2} & \frac{\sqrt{2}}{2} \\ 0 & \frac{\sqrt{2}}{2} & -\frac{\sqrt{2}}{2} \\ 0 & 0 & 1 \end{bmatrix}$

19. $Q = \begin{bmatrix} \frac{1}{\sqrt{3}} & -\frac{2}{\sqrt{15}} & \frac{3}{\sqrt{35}} \\ \frac{1}{\sqrt{3}} & \frac{1}{\sqrt{15}} & -\frac{4}{\sqrt{35}} \\ \frac{1}{\sqrt{3}} & \frac{1}{\sqrt{15}} & \frac{1}{\sqrt{35}} \\ 0 & \frac{3}{\sqrt{15}} & \frac{3}{\sqrt{35}} \end{bmatrix}$,

$R = \begin{bmatrix} \sqrt{3} & \frac{2}{\sqrt{3}} & \frac{2}{\sqrt{3}} \\ 0 & \frac{5}{\sqrt{15}} & \frac{2}{\sqrt{15}} \\ 0 & 0 & \frac{7}{\sqrt{35}} \end{bmatrix}$

Section 3.4

1. $\begin{bmatrix} 2 & 4 & 4 \\ 4 & 8 & 8 \\ 4 & 8 & 8 \end{bmatrix}\begin{bmatrix} x \\ y \\ z \end{bmatrix} = \begin{bmatrix} 4 \\ 8 \\ 8 \end{bmatrix}$

3. $y = 2x - \frac{2}{5}$, where x is a free variable.

5. $x = y = \frac{1}{2}$

7. $x = \frac{5}{2} - y$, where y is a free variable.

9. $y = -2x + \frac{3}{2} - z$, where x and z are free variable.

11. $y = x$, $z = -3x + 1$, where x is a free variable.

13. $x = \frac{11}{27}$, $y = \frac{10}{27}$

15. $x = -\frac{2}{15}$, $y = \frac{8}{15}$

17. $\frac{1}{3}\begin{bmatrix} 4 \\ 2 \\ 3 \\ 1 \end{bmatrix}$

19. $\frac{1}{7}\begin{bmatrix} 2 \\ 2 \\ 2 \\ 1 \\ 1 \end{bmatrix}$

21. $\frac{1}{2}\begin{bmatrix}1&0&0&1\\0&0&0&0\\0&0&2&0\\1&0&0&1\end{bmatrix}$

23. $\frac{1}{5}\begin{bmatrix}3&2&1&1\\2&3&-1&-1\\1&-1&2&2\\1&-1&2&2\end{bmatrix}$

Section 3.5

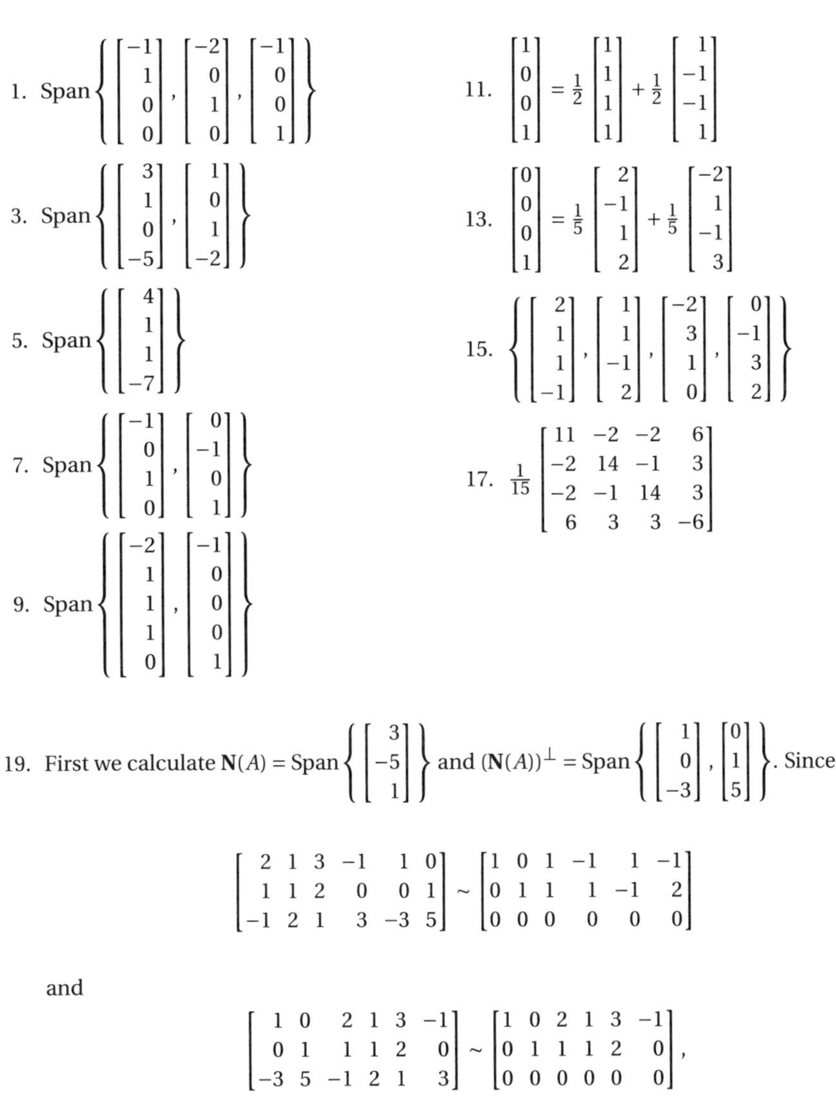

1. $\operatorname{Span}\left\{\begin{bmatrix}-1\\1\\0\\0\end{bmatrix},\begin{bmatrix}-2\\0\\1\\0\end{bmatrix},\begin{bmatrix}-1\\0\\0\\1\end{bmatrix}\right\}$

3. $\operatorname{Span}\left\{\begin{bmatrix}3\\1\\0\\-5\end{bmatrix},\begin{bmatrix}1\\0\\1\\-2\end{bmatrix}\right\}$

5. $\operatorname{Span}\left\{\begin{bmatrix}4\\1\\1\\-7\end{bmatrix}\right\}$

7. $\operatorname{Span}\left\{\begin{bmatrix}-1\\0\\1\\0\end{bmatrix},\begin{bmatrix}0\\-1\\0\\1\end{bmatrix}\right\}$

9. $\operatorname{Span}\left\{\begin{bmatrix}-2\\1\\1\\1\\0\end{bmatrix},\begin{bmatrix}-1\\0\\0\\0\\1\end{bmatrix}\right\}$

11. $\begin{bmatrix}1\\0\\0\\1\end{bmatrix}=\frac{1}{2}\begin{bmatrix}1\\1\\1\\1\end{bmatrix}+\frac{1}{2}\begin{bmatrix}1\\-1\\-1\\1\end{bmatrix}$

13. $\begin{bmatrix}0\\0\\0\\1\end{bmatrix}=\frac{1}{5}\begin{bmatrix}2\\-1\\1\\2\end{bmatrix}+\frac{1}{5}\begin{bmatrix}-2\\1\\-1\\3\end{bmatrix}$

15. $\left\{\begin{bmatrix}2\\1\\1\\-1\end{bmatrix},\begin{bmatrix}1\\1\\-1\\2\end{bmatrix},\begin{bmatrix}-2\\3\\1\\0\end{bmatrix},\begin{bmatrix}0\\-1\\3\\2\end{bmatrix}\right\}$

17. $\frac{1}{15}\begin{bmatrix}11&-2&-2&6\\-2&14&-1&3\\-2&-1&14&3\\6&3&3&-6\end{bmatrix}$

19. First we calculate $\mathbf{N}(A)=\operatorname{Span}\left\{\begin{bmatrix}3\\-5\\1\end{bmatrix}\right\}$ and $(\mathbf{N}(A))^{\perp}=\operatorname{Span}\left\{\begin{bmatrix}1\\0\\-3\end{bmatrix},\begin{bmatrix}0\\1\\5\end{bmatrix}\right\}$. Since

$$\begin{bmatrix}2&1&3&-1&1&0\\1&1&2&0&0&1\\-1&2&1&3&-3&5\end{bmatrix}\sim\begin{bmatrix}1&0&1&-1&1&-1\\0&1&1&1&-1&2\\0&0&0&0&0&0\end{bmatrix}$$

and

$$\begin{bmatrix}1&0&2&1&3&-1\\0&1&1&1&2&0\\-3&5&-1&2&1&3\end{bmatrix}\sim\begin{bmatrix}1&0&2&1&3&-1\\0&1&1&1&2&0\\0&0&0&0&0&0\end{bmatrix},$$

we have $(\mathbf{N}(A))^{\perp}=\mathbf{C}(A^T)$.

Chapter 4

1. 14
3. 63
5. 8
7. −25
9. −10
11. 7
13. 5

15. $$\det\begin{bmatrix} a_1 & b_1 & c_1 & d_1 & e_1 \\ a_2 & b_2 & 0 & d_2 & e_2 \\ a_3 & b_3 & 0 & d_3 & e_3 \\ a_4 & b_4 & 0 & d_4 & e_4 \\ a_5 & b_5 & 0 & d_5 & e_5 \end{bmatrix} = -\det\begin{bmatrix} c_1 & a_1 & b_1 & d_1 & e_1 \\ 0 & a_2 & b_2 & d_2 & e_2 \\ 0 & a_3 & b_3 & d_3 & e_3 \\ 0 & a_4 & b_4 & d_4 & e_4 \\ 0 & a_5 & b_5 & d_5 & e_5 \end{bmatrix}$$

17. $$\det\begin{bmatrix} a_1 & b_1 & c_1 & d_1 & e_1 \\ a_2 & b_2 & c_2 & d_2 & e_2 \\ 0 & 0 & 0 & d_3 & 0 \\ a_4 & b_4 & c_4 & d_4 & e_4 \\ a_5 & b_5 & c_5 & d_5 & e_5 \end{bmatrix} = \det\begin{bmatrix} 0 & 0 & 0 & d_3 & 0 \\ a_1 & b_1 & c_1 & d_1 & e_1 \\ a_2 & b_2 & c_2 & d_2 & e_2 \\ a_4 & b_4 & c_4 & d_4 & e_4 \\ a_5 & b_5 & c_5 & d_5 & e_5 \end{bmatrix} = -\det\begin{bmatrix} d_3 & 0 & 0 & 0 & 0 \\ d_1 & a_1 & b_1 & c_1 & e_1 \\ d_2 & a_2 & b_2 & c_2 & e_2 \\ d_4 & a_4 & b_4 & c_4 & e_4 \\ d_4 & a_5 & b_5 & c_5 & e_5 \end{bmatrix}$$

19. $$\det\begin{bmatrix} a_1 & b_1 & c_1 & d_1 \\ a_2 & b_2 & c_2 & d_2 \\ 0 & b_3 & c_3 & d_3 \\ 0 & b_4 & c_4 & d_4 \end{bmatrix} = \det\begin{bmatrix} a_1 & b_1 & c_1 & d_1 \\ 0 & b_2 - \frac{a_2}{a_1} b_1 & c_2 - \frac{a_2}{a_1} c_1 & d_2 - \frac{a_2}{a_1} d_1 \\ 0 & b_3 & c_3 & d_3 \\ 0 & b_4 & c_4 & d_4 \end{bmatrix}$$

21. $$\det\begin{bmatrix} a_1 & b_1 & c_1 & d_1 \\ a_2 & b_2 & c_2 & d_2 \\ 0 & b_3 & c_3 & d_3 \\ 0 & b_4 & c_4 & d_4 \end{bmatrix} = \det\begin{bmatrix} a_1 & b_1 & c_1 & d_1 \\ a_2 & b_2 & c_2 & d_2 \\ 0 & b_3 - \frac{a_3}{a_1} b_1 & c_3 - \frac{a_3}{a_1} c_1 & d_3 - \frac{a_3}{a_1} d_1 \\ 0 & b_4 & c_4 & d_4 \end{bmatrix}$$

23. $$\det\begin{bmatrix} a_1 & b_1 & c_1 & d_1 \\ a_2 & b_2 & c_2 & d_2 \\ a_3 & b_3 & c_3 & d_3 \\ a_4 & b_4 & c_4 & d_4 \end{bmatrix} = \det\begin{bmatrix} a_1 & b_1 & c_1 & d_1 \\ a_2 & b_2 & c_2 & d_2 \\ a_3 & b_3 & c_3 & d_3 \\ 0 & b_4 - \frac{a_4}{a_1} b_1 & c_4 - \frac{a_4}{a_1} c_1 & d_4 - \frac{a_4}{a_1} d_1 \end{bmatrix}$$

25. From the equality $AA^{-1} = I_n$ we get $\det A \det A^{-1} = 1$.

27. Use Theorem 4.14 and Theorem 4.15.

29. $$\det\begin{bmatrix} b_1 & a_1 & c_1 & d_1 \\ b_2 & a_2 & c_2 & d_2 \\ b_3 & a_3 & c_3 & d_3 \\ b_4 & a_4 & c_4 & d_4 \end{bmatrix} = -\det\begin{bmatrix} a_1 & b_1 & c_1 & d_1 \\ a_2 & b_2 & c_2 & d_2 \\ a_3 & b_3 & c_3 & d_3 \\ a_4 & b_4 & c_4 & d_4 \end{bmatrix}$$

31. $$\det\begin{bmatrix} a_3 & a_1 & a_2 & a_4 \\ b_3 & b_1 & b_2 & b_4 \\ c_3 & c_1 & c_2 & c_4 \\ d_3 & d_1 & d_2 & d_4 \end{bmatrix} = \det\begin{bmatrix} a_1 & a_2 & a_3 & a_4 \\ b_1 & b_2 & b_3 & b_4 \\ c_1 & c_2 & c_3 & c_4 \\ d_1 & d_2 & d_3 & d_4 \end{bmatrix} = \det\begin{bmatrix} a_1 & b_1 & c_1 & d_1 \\ a_2 & b_2 & c_2 & d_2 \\ a_3 & b_3 & c_3 & d_3 \\ a_4 & b_4 & c_4 & d_4 \end{bmatrix}$$

33. $y = 4$

35. $x = \frac{1}{7}$

37. $\frac{5}{9}$

39. $-\frac{5}{8}$

41. $\det\begin{bmatrix} B & \mathbf{0}_1 & C \\ \mathbf{0}_2 & a & \mathbf{0}_3 \\ D & \mathbf{0}_4 & E \end{bmatrix} = (-1)^m \det\begin{bmatrix} \mathbf{0}_2 & a & \mathbf{0}_3 \\ B & \mathbf{0}_1 & C \\ D & \mathbf{0}_4 & E \end{bmatrix} = (-1)^{m+p} \det\begin{bmatrix} a & \mathbf{0}_2 & \mathbf{0}_3 \\ \mathbf{0}_1 & B & C \\ \mathbf{0}_4 & D & E \end{bmatrix}$

Section 5.1

1. $\lambda^3 - 3\lambda^2 - 2\lambda + 2$

3. $\lambda^4 - 4\lambda^3 - \lambda^2 + 12\lambda$

5. 3 is an eigenvalue because, if we take $\lambda = 3$ in the characteristic polynomial $\begin{bmatrix} -\lambda+2 & 5 & 9 \\ 4 & -\lambda+5 & 1 \\ 4 & 2 & -\lambda+4 \end{bmatrix}$, we get $\det\begin{bmatrix} -1 & 5 & 9 \\ 4 & 2 & 1 \\ 4 & 2 & 1 \end{bmatrix} = 0.$

7. 2 is an eigenvalue because, if we take $\lambda = 2$ in the characteristic polynomial $\det\begin{bmatrix} -\lambda+3 & 2 & 5 & 1 \\ 1 & -\lambda+4 & 5 & 1 \\ 1 & 1 & -\lambda & 4 \\ 2 & 1 & 7 & -\lambda+3 \end{bmatrix}$, we get $\det\begin{bmatrix} 1 & 2 & 5 & 1 \\ 1 & 2 & 5 & 1 \\ 1 & 1 & -2 & 4 \\ 2 & 1 & 7 & 1 \end{bmatrix} = 0.$

9. 2 and 3.

11. 3 and 5.

13. 0, 1, and 2.

15. $\operatorname{Span}\left\{\begin{bmatrix} 5 \\ 7 \\ -15 \end{bmatrix}\right\}$

17. $\operatorname{Span}\left\{\begin{bmatrix} 2 \\ 0 \\ -3 \end{bmatrix}, \begin{bmatrix} 0 \\ 1 \\ -2 \end{bmatrix}\right\}$

19. $\operatorname{Span}\left\{\begin{bmatrix} 4 \\ -5 \\ -1 \\ 2 \end{bmatrix}\right\}$

21. $\operatorname{Span}\left\{\begin{bmatrix} 1 \\ -1 \\ 0 \\ 0 \end{bmatrix}, \begin{bmatrix} 1 \\ 0 \\ 1 \\ -1 \end{bmatrix}\right\}$

23. $\operatorname{Span}\left\{\begin{bmatrix} 1 \\ -1 \\ 0 \\ 0 \end{bmatrix}, \begin{bmatrix} 1 \\ 0 \\ -1 \\ 0 \end{bmatrix}, \begin{bmatrix} 1 \\ 0 \\ 0 \\ -1 \end{bmatrix}\right\}$

25. 1, 2, 7

27. 1 (double), 10

29. 2, 3, 4, 7

31. 1 (triple), 7

33. $D = \begin{bmatrix} 3 & 0 \\ 0 & 9 \end{bmatrix}, P = \begin{bmatrix} 1 & 1 \\ -1 & 5 \end{bmatrix}$

35. $D = \begin{bmatrix} 0 & 0 & 0 \\ 0 & 1 & 0 \\ 0 & 0 & 9 \end{bmatrix}$, $P = \begin{bmatrix} 4 & 1 & 1 \\ -1 & 2 & 2 \\ -1 & -2 & 2 \end{bmatrix}$

37. $D = \begin{bmatrix} 3 & 0 & 0 \\ 0 & 8 & 0 \\ 0 & 0 & 3 \end{bmatrix}$, $P = \begin{bmatrix} 2 & 1 & 2 \\ -1 & 1 & 0 \\ 0 & 1 & -1 \end{bmatrix}$

39. $D = \begin{bmatrix} 7 & 0 & 0 & 0 \\ 0 & 1 & 0 & 0 \\ 0 & 0 & 1 & 0 \\ 0 & 0 & 0 & 1 \end{bmatrix}$, $P = \begin{bmatrix} 1 & 1 & 1 & 1 \\ 1 & -1 & 0 & 0 \\ 1 & 0 & -1 & 0 \\ 3 & 0 & 0 & -1 \end{bmatrix}$

41. $\operatorname{Span}\left\{\begin{bmatrix} 2 \\ 3 \\ 1 \end{bmatrix}, \begin{bmatrix} 1 \\ 1 \\ 0 \end{bmatrix}\right\} \neq \mathbb{R}^3$

43. $\operatorname{Span}\left\{\begin{bmatrix} 1 \\ 1 \\ 0 \end{bmatrix}\right\} \neq \mathbb{R}^3$

Section 5.2

1. $A = \begin{bmatrix} 2 & 1 \\ 1 & -2 \end{bmatrix}\begin{bmatrix} 3 & 0 \\ 0 & 4 \end{bmatrix}\begin{bmatrix} 2/5 & 1/5 \\ 1/5 & -2/5 \end{bmatrix} = \begin{bmatrix} 16/5 & -2/5 \\ -2/5 & 19/5 \end{bmatrix}$

3. $A = \begin{bmatrix} x & y \\ y & -x \end{bmatrix}\begin{bmatrix} \lambda & 0 \\ 0 & \lambda \end{bmatrix}\begin{bmatrix} \frac{x}{x^2+y^2} & \frac{y}{x^2+y^2} \\ \frac{y}{x^2+y^2} & -\frac{x}{x^2+y^2} \end{bmatrix} = \begin{bmatrix} \lambda & 0 \\ 0 & \lambda \end{bmatrix}$

5. $A = \begin{bmatrix} 1 & 1 & 5 \\ 1 & -3 & -1 \\ 1 & 2 & -4 \end{bmatrix}\begin{bmatrix} 1 & 0 & 0 \\ 0 & 2 & 0 \\ 0 & 0 & 3 \end{bmatrix}\begin{bmatrix} 1/3 & 1/3 & 1/3 \\ 1/14 & -3/14 & 1/7 \\ 5/42 & -1/42 & -2/21 \end{bmatrix} = \begin{bmatrix} \frac{95}{42} & -\frac{19}{42} & -\frac{17}{21} \\ -\frac{19}{42} & \frac{71}{42} & -\frac{5}{21} \\ -\frac{17}{21} & -\frac{5}{21} & \frac{43}{21} \end{bmatrix}$

7. $A = \begin{bmatrix} 1 & 2 & 2 \\ 1 & 1 & -1 \\ 1 & 3 & -1 \end{bmatrix}\begin{bmatrix} 1 & 0 & 0 \\ 0 & 1 & 0 \\ 0 & 0 & 2 \end{bmatrix}\begin{bmatrix} 1/3 & 4/3 & -2/3 \\ 0 & -1/2 & 1/2 \\ 1/3 & -1/6 & -1/6 \end{bmatrix} = \begin{bmatrix} 5/3 & -1/3 & -1/3 \\ -1/3 & 7/6 & 1/6 \\ -1/3 & 1/6 & 7/6 \end{bmatrix}$

9. $A = \begin{bmatrix} 2 & 1 & 7 \\ 3 & 2 & -5 \\ 1 & 3 & 1 \end{bmatrix}\begin{bmatrix} 0 & 0 & 0 \\ 0 & 0 & 0 \\ 0 & 0 & 1 \end{bmatrix}\begin{bmatrix} \frac{17}{75} & \frac{4}{15} & -\frac{19}{75} \\ -\frac{8}{75} & -\frac{1}{15} & \frac{31}{75} \\ \frac{7}{75} & -\frac{1}{15} & \frac{1}{75} \end{bmatrix} = \begin{bmatrix} \frac{49}{75} & -\frac{7}{15} & \frac{7}{75} \\ -\frac{7}{15} & \frac{1}{3} & -\frac{1}{15} \\ \frac{7}{75} & -\frac{1}{15} & \frac{1}{75} \end{bmatrix}$

11. $A = \begin{bmatrix} 1 & 3 & 1 \\ 3 & -1 & 0 \\ 1 & 0 & -1 \end{bmatrix}\begin{bmatrix} 2 & 0 & 0 \\ 0 & 1 & 0 \\ 0 & 0 & 1 \end{bmatrix}\begin{bmatrix} \frac{1}{11} & \frac{3}{11} & \frac{1}{11} \\ \frac{3}{11} & -\frac{2}{11} & \frac{3}{11} \\ \frac{1}{11} & \frac{3}{11} & -\frac{10}{11} \end{bmatrix} = \begin{bmatrix} \frac{12}{11} & \frac{3}{11} & \frac{1}{11} \\ \frac{3}{11} & \frac{20}{11} & \frac{3}{11} \\ \frac{1}{11} & \frac{3}{11} & \frac{12}{11} \end{bmatrix}$

13. $A = \begin{bmatrix} 2 & 1 & 1 & 1 \\ 1 & 2 & 1 & 1 \\ 1 & 1 & 2 & 1 \\ 1 & 1 & 1 & -4 \end{bmatrix}\begin{bmatrix} 1 & 0 & 0 & 0 \\ 0 & 1 & 0 & 0 \\ 0 & 0 & 1 & 0 \\ 0 & 0 & 0 & 0 \end{bmatrix}\begin{bmatrix} \frac{14}{19} & -\frac{5}{19} & -\frac{5}{19} & \frac{1}{19} \\ -\frac{5}{19} & \frac{14}{19} & -\frac{5}{19} & \frac{1}{19} \\ -\frac{5}{19} & -\frac{5}{19} & \frac{14}{19} & \frac{1}{19} \\ \frac{1}{19} & \frac{1}{19} & \frac{1}{19} & -\frac{4}{19} \end{bmatrix}$

$= \begin{bmatrix} \frac{18}{19} & -\frac{1}{19} & -\frac{1}{19} & \frac{4}{19} \\ -\frac{1}{19} & \frac{18}{19} & -\frac{1}{19} & \frac{4}{19} \\ -\frac{1}{19} & -\frac{1}{19} & \frac{18}{19} & \frac{4}{19} \\ \frac{4}{19} & \frac{4}{19} & \frac{4}{19} & \frac{3}{19} \end{bmatrix}$

15. $A = \begin{bmatrix} 1 & 1 & 1 & 1 \\ 1 & -1 & 0 & 0 \\ 1 & 1 & -1 & 0 \\ 1 & 1 & 0 & -1 \end{bmatrix}\begin{bmatrix} 1 & 0 & 0 & 0 \\ 0 & 1 & 0 & 0 \\ 0 & 0 & 0 & 0 \\ 0 & 0 & 0 & 0 \end{bmatrix}\begin{bmatrix} 1/6 & 1/2 & 1/6 & 1/6 \\ 1/6 & -1/2 & 1/6 & 1/6 \\ 1/3 & 0 & -2/3 & 1/3 \\ 1/3 & 0 & 1/3 & -2/3 \end{bmatrix}$

$= \begin{bmatrix} 1/3 & 0 & 1/3 & 1/3 \\ 0 & 1 & 0 & 0 \\ 1/3 & 0 & 1/3 & 1/3 \\ 1/3 & 0 & 1/3 & 1/3 \end{bmatrix}$

17. $D = \begin{bmatrix} 20 & 0 \\ 0 & 25 \end{bmatrix}, P = \begin{bmatrix} \frac{2}{\sqrt{5}} & \frac{1}{\sqrt{5}} \\ -\frac{1}{\sqrt{5}} & \frac{2}{\sqrt{5}} \end{bmatrix}$

19. $D = \begin{bmatrix} 0 & 0 & 0 \\ 0 & 3 & 0 \\ 0 & 0 & 14 \end{bmatrix}, P = \begin{bmatrix} \frac{1}{\sqrt{46}} & \frac{1}{\sqrt{3}} & \frac{3}{\sqrt{14}} \\ -\frac{4}{\sqrt{46}} & -\frac{1}{\sqrt{3}} & \frac{2}{\sqrt{14}} \\ \frac{5}{\sqrt{46}} & -\frac{1}{\sqrt{3}} & \frac{1}{\sqrt{14}} \end{bmatrix}$

21. $D = \begin{bmatrix} 2 & 0 & 0 \\ 0 & 9 & 0 \\ 0 & 0 & 18 \end{bmatrix}$, $P = \begin{bmatrix} \frac{1}{\sqrt{2}} & \frac{1}{\sqrt{3}} & -\frac{1}{\sqrt{6}} \\ \frac{1}{\sqrt{2}} & -\frac{1}{\sqrt{3}} & \frac{1}{\sqrt{6}} \\ 0 & \frac{1}{\sqrt{3}} & \frac{2}{\sqrt{6}} \end{bmatrix}$

23. $D = \begin{bmatrix} 30 & 0 & 0 \\ 0 & 30 & 0 \\ 0 & 0 & -30 \end{bmatrix}$, $P = \begin{bmatrix} \frac{1}{\sqrt{5}} & \frac{2}{\sqrt{6}} & -\frac{2}{\sqrt{30}} \\ \frac{2}{\sqrt{5}} & -\frac{1}{\sqrt{6}} & \frac{1}{\sqrt{30}} \\ 0 & \frac{1}{\sqrt{6}} & \frac{5}{\sqrt{30}} \end{bmatrix}$

25. $D = \begin{bmatrix} 9 & 0 & 0 \\ 0 & 18 & 0 \\ 0 & 0 & 18 \end{bmatrix}$, $P = \begin{bmatrix} -\frac{2}{3} & \frac{1}{\sqrt{2}} & \frac{1}{\sqrt{18}} \\ \frac{1}{3} & 0 & \frac{4}{\sqrt{18}} \\ \frac{2}{3} & \frac{1}{\sqrt{2}} & -\frac{1}{\sqrt{18}} \end{bmatrix}$

27. $A = \begin{bmatrix} 1 \\ 1 \\ 0 \end{bmatrix} \begin{bmatrix} 1 & 1 & 0 \end{bmatrix} + 3 \begin{bmatrix} 1 \\ -1 \\ 1 \end{bmatrix} \begin{bmatrix} 1 & -1 & 1 \end{bmatrix} + 3 \begin{bmatrix} -1 \\ 1 \\ 2 \end{bmatrix} \begin{bmatrix} -1 & 1 & 2 \end{bmatrix}$

29. $A = 6 \begin{bmatrix} 1 \\ 2 \\ 0 \end{bmatrix} \begin{bmatrix} 1 & 2 & 0 \end{bmatrix} + 5 \begin{bmatrix} 2 \\ -1 \\ 1 \end{bmatrix} \begin{bmatrix} 2 & -1 & 1 \end{bmatrix} - \begin{bmatrix} -2 \\ 1 \\ 5 \end{bmatrix} \begin{bmatrix} -2 & 1 & 5 \end{bmatrix}$

31. $A = 5 \begin{bmatrix} 1 \\ 2 \end{bmatrix} \begin{bmatrix} 1 & 2 \end{bmatrix} + 4 \begin{bmatrix} 2 \\ -1 \end{bmatrix} \begin{bmatrix} 2 & -1 \end{bmatrix}$

33. $D = \begin{bmatrix} 0 & 0 & 0 & 0 \\ 0 & 1 & 0 & 0 \\ 0 & 0 & 2 & 0 \\ 0 & 0 & 0 & 3 \end{bmatrix}$, $P = \begin{bmatrix} 0 & \frac{1}{\sqrt{2}} & 0 & \frac{1}{\sqrt{2}} \\ 0 & -\frac{1}{\sqrt{2}} & 0 & \frac{1}{\sqrt{2}} \\ \frac{1}{\sqrt{2}} & 0 & \frac{1}{\sqrt{2}} & 0 \\ -\frac{1}{\sqrt{2}} & 0 & \frac{1}{\sqrt{2}} & 0 \end{bmatrix}$

35. $D = \begin{bmatrix} 1 & 0 & 0 & 0 \\ 0 & 1 & 0 & 0 \\ 0 & 0 & 1 & 0 \\ 0 & 0 & 0 & 4 \end{bmatrix}$, $P = \begin{bmatrix} 0 & \frac{1}{\sqrt{2}} & \frac{1}{\sqrt{6}} & \frac{1}{\sqrt{3}} \\ 0 & -\frac{1}{\sqrt{2}} & \frac{1}{\sqrt{6}} & \frac{1}{\sqrt{3}} \\ 0 & 0 & -\frac{2}{\sqrt{6}} & \frac{1}{\sqrt{3}} \\ 1 & 0 & 0 & 0 \end{bmatrix}$

37. $\alpha \begin{bmatrix} 1 & 3 & 1 \\ 3 & 9 & 3 \\ 1 & 3 & 1 \end{bmatrix} + \beta \begin{bmatrix} 1 & 1 & -4 \\ 1 & 1 & -4 \\ -4 & -4 & 16 \end{bmatrix} = \begin{bmatrix} \alpha+\beta & 3\alpha+\beta & \alpha-4\beta \\ 3\alpha+\beta & 9\alpha+\beta & 3\alpha-4\beta \\ \alpha-4\beta & 3\alpha-4\beta & \alpha+16\beta \end{bmatrix}$

Section 6.1

1. $f\left(\begin{bmatrix} x_1 \\ x_2 \\ x_3 \\ x_4 \end{bmatrix}\right) = [2 \;\; -3 \;\; 1 \;\; 2] \begin{bmatrix} x_1 \\ x_2 \\ x_3 \\ x_4 \end{bmatrix}$

3. $f\left(\begin{bmatrix} x \\ y \end{bmatrix}\right) = \begin{bmatrix} 1 & 0 \\ 1 & 1 \\ 0 & 2 \\ 1 & -1 \\ 1 & 1 \end{bmatrix} \begin{bmatrix} x \\ y \end{bmatrix}$

5. The standard matrix of f is $A = \begin{bmatrix} 2 & 1 \\ 3 & -2 \end{bmatrix}$ and the standard matrix of g is $B = \begin{bmatrix} 2 & 1 \end{bmatrix}$. Since $g\left(f\left(\begin{bmatrix} x \\ y \end{bmatrix}\right)\right) = 7x = \begin{bmatrix} 7 & 0 \end{bmatrix}\begin{bmatrix} x \\ y \end{bmatrix}$, the standard matrix of the linear transformation $g \circ f$ is $\begin{bmatrix} 7 & 0 \end{bmatrix}$. Note that the same matrix is obtained as the product of the standard matrices of g and f: $BA = \begin{bmatrix} 2 & 1 \end{bmatrix}\begin{bmatrix} 2 & 1 \\ 3 & -2 \end{bmatrix} = \begin{bmatrix} 7 & 0 \end{bmatrix}$.

7. The standard matrix of f is $A = \begin{bmatrix} 0 & 1 \\ 2 & -1 \end{bmatrix}$ and the standard matrix of g is $B = \begin{bmatrix} 2 & 0 \\ 1 & -1 \end{bmatrix}$. Since $g\left(f\left(\begin{bmatrix} x \\ y \end{bmatrix}\right)\right) = \begin{bmatrix} 2y \\ 2y - 2x \end{bmatrix} = \begin{bmatrix} 0 & 2 \\ -2 & 2 \end{bmatrix}\begin{bmatrix} x \\ y \end{bmatrix}$, the standard matrix of the linear transformation $g \circ f$ is $\begin{bmatrix} 0 & 2 \\ -2 & 2 \end{bmatrix}$. Note that the same matrix is obtained as the product of the standard matrices of g and f: $BA = \begin{bmatrix} 2 & 0 \\ 1 & -1 \end{bmatrix}\begin{bmatrix} 0 & 1 \\ 2 & -1 \end{bmatrix} = \begin{bmatrix} 0 & 2 \\ -2 & 2 \end{bmatrix}$.

9. $f^{-1}\left(\begin{bmatrix} x \\ y \end{bmatrix}\right) = \begin{bmatrix} -5x + 3y \\ 2x - y \end{bmatrix}$

11. $f^{-1}\left(\begin{bmatrix} x \\ y \\ z \end{bmatrix}\right) \begin{bmatrix} 2x + y - z \\ -x + z \\ -x - y + z \end{bmatrix}$

13. $f\left(\begin{bmatrix} x \\ y \end{bmatrix}\right) = \begin{bmatrix} 3 & 1 \\ 2 & 1 \end{bmatrix}\begin{bmatrix} 1 & 1 \\ 1 & 2 \end{bmatrix}^{-1}\begin{bmatrix} x \\ y \end{bmatrix} = \begin{bmatrix} 5 & -2 \\ 3 & -1 \end{bmatrix}\begin{bmatrix} x \\ y \end{bmatrix}$

15. $\begin{bmatrix} 1 & 2 & 1 \\ 2 & 2 & 1 \\ 1 & 1 & 1 \end{bmatrix}\begin{bmatrix} 1 & 1 & 1 \\ 0 & 1 & 1 \\ 0 & 0 & 1 \end{bmatrix}^{-1} = \begin{bmatrix} 1 & 2 & 1 \\ 2 & 2 & 1 \\ 1 & 1 & 1 \end{bmatrix}\begin{bmatrix} 1 & -1 & 0 \\ 0 & 1 & -1 \\ 0 & 0 & 1 \end{bmatrix} = \begin{bmatrix} 1 & 1 & -1 \\ 2 & 0 & -1 \\ 1 & 0 & 0 \end{bmatrix}$

17. $\begin{bmatrix} 1 & 1 & 1 \\ 1 & 0 & 2 \end{bmatrix}\begin{bmatrix} 2 & 1 & 1 \\ 1 & 2 & 1 \\ 1 & 1 & 2 \end{bmatrix}^{-1} = \begin{bmatrix} 1 & 1 & 1 \\ 1 & 0 & 2 \end{bmatrix}\begin{bmatrix} 3/4 & -1/4 & -1/4 \\ -1/4 & 3/4 & -1/4 \\ -1/4 & -1/4 & 3/4 \end{bmatrix} = \begin{bmatrix} 1/4 & 1/4 & 1/4 \\ 1/4 & -3/4 & 5/4 \end{bmatrix}$

19. If $\mathbf{x} = \begin{bmatrix} x_1 \\ \vdots \\ x_n \end{bmatrix}$ and $\{\mathbf{e}_1, \dots, \mathbf{e}_n\}$ is the standard basis in $\mathbb{R}^n$, then

$$f(\mathbf{x}) = f(x_1\mathbf{e}_1 + \dots + x_n\mathbf{e}_n) = x_1 f(\mathbf{e}_1) + \dots + x_n f(\mathbf{e}_n) = \mathbf{x} \cdot \begin{bmatrix} f(\mathbf{e}_1) \\ \vdots \\ f(\mathbf{e}_n) \end{bmatrix}.$$

Section 6.2

1. $\begin{bmatrix} 1/3 & 1/3 \\ 1/3 & -2/3 \end{bmatrix}\begin{bmatrix} 1 & -1 \\ 2 & 3 \end{bmatrix}\begin{bmatrix} 2 & 1 \\ 1 & -1 \end{bmatrix} = \begin{bmatrix} 8/3 & 1/3 \\ -13/3 & 4/3 \end{bmatrix}$

3. $\begin{bmatrix} 3/2 & -1/2 \\ -1/2 & 1/2 \end{bmatrix}\begin{bmatrix} 2 & 5 \\ 1 & 7 \end{bmatrix}\begin{bmatrix} 1 & 1 \\ 1 & 3 \end{bmatrix} = \begin{bmatrix} 13/2 & 29/2 \\ 1/2 & 5/2 \end{bmatrix}$

5. $\begin{bmatrix} 0 & 1 & -1 \\ 1 & 0 & 0 \\ -1 & 0 & 1 \end{bmatrix}\begin{bmatrix} 1 & 0 & 0 \\ 1 & 1 & 0 \\ 1 & 1 & 1 \end{bmatrix}\begin{bmatrix} 0 & 1 & 0 \\ 1 & 1 & 1 \\ 0 & 1 & 1 \end{bmatrix} = \begin{bmatrix} 0 & -1 & -1 \\ 0 & 1 & 0 \\ 1 & 2 & 2 \end{bmatrix}$

7. $\begin{bmatrix} 2 & 1 \\ 1 & 2 \end{bmatrix} \begin{bmatrix} 5 & 1 \\ 7 & 4 \end{bmatrix} \begin{bmatrix} 2 & 1 \\ 1 & 2 \end{bmatrix}^{-1} = \begin{bmatrix} 2 & 1 \\ 1 & 2 \end{bmatrix} \begin{bmatrix} 5 & 1 \\ 7 & 4 \end{bmatrix} \begin{bmatrix} 2/3 & -1/3 \\ -1/3 & 2/3 \end{bmatrix} = \begin{bmatrix} 28/3 & -5/3 \\ 29/3 & -1/3 \end{bmatrix}$

9. $\begin{bmatrix} 2 & 1 & 1 \\ 1 & 1 & -3 \\ 1 & 2 & 1 \end{bmatrix} \begin{bmatrix} 3 & 0 & 0 \\ 0 & 4 & 0 \\ 0 & 0 & 1 \end{bmatrix} \begin{bmatrix} 7/11 & 1/11 & -4/11 \\ -4/11 & 1/11 & 7/11 \\ 1/11 & -3/11 & 1/11 \end{bmatrix} = \begin{bmatrix} \frac{27}{11} & \frac{7}{11} & \frac{5}{11} \\ \frac{2}{11} & \frac{16}{11} & \frac{13}{11} \\ -\frac{10}{11} & \frac{8}{11} & \frac{45}{11} \end{bmatrix}$

11. $\begin{bmatrix} 1 & 2 & 4 \\ 1 & 1 & -5 \\ 1 & -3 & 1 \end{bmatrix} \begin{bmatrix} 2 & 0 & 0 \\ 0 & 1 & 0 \\ 0 & 0 & 2 \end{bmatrix} \begin{bmatrix} 1/3 & 1/3 & 1/3 \\ 1/7 & 1/14 & -3/14 \\ 2/21 & -5/42 & 1/42 \end{bmatrix} = \begin{bmatrix} 12/7 & -1/7 & 3/7 \\ -1/7 & 27/14 & 3/14 \\ 3/7 & 3/14 & 19/14 \end{bmatrix}$

   The matrix is symmetric because the vectors $\mathbf{u}_1, \mathbf{u}_2, \mathbf{u}_3$ are orthogonal.

13. $\begin{bmatrix} 2 & 1 & 0 \\ 1 & 2 & 0 \\ 0 & 1 & 1 \end{bmatrix}^{-1} \begin{bmatrix} 1 & 1 \\ 1 & -1 \\ 2 & 0 \end{bmatrix} \begin{bmatrix} 1 & 1 \\ 3 & -1 \end{bmatrix} = \begin{bmatrix} 2/3 & -1/3 & 0 \\ -1/3 & 2/3 & 0 \\ 1/3 & -2/3 & 1 \end{bmatrix} \begin{bmatrix} 1 & 1 \\ 1 & -1 \\ 2 & 0 \end{bmatrix} \begin{bmatrix} 1 & 1 \\ 3 & -1 \end{bmatrix} = \begin{bmatrix} 10/3 & -2/3 \\ -8/3 & 4/3 \\ 14/3 & 2/3 \end{bmatrix}$

15. $\begin{bmatrix} 2 & 1 & 0 \\ 1 & 2 & 1 \\ 1 & -2 & 1 \end{bmatrix} \begin{bmatrix} 1 & 2 \\ 3 & 1 \\ 2 & -1 \end{bmatrix} \begin{bmatrix} 3 & -2 \\ -1 & 1 \end{bmatrix} = \begin{bmatrix} 10 & -5 \\ 24 & -15 \\ -8 & 5 \end{bmatrix}$

17. $\begin{bmatrix} 2 & -1 \\ 1 & 2 \end{bmatrix} \begin{bmatrix} 2 & 0 & 1 \\ 0 & 7 & 1 \end{bmatrix} \begin{bmatrix} 1/11 & 7/11 & 3/11 \\ 1/11 & 3/22 & -5/22 \\ 2/11 & -5/22 & 1/22 \end{bmatrix} = \begin{bmatrix} -1/11 & 15/11 & 30/11 \\ 2 & 5/2 & -5/2 \end{bmatrix}$

Chapter 7

1. The characteristic polynomial is $(\lambda-4)^2$ and we have

   $\mathbf{N}(A-4I_2) = \mathbf{N}\left(\begin{bmatrix} -6 & 9 \\ -4 & 6 \end{bmatrix}\right) = \operatorname{Span}\left\{\begin{bmatrix} 3 \\ 2 \end{bmatrix}\right\}$ and $\mathbf{N}((A-4I_2)^2) = \mathbf{N}\left(\begin{bmatrix} 0 & 0 \\ 0 & 0 \end{bmatrix}\right) = \mathbb{R}^2$.

   Since the vector $\begin{bmatrix} 1 \\ 1 \end{bmatrix}$ is not in $\operatorname{Span}\left\{\begin{bmatrix} 3 \\ 2 \end{bmatrix}\right\}$ and $\begin{bmatrix} -6 & 9 \\ -4 & 6 \end{bmatrix}\begin{bmatrix} 1 \\ 1 \end{bmatrix} = \begin{bmatrix} 3 \\ 2 \end{bmatrix}$, we can take $P = \begin{bmatrix} 3 & 1 \\ 2 & 1 \end{bmatrix}$ and $J = \begin{bmatrix} 4 & 1 \\ 0 & 4 \end{bmatrix}$.

3. The characteristic polynomial is $(\lambda-4)^3$ and we have

   $\mathbf{N}(A-4I_3) = \mathbf{N}\left(\begin{bmatrix} -6 & 16 & -4 \\ -3 & 8 & -2 \\ -3 & 8 & -2 \end{bmatrix}\right) = \operatorname{Span}\left\{\begin{bmatrix} 8 \\ 3 \\ 0 \end{bmatrix}, \begin{bmatrix} -2 \\ 0 \\ 3 \end{bmatrix}\right\}$

   and $\mathbf{N}((A-4I_3)^2) = \mathbb{R}^3$. Now we use Theorem 7.5.

   Since $\begin{bmatrix} -6 & 16 & -4 \\ -3 & 8 & -2 \\ -3 & 8 & -2 \end{bmatrix}\begin{bmatrix} 1 \\ 0 \\ 0 \end{bmatrix} = \begin{bmatrix} -6 \\ -3 \\ -3 \end{bmatrix}$, we can take $P = \begin{bmatrix} -6 & 1 & 8 \\ -3 & 0 & 3 \\ -3 & 0 & 0 \end{bmatrix}$ and $J = \begin{bmatrix} 4 & 1 & 0 \\ 0 & 4 & 0 \\ 0 & 0 & 4 \end{bmatrix}$.

5. We calculate $\mathbf{N}(A-3I_3) = \mathbf{N}\left(\begin{bmatrix} 10 & -32 & 12 \\ 9 & -28 & 10 \\ 13 & -40 & 14 \end{bmatrix}\right) = \operatorname{Span}\left\{\begin{bmatrix} 2 \\ 1 \\ 1 \end{bmatrix}\right\}$,

   $\mathbf{N}((A-3I_3)^2) = \mathbf{N}\left(\begin{bmatrix} -32 & 96 & -32 \\ -32 & 96 & -32 \\ -48 & 144 & -48 \end{bmatrix}\right) = \operatorname{Span}\left\{\begin{bmatrix} 3 \\ 1 \\ 0 \end{bmatrix}, \begin{bmatrix} -1 \\ 0 \\ 1 \end{bmatrix}\right\}$,

   and $\mathbf{N}(A+I_3) = \mathbf{N}\left(\begin{bmatrix} 14 & -32 & 12 \\ 9 & -24 & 10 \\ 13 & -40 & 18 \end{bmatrix}\right) = \operatorname{Span}\left\{\begin{bmatrix} 2 \\ 2 \\ 3 \end{bmatrix}\right\}$. Since the vector $\begin{bmatrix} -1 \\ 0 \\ 1 \end{bmatrix}$

is not in $\mathbf{N}(A-3I_3)$ and $\begin{bmatrix} 10 & -32 & 12 \\ 9 & -28 & 10 \\ 13 & -40 & 14 \end{bmatrix}\begin{bmatrix} -1 \\ 0 \\ 1 \end{bmatrix} = \begin{bmatrix} 2 \\ 1 \\ 1 \end{bmatrix}$, we can take

$$P = \begin{bmatrix} 2 & -1 & 2 \\ 1 & 0 & 2 \\ 1 & 1 & 3 \end{bmatrix} \text{ and } J = \begin{bmatrix} 3 & 1 & 0 \\ 0 & 3 & 0 \\ 0 & 0 & -1 \end{bmatrix}.$$

7. We calculate $\mathbf{N}(A-2I_3) = \mathbf{N}\left(\begin{bmatrix} -4 & 5 & -2 \\ -6 & 8 & -4 \\ -5 & 7 & -4 \end{bmatrix}\right) = \text{Span}\left\{\begin{bmatrix} 2 \\ 2 \\ 1 \end{bmatrix}\right\}$,

$\mathbf{N}((A-2I_3)^2) = \mathbf{N}\left(\begin{bmatrix} -4 & 6 & -4 \\ -4 & 6 & -4 \\ -2 & 3 & -2 \end{bmatrix}\right) = \text{Span}\left\{\begin{bmatrix} 3 \\ 2 \\ 0 \end{bmatrix}, \begin{bmatrix} -1 \\ 0 \\ 1 \end{bmatrix}\right\}$, $\mathbf{N}((A-2I_3)^2) = \mathbb{R}^3$.

Since the vector $\begin{bmatrix} 1 \\ 0 \\ 0 \end{bmatrix}$ is not in $\mathbf{N}((A-2I_3)^2)$, $\begin{bmatrix} -4 & 5 & -2 \\ -6 & 8 & -4 \\ -5 & 7 & -4 \end{bmatrix}\begin{bmatrix} 1 \\ 0 \\ 0 \end{bmatrix} = \begin{bmatrix} -4 \\ -6 \\ -5 \end{bmatrix}$,

and $\begin{bmatrix} -4 & 6 & -4 \\ -4 & 6 & -4 \\ -2 & 3 & -2 \end{bmatrix}\begin{bmatrix} 1 \\ 0 \\ 0 \end{bmatrix} = \begin{bmatrix} -4 \\ -4 \\ -2 \end{bmatrix}$, we can take $P = \begin{bmatrix} -4 & -4 & 1 \\ -4 & -6 & 0 \\ -2 & -5 & 0 \end{bmatrix}$ and

$$J = \begin{bmatrix} 2 & 1 & 0 \\ 0 & 2 & 1 \\ 0 & 0 & 2 \end{bmatrix}.$$

9. $\mathbf{N}(A+I_3) = \mathbf{N}\left(\begin{bmatrix} 5 & -3 & 2 \\ 5 & -3 & 2 \\ -5 & 3 & -2 \end{bmatrix}\right) = \text{Span}\left\{\begin{bmatrix} 3 \\ 5 \\ 0 \end{bmatrix}, \begin{bmatrix} -2 \\ 0 \\ 5 \end{bmatrix}\right\}$ and $\mathbf{N}((A+I_3)^2) = \mathbb{R}^3$. Since the vector $\begin{bmatrix} 0 \\ 0 \\ 1 \end{bmatrix}$ is not in $\mathbf{N}(A+I_3)$ and $\begin{bmatrix} 5 & -3 & 2 \\ 5 & -3 & 2 \\ -5 & 3 & -2 \end{bmatrix}\begin{bmatrix} 0 \\ 0 \\ 1 \end{bmatrix} = \begin{bmatrix} 2 \\ 2 \\ -2 \end{bmatrix}$, we can take

$$P = \begin{bmatrix} 2 & 0 & 3 \\ 2 & 0 & 5 \\ -2 & 1 & 0 \end{bmatrix} \text{ and } J = \begin{bmatrix} -1 & 1 & 0 \\ 0 & -1 & 0 \\ 0 & 0 & -1 \end{bmatrix}.$$

11. $\mathbf{N}(A-I_4) = \mathbf{N}\left(\begin{bmatrix} 2 & -2 & -3 & 3 \\ 4 & -8 & -9 & 13 \\ 2 & -3 & -4 & 5 \\ 4 & -8 & -9 & 13 \end{bmatrix}\right) = \text{Span}\left\{\begin{bmatrix} 1 \\ 1 \\ 1 \\ 1 \end{bmatrix}\right\}$,

$\mathbf{N}((A-I_4)^2) = \mathbf{N}\left(\begin{bmatrix} 2 & -3 & -3 & 4 \\ 10 & -21 & -21 & 32 \\ 4 & -8 & -8 & 12 \\ 10 & -21 & -21 & 32 \end{bmatrix}\right) = \text{Span}\left\{\begin{bmatrix} 1 \\ 2 \\ 0 \\ 1 \end{bmatrix}, \begin{bmatrix} 0 \\ 1 \\ -1 \\ 0 \end{bmatrix}\right\}$,

$\mathbf{N}(A-2I_4) = \mathbf{N}\left(\begin{bmatrix} 1 & -2 & -3 & 3 \\ 4 & -9 & -9 & 13 \\ 2 & -3 & -5 & 5 \\ 4 & -8 & -9 & 12 \end{bmatrix}\right) = \text{Span}\left\{\begin{bmatrix} -1 \\ 1 \\ 0 \\ 1 \end{bmatrix}\right\}$,

and $\mathbf{N}(A-3I_4) = \mathbf{N}\left(\begin{bmatrix} 0 & -2 & -3 & 3 \\ 4 & -10 & -9 & 13 \\ 2 & -3 & -6 & 5 \\ 4 & -8 & -9 & 11 \end{bmatrix}\right) = \text{Span}\left\{\begin{bmatrix} 0 \\ 3 \\ 1 \\ 3 \end{bmatrix}\right\}$. Since the vector

$\begin{bmatrix} 0 \\ 1 \\ -1 \\ 0 \end{bmatrix}$ is not in $\mathbf{N}(A - I_4)$ and $\begin{bmatrix} 2 & -2 & -3 & 3 \\ 4 & -8 & -9 & 13 \\ 2 & -3 & -4 & 5 \\ 4 & -8 & -9 & 13 \end{bmatrix}\begin{bmatrix} 0 \\ 1 \\ -1 \\ 0 \end{bmatrix} = \begin{bmatrix} 1 \\ 1 \\ 1 \\ 1 \end{bmatrix}$, we can take

$$P = \begin{bmatrix} 1 & 0 & -1 & 0 \\ 1 & 1 & 1 & 3 \\ 1 & -1 & 0 & 1 \\ 1 & 0 & 1 & 3 \end{bmatrix} \text{ and } J = \begin{bmatrix} 1 & 1 & 0 & 0 \\ 0 & 1 & 0 & 0 \\ 0 & 0 & 2 & 0 \\ 0 & 0 & 0 & 3 \end{bmatrix}.$$

13. $\mathbf{N}(A+2I_4) = \mathbf{N}\left(\begin{bmatrix} -3 & 0 & 4 & 3 \\ 0 & -2 & 2 & 4 \\ -3 & 1 & 3 & 1 \\ 2 & -2 & -1 & 2 \end{bmatrix}\right) = \text{Span}\left\{\begin{bmatrix} 1 \\ 2 \\ 0 \\ 1 \end{bmatrix}\right\},$

$\mathbf{N}((A+2I_4)^2) = \mathbf{N}\left(\begin{bmatrix} 2 & -3 & -3 & 4 \\ 10 & -21 & -21 & 32 \\ 4 & -8 & -8 & 12 \\ 10 & -21 & -21 & 32 \end{bmatrix}\right) = \text{Span}\left\{\begin{bmatrix} 1 \\ 2 \\ 0 \\ 1 \end{bmatrix}, \begin{bmatrix} 1 \\ 0 \\ 1 \\ 0 \end{bmatrix}\right\},$

and $\mathbf{N}((A+2I_4)^3) = \mathbf{N}\left(\begin{bmatrix} 2 & -1 & -2 & 0 \\ 4 & -2 & -4 & 0 \\ 0 & 0 & 0 & 0 \\ 2 & -1 & -2 & 0 \end{bmatrix}\right) = \text{Span}\left\{\begin{bmatrix} 1 \\ 2 \\ 0 \\ 0 \end{bmatrix}, \begin{bmatrix} 0 \\ -2 \\ 1 \\ 0 \end{bmatrix}, \begin{bmatrix} 0 \\ 0 \\ 0 \\ 1 \end{bmatrix}\right\}.$

Now we use Theorem 7.11. Since the vector $\begin{bmatrix} 1 \\ 0 \\ 0 \\ 0 \end{bmatrix}$ is not in $\mathbf{N}((A+2I_4)^3)$,

$\begin{bmatrix} -3 & 0 & 4 & 3 \\ 0 & -2 & 2 & 4 \\ -3 & 1 & 3 & 1 \\ 2 & -2 & -1 & 2 \end{bmatrix}\begin{bmatrix} 1 \\ 0 \\ 0 \\ 0 \end{bmatrix} = \begin{bmatrix} -3 \\ 0 \\ -3 \\ 2 \end{bmatrix}, \begin{bmatrix} 3 & -2 & -3 & 1 \\ 2 & -2 & -2 & 2 \\ 2 & -1 & -2 & 0 \\ 1 & -1 & -1 & 1 \end{bmatrix}\begin{bmatrix} 1 \\ 0 \\ 0 \\ 0 \end{bmatrix} = \begin{bmatrix} 3 \\ 2 \\ 2 \\ 1 \end{bmatrix}$, and

$\begin{bmatrix} 2 & -1 & -2 & 0 \\ 4 & -2 & -4 & 0 \\ 0 & 0 & 0 & 0 \\ 2 & -1 & -2 & 0 \end{bmatrix}\begin{bmatrix} 1 \\ 0 \\ 0 \\ 0 \end{bmatrix} = \begin{bmatrix} 2 \\ 4 \\ 0 \\ 2 \end{bmatrix}$, we can take $P = \begin{bmatrix} 2 & 3 & -3 & 1 \\ 4 & 2 & 0 & 0 \\ 0 & 2 & -3 & 0 \\ 2 & 1 & 2 & 0 \end{bmatrix}$ and

$$J = \begin{bmatrix} -2 & 1 & 0 & 0 \\ 0 & -2 & 1 & 0 \\ 0 & 0 & -2 & 1 \\ 0 & 0 & 0 & -2 \end{bmatrix}.$$

14. We can take $P = \begin{bmatrix} 1 & 2 & -1 & 1 \\ 1 & 0 & 1 & 2 \\ 2 & 4 & -1 & 2 \\ 3 & 5 & 1 & 4 \end{bmatrix}$ and $J = \begin{bmatrix} 4 & 1 & 0 & 0 \\ 0 & 4 & 0 & 0 \\ 0 & 0 & 4 & 1 \\ 0 & 0 & 0 & 4 \end{bmatrix}.$

15. Let $\mathbf{u}$ be a vector in $\mathbf{N}((A-\alpha I_5)^3) = \mathbb{R}^5$ which is not in $\mathbf{N}((A-\alpha I_5)^2)$ and let $\mathbf{v}$ be a vector in $\mathbf{N}((A-\alpha I_5)^2)$ which is not in $\text{Span}\{(A-\alpha I_5)\mathbf{u}, \mathbf{N}(A-\alpha I_5)\}$. The vector $(A-\alpha I_5)\mathbf{u}$ is in $\mathbf{N}((A-\alpha I_5)^2)$ and the vectors $(A-\alpha I_5)^2\mathbf{u}$ and $(A-\alpha I_5)\mathbf{v}$ are in $\mathbf{N}(A-\alpha I_5)$. We will show that $\{(A-\alpha I_5)^2\mathbf{u}, (A-\alpha I_5)\mathbf{u}, \mathbf{u}, (A-\alpha I_5)\mathbf{v}, \mathbf{v}\}$ is the desired basis.

From the equality

$$x_1(A-\alpha I_5)^2\mathbf{u} + x_2(A-\alpha I_5)\mathbf{u} + x_3\mathbf{u} + x_4(A-\alpha I_5)\mathbf{v} + x_5\mathbf{v} = \mathbf{0} \tag{12.1}$$

we get $x_3(A-\alpha I_5)^2\mathbf{u}=\mathbf{0}$ and consequently $x_3=0$. Next we get $x_2(A-\alpha I_5)^2\mathbf{u}+x_5(A-\alpha I_5)\mathbf{v}=\mathbf{0}$, and hence the vector $x_2(A-\alpha I_5)\mathbf{u}+x_5\mathbf{v}$ is in $\mathbf{N}(A-\alpha I_5)$ and the vector $x_5\mathbf{v}$ is in $\operatorname{Span}\{(A-\alpha I_5)\mathbf{u},\mathbf{N}(A-\alpha I_5)\}$. We must have $x_5=0$ and then $x_2=0$. Now (12.1) becomes $x_1(A-\alpha I_5)^2\mathbf{u}+x_4(A-\alpha I_5)\mathbf{v}=0$ which gives us $x_1=x_4=0$, as before. Now we have $A(A-\alpha I_5)^2\mathbf{u}=(A-\alpha I_5)^3\mathbf{u}+\alpha(A-\alpha I_5)^2\mathbf{u}=\alpha(A-\alpha I_5)^2\mathbf{u}$, $A(A-\alpha I_5)\mathbf{u}=(A-\alpha I_5)^2\mathbf{u}+\alpha(A-\alpha I_5)\mathbf{u}$, $A\mathbf{u}=(A-\alpha I_5)\mathbf{u}+\alpha\mathbf{u}$, $A(A-\alpha I_5)\mathbf{v}=(A-\alpha I_5)^2\mathbf{v}+\alpha(A-\alpha I_5)\mathbf{v}=\alpha(A-\alpha I_5)\mathbf{v}$, $A\mathbf{v}=(A-\alpha I_5)\mathbf{v}+\alpha\mathbf{v}$.

17. Let $\mathbf{u}$ and $\mathbf{v}$ be linearly independent vectors in $\mathbf{N}((A-\alpha I_5)^2)$ which are not in $\mathbf{N}(A-\alpha I_5)$. Then the vectors $(A-\alpha I_5)\mathbf{u}$ and $(A-\alpha I_5)\mathbf{v}$ are in $\mathbf{N}(A-\alpha I_5)$. Let $\mathbf{w}$ be a vector $\neq\mathbf{0}$ in $\mathbf{N}(A-\alpha I_5)$ which is not in $\operatorname{Span}\{(A-\alpha I_5)\mathbf{u},(A-\alpha I_5)\mathbf{v}\}$. We will show that $\{(A-\alpha I_5)\mathbf{u},\mathbf{u},(A-\alpha I_5)\mathbf{v},\mathbf{v},\mathbf{w}\}$ is the desired basis.

    From the equality $x_1(A-\alpha I_5)\mathbf{u}+x_2\mathbf{u}+x_3(A-\alpha I_5)\mathbf{v}+x_4\mathbf{v}+x_5\mathbf{w}=\mathbf{0}$ we get $x_2(A-\alpha I_5)\mathbf{u}+x_4(A-\alpha I_5)\mathbf{v}=\mathbf{0}$ which implies that the vector $x_2\mathbf{u}+x_4\mathbf{v}$ is in $\mathbf{N}(A-\alpha I_5)$ and consequently $x_2=x_4=0$. Next we get $x_1(A-\alpha I_5)\mathbf{u}+x_3(A-\alpha I_5)\mathbf{v}+x_5\mathbf{w}=\mathbf{0}$ which shows that $x_5=0$, and thus $x_1=x_3=0$, as before. Now we have $A(A-\alpha I_5)\mathbf{u}=(A-\alpha I_5)^2\mathbf{u}+\alpha(A-\alpha I_5)\mathbf{u})=\alpha(A-\alpha I_5)\mathbf{u}$, $A(\mathbf{u})=(A-\alpha I_5)\mathbf{u}+\alpha\mathbf{u}$, $A(A-\alpha I_5)\mathbf{v}=(A-\alpha I_5)^2\mathbf{v}+\alpha(A-\alpha I_5)\mathbf{v}=\alpha(A-\alpha I_5)\mathbf{v}$, $A\mathbf{v}=(A-\alpha I_5)\mathbf{v}+\alpha\mathbf{v}$, $A\mathbf{w}=\alpha\mathbf{w}$.

19. $\mathbf{N}(A-2I_5)=\mathbf{N}\left(\begin{bmatrix}5&5&1&-22&6\\6&6&1&-27&8\\4&4&1&-18&5\\4&4&1&-18&5\\5&5&2&-23&6\end{bmatrix}\right)=\operatorname{Span}\left\{\begin{bmatrix}-1\\1\\0\\0\\0\end{bmatrix},\begin{bmatrix}3\\0\\1\\1\\1\end{bmatrix}\right\}$,

    $\mathbf{N}((A-2I_5)^2)=\mathbf{N}\left(\begin{bmatrix}1&1&1&-5&1\\2&2&2&-10&2\\1&1&1&-5&1\\1&1&1&-5&1\\1&1&1&-5&1\end{bmatrix}\right)=\operatorname{Span}\left\{\begin{bmatrix}-1\\1\\0\\0\\0\end{bmatrix},\begin{bmatrix}-1\\0\\1\\0\\0\end{bmatrix},\begin{bmatrix}5\\0\\0\\1\\0\end{bmatrix},\begin{bmatrix}-1\\0\\0\\0\\1\end{bmatrix}\right\}$, and $\mathbf{N}((A-2I_5)^3)=\mathbf{N}\left(\begin{bmatrix}0&0&0&0&0\\0&0&0&0&0\\0&0&0&0&0\\0&0&0&0&0\\0&0&0&0&0\end{bmatrix}\right)=\mathbb{R}^5$. Now we use the result from Exercise 16.

    Since $\begin{bmatrix}5&5&1&-22&6\\6&6&1&-27&8\\4&4&1&-18&5\\4&4&1&-18&5\\5&5&2&-23&6\end{bmatrix}\begin{bmatrix}1\\0\\0\\0\\0\end{bmatrix}=\begin{bmatrix}5\\6\\4\\4\\5\end{bmatrix}$, $\begin{bmatrix}1&1&1&-5&1\\2&2&2&-10&2\\1&1&1&-5&1\\1&1&1&-5&1\\1&1&1&-5&1\end{bmatrix}\begin{bmatrix}1\\0\\0\\0\\0\end{bmatrix}=\begin{bmatrix}1\\2\\1\\1\\1\end{bmatrix}$, and $\begin{bmatrix}5&5&1&-22&6\\6&6&1&-27&8\\4&4&1&-18&5\\4&4&1&-18&5\\5&5&2&-23&6\end{bmatrix}\begin{bmatrix}1\\0\\-1\\0\\0\end{bmatrix}=\begin{bmatrix}4\\5\\3\\3\\3\end{bmatrix}$, we can take $P=\begin{bmatrix}1&5&1&4&1\\2&6&0&5&0\\1&4&0&3&-1\\1&4&0&3&0\\1&5&0&3&0\end{bmatrix}$.

Section 8.1

1. $\mathbf{u}\mathbf{v}^T=\begin{bmatrix}-20&45\\-8&18\end{bmatrix}$, $\mathbf{v}\mathbf{u}^T=\begin{bmatrix}-20&-8\\45&18\end{bmatrix}$

3. $\mathbf{u}\mathbf{v}^T = \begin{bmatrix} 2 & 1 & 3 & 7 \\ 4 & 2 & 6 & 14 \\ 10 & 5 & 15 & 35 \end{bmatrix}$, $\mathbf{v}\mathbf{u}^T = \begin{bmatrix} 2 & 4 & 10 \\ 1 & 2 & 5 \\ 3 & 6 & 15 \\ 7 & 14 & 35 \end{bmatrix}$

5. $A = 2\sqrt{2}\mathbf{u}_1\mathbf{v}_1^T + \sqrt{2}\mathbf{u}_2\mathbf{v}_2^T$, where $\mathbf{v}_1 = \frac{1}{\sqrt{2}}\begin{bmatrix} 1 \\ -1 \end{bmatrix}$, $\mathbf{v}_2 = \frac{1}{\sqrt{2}}\begin{bmatrix} 1 \\ 1 \end{bmatrix}$, $\mathbf{u}_1 = \begin{bmatrix} -1 \\ 0 \end{bmatrix}$, $\mathbf{u}_2 = \begin{bmatrix} 0 \\ 1 \end{bmatrix}$

7. $A = \sqrt{33}\mathbf{u}_1\mathbf{v}_1^T$, where $\mathbf{v}_1 = \frac{1}{\sqrt{11}}\begin{bmatrix} 3 \\ 1 \\ 1 \end{bmatrix}$, $\mathbf{u}_1 = \frac{1}{\sqrt{33 \cdot 11}}\begin{bmatrix} 11 \\ 11 \\ 11 \end{bmatrix} = \frac{1}{\sqrt{3}}\begin{bmatrix} 1 \\ 1 \\ 1 \end{bmatrix}$

9. $A = \sqrt{15}\mathbf{u}_1\mathbf{v}_1^T + \sqrt{5}\mathbf{u}_2\mathbf{v}_2^T$, where $\mathbf{v}_1 = \frac{1}{\sqrt{30}}\begin{bmatrix} 3 \\ 4 \\ \sqrt{5} \end{bmatrix}$, $\mathbf{v}_2 = \frac{1}{\sqrt{10}}\begin{bmatrix} 1 \\ -2 \\ \sqrt{5} \end{bmatrix}$, $\mathbf{u}_1 = \frac{1}{\sqrt{2}}\begin{bmatrix} 1 \\ 1 \end{bmatrix}$,

$\mathbf{u}_2 = \frac{1}{\sqrt{2}}\begin{bmatrix} 1 \\ -1 \end{bmatrix}$

11. $A = \sqrt{150}\mathbf{u}_1\mathbf{v}_1^T + \sqrt{20}\mathbf{u}_2\mathbf{v}_2^T$, where $\mathbf{v}_1 = \frac{1}{\sqrt{5}}\begin{bmatrix} -2 \\ 1 \end{bmatrix}$, $\mathbf{v}_2 = \frac{1}{\sqrt{5}}\begin{bmatrix} 1 \\ 2 \end{bmatrix}$, $\mathbf{u}_1 = \frac{1}{\sqrt{30}}\begin{bmatrix} -1 \\ -4 \\ -3 \\ 2 \end{bmatrix}$, $\mathbf{u}_2 =$

$\frac{1}{2}\begin{bmatrix} 1 \\ 1 \\ -1 \\ 1 \end{bmatrix}$

13. $A = \sqrt{14}\mathbf{u}_1\mathbf{v}_1^T + 2\mathbf{u}_2\mathbf{v}_2^T$, where $\mathbf{v}_1 = \frac{1}{\sqrt{10}}\begin{bmatrix} 3 \\ 1 \end{bmatrix}$, $\mathbf{v}_2 = \frac{1}{\sqrt{10}}\begin{bmatrix} 1 \\ -3 \end{bmatrix}$, $\mathbf{u}_1 = \frac{1}{\sqrt{35}}\begin{bmatrix} 5 \\ 2 \\ 2 \\ \sqrt{2} \end{bmatrix}$,

$\mathbf{u}_2 = \frac{1}{\sqrt{10}}\begin{bmatrix} 0 \\ -1 \\ -1 \\ 2\sqrt{2} \end{bmatrix}$

15. $A = \sqrt{21}\mathbf{u}_1\mathbf{v}_1^T + 2\mathbf{u}_2\mathbf{v}_2^T$, where $\mathbf{v}_1 = \frac{1}{\sqrt{17}}\begin{bmatrix} 4 \\ 1 \end{bmatrix}$, $\mathbf{v}_2 = \frac{1}{\sqrt{17}}\begin{bmatrix} 1 \\ -4 \end{bmatrix}$, $\mathbf{u}_1 = \frac{1}{\sqrt{21 \cdot 17}}\begin{bmatrix} -5 \\ 3 \\ 5 \\ -3 \\ 17 \end{bmatrix} =$

$\frac{1}{\sqrt{357}}\begin{bmatrix} -5 \\ 3 \\ 5 \\ -3 \\ 17 \end{bmatrix}$, $\mathbf{u}_2 = \frac{1}{2\sqrt{17}}\begin{bmatrix} 3 \\ 5 \\ -3 \\ -5 \\ 0 \end{bmatrix}$

17. $A = \sqrt{20}\mathbf{u}_1\mathbf{v}_1^T + \sqrt{15}\mathbf{u}_2\mathbf{v}_2^T$, where $\mathbf{v}_1 = \frac{1}{\sqrt{20\cdot 5}}\begin{bmatrix} 5 \\ 5 \\ 0 \\ 5 \\ -5 \end{bmatrix} = \frac{1}{2}\begin{bmatrix} 1 \\ 1 \\ 0 \\ 1 \\ -1 \end{bmatrix}$, $\mathbf{v}_2 = \frac{1}{\sqrt{15\cdot 5}}\begin{bmatrix} -5 \\ 0 \\ -5 \\ 0 \\ -5 \end{bmatrix} =$

$\frac{1}{\sqrt{3}}\begin{bmatrix} -1 \\ 0 \\ -1 \\ 0 \\ -1 \end{bmatrix}$, $\mathbf{u}_1 = \frac{1}{\sqrt{5}}\begin{bmatrix} 1 \\ 2 \end{bmatrix}$, $\mathbf{u}_2 = \frac{1}{\sqrt{5}}\begin{bmatrix} -2 \\ 1 \end{bmatrix}$

19. Assume that the outer product expansion of A is $A = \sigma_1\mathbf{u}_1\mathbf{v}_1^T + \sigma_2\mathbf{u}_2\mathbf{v}_2^T$ and that $\{\mathbf{v}_1, \mathbf{v}_2, \mathbf{v}_3\}$ is an orthonormal basis of eigenvectors of the matrix A^TA. A vector $\mathbf{x}$ in $\mathbb{R}^3$ is a least square solution of the equation $A\mathbf{x} = \mathbf{b}$ if it satisfies the equation $A\mathbf{x} = \text{proj}_{\mathbf{C}(A)}(\mathbf{b})$. Since $\{\mathbf{v}_1, \mathbf{v}_2, \mathbf{v}_3\}$ is a basis in $\mathbb{R}^3$, we can write $\mathbf{x} = x_1\mathbf{v}_1 + x_2\mathbf{v}_2 + x_3\mathbf{v}_3$. Then the equation $A\mathbf{x} = \text{proj}_{\mathbf{C}(A)}(\mathbf{b})$ becomes $A(x_1\mathbf{v}_1 + x_2\mathbf{v}_2 + x_3\mathbf{v}_3) = \text{proj}_{\mathbf{C}(A)}(\mathbf{b})$. Since $A(x_1\mathbf{v}_1) = x_1\sigma_1\mathbf{u}_1$, $A(x_2\mathbf{v}_2) = x_2\sigma_2\mathbf{u}_2$, and $A(x_3\mathbf{v}_3) = \mathbf{0}$, we have $x_1\sigma_1\mathbf{u}_1 + x_2\sigma_2\mathbf{u}_2 = \text{proj}_{\mathbf{C}(A)}(\mathbf{b})$.

Now, since $\{\mathbf{u}_1, \mathbf{u}_2\}$ is an orthonormal basis in $\mathbf{C}(A)$, $x_1\sigma_1 = \mathbf{b}\cdot\mathbf{u}_1 = \mathbf{u}_1^T\mathbf{b}$, and $x_2\sigma_2 = \mathbf{b}\cdot\mathbf{u}_2 = \mathbf{u}_2^T\mathbf{b}$, we can conclude that $\mathbf{x} = \frac{1}{\sigma_1}(\mathbf{u}_1^T\mathbf{b})\mathbf{v}_1 + \frac{1}{\sigma_2}(\mathbf{u}_2^T\mathbf{b})\mathbf{v}_2 + x_3\mathbf{v}_3$.

From the Pythagorean Theorem we get
$\left\|\frac{1}{\sigma_1}(\mathbf{u}_1^T\mathbf{b})\mathbf{v}_1 + \frac{1}{\sigma_2}(\mathbf{u}_2^T\mathbf{b})\mathbf{v}_2 + x_3\mathbf{v}_3\right\|^2 = \frac{1}{\sigma_1^2}|\mathbf{u}_1^T\mathbf{b}|^2 + \frac{1}{\sigma_2^2}|\mathbf{u}_2^T\mathbf{b}|^2 + x_3^2$.
It is clear that the norm is minimized if $x_3 = 0$. Consequently, $\mathbf{x} = \frac{1}{\sigma_1}(\mathbf{u}_1^T\mathbf{b})\mathbf{v}_1 + \frac{1}{\sigma_2}(\mathbf{u}_2^T\mathbf{b})\mathbf{v}_2$ is the unique least square solution of minimal length. Moreover, since
$(\mathbf{u}^T\mathbf{b})\mathbf{v} = (\mathbf{u}\cdot\mathbf{b})\mathbf{v} = \mathbf{v}(\mathbf{u}\cdot\mathbf{b}) = \mathbf{v}(\mathbf{u}^T\mathbf{b}) = (\mathbf{v}\mathbf{u}^T)\mathbf{b}$,
for any $\mathbf{u}$, $\mathbf{b}$ in $\mathbb{R}^m$ and $\mathbf{v}$ in $\mathbb{R}^n$, we have
$\frac{1}{\sigma_1}(\mathbf{u}_1^T\mathbf{b})\mathbf{v}_1 + \frac{1}{\sigma_2}(\mathbf{u}_2^T\mathbf{b})\mathbf{v}_2 = \left(\frac{1}{\sigma_1}\mathbf{v}_1\mathbf{u}_1^T + \frac{1}{\sigma_2}\mathbf{v}_2\mathbf{u}_2^T\right)\mathbf{b}$.

21. $\begin{bmatrix} 3/20 & 1/20 \\ -3/20 & -1/20 \end{bmatrix}$

23. $\begin{bmatrix} \frac{1}{15} & \frac{4}{75} & -\frac{22}{75} \\ \frac{2}{15} & \frac{31}{150} & \frac{17}{150} \end{bmatrix}$

25. $\begin{bmatrix} \frac{27}{100} & \frac{2}{25} & \frac{11}{100} & -\frac{3}{50} \\ \frac{1}{100} & \frac{1}{25} & -\frac{7}{100} & \frac{11}{50} \end{bmatrix}$

27. $\begin{bmatrix} \frac{1}{45} & \frac{2}{45} & \frac{4}{9} \\ \frac{4}{45} & \frac{8}{45} & -\frac{2}{9} \\ -\frac{1}{45} & -\frac{2}{45} & \frac{5}{9} \end{bmatrix}$

29. If $A = \sigma_1\mathbf{u}_1\mathbf{v}_1^T + \cdots + \sigma_r\mathbf{u}_r\mathbf{v}_r^T$, then $A^+ = \frac{1}{\sigma_1}\mathbf{v}_1\mathbf{u}_1^T + \cdots + \frac{1}{\sigma_r}\mathbf{v}_r\mathbf{u}_r^T$ and
$AA^+ = \left(\sigma_1\mathbf{u}_1\mathbf{v}_1^T + \cdots + \sigma_r\mathbf{u}_r\mathbf{v}_r^T\right)\left(\frac{1}{\sigma_1}\mathbf{v}_1\mathbf{u}_1^T + \cdots + \frac{1}{\sigma_r}\mathbf{v}_r\mathbf{u}_r^T\right) = \mathbf{u}_1\mathbf{u}_1^T + \cdots + \mathbf{u}_r\mathbf{u}_r^T = \text{proj}_{\mathbf{C}(A)}$.

31. If $A = \sigma_1\mathbf{u}_1\mathbf{v}_1^T + \cdots + \sigma_r\mathbf{u}_r\mathbf{v}_r^T$, then $A^T = \sigma_1\mathbf{v}_1\mathbf{u}_1^T + \cdots + \sigma_r\mathbf{v}_r\mathbf{u}_r^T$ and $A^+ = \frac{1}{\sigma_1}\mathbf{v}_1\mathbf{u}_1^T + \cdots + \frac{1}{\sigma_r}\mathbf{v}_r\mathbf{u}_r^T$. Consequently $A^TA = \sigma_1^2\mathbf{v}_1\mathbf{v}_1^T + \cdots + \sigma_r^2\mathbf{v}_r\mathbf{v}_r^T$ and
$A^TAA^+ = \left(\sigma_1^2\mathbf{v}_1\mathbf{v}_1^T + \cdots + \sigma_r^2\mathbf{v}_r\mathbf{v}_r^T\right)\left(\frac{1}{\sigma_1}\mathbf{v}_1\mathbf{u}_1^T + \cdots + \frac{1}{\sigma_r}\mathbf{v}_r\mathbf{u}_r^T\right) = \sigma_1\mathbf{v}_1\mathbf{u}_1^T + \cdots + \sigma_r\mathbf{v}_r\mathbf{u}_r^T = A^T$.

Section 8.2

1. $A = \begin{bmatrix} \mathbf{u}_1 & \mathbf{u}_2 & \mathbf{u}_3 & \mathbf{u}_4 \end{bmatrix} \begin{bmatrix} \sigma_1 & 0 & 0 & 0 \\ 0 & \sigma_2 & 0 & 0 \\ 0 & 0 & \sigma_3 & 0 \\ 0 & 0 & 0 & 0 \end{bmatrix} \begin{bmatrix} \mathbf{v}_1^T \\ \mathbf{v}_2^T \\ \mathbf{v}_3^T \\ \mathbf{v}_4^T \end{bmatrix}$

3. $A = \begin{bmatrix} \mathbf{u}_1 & \mathbf{u}_2 & \mathbf{u}_3 & \mathbf{u}_4 & \mathbf{u}_5 \end{bmatrix} \begin{bmatrix} \sigma_1 & 0 \\ 0 & \sigma_2 \\ 0 & 0 \\ 0 & 0 \\ 0 & 0 \end{bmatrix} \begin{bmatrix} \mathbf{v}_1^T \\ \mathbf{v}_2^T \end{bmatrix}$

5. $A = \begin{bmatrix} \frac{2\sqrt{5}}{5} & \frac{2\sqrt{5}}{10} \\ -\frac{2\sqrt{5}}{10} & \frac{2\sqrt{5}}{5} \end{bmatrix} \begin{bmatrix} 2\sqrt{10} & 0 \\ 0 & \sqrt{10} \end{bmatrix} \begin{bmatrix} \frac{\sqrt{2}}{2} & \frac{\sqrt{2}}{2} \\ -\frac{\sqrt{2}}{2} & \frac{\sqrt{2}}{2} \end{bmatrix}$

7. $A = \begin{bmatrix} \frac{3}{\sqrt{10}} & \frac{1}{\sqrt{10}} \\ \frac{1}{\sqrt{10}} & -\frac{3}{\sqrt{10}} \end{bmatrix} \begin{bmatrix} \sqrt{20} & 0 \\ 0 & 0 \end{bmatrix} \begin{bmatrix} \frac{1}{\sqrt{2}} & -\frac{1}{\sqrt{2}} \\ \frac{1}{\sqrt{2}} & \frac{1}{\sqrt{2}} \end{bmatrix}$

9. $A = \begin{bmatrix} \frac{\sqrt{5}}{5} & -\frac{2\sqrt{5}}{5} \\ \frac{2\sqrt{5}}{5} & \frac{\sqrt{5}}{5} \end{bmatrix} \begin{bmatrix} \sqrt{15} & 0 & 0 \\ 0 & \sqrt{10} & 0 \end{bmatrix} \begin{bmatrix} \frac{\sqrt{3}}{3} & \frac{7\sqrt{3}}{15} & -\frac{\sqrt{3}}{15} \\ 0 & \frac{\sqrt{2}}{10} & \frac{7\sqrt{2}}{10} \\ \frac{\sqrt{6}}{3} & -\frac{7\sqrt{6}}{30} & \frac{\sqrt{6}}{30} \end{bmatrix}$

11. $A = \begin{bmatrix} \frac{\sqrt{2}}{2} & -\frac{\sqrt{2}}{2} & 0 \\ \frac{\sqrt{2}}{2} & \frac{\sqrt{2}}{2} & 0 \\ 0 & 0 & 1 \end{bmatrix} \begin{bmatrix} 4 & 0 & 0 \\ 0 & 2\sqrt{3} & 0 \\ 0 & 0 & \sqrt{3} \end{bmatrix} \begin{bmatrix} -\frac{\sqrt{2}}{2} & 0 & \frac{\sqrt{2}}{2} \\ \frac{\sqrt{6}}{6} & -\frac{\sqrt{6}}{3} & \frac{\sqrt{6}}{6} \\ \frac{\sqrt{3}}{3} & \frac{\sqrt{3}}{3} & \frac{\sqrt{3}}{3} \end{bmatrix}$

13. $A = \begin{bmatrix} 1/6\sqrt{6} & -1/3\sqrt{3} & \frac{\sqrt{2}}{2} \\ -1/3\sqrt{6} & -1/3\sqrt{3} & 0 \\ 1/6\sqrt{6} & -1/3\sqrt{3} & -\frac{\sqrt{2}}{2} \end{bmatrix} \begin{bmatrix} 2\sqrt{3} & 0 & 0 \\ 0 & 3 & 0 \\ 0 & 0 & 0 \end{bmatrix} \begin{bmatrix} \frac{\sqrt{2}}{2} & 0 & \frac{\sqrt{2}}{2} \\ -1/3\sqrt{3} & 1/3\sqrt{3} & 1/3\sqrt{3} \\ -1/6\sqrt{6} & -1/3\sqrt{6} & 1/6\sqrt{6} \end{bmatrix}$

15. $A = \begin{bmatrix} \frac{3\sqrt{2}}{10} & -\frac{4}{5} & -\frac{\sqrt{6}}{6} & -\frac{\sqrt{3}}{15} \\ \frac{\sqrt{2}}{5} & -\frac{1}{5} & \frac{\sqrt{6}}{3} & -\frac{4\sqrt{3}}{15} \\ -\frac{\sqrt{2}}{10} & -\frac{2}{5} & \frac{\sqrt{6}}{6} & \frac{7\sqrt{3}}{15} \\ 3\sqrt{2} & \frac{2}{5} & 0 & \frac{\sqrt{3}}{5} \end{bmatrix} \begin{bmatrix} 2\sqrt{5} & 0 \\ 0 & \sqrt{10} \\ 0 & 0 \\ 0 & 0 \end{bmatrix} \begin{bmatrix} \frac{\sqrt{10}}{10} & \frac{3\sqrt{10}}{10} \\ -\frac{3\sqrt{10}}{10} & \frac{\sqrt{10}}{10} \end{bmatrix}$

17. $A = \begin{bmatrix} \frac{\sqrt{10}}{10} & -\frac{3\sqrt{10}}{10} \\ \frac{3\sqrt{10}}{10} & \frac{\sqrt{10}}{10} \end{bmatrix} \begin{bmatrix} 2\sqrt{5} & 0 & 0 & 0 \\ 0 & \sqrt{10} & 0 & 0 \end{bmatrix} \begin{bmatrix} \frac{3\sqrt{2}}{10} & \frac{\sqrt{2}}{5} & -\frac{\sqrt{2}}{10} & 3\sqrt{2} \\ -\frac{4}{5} & -\frac{1}{5} & -\frac{2}{5} & \frac{2}{5} \\ -\frac{\sqrt{6}}{6} & \frac{\sqrt{6}}{3} & \frac{\sqrt{6}}{6} & 0 \\ -\frac{\sqrt{3}}{15} & -\frac{4\sqrt{3}}{15} & \frac{7\sqrt{3}}{15} & \frac{\sqrt{3}}{5} \end{bmatrix}$

19. (a) $\mathbf{C}(A)$ is a subspace of $\mathbb{R}^8$ of dimension 3,

(b) $\mathbf{C}(A^T)$ is a subspace of $\mathbb{R}^5$ of dimension 3,

(c) $\mathbf{N}(A)$ is a subspace of $\mathbb{R}^5$ of dimension 2, and

(d) $\mathbf{N}(A^T)$ is a subspace of $\mathbb{R}^8$ of dimension 5.

21. $\mathbf{C}(A) = \text{Span}\{\mathbf{u}_1\}, \mathbf{N}(A^T) = \text{Span}\{\mathbf{u}_2, \mathbf{u}_3, \mathbf{u}_4, \mathbf{u}_5\}, \mathbf{C}(A^T) = \text{Span}\{\mathbf{v}_1\}, \mathbf{N}(A) = \text{Span}\{\mathbf{v}_2, \mathbf{v}_3, \mathbf{v}_4\}$.

Chapter 9

1. A is indefinite because the eigenvalues are $3+\sqrt{10}$ and $3-\sqrt{10}$.
3. A is positive definite because the eigenvalues are 10 and 1.
5. A is indefinite because the eigenvalues are -2, $3-\sqrt{3}$, and $3+\sqrt{3}$.
7. A is positive semidefinite because the eigenvalues are 0, 6, and 12.
9. A is negative definite because the eigenvalues are -3, -6, and -9.

11. $\begin{bmatrix} \sqrt{5} & 0 \\ \frac{\sqrt{5}}{5} & \frac{2\sqrt{5}}{5} \end{bmatrix}$

13. $\begin{bmatrix} \sqrt{7} & 0 \\ -\frac{5\sqrt{7}}{7} & \frac{\sqrt{21}}{7} \end{bmatrix}$

15. $\begin{bmatrix} \sqrt{5} & 0 & 0 \\ \frac{2\sqrt{5}}{5} & \frac{\sqrt{5}}{5} & 0 \\ \frac{\sqrt{5}}{5} & \frac{3\sqrt{5}}{5} & \sqrt{2} \end{bmatrix}$

17. $\begin{bmatrix} 1 & 0 & 0 \\ -2 & \sqrt{3} & 0 \\ 1 & \sqrt{3} & 1 \end{bmatrix}$

19. In the proof of Theorem 9.13 we show that $a_{11} > 0$. Since $a_{22} - \frac{a_{21}^2}{a_{11}} = \frac{1}{a_{11}} \det \begin{bmatrix} a_{11} & a_{21} \\ a_{21} & a_{22} \end{bmatrix}$, the second inequality follows.

21. With the notations from the proof of Theorem 9.13, we have

$$\begin{aligned} 0 < b_{22} - \frac{b_{21}^2}{b_{11}} &= \frac{(a_{11}a_{22} - a_{21}^2)(a_{11}a_{33} - a_{31}^2) - (a_{11}a_{32} - a_{21}a_{31})^2}{a_{11}} \\ &= \frac{a_{11}(a_{11}a_{22}a_{33} - a_{22}a_{31}^2 - a_{33}a_{21}^2 - a_{11}a_{32}^2 + 2a_{21}a_{31}a_{32})}{a_{11}} \\ &= a_{11}a_{22}a_{33} - a_{22}a_{31}^2 - a_{33}a_{21}^2 - a_{11}a_{32}^2 + 2a_{21}a_{31}a_{32} \\ &= \det \begin{bmatrix} a_{11} & a_{21} & a_{31} \\ a_{21} & a_{22} & a_{32} \\ a_{31} & a_{32} & a_{33} \end{bmatrix}. \end{aligned}$$

Now the proof is a consequence of the proof of Theorem 9.13.

22. The matrix is positive definite because

$1 > 0$, $\det \begin{bmatrix} 1 & -2 \\ -2 & 7 \end{bmatrix} = 3$, and $\det \begin{bmatrix} 1 & -2 & 1 \\ -2 & 7 & 1 \\ 1 & 1 & 5 \end{bmatrix} = 3$.

23. We have $6x^2 + 3y^2 + 4xy = \begin{bmatrix} x & y \end{bmatrix} \begin{bmatrix} 6 & 2 \\ 2 & 3 \end{bmatrix} \begin{bmatrix} x \\ y \end{bmatrix}$ and $\begin{bmatrix} 6 & 2 \\ 2 & 3 \end{bmatrix} = P \begin{bmatrix} 7 & 0 \\ 0 & 2 \end{bmatrix} P^T$, where $P = \begin{bmatrix} \frac{2}{\sqrt{5}} & -\frac{1}{\sqrt{5}} \\ \frac{1}{\sqrt{5}} & \frac{2}{\sqrt{5}} \end{bmatrix}$. In terms of the new variables x' and y', the quadratic form becomes $7(x')^2 + 2(y')^2$.

25. We have $6x^2+3y^2+4xy = \begin{bmatrix} x & y \end{bmatrix}\begin{bmatrix} -11 & 3 \\ 3 & -19 \end{bmatrix}\begin{bmatrix} x \\ y \end{bmatrix}$ and $\begin{bmatrix} -11 & 3 \\ 3 & -19 \end{bmatrix} = P\begin{bmatrix} -10 & 0 \\ 0 & -20 \end{bmatrix}P^T$, where $P = \begin{bmatrix} \frac{3}{\sqrt{10}} & -\frac{1}{\sqrt{10}} \\ \frac{1}{\sqrt{10}} & \frac{3}{\sqrt{10}} \end{bmatrix}$. In terms of the new variables x' and y', the quadratic form becomes $-10(x')^2-20(y')^2$.

27. We have $7x^2+7y^2+10z^2+2xy+4xz+4yz = \begin{bmatrix} x & y & z \end{bmatrix}\begin{bmatrix} 7 & 1 & 2 \\ 1 & 7 & 2 \\ 2 & 2 & 10 \end{bmatrix}\begin{bmatrix} x \\ y \\ z \end{bmatrix}$ and $\begin{bmatrix} 7 & 1 & 2 \\ 1 & 7 & 2 \\ 2 & 2 & 10 \end{bmatrix} = P\begin{bmatrix} 6 & 0 & 0 \\ 0 & 6 & 0 \\ 0 & 0 & 12 \end{bmatrix}P^T$, where $P = \begin{bmatrix} \frac{1}{\sqrt{2}} & \frac{1}{\sqrt{3}} & \frac{1}{\sqrt{6}} \\ -\frac{1}{\sqrt{2}} & \frac{1}{\sqrt{3}} & \frac{1}{\sqrt{6}} \\ 0 & -\frac{1}{\sqrt{3}} & -\frac{2}{\sqrt{6}} \end{bmatrix}$. In terms of the new variables x', y' and z', the quadratic form becomes $6(x')^2+6(y')^2+12(z')^2$.

29. We have $\begin{bmatrix} 2 & 3 & -3 \\ 3 & 10 & 1 \\ -3 & 1 & 10 \end{bmatrix} = P\begin{bmatrix} 0 & 0 & 0 \\ 0 & 11 & 0 \\ 0 & 0 & 11 \end{bmatrix}P^T$, where $P = \begin{bmatrix} \frac{3}{\sqrt{11}} & \frac{1}{\sqrt{10}} & \frac{3}{\sqrt{110}} \\ -\frac{1}{\sqrt{11}} & \frac{3}{\sqrt{10}} & -\frac{1}{\sqrt{110}} \\ \frac{1}{\sqrt{11}} & 0 & -\frac{10}{\sqrt{110}} \end{bmatrix}$. In terms of the new variables x', y' and z', the quadratic form becomes $11(y')^2+11(z')^2$.

31. Suppose that the matrix A is positive definite and has the Cholesky decomposition

$$A = \begin{bmatrix} m_{11} & 0 & 0 \\ m_{21} & m_{22} & 0 \\ m_{31} & m_{32} & m_{33} \end{bmatrix}\begin{bmatrix} m_{11} & m_{21} & m_{31} \\ 0 & m_{22} & m_{32} \\ 0 & 0 & m_{33} \end{bmatrix}.$$

Then

$$\begin{aligned} A &= \begin{bmatrix} 1 & 0 & 0 \\ m_{21}m_{11}^{-1} & 1 & 0 \\ m_{31}m_{11}^{-1} & m_{32}m_{22}^{-1} & 1 \end{bmatrix}\begin{bmatrix} m_{11}^2 & m_{21}m_{11} & m_{31}m_{11} \\ 0 & m_{22}^2 & m_{32}m_{22} \\ 0 & 0 & m_{33}^2 \end{bmatrix} \\ &= \begin{bmatrix} 1 & 0 & 0 \\ m_{21}m_{11}^{-1} & 1 & 0 \\ m_{31}m_{11}^{-1} & m_{32}m_{22}^{-1} & 1 \end{bmatrix}\begin{bmatrix} m_{11}^2 & 0 & 0 \\ 0 & m_{22}^2 & 0 \\ 0 & 0 & m_{33}^2 \end{bmatrix}\begin{bmatrix} 1 & m_{21}m_{11}^{-1} & m_{31}m_{11}^{-1} \\ 0 & 1 & m_{32}m_{22}^{-1} \\ 0 & 0 & 1 \end{bmatrix}. \end{aligned}$$

33. The Cholesky decomposition of A is $A = \begin{bmatrix} \sqrt{d_1} & 0 & 0 \\ l_{21}\sqrt{d_1} & \sqrt{d_2} & 0 \\ l_{31}\sqrt{d_1} & l_{32}\sqrt{d_2} & \sqrt{d_3} \end{bmatrix}\begin{bmatrix} \sqrt{d_1} & l_{21}\sqrt{d_1} & l_{31}\sqrt{d_1} \\ 0 & \sqrt{d_2} & l_{32}\sqrt{d_2} \\ 0 & 0 & \sqrt{d_3} \end{bmatrix}$, which also means that the matrix A is positive definite.

Section 10.1

1. We have $C = A + B$.

3. We have $C = 2A - B$.

5. $s = 14$

7. $s = \frac{2}{5}$, $t = \frac{1}{5}$

9. The subspace is Span$\left\{\begin{bmatrix} 1 & 0 \\ 0 & 1 \end{bmatrix}, \begin{bmatrix} 0 & 1 \\ 1 & 0 \end{bmatrix}\right\}$ and $\left\{\begin{bmatrix} 1 & 0 \\ 0 & 1 \end{bmatrix}, \begin{bmatrix} 0 & 1 \\ 1 & 0 \end{bmatrix}\right\}$ is a basis in this subspace.

11. The vectors are linearly independent.

13. $s = \frac{9}{8}$

15. We have $\mathbf{p} = x_1\mathbf{a} + x_2\mathbf{b} + x_3\mathbf{c} + x_4\mathbf{d}$. Now $x_1 \neq 0$ or $x_2 \neq 0$ or $x_3 \neq 0$ or $x_4 \neq 0$, because the vector $\mathbf{p}$ is nonzero.
If $x_1 \neq 0$, then $\mathbf{a} = \frac{1}{x_1}\mathbf{p} - \frac{x_2}{x_1}\mathbf{b} - \frac{x_3}{x_1}\mathbf{c} - \frac{x_4}{x_1}\mathbf{d}$ and the vector $\mathbf{a}$ is in Span$\{\mathbf{b}, \mathbf{c}, \mathbf{d}, \mathbf{p}\}$.
We show in a similar way that, if $x_2 \neq 0$, then the vector $\mathbf{b}$ is in Span$\{\mathbf{a}, \mathbf{c}, \mathbf{d}, \mathbf{p}\}$, if $x_3 \neq 0$, then the vector $\mathbf{c}$ is in Span$\{\mathbf{a}, \mathbf{b}, \mathbf{d}, \mathbf{p}\}$, and if $x_4 \neq 0$, then the vector $\mathbf{d}$ is in Span$\{\mathbf{a}, \mathbf{b}, \mathbf{c}, \mathbf{p}\}$.

17. We have $\mathbf{r} = x_1\mathbf{a} + x_2\mathbf{b} + x_3\mathbf{p} + x_4\mathbf{q}$. Now $x_1 \neq 0$ or $x_2 \neq 0$, because the vectors $\mathbf{p}, \mathbf{q}$ and $\mathbf{r}$ are linearly independent.
If $x_1 \neq 0$, then $\mathbf{a} = \frac{1}{x_1}\mathbf{r} - \frac{x_2}{x_1}\mathbf{b} - \frac{x_3}{x_1}\mathbf{p} - \frac{x_4}{x_1}\mathbf{q}$ and the vector $\mathbf{a}$ is in Span$\{\mathbf{b}, \mathbf{p}, \mathbf{q}, \mathbf{r}\}$.
If $x_2 \neq 0$, then $\mathbf{b} = \frac{1}{x_2}\mathbf{r} - \frac{x_1}{x_2}\mathbf{a} - \frac{x_3}{x_2}\mathbf{p} - \frac{x_4}{x_2}\mathbf{q}$ and the vector $\mathbf{b}$ is in Span$\{\mathbf{a}, \mathbf{p}, \mathbf{q}, \mathbf{r}\}$.

19. In an introductory differential equations course we learn that the space of solutions is Span$\{f, g\}$ from the calculus course. The set $\{f, g\}$ is a basis in this subspace because the functions f and g are linearly independent. Indeed, the equation $ae^{3t} + bte^{4t} = 0$ for $t = 0$ gives us $a + b = 0 = 0$. Now we differentiate the equation and get $3ae^{3t} + 4bte^{4t} = 0$. Using $t = 0$ in this equation we get $3a + 4b = 0$. The equalities $a + b = 0$ and $3a + 4b = 0$ give us $a = b = 0$.

21. In an introductory differential equations course we learn that the space of solutions is Span$\{f, g\}$. The set $\{f, g\}$ is a basis in this subspace because the functions f, g are linearly independent. Indeed, the equation $ae^{2t} + bte^{2t} = 0$ gives us $a = 0$ for $t = 0$ and then for $b = 0$ for $t = 1$.

23. In an introductory differential equations course we learn that the space of solutions is Span$\{f, g\}$. The set $\{f, g\}$ is a basis in this subspace because the functions f, g are linearly independent. Indeed, the equation $ae^t \cos t + be^t \sin t = 0$ gives us for $a = 0$ for $t = 0$ and $b = 0$ for $t = \frac{\pi}{2}$.

25. We have
$(f+g)(\mathbf{v}+\mathbf{w}) = f(\mathbf{v}+\mathbf{w}) + g(\mathbf{v}+\mathbf{w}) = f(\mathbf{v}) + f(\mathbf{w}) + g(\mathbf{v}) + g(\mathbf{w}) = f(\mathbf{v}) + g(\mathbf{v}) + f(\mathbf{w}) + g(\mathbf{w}) = (f+g)(\mathbf{v}) + (f+g)(\mathbf{w})$
and
$(f+g)(c\mathbf{v}) = f(c\mathbf{v}) + g(c\mathbf{v}) = cf(\mathbf{v}) + cg(\mathbf{v}) = c(f+g)(\mathbf{v})$.

27. We have
$h(\mathbf{v}+\mathbf{w}) = g(f(\mathbf{v}+\mathbf{w})) = g(f(\mathbf{v}) + f(\mathbf{w})) = g(f(\mathbf{v})) + g(f(\mathbf{w})) = h(\mathbf{v}) + h(\mathbf{w})$
and
$h(c\mathbf{v}) = g(f(c\mathbf{v})) = g(cf(\mathbf{v})) = cg(f(\mathbf{v})) = ch(\mathbf{v})$.

29. Let $\mathbf{v}$ and $\mathbf{w}$ be vectors in $\ker f$ and let c be a real number. Then
$f(\mathbf{v}+\mathbf{w}) = f(\mathbf{v}) + f(\mathbf{w}) = \mathbf{0} + \mathbf{0} = \mathbf{0}$ and $f(c\mathbf{v}) = cf(\mathbf{v}) = c \cdot \mathbf{0} = \mathbf{0}$.

31. The fact that f is a linear transformation can be shown directly or using Exercise 27 and Example **??**. Since every solution of the equation $y''' = 0$ is of the form $y(t) = at^2+bt+c$, for arbitrary real numbers a, b, c, these functions constitute the kernel of the linear transformation f.

33. Let $\{\mathbf{u}_1, \mathbf{u}_2, \mathbf{u}_3\}$ be a basis in $f(\mathscr{L})$ and let $\mathbf{v}_1$, $\mathbf{v}_2$, and $\mathbf{v}_3$ be such that $f(\mathbf{v}_1) = \mathbf{u}_1$, $f(\mathbf{v}_1) = \mathbf{u}_1$ and $f(\mathbf{v}_1) = \mathbf{u}_1$. Let $\{\mathbf{w}_1, \mathbf{w}_2\}$ be a basis in $\ker f$. We will show that $\{\mathbf{v}_1, \mathbf{v}_2, \mathbf{v}_3, \mathbf{w}_1, \mathbf{w}_2\}$ is a basis in $\mathscr{L}$.

    Let $\mathbf{v}$ be a vector of $\mathscr{L}$. Then

    $f(\mathbf{v}) = x_1\mathbf{u}_1 + x_2\mathbf{u}_2 + x_3\mathbf{u}_3 = x_1 f(\mathbf{v}_1) + x_2 f(\mathbf{v}_2) + x_3 f(\mathbf{v}_3) = f(x_1\mathbf{v}_1 + x_2\mathbf{v}_2 + x_3\mathbf{v}_3)$.

    Hence

    $f(\mathbf{v} - (x_1\mathbf{v}_1 + x_2\mathbf{v}_2 + x_3\mathbf{v}_3)) = f(\mathbf{v}) - f(x_1\mathbf{v}_1 + x_2\mathbf{v}_2 + x_3\mathbf{v}_3) = \mathbf{0}$

    and, consequently, the vector $\mathbf{v} - (x_1\mathbf{v}_1 + x_2\mathbf{v}_2 + x_3\mathbf{v}_3)$ is in $\ker f$ and we have

    $\mathbf{v} = x_1\mathbf{v}_1 + x_2\mathbf{v}_2 + x_3\mathbf{v}_3 + y_1\mathbf{w}_1 + y_2\mathbf{w}_2$.

    Next we show that the vectors $\mathbf{v}_1, \mathbf{v}_2, \mathbf{v}_3, \mathbf{w}_1, \mathbf{w}_2$ are linearly independent. If

    $x_1\mathbf{v}_1 + x_2\mathbf{v}_2 + x_3\mathbf{v}_3 + y_1\mathbf{w}_1 + y_2\mathbf{w}_2 = \mathbf{0}$,

    then

    $x_1 f(\mathbf{v}_1) + x_2 f(\mathbf{v}_2) + x_3 f(\mathbf{v}_3) + y_1 f(\mathbf{w}_1) + y_2 f(\mathbf{w}_2) = \mathbf{0}$,

    which means that

    $x_1\mathbf{u}_1 + x_2\mathbf{u}_2 + x_3\mathbf{u}_3 + y_1 \cdot \mathbf{0} + y_2 \cdot \mathbf{0} = \mathbf{0}$,

    which gives us $x_1 = x_2 = x_3 = 0$, because the vectors $\mathbf{u}_1$, $\mathbf{u}_2$, and $\mathbf{u}_3$ are linearity independent. Now, using linear independence of the vectors $\mathbf{w}_1$ and $\mathbf{w}_2$, we get $y_1 = y_2 = 0$. This shows that the vectors $\mathbf{v}_1, \mathbf{v}_2, \mathbf{v}_3, \mathbf{w}_1, \mathbf{w}_2$ are linearly independent.

Section 10.2

1. We have

$$\left\langle \begin{bmatrix} u_1 & u_2 \\ u_3 & u_4 \end{bmatrix}, \begin{bmatrix} v_1 & v_2 \\ v_3 & v_4 \end{bmatrix} \right\rangle = u_1v_1 + u_2v_2 + u_3v_3 + u_4v_4 = \left\langle \begin{bmatrix} v_1 & v_2 \\ v_3 & v_4 \end{bmatrix}, \begin{bmatrix} u_1 & u_2 \\ u_3 & u_4 \end{bmatrix} \right\rangle,$$

$$\left\langle \begin{bmatrix} u_1 & u_2 \\ u_3 & u_4 \end{bmatrix}, \begin{bmatrix} v_1+w_1 & v_2+w_2 \\ v_3+w_3 & v_4+w_4 \end{bmatrix} \right\rangle = u_1(v_1+w_1)+u_2(v_2+w_2)+u_3(v_3+w_3)+u_4(v_4+w_4)$$

$$= \left\langle \begin{bmatrix} u_1 & u_2 \\ u_3 & u_4 \end{bmatrix}, \begin{bmatrix} v_1 & v_2 \\ v_3 & v_4 \end{bmatrix} \right\rangle + \left\langle \begin{bmatrix} u_1 & u_2 \\ u_3 & u_4 \end{bmatrix}, \begin{bmatrix} w_1 & w_2 \\ w_3 & w_4 \end{bmatrix} \right\rangle,$$

$$\left\langle \begin{bmatrix} cu_1 & cu_2 \\ cu_3 & cu_4 \end{bmatrix}, \begin{bmatrix} v_1 & v_2 \\ v_3 & v_4 \end{bmatrix} \right\rangle = c(u_1v_1 + u_2v_2 + u_3v_3 + u_4v_4) = c\left\langle \begin{bmatrix} u_1 & u_2 \\ u_3 & u_4 \end{bmatrix}, \begin{bmatrix} v_1 & v_2 \\ v_3 & v_4 \end{bmatrix} \right\rangle,$$

$$\left\langle \begin{bmatrix} u_1 & u_2 \\ u_3 & u_4 \end{bmatrix}, \begin{bmatrix} u_1 & u_2 \\ u_3 & u_4 \end{bmatrix} \right\rangle = u_1^2 + u_2^2 + u_3^2 + u_4^2 \ge 0,$$

and $\left\langle \begin{bmatrix} u_1 & u_2 \\ u_3 & u_4 \end{bmatrix}, \begin{bmatrix} u_1 & u_2 \\ u_3 & u_4 \end{bmatrix} \right\rangle = 0$ if and only if $u_1^2+u_2^2+u_3^2+u_4^2 = 0$ if and only if $\begin{bmatrix} u_1 & u_2 \\ u_3 & u_4 \end{bmatrix} = \begin{bmatrix} 0 & 0 \\ 0 & 0 \end{bmatrix}$.

3. We have

$$\begin{aligned}
\langle f,g\rangle &= \int_0^1 f(t)g(t)t\,dt = \int_0^1 g(t)f(t)t\,dt = \langle g,f\rangle,\\
\langle f,g+h\rangle &= \int_0^1 f(t)(g+h)(t)t\,dt = \int_0^1 f(t)g(t)t\,dt + \int_0^1 f(t)h(t)t\,dt = \langle f,g\rangle + \langle f,h\rangle,\\
\langle cf,g\rangle &= \int_0^1 f(t)g(t)t\,dt = \int_0^1 cf(t)g(t)t\,dt = c\langle f,g\rangle,\\
\langle f,f\rangle &= \int_0^1 (f(t))^2 t\,dt \ge 0.
\end{aligned}$$

$\langle f,f\rangle = 0$ is equivalent to $\int_0^1 (f(t))^2 t\,dt = 0$ and this is equivalent to $f(t) = 0$ for every t in the interval $[0,1]$ as a consequence of the continuity of the function f.

5. $\langle \mathbf{u}+\mathbf{v},\mathbf{w}\rangle = \langle \mathbf{w},\mathbf{u}+\mathbf{v}\rangle = \langle \mathbf{w},\mathbf{u}\rangle + \langle \mathbf{w},\mathbf{v}\rangle = \langle \mathbf{u},\mathbf{w}\rangle + \langle \mathbf{v},\mathbf{w}\rangle$

7. $\|c\mathbf{v}\|^2 = \langle c\mathbf{v},c\mathbf{v}\rangle = c^2\|\mathbf{v}\|^2$

9. $\langle t,e^t\rangle = \displaystyle\int_0^1 te^{2t}\,dt = \frac{e^2}{2}$

11. $\|t\| = \displaystyle\int_{-1}^1 t^4\,dt = \frac{2}{5}$

13. $\left\langle \mathbf{u},\mathbf{v}-\dfrac{\langle \mathbf{v},\mathbf{u}\rangle}{\langle \mathbf{u},\mathbf{u}\rangle}\mathbf{u}\right\rangle = \langle \mathbf{u},\mathbf{v}\rangle - \left\langle \mathbf{u},\dfrac{\langle \mathbf{v},\mathbf{u}\rangle}{\langle \mathbf{u},\mathbf{u}\rangle}\mathbf{u}\right\rangle = \langle \mathbf{u},\mathbf{v}\rangle - \dfrac{\langle \mathbf{v},\mathbf{u}\rangle}{\langle \mathbf{u},\mathbf{u}\rangle}\langle \mathbf{u},\mathbf{u}\rangle = 0$

15. $\left\|\dfrac{\langle \mathbf{v},\mathbf{u}\rangle}{\langle \mathbf{u},\mathbf{u}\rangle}\mathbf{u}\right\|^2 = \left(\dfrac{\langle \mathbf{v},\mathbf{u}\rangle}{\langle \mathbf{u},\mathbf{u}\rangle}\right)^2\|\mathbf{u}\|^2 = \left(\dfrac{\langle \mathbf{v},\mathbf{u}\rangle}{\|\mathbf{u}\|^2}\right)^2\|\mathbf{u}\|^2 = \dfrac{|\langle \mathbf{v},\mathbf{u}\rangle|^2}{\|\mathbf{u}\|^2}$

17. From Exercise 16 we get
$\|\mathbf{u}+\mathbf{v}\|^2 = \langle \mathbf{u}+\mathbf{v},\mathbf{u}+\mathbf{v}\rangle = \|\mathbf{u}\|^2+\|\mathbf{v}\|^2+2\langle \mathbf{u},\mathbf{v}\rangle \le \|\mathbf{u}\|^2+\|\mathbf{v}\|^2+2\|\mathbf{u}\|\|\mathbf{v}\| = (\|\mathbf{u}\|+\|\mathbf{v}\|)^2$

19. $\dfrac{2(\alpha+2)-3(\alpha+1)}{6(\alpha+1)(\alpha+2)} + \dfrac{12(2(\alpha+1)-(\alpha+2))}{2(\alpha+1)(\alpha+2)}t$

21. This is a particular case of the system from Theorem **??**.

23. $\dfrac{3}{4}\cos t + \dfrac{1}{4}\cos 3t$

25. $\dfrac{\pi^2}{3} - 4\cos t + \cos 2t$

27. $\dfrac{3\pi}{4} - \dfrac{2}{\pi}\cos t - \dfrac{2}{9\pi}\cos 3t - \sin t - \dfrac{1}{2}\sin 2t - \dfrac{1}{3}\sin 3t$

29. $\dfrac{e^{2\pi}-1}{2\pi} + \dfrac{e^{2\pi}-1}{2\pi}\cos t - \dfrac{e^{2\pi}-1}{2\pi}\sin t$

31. We take $q_1 = t - \dfrac{\int_0^1 t\,dt}{\int_0^1 dt} = t - \dfrac{1}{2}$ and

$$q_2 = t^2 - \frac{\int_0^1 t^2\,dt}{\int_0^1 dt} - \frac{\int_0^1 t^2(t-\frac{1}{2})\,dt}{\int_0^1 \left(t-\frac{1}{2}\right)^2 dt}\left(t-\frac{1}{2}\right) = t^2 - \frac{1}{3} - \left(t-\frac{1}{2}\right) = t^2 - t + \frac{1}{6}.$$

Then $\langle 1, t-\frac{1}{2}\rangle = 0$, $\langle 1, t^2-t+\frac{1}{6}\rangle = 0$, and $\langle t-\frac{1}{2}, t^2-t+\frac{1}{6}\rangle = \langle t, t^2-t+\frac{1}{6}\rangle = 0$.

Bibliography

[1] D. Atanasiu and P. Mikusiński, A Bridge to Linear Algebra, World Scientific, 2019.

[2] V. Blanloeil, Une Introduction Moderne à l'Algèbre Linéaire, Ellipses, 2012.

[3] T. W. Körner, Vectors Pure and Applied, Cambridge University Press, 2013.

[4] S. Lang, Linear Algebra, 3rd edition, Springer, 1987.

[5] R. Larson, Elementary Linear Algebra, 8th edition, Cengage, 2017.

[6] D. Lay and S. Lay, Elementary Linear Algebra, 5th edition, Pearson, 2015.

[7] J. R. Munkres, Analysis on manifolds, Addison Wesley, 1991.

[8] L. Spence, A. Insel, and S. Friedberg, Elementary Linear Algebra, 2nd edition, Pearson, 2007.

[9] G. Strang, Introduction to Linear Algebra, 5th edition, Wellesley-Cambridge Press, 2016.

[10] S. Weintraub, Jordan canonical form: Theory and Practice, Morgan & Claypool Publishers, 2009.

Index